ENCYCLOPAEDIA OF COMMERCE

Ashu Ahuja

ANMOL PUBLICATIONS PVT. LTD.
NEW DELHI - 110 002 (INDIA)

ANMOL PUBLICATIONS PVT. LTD.
H.O.: 4374/4B, Ansari Road, Daryaganj
New Delhi - 110 002
Ph.: 23278000, 23261597
B.O.: No. 1015, Ist Main Road, BSK III Stage
III Phase, III Block
Bangalore - 560 085 (India)
Visit us at: www.anmolpublications.com

Encyclopaedia of Commerce

First Published, 2006

ISBN 81-261-2863-1

PRINTED IN INDIA

Printed at Mehra Offset Press, Delhi.

Preface

The aim of this dictionary is to provide teachers, students and laymen interested in commerce with clear, concise, and correct definitions and descriptions of the commerce terms, throughout. In the volume all the branches are represented. The great field that must be compressed within the limits of a small volume makes omissions inevitable. I trust that the present volume will serve as reliably as the chronometer of today, in the time-pattern of the commercial world.

The explanations of entries range from brief definitions of a few lines to expository discussions of various lengths. Several goals motivated the treatment. The dictionary is to satisfy the immediate technical needs of the user. Commerce is to be presented as a unified, ever-growing, and stimulating body of knowledge. The reader who is a non specialist in commerce is to be served in commerce and in related fields (that are significant for the future). In terms of its broader objectives, the present book may be viewed as an outline or compendium of commerce ideas correlated with a large collection of specialized items. There are concern and activity in the educational field with regard to revitalizing and modernizing scientific training and creating a better understanding of this field; it is hoped that this volume will contribute to these aims in commerce.

The aim of this volume is to provide a concise repository of number of major concepts in this field. In preparation of this book, the author has freely consulted large number of books and journals so no authenticity is claimed. Author is specially thankful to M/S Anmol Publications for shaping this book in its final form.

Suggestions for further improvement of this book are not only welcome but also greatly appreciated.

Author

Aad referendum Denoting a contract that has been signed although minor points remain to be decided.

Aaptive exponential smoothing A quantitative forecasting method in which averages derived from historical data are smoothed by a coefficient, which is allowed to fluctuate with time in relation to changes in the demand pattern. The larger the coefficient, the greater the smoothing effect. *See*quantitative forecasting techniques.

Abandonment A process by which the court releases property from its control. This occurs when property of the estate is of little or no value to the estate.

If the debtor's real estate is worth Rs.100,000, but has mortgages of Rs.110,000 against it, the trustee may "abandon" the property rather than administer it.

Abandonment option The option of terminating an investment earlier than originally planned.

Abbreviated accounts A company qualifying as a small or medium-sized company may file abbreviated accounts with the Registrar of Companies instead of the full report and accounts. These accounts, previously known as modified accounts, are an additional set of financial statements drawn from the full financial statements, specifically for the purpose of delivery to the Registrar; they thus become a public document and must be accompanied by a special report of the auditors.

Ability to pay Refers to the borrower's ability to make interest and principal payments on debts. In context of municipal bonds, refers to the issuer's present and future ability to create sufficient tax revenue to fulfill its contractual obligations, accounting for municipal income and property values. In context of taxation, notion that tax rates should be determined according to income or wealth.

Abnormal returns Part of the return that is not due to systematic influences (market wide influences). In other words, abnormal returns are above those predicted by the market movement alone. Related: excess returns.

Absolute advantage A person, company or country has an absolute advantage if its output per unit of input of all goods and services produced is higher than that of another person, company or country.

Absolute form of purchasing power parity A theory that prices of products of two different countries should be equal when measured by a common currency. Also called the "law of one price."

Absolute performance standard A standard set at the theoretical limit of performance. For example, in manufacturing, the theoretical quality standard of 'zero defects', which it is impossible to improve on, might be set as the absolute performance standard. Such standards may be achievable in practice or may form an ideal against which an organization may judge its progress.

Absorbed Used in context of general equities. Securities are "absorbed" as long as there are corresponding orders to buy and sell. The market has reached the absorption point when further assimilation is impossible without an adjustment in price.

Absorption costing The cost accounting system in which the overheads of an organization are charged to the production by means of the process of absorption. Costs are first apportioned to cost centres, where they are absorbed using absorption rates.

Absorption rate (overhead absorption rate; recovery rate) The rate or rates calculated in an absorption costing system in advance of an accounting period for the purpose of charging the overheads to the production of that period. Absorption rates are calculated for the accounting period in question using the following formula: In absorption costing production may be expressed in a number of different ways; the way chosen to express production will determine the absorption rate to be used. The seven major methods of measuring production, together with their associated absorption rate, are given in the table. The rate is used during the accounting period to obtain the.

Abstention In certain circumstances, the court may choose to abstain from control of the case or of an adversary proceeding within the case.

When an existing lawsuit between the debtor and another party is being heard in another court and someone tries to bring that dispute to bankruptcy court, the bankruptcy judge may decide to "abstain" and allow it to continue in the other court.

Abstract of title A document used in conveyancing land that is not registered to show how the vendor derived good title. It consists of a summary of certain documents, such as conveyances of the land, and recitals of certain events, such as marriages and deaths of previous owners. The purchaser will check the abstract against title deeds, grants of probate, etc. This document is not needed when registered land is being conveyed, as the land certificate shows good title.

Accelerated cost recovery system (ACRS) Schedule of depreciation rates allowed for tax purposes.

Accelerated depreciation Any depreciation method that produces larger deductions for depreciation in the early years of a project's life. Accelerated cost recovery system (ACRS), which is a depreciation schedule allowed for tax purposes, is one such example.

Acceptance Contractual agreement instigated when the drawee of a time draft "accepts" the draft by writing the word "accepted" thereon. The drawee assumes responsibility as the acceptor and for payment at maturity.

Acceptance credit A means of financing the sale of goods, particularly in international trade. It involves a commercial bank or merchant bank extend- ing credit to a foreign importer, whom it deems creditworthy. An acceptance credit is opened against which the exporter can draw a bill of exchange. Once accepted by the bank the bill can be discounted on the money market or allowed to run to maturity. In return for this service the exporter pays the bank a fee known as an acceptance commission.

Acceptance sampling A statistical quality-control technique that uses inspection of a sample from a batch to decide whether to accept or reject the whole batch. In deciding the level of a balance has to be struck between achieving the desired quality levels and the cost of undertaking the sampling.

Access time The time taken to obtain information from a computer memory, or to write information to the memory. Access times can vary from a thousand millionth of a second (a nanosecond) in a fast electronic memory to a second or longer with magnetic-tape memories.

Accident insurance An insurance policy that pays a specified amount of money to the policyholder in the event of the loss of one or more eyes or limbs in any type of accident It also pays a sum to the dependants of the policyholder in the event of his or her death. These policies first appeared in the early days of railway travel, when passengers felt a train journey was hazardous and they needed some protection for their dependants if they were to be killed or injured.

Accommodation bill A bill of exchange signed by a person (the accommodation party) who acts as a guarantor. The accommodation party is liable for the bill should the acceptor fail to pay at maturity. Accommodation bills are sometimes known as windbills or windmills.

Accommodation endorser A person or a bank that endorses a loan to another party; for example, a parent company may endorse a bank loan to a subsidiary. The endorser becomes a guarantor and is secondarily liable in case of default. Banks may endorse other banks' acceptance notes, which can then be traded on the secondary market.

Accommodative monetary policy Federal Reserve System policy to increase the amount of money available to banks for lending. See: Monetary policy.

Accord and satisfaction A device enabling one party to a contract to avoid an obligation that arises under the contract, provided that the other party agrees. The accord is the agreement by which the contractual obligation is discharged and the satisfaction is the consideration making the agreement legally operative. Such an agreement only discharges the contractual obligation if it is accompanied by consideration. For example, under a contract of sale the seller of goods may discharge the contractual obligation by delivering goods of different quality to that specified in the contract, provided there is agreement with the buyer (the accord) and the seller offers a reduction in the contract price (the satisfaction). The seller has therefore 'purchased' the release from the obligation. Accord and satisfaction refer to the discharge of an obligation arising under the law of tort.

Account Ad Valorem Duty An imported merchandise tax expressed as a percentage.

Account balance Credits minus debits at the end of a reporting period.

Account In the context of bookkeeping, refers to the ledger pages upon which various assets, liabilities, income, and expenses are represented.

The context of investment banking, refers to the status of securities sold and owned or the relationship between parties to an underwriting syndicate. In the context of securities, the relationship between a client and a broker/dealer firm allowing the firm's employee to be the client's buying and selling agent. See: Account executive; account statement.

Account management group 1. A group within an advertising, marketing, or public-relations agency responsible for planning, supervising, and coordinating all the work done on behalf of a client. In large agencies handling large accounts the group might consist of an account director, account manager, account or media planner, and account executive.

2. A group in the sales department of an

organization that is responsible for managing the relationship with existing clients.

Account of profits A legal remedy available as an alternative to damages in certain circumstances, especially inbreach of copyright cases. The person whose copyright has been breached sues the person who breached it for a sum of money equal to the gain made as a result of the breach.

Account Party Party who applies to open a bank for the issuance of a letter of credit.

Account payee only Words that, in accordance with the *Cheque Act*, make the cheque non-transferable. This is to avoid cheques being endorsed and paid into an account other than that of the payee, although it should be noted that banks may argue in some circumstances that they acted in good faith and without negligence if an endorsed cheque is honoured by the bank. In spite of this most cheques are now overprinted 'account payee only', and the words 'not negotiable' are sometimes added.

Account reconciliation The reviewing and adjusting of the balance in a personal checkbook to match your bank statement.

Account statement In the context of banking, refers to a summary of all balances. In the context of securities, a summary of all transactions and positions (long and short) between a broker/dealer and a client.

Accountability An obligation to give an account. For limited companies, it is assumed that the directors of the company are accountable to the shareholders and that this responsibility is discharged, in part, by the directors providing an annual report and accounts. In an accountability relationship will be at least one principal and at least one agent. This forms the basis of an agency relationship.

Accountant A person who has passed the accountancy examinations of one of the recognized accountancy bodies and completed the required work experience. Each of the bodies varies in the way they train their students and the type of work expected to be undertaken. For example, accountants who are members of the Chartered Institute of Public Finance and Accountancy generally work in local authorities, the National Health Service, or other similar public bodies, while members of the Chartered Institute of Management Accountants work in industry. Wherever accountants work, their responsibilities centre on the collating, recording, and communicating of financial information and the preparation of analyses for decision-making purposes.

Accountant's opinion A signed statement from an independent public accountant after examination of a firm's records and accounts. The opinion may be unqualified or qualified.

Accounting code (cost code; expenditure code; income code) In modern accounting systems, a numerical reference given to each account to facilitate the recording of voluminous accounting transactions by computer.

Accounting earnings Earnings of a firm as reported on its income statement.

Accounting entity A unit for which accounting records are maintained and for which financial statements are prepared. As an accounting concept, it is assumed that the financial records are prepared for a particular unit or entity. By law, limited companies constitute the accounting entity. For sole traders and partnerships accounts are also prepared to reflect the transactions of the business as an accounting entity, not those of the owner(s) of the business. Changing the boundaries of the accounting entity can have a significant impact on the accounts themselves, as these will reflect the purpose of the accounts and for whom they are prepared.

Accounting event A transaction or change

(internal or external) recognized by the accounting recording system. Events are recorded as debit and credit entries. For example, when a sale is made for cash the double entry for the sales transaction would be debit bank, credit sales. accounting period 1. The period for which a business prepares its accounts. Internally, management accounts may be produced monthly or quarterly. Externally, financial accounts are produced for a period of 12 months, although this may vary when a business is set up or ceases or if it changes its accounting year end. 2. (chargeable account period) A period in respect of which a corporation tax assessment is raised. It cannot be more than 12 months in length. An accounting period starts when a company begins to trade or immediately after a previous accounting period ends. An accounting period ends at the earliest of:

12 months after the start date,

- at the end of the company's period of account,
- the start of a winding-up.
- on ceasing to be UK resident.

Accounting exposure The change in the value of a firm's foreign currency-denominated accounts due to a change in exchange rates.

Accounting insolvency Total liabilities exceed total assets. A firm with a negative net worth is insolvent on the books.

Accounting liquidity The ease and quickness with which assets can be converted to cash.

Accounting policies The specific accounting bases adopted and consistently followed by an organization in the preparation of its financial statements. These bases will have been determined by the organization to be the most appropriate for presenting fairly its financial results and operations; they will concentrate on such specific topics as pension schemes, goodwill, research and development costs, and foreign exchange. Under Statement of Standard Accounting Practice 2, companies are required to disclose their accounting policies in their annual accounts.

Accounting records The records kept by a company to comply with the *Companies Act*, which requires companies to keep accounting records sufficient to show and explain their transactions and to prepare accounts that give a true and fair view of their activities. Accounting records take the form of a manual or computerized ledgers, journals, and the supporting documentation.

Accounts receivable financing A short-term financing method in which accounts receivable are collateral for cash advances.

Accounts receivable turnover The ratio of net credit sales to average accounts receivable, a measure of how quickly customers pay their bills.

Accreting Swap An interest rate swap in which the notional principal amount increases over time, for example as with a construction loan provided in tranches as each stage of the project is completed.

Accrual accounting A system of accounting in which revenue is recognized when it is earned and expenses are recognized as they are incurred. Accrual accounting is a basic accounting concept used in the preparation of the profit and loss account and balance sheet of a business. It differs from cash accounting, which recognizes transactions when cash has been received or paid. In preparing financial statements for an accounting period using accrual accounting, there will inevitably be some estimation and uncertainty in respect of transactions. The reader of the financial statements therefore cannot have the same high level of confidence in these statements as in those using cash accounting. Income and expenses should be matched with one another, as far as their relationship can be established or justifiably assumed and dealt with in the profit and loss account of the period to which they relate. However, if there is a conflict between the accruals concept and the prudence concept, the latter prevails. Accruals and prepayments are examples of the application of the accruals concept in practice. For example, if a rates bill for both a current and future period is paid, that part relating to the future period is carried forward as a current asset (a prepayment) until it can be matched to the future periods.accrued benefits Benefits due under a pension scheme in respect of service up to a given time, irrespective of whether the rights to the benefits are vested or not. Accrued benefits may be calculated in relation to current earnings of protected final earnings.

The benefits that the members assumed to be in service on the given date will receive for service up to that date only Allowance may be made for expected increases in earnings after the given date, and for additional pension increases not promised by the rules. The given date may be a current or future date. The further into the future the adopted date lies, the closer the results will be to those obtained by a prospective benefits valuation method.

Accrual Accounting Convention An accounting system that tries to match the recognition of revenues earned with the expenses incurred in generating those revenues. It ignores the timing of the cash flows associated with revenues and expenses.

Accrual bond A bond on which interest accrues but is not paid to the investor during the time of accrual. The amount of accrued interest is added to the remaining principal of the bond and is paid at maturity.

Accrued discount Interest that accumulates on savings bonds from the date of purchase until the date of redemption or final maturity, whichever comes first. Series A, B, C, D, E, EE, F, I, and J are discount or accrual bonds, meaning principal and interest are paid when the bonds are redeemed. Series G, H, HH, and K are income bonds, and the semiannual interest paid to their holders is not included in accrued discount.

Accrued interest Applies mainly to convertible securities. Interest that has accumulated between the most recent payment and the sale of a bond or other fixed-income security. At the time of sale, the buyer pays the seller the bond's price plus "accrued interest," calculated by multiplying the coupon rate by the fraction of the coupon period that has elapsed since the last payment. (If a bondholder receives Rs.40 in coupon payments per bond semiannually and sells the bond one-quarter of the way into the coupon period, the buyer pays the seller Rs.10 as the latter's proportion of earned.)

Accumulate Broker/analyst recommendation that could mean slightly different things depending on the broker/analyst. In general, it means to increase the number of shares of a particular security over the near term, but not to liquidate other parts of the portfolio to buy a security that might skyrocket. A buy recommendation, but not an urgent buy.

Accumulated Benefit Obligation (ABO) An approximate measure of the liability of a plan in the event of a termination at the date the calculation is performed. Related: projected benefit obligation.

Accumulated dividend A dividend that has not been paid to a holder of cumulative preference shares and is carried forward (i.e. accumulated) to the next accounting period. It represents a liability to the company. The *Companies Act* requires that where any fixed cumulative dividends on a company's shares are in arrears, both the amount of the arrears and the period(s) in arrears must be disclosed for each class of shares.

Accumulated profits The amount showing in

the appropriation of profits account that can be carried forward to the next year's accounts, i.e. after paying dividends, taxes, and putting some to reserve.

Accumulating shares Ordinary shares issued to holders of ordinary shares in a company, instead of a dividend. Accumulating shares are a way of replacing annual income with capital growth; they avoid income tax but not capital gains tax. Usually tax is deducted by the company from the declared dividend in the usual way, and the net dividend is then used to buy additional ordinary shares for the shareholder.

Accumulation unit A unit in an investment trust in which dividends are ploughed back into the trust, after deducting income tax, enabling the value of the unit to increase. It is usually linked to a life-assurance policy.

Achievement motivation theory The theory that establishes a relationship between personal characteristics, social background and achievement. A person with a strong need for achievement tends to exhibit such characteristics as:Regarding the task as more important than any relationship;

Having a preference for tasks over which they have control and responsibility;

Needing to identify closely, and be identified closely, with the successful outcomes of their actions; Seeking tasks that are sufficiently difficult to be challenging, to be capable of demonstrating expertise, and to gain recognition from others, while also being sufficiently easy to be capable of achievement; Avoiding the likelihood and consequences of failure; Requiring feedback on achievements to ensure that success is recognized; Needing opportunities for promotion.

The need for achievement is based on a combination of an intrinsic motivation (drives from within the individual) and an extrinsic motivation (pressures and expectations exerted by an organization peers, and society). Achievement is also clearly influence by education, social awareness, cultural background, and values.

Accumulation In the context of corporate finance, refers to profits that are added to the capital base of the company rather than paid out as dividends. See: Accumulated profits tax. In the context of investments, refers to the purchase by an institutional broker of a large number of shares over a period of time in order to avoid pushing the price of that share up. In the context of mutual funds, refers to the regular investing of a fixed amount while reinvesting dividends and capital gains.

Acquired surplus The surplus acquired when a company is purchased in a pooling of interests combination, i.e. the net worth not considered to be capital stock.

Acquisition cost Refers to the price (including the closing costs) to purchase another company or property.

In the context of investments, refers to price plus brokerage commissions, of a security, or the sales charge applied to load funds.

Acquisition of stock A merger or consolidation in which an acquirer purchases the acquiree's stock.

Across the board Movement or trend in the stock market that causes all stocks in all sectors to move in the same direction.

Active Management The pursuit of investment returns in excess of a specified benchmark.

Active portfolio strategy A strategy that uses available information and forecasting techniques to seek better performance than a buy and hold portfolio. Related: Passive portfolio strategy.

A-D Advance-Decline, or measurement of the number of issues trading above their previous closing prices less the number trading below their previous closing prices over a particular period. As a technical measure of market breadth, the steepness of the AD line indicates whether a strong bull or bear market is under way.

Ad valorem tax A type of tax calculated based on percentage of gross or stated value. For example, VAT.

Adequacy of coverage A test that measures the extent to which the value of an asset is protected from potential loss either through insurance or hedging.

Adequate protection Relief fashioned to protect the value of a secured creditor's lien so it does not diminish during the bankruptcy proceeding. This relief can take the form of periodic payments, interest payments, a replacement lien on other property, or any other form determined by the court to be "adequate."

Adjustable rate Applies mainly to convertible securities. Refers to interest rate or dividend that is adjusted periodically, usually according to a standard market rate outside the control of the bank or savings institution, such as that prevailing on Treasury bonds or notes. Typically, such issues have a set floor or ceiling, called caps and collars that limits the adjustment.

Adjustable-rate mortgage (ARM) A mortgage that features predetermined adjustments of the loan interest rate at regular intervals based on an established index. The interest rate is adjusted at each interval to a rate equivalent to the index value plus a predetermined spread, or margin, over the index, usually subject to per-interval and to life-of-loan interest rate and/or payment rate caps.

Adjusted balance method Method of calculating finance charges that uses the account balance remaining after adjusting for all transactions posted during the given billing period as its basis. Related: Average daily balance method, previous balance method, past due balance method.

Adjusted basis Price from which to calculate and derive capital gains or losses upon sale of an asset. Account actions such as any stock splits that have occurred since the initial purchase must be accounted for.

Adjusted debit balance (ADB) The account balance for a margin account that is calculated by combining the balance owed to a broker with any outstanding balance in the special miscellaneous account, and any paper profits on short accounts.

Adjusted gross income (AGI) Gross income less allowable adjustments, which is the income on which an individual is taxed by the federal government.

Adjusted present value (APV) The net present value analysis of an asset if financed solely by equity (present value of un-levered cash flows), plus the present value of any financing decisions (levered cash flows). In other words, the various tax shields provided by the deductibility of interest and the benefits of other investment tax credits are calculated separately. This analysis is often used for highly leveraged transactions such as a leverage buy-out.

Adjusting events (post-balancesheet events) Events that occur between a balance-sheet date and the date on which financial statements are approved, providing additional evidence of conditions existing at the balancesheet date. For example, a valuation of a property held at the balance-sheet date that provides evidence of a permanent diminution in value would need to be adjusted in the financial statements. Such events include those that, because of statutory or conventional requirements, are reflected in financial statements. Statement of Standard Accounting Practice 17, Accounting for Post Balance Sheet Events', requires that such material events should be reflected in the actual account balances in the financial accounts, where they purport to give a true and fair view.A non-adjusting event concerns conditions that did not exist at the balancesheet date; such events may need to be disclosed in the notes to the accounts. An issue of shares after the post-balancesheet period would be an example of a non-adjusting event. If an event would be classed as non-adjusting, but the application of the going-concern concept to the whole, or a material part, of the company is not appropriate, it should be treated as an adjusting event. For example, if serious industrial action has occurred, which if it continues could threaten the continued existence of the business, an appropriate provision should be made in the accounts.administered strategic alliance (administered vertical marketing system; AVMS) A vertical marketing system:(retailer?distributor?manufacturer) in which a powerful channel captain (quite often a major retailing chain) coordinates marketing activities at all levels in the channel, including planning and management, even though they do not directly own the other channel members.administration order 1. An order made in a county court for the administration of the estate of a judgement debtor. The order normally requires the debtor to pay all debts by instalments; so long as he does so, the creditors referred to in the order cannot enforce their individual claims by other methods without the leave of the court. Administration orders are issued when the debtor has multiple debts but it is thought that bankruptcy can be avoided. 2. An order of the court under the made in relation to a company in financial difficulties with a view to securing its survival as a going concern or, failing that, to achieving a more favourable realization of its assets than would be possible on a liquidation. While the order is in force, the affairs of the company are managed by an administrator.administrative management (classical school of management) The traditional view of management that centres on how a business should be organized and the practices an effective manager should follow. The two major contributors to this school of thought were Henri Fayol (1930) and Max Weber (1922). Fayol's 14 principles of management are still relevant, while Weber's bureaucracy model still has some relevance in medium and large organizations.administrative receiver A receiver appointed by the holder of a floating charge covering the whole, or substantially all, of a company's assets in order to recover money due to a secured creditor. The administrative receiver has the power to sell the assets that are secured by the charge or to carry on the company's business.administrator 1. Any person appointed by the courts, or by private arrangement, to manage the property of another. 2. Any person appointed by the courts to take charge of the affairs of a deceased person, who died without making a will. This includes collection of assets, payment of debts, and distribution of the surplus to those persons entitled to inherit, according to the laws of intestacy. The administrator must be in possession of letters of administration as proof of the authority vested by the courts.adoption of innovations It has been suggested that those consumers who eventually accept an innovation can fall into the five groups shown in the diagram. Understanding each of these groups should help a company in devising effective marketing strategies.

Innovators — the first to buy and use new products, i.e. they put the innovation on show to create the image of being venturesome. They are likely to communicate with and persuade others to try the product. This group is defined as the first 2 ½% to adopt the new product.

Early adopters — tend to be opinion leaders and to adopt new ideas early but carefully. This group is defined as the next 13 ½% of the adopting consumers.

Early majority — people regarded as being deliberate in their decisions, who are rarely leaders. These form the next 34% of the adopting consumers.

Late majority – skeptics who only adopt an innovation after most other people have tried it. These form the next 34% of the adopting consumers.

Laggards – the most tradition-bound, who are suspicious of changes and innovations. They tend only to adopt the innovation when it has become widely accepted (i.e. no longer an innovation). The innovators are clearly critical to the process of adoption. Without their support an innovation is unlikely to be successful. *See also* product life cycle.

Adoption process The mental and behavioral stages through which an individual passes before making a purchase or placing an order. The stages are: awareness?interest?evaluation?trial, and finally adoption of the product or service.

Adudication 1. The judgement or decision of a court, especially in bankruptcy proceedings.

2. An assessment by the Commissioners of Inland Revenue of the amount of stamp duty due on a document. A document sent for adjudication will either be stamped as having no duty to pay or the taxpayer will be advised how much is due. An appeal may be made to the High Court if the taxpayer disagrees with the adjudication.

Adustable-rate mortgage (ARM) A mortgage in which the interest rate is adjusted at periodic intervals, usually to reflect the prevailing rate of interest in the money markets. Borrowers are often protected by a cap, or ceiling, above which the interest rate is not allowed to rise. Such a cap may be reviewed annually or may cover the whole term of the mortgage. Such variable-rate mortgages usually start with a lower rate than a fixed-rate mortgage in order to attract borrowers. In the USA this low initial rate is often known as a teaser rate.

Advance commitment A promise to sell an asset before the seller has lined up purchase of the asset. This seller can offset risk by purchasing a futures contract to fix the sales price approximately.

Advance Computerized Execution System (ACES) Refers to the Advance Computerized Execution System, run by Nasdaq. ACES automates trades between order entry and market maker firms that have established trading relationships with each other. Securities are designated as specified for automatic execution.

Advance refunding In the context of municipal bonds, refers to the sale of new bonds (the refunding issue) before the first call date of old bonds (the issue to be refunded). The refunding issue usually specifies a rate lower than the issue to be refunded, and the proceeds are invested, usually in government securities, until the higher-rate bonds become callable.

Advertising A communication that is paid for by an identified sponsor with the object of promoting ideas, goods, or services. It is intended to persuade and sometimes to inform. The two basic aspects of advertising are the message and the medium. The media that carry advertising range from the press, television, cinema, radio, and posters to company logos on apparel. Advertising creates awareness of a product, extensive advertising creates confidence in the product, and good advertising creates a desire to buy the product. Advertising is a part of an organization's total marketing communications programme (i.e. its promotion mix).

Advertising agency A business organization specializing in planning and handling advertising on behalf of clients. A full-service agency provides a range of services to clients, including booking advertising space, designing and producing advertisements, devising media schedules, commissioning research, providing sales promotion advice, and acting as a marketing consultant. The departments within an agency include research, planning, creative design, media bookings, production, and accounts. Most advertising agents work on the basis of a commission on the total sums spent by the client.

Advertising brief An agreement between an advertising agency and a client on the objectives of an advertising campaign. It is important that the client knows exactly what the objectives are, helps to plan the overall strategy, and sets the budgets. Once the brief has been agreed the agency can prepare and evaluate the advertisements themselves and develop the media plan.

Advertising evaluation An attempt to measure the effectiveness of an advertising campaign. Generally this involves an assessment of the communications and sales effects of the advertisement as well as copy testing. This testing can be done before or after an advertisement is broadcast or published. Testing before would include direct rating, portfolio tests, and laboratory tests. Testing after would include recall tests (aided and unaided) as well as recognition tests. With communication and sales effectiveness, clearly the intention is to make consumers aware of and interested in the product, leading to the act of buying and producing satisfaction. It is recognized, however, that some potential, buyers are unlikely to follow this course, particularly if the advertising is ineffective and the product does not meet the buyers' needs.

Advertising objective (communication goals) A specific communications (advertising) task to be achieved, with a specific target audience, during a specific period. In the context of marketing, the objectives will be to gain attention, to be understood, to be believed, and to be remembered (i.e. to inform, persuade, compare, and remind).

Advertising theme An advertising appeal used in several different advertisements to give continuity to an advertising campaign. For example, Visa credit cards are advertised as the 'most widely accepted card', while the theme of American Express card advertising is 'the card has no spending limits'.

Advice note A note sent to a customer by a supplier of goods to advise him that an order has been fulfilled. The advice note may either accompany the goods or be sent separately, thus preceding the invoice and any delivery note. The advice note refers to a particular batch of goods, denoting them by their marks and numbers (if more than one package); it also details the date and method of dispatch.

Advocacy promotion A firm's attempt to build customer goodwill and develop a better public image by promoting causes that are not naturally tied in to its products. A breakfast cereal producer supporting the Worldwide Fund for Nature (WWF) might be an example.

Affidavit A sworn written statement by a person (the deponent), who signs it in the presence of a commissioner for oaths. It sets out facts known to the deponent. In certain cases, particularly proceedings in the Chancery division of the High Court, evidence may be taken by affidavit rather than by the witness appearing in person.

Affidavit of Loss A sworn statement describing the particulars and circumstances of the loss of securities. This affidavit is required before a Bond of Indemnity can be issued and the securities replaced.

Affirmation of contract Treating a contract as being valid, rather than exercising a right to rescind it for a good reason. Affirmation can only occur if it takes place with full knowledge of the facts. It may take the form of a declaration of intention, be inferred from such conduct as selling goods purchased under the contract, or allowing time to pass without seeking a remedy.

Affirmative obligation A New York Stock Exchange rule that governs the behavior of specialists. Affirmative obligation is the mandate of the specialists to step in and act as either the buyer or the seller when public investor orders exist do not match up naturally.

Affordability index An index that measures the financial ability of consumers to purchase a home.

Afghani (AF) The standard monetary unit of Afghanistan, divided into 100 puls.

Afloat Denoting goods, especially commodities, that are on a ship from their port of origin to a specified port of destination; for example, "afloat Rotterdam" means the goods are on their way to Rotterdam. The price of such goods will usually be between the price of spot goods and goods for immediate shipment from origin.

After acquired clause A contractual clause in a mortgage agreement stating that any additional mortgageable property attained by the borrower after the mortgage is signed will be regarded as additional security for the obligation addressed in the mortgage.

After-hours deals Transactions made on the London Stock Exchange after its official close at the end of its mandatory quote period. These deals are recorded as part of the next day's trading and referred to as early bargains. The mandatory quote period on the London Stock Exchange for the Stock Exchange Automated Quotations System (SEAQ) is 8.30 am to 4.30 pm, and on SEAQ International, 9.30 am to 4.00 pm.

After-sales service Maintenance of products by the manufacturers or their agents after purchase. This often takes the form of a guarantee, which is effective for a stated period during which the service is free in respect of both parts and labour, followed by a maintenance contract for which the buyer of the product has to pay. Efficient and effective after-sales service is an essential component of good marketing policy, especially for such consumer durables as cars and computers; in the case of exported goods it is of overriding importance.

Age analysis A listing of debtors' accounts (i.e. the amounts owing to a business), usually produced monthly, which analyses the age of the debts by splitting them into such categories as those up to one month old, two months old, and more than two months old. As a basic part of the credit control system, the analysis should be regularly examined so that any appropriate follow-up action may be taken.

Aged fail An account between two broker/dealers that remains intact 30 days after the settlement date. The receiving firm must adjust its capital as it can no longer treat this account as an asset.

Agency bank A form of organization commonly used by foreign banks to enter the US market. An agency bank cannot accept deposits or extend loans in its own name; it acts as agent for the parent bank. It is also the financial_institution that issues ADRs to the general market.

Agency basis A means of compensating the broker of a program trade solely on the basis of commission established through bids submitted by various brokerage firms. agency incentive arrangement. A means of compensating the broker of a program trade using benchmark prices for issues to be traded in determining commissions or fees.

Agency In context of general equities, buying or selling for the account and risk of a customer. Generally, an agent, or broker, acts as intermediary between buyer and seller, taking no financial risk personally or as a firm, and charging a commission for the service. The broker represents a customer buyer/seller to a customer seller/buyer and does not act as principal for the firm's own trading account. Antithesis of principal. See: Dealer.

Agency loan A loan available for localgovernment authorities, public organizations, etc., from the European Investment Bank.

Agenda The list of items to be discussed at a business meeting. For the annual general meeting or an extraordinary general meeting of a company, the agenda is usually sent to shareholders in advance. *See also*order of business.

Agent A party appointed to act on behalf of a principal entity or person. In context of project financing, refers to the bank in charge of administering the project financing.

Aggregate demand The sum of demands for all the goods and services in an economy at any particular time. Made a central concept in macroeconomics by J M Keynes (1883-1946), it is usually defined as the sum of consumers' expenditure, investment, government expenditure, and imports less exports. Keynesian theory proposes that the free market will not always maintain a sufficient level of aggregate demand to ensure full employment and that at such times the government should seek to stimulate aggregate demand, However, many macroeconomists have questioned the feasibility of such policies and this remains a critical issue in macroeconomics.

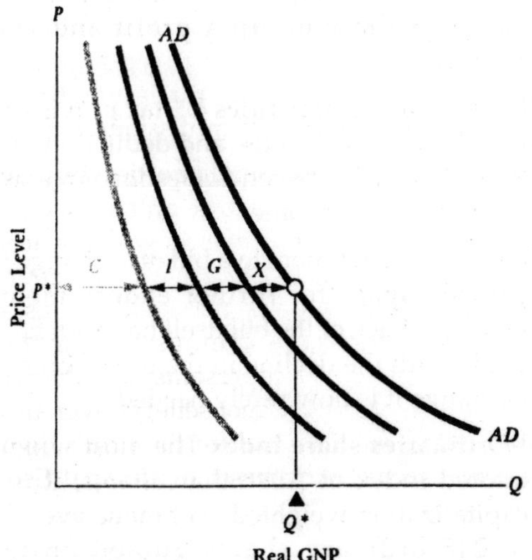

Fig. Aggregate demand (Components)

Aggregate exercise price The exercise price multiplied by the number of shares in a put or call contract. The option premium is excluded in the aggregate exercise price. In the case of options traded on debt instruments, the aggregate exercise price is the exercise price of the underlying security multiplied by its face value.

Aggregate planning An approach to planning that enables overall output levels and the appropriate resource input mix to be set for related groups of products over the near to medium term. Thus the aggregate plan for a car factory would consider the total number of units produced per month, but would not be concerned with the scheduling of individual models, colours, standards of finish, engine sizes, etc. It would be concerned with the labour and machine hours available in a given period but not with what each machine would be doing at a particular time. Aggregate planning assumes that the product/service mix remains stable over the period of the plan and that the capacity remains fixed, i.e. that problems cannot be solved by bringing new plant on stream within the timescale of the plan. Aggregate planning may also seek to influence demand through such variables as price, advertising, and the product mix.

Aggregation Process in corporate financial planning whereby the smaller investment proposals of each of the firm's operational units are added up and in effect treated as a big picture.

Aggressively Used in context of general equities. For a customer it means working to buy or sell one's stock, with an emphasis on execution over price. For a trader it means acting in a way that puts the firm's capital at higher risk through paying a higher price, selling cheaper, or making a larger short sale or purchase than the trader would under normal circumstances.

Agreement among underwriters A contract among participating members of a syndicate that defines the members' proportionate liability, which is usually limited to and based on the participants' level of involvement. The contract outlines the payment schedule on the settlement date.

Agreement corporation Corporation chartered by a state to engage in international banking: so named because the corporation enters into an "agreement" with the Fed's Board of Governors that it will limit its activities to those permitted by an Edge Act Corporation.

Agregate supply The total supply of all the goods and services in an economy. J M Keynes made aggregate demand the focus of macroeconomics; however, since the 1970s many economists have questioned the importance of aggregate demand in determining the health of an economy, suggesting instead that governments should concentrate on establishing conditions to encourage the supply of goods and services. This could entail deregulation, encouraging competition, and removing restrictive practices in the labour market.

Air Freight Consolidator An air freight carrier that does not own or operate its own aircraft but ships its cargo with actual equipment operating carriers. Consolidators issue house air waybills to their customers and receive master air waybills from the actual carriers.

Alienation of assets The sale by a borrower of some or all of the assets that form the actual or implied security for a loan. It is therefore common practice to include a clause in the document setting up a loan, which restricts the disposal of the borrower's assets to specific circumstances.

All or none order (AON) Used in context of general equities. A limited price order that is to be executed in its entirety or not at all (no partial transaction), and thus is testing the strength/conviction of the counterparty. Unlike an FOK order, an AON order is not to be treated as cancelled if not executed as soon as it is represented in the trading crowd, but instead remains alive until executed or cancelled. The making of "all or none" bids or offers in stocks is prohibited, and the making of "all or none" bids or offers in bonds is subject to the restrictions of Rules. AON orders are not shown on the specialist's book because they cannot be traded in pieces. Antithesis of any-part-of order.

Alligator spread The term used to describe a spread in the options market that generates such a large commission that the client is unlikely to make a profit even if the markets move as the investor anticipated.

All-inclusive income concept A concept used in drawing up a profit and loss account, in which all items of profit and loss are included in the statement to arrive at a figure of earnings; this is the approach adopted in the UK and the USA. Although it is claimed that this basis gives the fullest picture of the operation of an enterprise, it does lead to a volatility in earnings figures as oneoff costs, such as redundancies and sale of assets, will be included. To assist prediction of future profits, users are often more interested in the sustainable profits, which are shown using reserve accounting, which is the alternative basis for drawing up a profit and loss account.

Allocation-of-income rules US tax provisions that define how income and deductions are to be allocated between domestic source and foreign source income.

Allonge An attachment to a bill of exchange to provide space for further endorsements when the back of the bill itself has been fully used. With the decline in the use of bills of exchange it is now rarely needed.

All-Ordinaries share index The most widely quoted index of Australian shares; it is a capitalization-weighted arithmetic average of 245 ordinary shares quoted on the Australian Stock Exchange. The index encompasses two others, which are sometimes quoted independently: the All-Mining index (accounting for 17.6% of the AllOrdinaries) and the All-Resources (34.1%).

All-or-none underwriting An arrangement whereby a security issue is cancelled if the underwriter is unable to resell the entire issue.

Allowance 1. A tax-free amount that may be deducted before a particular tax is calculated. Typical allowances in assessing a person's liability to income tax are the personal allowances, e.g. the single-person's allowance and married man's allowance. It also includes tax-free annual amounts, such as the annual allowances available in respect of capital gains tax or inheritance tax.

2. Money paid to an employee for expenses incurred in the course of business. **3.** A deduction from an invoice for a specified purpose, such as substandard quality of goods, late delivery, etc.

4. A price reduction or rebate given to a customer or intermediary on a large order or for some other specific reason.

5. An agreed time used in work measurement; it is added to the basic time in calculating the standard time (*see* standard performance) for a particular job. These allowances provide time for rest, relaxation, and the workers, personal needs (e.g. toilet breaks).

Alotment A method of distributing previously unissued shares in a limited company in exchange for a contribution of capital. An application for such shares will often be made after the issue of a prospectus on the flotation of a public company or on the privatization of a state-owned industry. The company accepts the application by dispatching a letter of allotment to the applicant stating how many shares have been allotted; the applicant then has an unconditional right to be entered in the register of members in respect of those shares. If the number of shares applied for exceeds the number available (oversubscription), allotment is made by a random draw or by a proportional allocation. An applicant allotted fewer shares than applied for is sent a cheque for the unallotted balance (an application must be accompanied by a cheque for the full value of the shares applied for). *See also*multiple application.

Alpha A measure of selection risk (also known as residual risk) of a mutual fund in relation to the market. A positive alpha is the extra return awarded to the investor for taking a risk, instead of accepting the market return. For example, an alpha of 0.4 means the fund outperformed the market-based return estimate by 0.4%. An alpha of -0.6 means a fund's monthly return was 0.6% less than would have been predicted from the change in the market alone. In a Jensen Index, it is factor to represent the portfolio's performance that diverges from its beta, representing a measure of the manager's performance.

Alpha equation The alpha of a fund is determined as follows: [(sum of y) -((b)(sum of x))] / n where: n =number of observations (36 months) b = beta of the fund x = rate of return for the S&P 500 y = rate of return for the fund.

Alphabet stock Categories of common stock of a corporation associated with a particular subsidiary resulting from acquisitions and restructuring. The various alphabetical categories have different voting rights and pay dividends tied to the operating performance of the particular divisions. See also: Tracking stocks.

Alteration of share capital An increase, reduction, or any other change in the authorized capital of a company (*see*share capital (e.g. by consolidating 100 shares of £1 into 25 shares of £4 or by subdividing

100 shares of £1 into 200 of 50p) and cancel unissued shares. These are reserved powers, passed — unless the articles of association provide otherwise — by an ordinary resolution.

Alternative investments Usually refers to investments in hedge funds. Many hedge funds pursue strategies that are uncommon relative to mutual funds. Examples of alternative investment strategies are: long—short equity, event driven, statistical arbitrage, fixed income arbitrage, convertible arbritage, short bias, global macro, and equity market neutral. May also refer to the high frequency style of commodity trading advisors who often employ technical and quantitative tools for intraday investments.

Alternative mortgage instruments Variations of mortgage instruments such as adjustable-rate and variable-rate mortgages, graduated-payment mortgages, reverse-annuity mortgages, and several seldom-used variations.

Alternative order Used in context of general equities. Order giving a broker a choice between two courses of action, either to buy or sell, never both. Execution of one course automatically eliminates the other. An example is a combination buy limit/ buy stop order, where the buy limit is below the current market and the buy stop is above. If the order is for one unit of trading, when one part of the order is executed on the occurrence of one alternative, the order on the other alternative is to be treated as cancelled. If the order is for an amount of more than one unit of trading, the number of units executed determines the amount of the alternative order to be treated as cancelled. Sometimes known as One Cancels the Other. Also see: Either-or order.

Amalgamation The combination of two or more companies. The combination may be effected by one company acquiring others, by the merging of two or more companies, or by existing companies being dissolved and a new company formed to take over the combined business.

Ambulance stocks High performance stocks recommended by a broker to a client whose portfolio has not fulfilled expectations. They either refresh the portfolio — and the relationship between broker and client — or they confirm the client's worst fears.

Amman Stock Exchange The only agency authorized as a formal market for trading securities in Jordan.

Amortization 1. The process of treating as an expense the annual amount deemed to waste away from a fixed asset. The concept is particularly applied to leases, which are acquired for a given sum for a specified term at the end of which the lease will have no value. It is customary to divide the cost of the lease by the number of years of its term and treat the result as an annual charge against profit. While this method does not necessarily reflect the value of the lease at any given time, it is an equitable way of allocating the original cost between periods. *Compare*depreciation

Goodwill may also be amortized. Statement of Standard Accounting Practice 22 recommends as its preferred method the writing-off in the year of purchase of all purchased goodwill. The charge should be to the reserves and not to the profit and loss account. However the standard also permits the writingoff of goodwill to the profit and loss account in regular instalments over the period of its economic life. Home-grown goodwill, if it is in the balance sheet at all, should be similarly dealt with by one of the two methods above.

2. The repayment of debt by a borrower in a series of instalments over a period. Each payment includes interest and part repayment of the capital.

3. The spreading of the front-end fee charged on taking out a loan over the life of a loan for accounting purposes.

4. In the USA, another word for depreciation.

Amortization factor The pool factor implied by the scheduled amortization assuming no prepayemts.

Amortizing mortgage A mortgage in which all the principal and all the interest has been repaid by the end of the mortgage agreement period. Although equal payments may be made during the term of the mortgage, the sums are divided, on a sliding scale, between interest payments and repayments of the principal. In the early years most of the payments go towards the interest charges, while in later years more repays the sum borrowed, until this sum is reduced to zero with the last payment. *Compare*balloon mortgage.

Amount outstanding and in circulation All currency issued by the Bureau of the Mint and intended as a medium of exchange. Coins sold by the Bureau of the Mint at premium prices are not included; uncirculated coin sets sold at face value plus handling charge are included.

Amounts differ The words stamped or written on a cheque or bill of exchange by a banker who returns it unpaid because the amount in words differs from that in figures. Banks usually make a charge for returning unpaid cheques.

Analyst Employee of a brokerage or fund management house who studies companies and makes buy-and-sell recommendations on stocks of these companies. Most specialize in a specific industry.

Ancillary credit business A business involved in credit brokerage, debt adjusting, debt counseling, debt collecting, or the operation of a credit-reference agency. Credit brokerage includes the effecting of introductions of individuals wishing to obtain credit to persons carrying on a consumer-credit business. Debt adjusting is the process by which a third party negotiates terms for the discharge of a debt due under consumer credit agreements or consumer-hire agreements with the creditor or owner on behalf of the debtor or hirer. The latter may also pay a third party to take over an obligation to discharge a debt or to undertake any similar activity concerned with its liquidation. Debt counseling is the giving of advice (other than by the original creditor and certain others) to debtors or hirers about the liquidation of debts due under consumer-credit agreements or consumer hire agreements. A credit-reference agency collects information concerning the financial standing of individuals and supplies this information to those seeking it.

Andean Pact A regional trade pact that includes Venezuela, Colombia, Ecuador, Peru, and Bolivia.

Announcement date Date on which particular news concerning a given company is announced to the public. Used in event studies, which researchers use to evaluate the economic impact of events of interest.

Annual accounts (annual report; report and accounts) The financial statements of an organization, generally published annually. In the UK, incorporated bodies have a legal obligation to publish annual accounts and file them at Companies House. Annual accounts consist of a profit and loss account, balance sheet, cash-flow statement (if required), and statement of total recognized gains and losses, together with supporting notes and the directors' report and auditors' report. Companies falling into the legally defined small companies and medium-sized companies categories may file abbreviated accounts that may not have been audited. Some bodies are regulated by other statutes; for example, many financial institutions and their accounts will have to comply with their own regulations. Non-incorporated bodies, such as partnerships, are not legally obliged to produce accounts but may do so for their own information, for their banks if funding is being sought, and for the Inland Revenue for taxation purposes.

Annual fund operating expenses For investment companies, the management fee and "other expenses," including the expenses for maintaining shareholder records, providing shareholders with financial statements, and providing custodial and accounting services. For 12b-1 funds, selling and marketing costs are included.

Annual general meeting (AGM) An annual meeting of the shareholders of a company, which must be held every year; the meetings may not be more than 15 months apart. Shareholders must be given 21 days' notice of the meeting. The usual business transacted at an AGM is the presentation of the audited accounts, the appointment of directors and auditors, the fixing of their remuneration, and recommendations for the payment of dividends. Other business may be transacted if notice of it has been given to the shareholders.

Annual hours work plan A contractual arrangement in which staff agree to work a certain number of hours per year, rather than per week or month. This is particularly useful when there is a significant fluctuation in the demand for a product or service; it avoids the costs of overtime, part-time, or casual working, etc., to cover the peaks.

Annual percentage rate (APR) In the context of credit cards, the periodic rate times the number of periods in a year. For example, a 1.5% monthly rate has an APR of 18%. In the context of consumer lending, the APR takes into account more than the interest rate applied to the principal per period. Under the Truth in Lending Act, it has a specific definition and includes all the costs paid by a non-exempt consumer borrower that are considered a "finance charge," including fees paid to third parties by the lender if not properly disclosed and excluded from the finance charge (such as credit insurance).

Annual percentage yield (APY) The effective, or true annual rate of return. The APY is the rate actually earned or paid in one year, taking into account the effect of compounding. The APY is calculated by taking one plus the periodic rate, raising it to the number of periods in a year and then subtracting one. For example, a 1% per month rate has an APY of 12.68% .

Annual rate of return There are many ways of calculating the annual rate of return. If the rate of return is calculated on a monthly basis, we sometimes multiply this by 12 to express an annual rate of return. This is often called the annual percentage rate (APR). The annual percentage yield (APY) includes the effect of compounding interest.

Annual report Yearly record of a publicly held company's financial condition. It includes a description of the firm's operations, as well as balance sheet, income statement, and cash flow statement information. SEC rules require that it be distributed to all shareholders. A more detailed version is called a 10-K.

Annual return A document that must be filed with the Registrar of Companies within 14 days of the annual general meeting of a company. Information required on the annual return includes the address of the registered office of the company and the names, addresses, nationality, and occupations of its directors. The financial statements, directors' report, and auditors' report must be annexed to the return. Legally defined small companies and mediumsized companies may file abbreviated accounts. Unlimited companies are exempt from filing the financial statements and dormant companies may not have to be audited. There are penalties for late filing of accounts.

Annuitize To commence a series of payments from the capital that has accumulated in an annuity. The payments may be a fixed amount, for a fixed period of time, or for a lifetime..

Annuity A regular periodic payment made by an insurance company to a policyholder for a specified period of time.

Annuity certain An annuity that pays a specific amount on a monthly basis for a set amount of time.

Annuity in arrears An annuity with a first payment one full period hence, rather than immediately.

Anticipation Arrangements whereby customers who pay before the final date may be entitled to deduct a normal rate of interest.

Antigreenmail Greenmail refers to the agreement between a large shareholder and a company in which the shareholder agrees to sell his stock back to the company, usually at a premium, in exchange for the promise not to seek control of the company for a specified period of time. Antigreenmail provisions prevent such arrangements unless the same repurchase offer is made to all shareholders or approved by shareholder vote. There are some states that have antigreenmail laws.

Anti-Persistence In R/S Analysis, an anti-persistent time series reverses itself more often than a random series would. If the system had been up in the previous period, it is more likely that it will be down in the next period and vice versa. Also called pink noise, or 1/f noise.

Antitrust laws Legislation established by the federal government to prevent the formation of monopolies and to regulate trade.

Any-or-all bid Often used in risk arbitrage. bid in which the acquirer offers to pay a set price for all outstanding shares of the target company, or any part thereof; contrasts with two-tier bid.

APEX Abbreviation for AdvancePurchase Excursion. It refers to a form of international return airline ticket offered at a discount to the standard fare, provided that bookings both ways are made 21 days in advance for international flights and 7-14 days for European flights, with no facilities for stopovers or cancellations.

Appeal A request to the U.S. District Court (or the bankruptcy appellate panel if there is one in the circuit) to review a decision of the bankruptcy court. A request to the Circuit Court of Appeals to review a decision of the U.S. District Court (or the bankruptcy appellate panel). A request to the U.S. Supreme Court to review a decision of a Circuit Court of Appeals.

Appellant A person or organization that appeals against the decision of a court. The party resisting the appeal is called the respondent.

Application for listing The process by which a company applies to a stock exchange for its securities to be traded on that exchange. In obtaining the listing a company will be required to abide by the rules of the exchange. The advantage for a company in obtaining a listing is that it will be able to raise funds by issuing shares on the stock exchange and the marketability of the shares it issues will attract investors.

Application form A form, issued by a newly floated company with its prospectus, on which members of the public apply for shares in the company.

Appraisal rights A right of shareholders in a merger to demand the payment of a fair price for their shares, as determined independently.

Appreciation 1. An increase in the value of an asset, usually as a result of inflation. This usually occurs with land and buildings; the directors of a company have an obligation to adjust the nominal value of land and buildings and other assets in balance sheets to take account of appreciation. *See*asset stripping

2. An increase in the value of a currency with a floating exchange rate relative to another currency.

Aquisition accounting The accounting procedures followed when one company is taken over by another. The fair value of the purchase consideration should, for the purpose of consolidated financial statements, be allocated between the underlying net tangible and intangible assets, other than goodwill, on the basis of the fair value to the acquiring company. Any difference between the fair value of the consideration and the aggregate of the fair values of the separable net assets (including identifiable intangibles, such as patents, licences, and trademarks) will represent goodwill. The results of the acquired company should be brought into the consolidated profit and loss account from the date of acquisition only. n certain circumstances merger accounting may be used when accounting for a business combination. Acquisition accounting differs from merger accounting in that shares issued as purchase consideration are valued at their market price, not par value (*see* share premium account), a goodwill figure may arise on consolidation, and pre-acquisition profits are not distributable. Merger accounting treats both parties as if they had always been combined, and values the purchase consideration at par.

Ar freight 1. The transport of goods y aircraft, either in a scheduled airliner or chartered airliner carrying passengers (an all-traffic service) or in a freight plane (an all-freight service). Air cargo usually consists of goods that have a high value compared to their weight.

2. The cost of transporting goods by aircraft. It is usually quoted on the basis of a price per kilogram.airspace

1. The space that lies above a state's land and sea territory and is subject to its exclusive jurisdiction.

2. The space above a piece of land. The owner of the land is entitled to the ownership and possession of the airspace above the land. This is not exclusive, however, and is limited by the reasonable and necessary use of that airspace by any neighbours as well as for the flight of aircraft.airtime

1. The amount of time allocated to an advertisement on radio or television.

2. The time of transmission of an advertisement on radio or television.air waybill (air consignment note) A document made out by a consignor of goods by air freight to facilitate swift delivery of the goods to the consignee. It gives the name of the consignor and the loading airport, the consignee and the airport of destination, a description of the goods, the value of the goods, and the marks, number, and dimensions of the packages.Aktb Abbreviation for *Aktiebolaget*. It appears after the name of a Swedish joint-stock company.Aktiengesellschaft A German, Austrian, or Swiss public limited company. A set of well-defined rules for solving a problem in a finite number of steps. Algorithms are extensively used in computer science. The steps in the algorithm are translated into a series of instructions that the computer can understand. These instructions form the computer program.alienation A feeling of detachment that causes employees to believe that their work is neither a relevant nor an important part of their life. Because conflicts and disputes, leading to underperformance, can be caused by alienation, managers need to be aware of its existence and its causes if they are to improve relationships within a workforce. Some of the principal causes of alienation are:

Powerlessness — the inability to influence work conditions, quality, volume, etc.;

Meaninglessness — an absence of recognition of the contribution made by the individual to the output of work;

Isolation— the absence of human interaction during working hours; this may result from the nature of the location or a psychological gap between individuals and their supervisors and managers;

Low self-esteem – a reflection of the lack of value placed on individuals by an organization and its managers;

Lack of prospects – a feeling of frustration at being trapped in a situation that offers little prospect of advancement;

Lack of equality – a result of strict differentiation between the grades and levels in a hierarchical organization.

Only by addressing these causes can managers achieve groups of people who feel pride in their work, supported in solving their problems, and rewarded for their skills and effort.

Arbitrage bonds Municipality issued bonds issued intended to gain an interest rate advantage by refunding a higher-rate bond in ahead of their call date. Lower-rate refunding issue proceeds are invested in Treasuries until the first call date of the higher-rate issue.

Arbitrage Pricing Theory (APT) An alternative model to the capital asset pricing model developed by Stephen Ross and based purely on arbitrage arguments. The APT implies that there are multiple risk factors that need to be taken into account when calculating risk-adjusted performance or alpha.

Arbitrage The non-speculative transfer of funds from one market to another to take advantage of differences in interest rates, exchange rates, or commodity prices between the two markets. It is non-speculative because an arbitrageur will only switch from one market to another knowing exactly what the rates or prices are in both markets and will only make the switch if the profit to be gained outweighs the costs of the operation. Thus, a large stock of a commodity in a user country may force its price below that in a producing country; if the difference is greater than the cost of shipping the goods back to the producing country, this could provide a profitable opportunity for arbitrage. Similar opportunities arise with bills of exchange and foreign currencies.

Arbitrageur One who profits from the differences in price when the same, or extremely similar, security, currency, or commodity is traded on two or more markets. The arbitrageur profits by simultaneously purchasing and selling these securities to take advantage of pricing differentials (spreads) created by market conditions. See: Risk arbitrage, convertible arbitrage, index arbitrage, and international arbitrage.

Architectural innovation An innovation that creates an improvement in the ways in which components, at least some of which may not in themselves be innovative, are put together. Examples include flexible manufacturing systems and networked computer systems.

Arithmetic average (mean) rate of return Arithmetic mean return.

Arithmetic mean (arithmetic average) An average obtained by adding together the individual numbers concerned and dividing the total by their number. For example, the arithmetic mean of 7, 20, 107, and 350 is 484/4 = 121. This value, however, gives no idea of the spread of numbers. *Compare*geometric mean; weighted average.

Arm's length price The price at which a willing buyer and a willing unrelated seller would freely agree to transact or a trade between related parties that is conducted as if they were unrelated, so that there is no conflict of interest in the transaction.

ARMs Adjustable rate mortgage. A mortgage that features predetermined adjustments of the loan interest rate at regular intervals based on an established index. The interest rate is adjusted at each interval to a rate equivalent to the index value plus a predetermined spread, or margin, over the index, usually subject to per-interval and to life-of-loan interest rate and/or payment rate caps.

Arms index. The index is usually calculated as the number of advancing issues divided by the number of declining issues. This, in turn, is divided by the advancing volume divided by the declining volume. If there is considerably more advancing volume relative to declining volume this will tend to reduce the index (i.e. increase the denominator). Hence, a value less than 1.0 is bullish while values greater than 1.0 indicate bearish demand. The index often is smoothed with a simple moving average.

Arrangement A method of enabling a debtor to enter into an agreement with any creditors (either privately or through the courts) to discharge his or her debts by partial payment, as an alternative to bankruptcy. This is generally achieved by a scheme of arrangement which involves applying the assets and income of the debtor in proportionate payment to the creditors. For instance, a scheme of arrangement may stipulate that the creditors will receive 20 pence for every pound that is owed to them. This is sometimes also known as composition. Once a scheme of arrangement has been agreed a deed of arrangement is drawn up, which must be registered with the department of Trade and Industry within seven days.

Arranger The senior tier of a syndication. This implies the entity that agreed and negotiated the project financing structure. Also refers to the bank or underwriter entitled to syndicate the loan or bond issue. Also known as the lead underwriter.

Arrears A liability that has not been settled by the due date. For example, cumulative preference shares entitle the shareholders to receive an annual fixed dividend. If this is not paid, the dividend is said to be in arrears and this fact must be disclosed in the notes to the financial statements.

Article of incorporation An official document that details a company's existence. It is similar to the UK memorandum of association.

Articles of association The document that governs the running of a company. It sets out voting rights of shareholders, conduct of shareholders' and directors' meetings, powers of the management, etc. Either the articles are submitted with the memorandum of association when application is made for incorporation or the relevant model articles contained in the Companies Regulations (Tables A to F) are adopted. Table A contains the model articles for companies limited by shares. The articles constitute a contract between the company and its members but this applies only to the rights of shareholders in their capacity as members. Therefore directors or company solicitors (for example) cannot use the articles to enforce their rights. The articles may be altered by a special resolution of the members in a general meeting.

Artificial person A person whose identry is recognized by the law but who is not an individual For example, a company is a person in the sense that it can sue and be sued, hold property, etc. in its own name. It is not, however, an individual or real person.

Asia-Pacific Economic Cooperation Pact (APEC) A loose economic affiliation of Southeast Asian and Far Eastern nations. The most prominent members are China, Japan, and Korea.

Asked price In context of general equities, price at which a security or commodity is offered for sale on an exchange or in the OTC Market.

Aspirin Australian Stock Price Riskless Indexed Notes. Zero-coupon four-year bonds repayable at face value plus the percentage increase by which the Australian stock index of all ordinaries (common stocks) rises above a predefined level during the given period.

Assay Metal purity test to confirm that the metal meets the standards for trading on a commodities exchange (commodities exchange center).

Assented stock A security, usually an ordinary share, the owner of which has agreed to the terms of a takeover bid. During the takeover negotiations, different prices may be quoted for assented and non-assented stock.

Assertiveness training Training courses designed to help employees to develop their abilities, exercise initiative, present themselves convincingly, translate ideas into action, and generally maximize their potential. Assertiveness training aims to raise trainees' selfesteem to enable them to accept that as individuals they have a right to express themselves, to be listened to, and to be taken seriously.

Assessed valuation The value assigned to property by a municipality for the purpose of tax assessment. Such an assessed valuation is important to investors in municipal bonds that are backed by property taxes.

Assessment centre A part of the training department of an organization that uses a range of tests in selecting new employees and considering current employees for promotion. The tests include questionnaires and interviews that attempt to identify the skills and qualities required for a particular job as well as the character traits of the individual. The overall purpose of the assessment centre is to provide an organization with a process of assessment that is consistent, free of prejudice, and fair.

Asset allocation mutual fund A mutual fund that rotates among stocks, bonds, and money market securities to maximize return on investment and minimize risk.

Asset Any object, tangible or intangible, that is of value to its possessor. In most cases it either is cash or can be turned into cash; exceptions include prepayments, which may represent payments made for rent, rates, or motor licences, in cases in which the time paid for has not yet expired. Tangible assets include land and buildings, plant and machinery, fixtures and fittings, trading stock, investments, debtors, and cash; intangible assets include goodwill, patents, copyrights, and trademarks.

Asset classes Categories of assets, such as stocks, bonds, real estate, and foreign securities.

Asset for asset swap Creditors exchange the debt of one defaulting borrower for the debt of another defaulting borrower.

Asset management account Account at a brokerage house, bank, or savings institution that integrates banking services and brokerage features.

Asset stripper A corporate raider (company A) that takes over a target company (company B) in order to sell large assets of company B to repay debt. Company A calculates that the net, selling off the assets and paying off the debt, will leave the raider with assets that are worth more than what it paid for company B.

Asset swap An interest rate swap used to alter the cash flow characteristics of an institution's assets in order to provide a better match with its liabilities.

Asset turnover Measure of operational efficiency - shows how much revenue is produced per £ of assets available to the business. (sales revenue/total assets less current liabilities).

Asset valuation An assessment of the value at which the assets of an organization, usually the capital assets, should be entered into its balance sheet. The valuation may be arrived at in a number of ways; for example, a revaluation of land and buildings would often involve taking professional advice.

Asset value (per share) The total value of the assets of a company less its liabilities, divided by the number of ordinary shares in issue. This represents in theory, although probably not in practice, the amount attributable to each share if the company was wound up. The asset value may not necessarily be the total of the values shown by a company's balance sheet, since it is not the function of balance sheets to value assets. It may, therefore, be necessary to substitute the best estimate that can be made of the market values of the assets (including goodwill) for the values shown in the balance sheet. If there is more than one class of share, it may be necessary to deduct amounts due to shareholders with a priority on winding up before arriving at the amounts attributable to shareholders with a lower priority. If a company goes into liquidation or receivership, the break-up value of the assets may well be below their book value, especially if the assets include specialized or obsolete machinery saleable only for its scrap value.

Asset/liability management Also called surplus management, the task of managing funds of a financial institution to accomplish the two goals of a financial institution:

(1) to earn an adequate return on funds invested and.

(2) to maintain a comfortable surplus of assets beyond liabilities.

Asset-backed fund A fund in which the money is invested in tangible or corporate assets, such as property or shares, rather than being treated as savings loaned to a bank or other institution. Asset-backed funds can be expected to grow with inflation in a way that bank savings cannot.

Asset-based financing Methods of financing in which lenders and equity investors look principally to the cash flow from a particular asset or set of assets for a return on, and the return of, their financing.

Assets Anything owned by the company having a monetary value; eg, 'fixed' assets like buildings, plant and machinery, vehicles (these are not assets if rentedand not owned) and potentially including intangibles like trade marks and brand names, and 'current' assets, such as stock, debtors and cash.

Assets requirements A common element of a financial plan that describes projected capital spending and the proposed uses of net working capital.

Assignment of lease The transfer of a lease by the tenant (assignor) to some other person (assignee). Leases are freely transferable at common law although it is common practice to restrict assignment by conditions (covenants) in the lease. An assignment that takes place in breach of such a covenant is valid but it may entitle the landlord to put an end to the lease and re-enter the premises. An assignment of a legal lease must be by deed. An assignment puts the assignee into the shoes of the assignor, so that there is 'privity of estate' between the landlord and the new tenant. This is important with regard to the enforceability of covenants in the lease. An assignment transfers the assignor's whole estate to the assignee, unlike a sub-lease.

Assignment The receipt of an exercise notice by an options writer that requires the writer to sell (in the case of a call) or purchase (in the case of a put) the underlying security at the specified strike price.

Association of South East Asian Nations (ASEAN) A political and economic grouping of the capitalist nations of South East Asia, formed in 1967 and comprising: Thailand, Malaysia, Singapore, Philippines, Indonesia, and Brunei. The countries are very diverse. For example the per capita income of Singapore in 1986 was some 12 times that of Indonesia; interests often diverge accordingly. While committed to strengthening economic ties, progress has been limited. There has also been political cooperation, for example over policy towards Indochina. There are regular consultations between ASEAN and the major industrialized countries.

Asymmetry A lack of equivalence between two things, such as the unequal tax treatment of interest expense and dividend payments.

At limit An instruction to a broker to buy or sell shares, stocks, commodities, currencies, etc, as specified, at a stated limiting price (i.e. not above a stated price if buying or not below a stated price if selling). When issuing such an instruction the principal should also state for how long the instruction stands, e.g. for a day, a week, etc.

At risk The exposure to the danger of economic loss. Frequently used in the context of claiming tax deductions. For example, a person can claim a tax deduction in a limited partnership if the taxpayer can show it is at risk of never realizing a profit and of losing its initial investment.

At the opening order In context of general equities, market order or limited price order that is to be executed at the opening (and corresponding price) of the stock or not at all, and any such order or portion thereof not so executed is to be treated as cancelled.

Athens Stock Exchange Greece's only major securities market. Greek language only.

Activity sampling (work sampling) A technique in which a large number of observations are made over a period of a group of machines, processes, or workers. Each observation records what is happening at that instant; the percentage of observations recorded for a particular activity enables the total time during which that activity is occurring to be predicted. Traditionally carried out by industrial engineers using manual recording systems, this technique used to be dependent on the correct calculation of the number of observations required to give statistically valid data. Increasing use of on-line computerized data now enables managers to account for every second of their staff's time. The technique is particularly suitable for such processes as phone selling, customer services, etc. It can, however, also be applied in manufacturing systems and to groups, e.g. by requiring teams to bar-code themselves into a work area.

Activity-based management The use made by the management of an organization of activity-based costing. The identification of activities and cost drivers encourages the management to review how projected cost levels compare with the activity levels achieved.

Attest To bear witness to an act or event. The law requires that some documents are only valid and binding if the signatures on them have been attested to by a third party. This also requires the third party's signature on the document. For instance, the signature of the purchaser of land under a contract must be attested to by a witness.

At-the-money An option is at the money if the strike price of the option is equal to the market price of the underlying security. For example, if xyz stock is trading at 54, then the xyz 54 option is at the money.

Attitude The way in which a person views and evaluates something or someone. Attitudes determine whether people like or dislike things — and therefore how they behave towards them. In recent years there has been a change in the link between the cognitive, affective, and behavioural components. The emphasis is now on defining attitude in terms of affect — the person's feelings towards the object, brand, etc. Is the brand good or bad? Is it likeable? In this shift, the importance of the cognitive and behavioural components is still accepted, but they are no longer regarded as critical components.

Attribute bias The tendency of stocks preferred by the dividend discount model to share certain equity attributes such as low price-earnings ratios, high dividend yield, high book-value ratio or membership in a particular industry sector.

Auction A method of sale in which goods are sold in public to the highest bidder. Auctions are used for any property for which there are likely to be a number of competing buyers, such as houses, second-hand and antique furniture, works of art, etc., as well as for certain commodities, such as tea, bristles, wool, furs, etc., which must be sold as individual lots, rather than on the basis of a standard sample or grading procedure. In most auctions the goods to be sold are available for viewing before the sale and it is usual for the seller to put a reserve price on the articles offered, i.e. the articles are withdrawn from sale unless more than a specified price is bid. The auctioneer acts as agent for the seller in most cases and receives a commission on the sale price. An auctioneer is an agent of the seller, who must have the authority of the seller to sell, and must know of no defect in the seller's title to the goods, without promising that a buyer will receive good title for a specific object. An advertisement that an auction will be held does not bind the auctioneer to hold it. It is illegal for a dealer (a person who buys at auction for subsequent resale) to offer a person a reward not to bid at an auction.

Auction Market Preferred Stock (AMPS) A type of Dutch Auction Preferred Stock (A Merrill Lynch product).

Auction markets Markets in which the prevailing price is determined through the free interaction of prospective buyers and sellers, as on the floor of the stock exchange.

Auction rate preferred stock (ARPS) Floating rate preferred stock, whose dividend is adjusted every seven weeks through a Dutch auction.

Audience research Research to establish readership, audience, and circulation data, which is vital information in advertising. Research into television audiences is undertaken by BARB (Broadcasters' Audience Research Board) into radio audiences by JICRAR (Joint Industry Committee for Radio Audience Research, and into national readership by JICNAR (Joint Industry Commitee for National Readership).

Audit An examination of a company's accounting records and books conducted by an outside professional in order to determine whether the company is maintaining records according to generally accepted accounting principles.

Audit evidence The evidence required by an auditor on which to base an audit opinion on the financial statements of the company whose accounts are being audited. Sources of information include the accounting systems and the underlying documentation of the enterprise, its tangible assets, management and employees, its customers, suppliers, and any other third parties who have dealings with, or knowledge of, the enterprise or its business. The evidence will be obtained by means of compliance tests (tests of controls) and substantive tests (tests of details and analytical review). Techniques involved in gathering the evidence include inspection, observation, enquiry, computation, and analytical work

Audit trail Resolves the validity of an accounting entry by a step-by-step record by which accounting data can be traced to their source.

Auditor's report A section of an annual report containing the auditor's opinion about the veracity of the financial statements.

Australian Stock Exchange (ASX) Australia's major securities market, formed when the six state stock exchanges (Adelaide, Brisbane, Hobart, Melbourne, Perth, and Sydney stock exchanges) were merged in 1987.

Autex Video communication network through

which brokerage houses alert institutional investors of their desire to transact block business (a purchase or sale) in a given security. Indications transmit small, medium, and large sizes only, with occasional limits mentioned. Supers are messages with specific size and price included. Both "indications" and "supers" can be only seen by customers (institutional subscribers to Autex). Trade recaps, advertised block trades entered by the dealer/subscribers, are also displayed, but can be seen by both institutions and dealers.

Authority bond A bond issued by a government agency or a corporation created to manage a revenue-producing public enterprise. The difference between an authority bond and a municipal bond is that margin protections may be incorporated in the authority bond contract as well as in the legislation that enables the authority.

Authorized share capital (nominal share capital; nominal capital; regis-tered capital) tered capital) The maximum amount of share capital that may be issued by a company, as detailed in the company's memorandum of association. The authorized share capital must be disclosed on the face of the balance sheet or alternatively in the notes to the accounts. *See also*authorized minimum share capital; issued share capital.

Automated Clearing House (ACH) A collection of 32 regional electronic interbank networks used to process transactions electronically with a guaranteed one-day bank collection float.

Automated guided vehicle (AGV) A self-propelled vehicle guided by cables set in the floor, by radio or microwave transmitters, by onboard computer, or by optical guidance systems. AGVs are used to transport materials or components without human intervention. They are most often used in manufacturing plants, especially if these involve hazardous chemicals or structures, where AGVs can form an integrated part of the process design, replacing conveyor belts, etc.

Automated Order System (AOS) Investment bank computerized order entry system that sends single order entries to DOT (Odd-Lot) or to investment banks floor brokers on the exchange. See: Round lot, GTC orders.

Automated teller machine (ATM) Computer-controlled terminal located on the premises of financial institutions or elsewhere, though which customers may make deposits, withdrawals or other transactions as they would through a bank teller. Other terms sometimes used to describe such terminals are customer-bank communications terminal (CBCT) and remote service unit (RSU). Groups of banks sometimes share ATMs. Sometimes called Automated Banking Machines.

Automatic Data Processing (ADP) A private company that acts as an intermediary to perform proxy services for several banks and brokers. Distributes proxy material to beneficial owners, tabulates the returned proxies, and provides the Corporation or its tabulator compiled reports of the tabulation results. ADP also distributes quarterly reports and other corporate information to the beneficial owners.

Automatic investment program A program in which an investor can invest or withdraw funds automatically. A mutual fund, for example, automatically withdraws a pre determined specified amount from the investor's bank account on a regular basis.

Auto-Regressive (AR) Process A stationary stochastic process where the current value of the time series is related to the past p values, where p is any integer, is called an AR(p) process. When the current value is related to the previous two values, it is an AR(2) process. An AR(1) process has an infinite memory.

Average 1. A single number used to represent a set of numbers; mean. *See* arithmetic mean; geometric mean; median; weighted average

2. A partial loss in marine insurance (from French: *avarie,* damage). In general average (GA) , a loss resulting from a deliberate act of the master of the ship (such as throwing overboard all or part of the cargo to save the ship) is shared by all the parties involved, i.e. by the shipowners and all the cargo owners. *See*average adjuster; average bond. In a particular average (PA) , an accidental loss is borne by the owners of the particular thing lost or damaged, e.g. the ship, an individual cargo, etc. Cargo can be insured either free of particular average (FPA) or with average (WA) . An FPA policy covers the cargo against loss by perils of the sea, fire, or collision and includes cover for any contribution payable in the event of a general average A WA policy gives better cover as it also includes damage by heavy seas and sea-water damage. In addition, marine cargo can be covered by an all-risks policy. *See also*free of all averages ; Institute cargo clause.

3. A method of sharing losses in property insurance to combat underinsurance. This is usually applied in an average clause in a fire insurance policy, in which it is stated that the sum payable in the event of a claim shall not be more than the proportion that the insured value of an item bears to its actual value.

Average bond A promise to pay general average contributions, if required. If a general-average loss occurs during a marine voyage the carrier has a right to take part of the cargo as payment of the cargo owners' contribution to the loss. As an alternative to the possibility of losing part of the cargo, the cargo owner may take out an average bond with insurers, who agree to pay any losses arising in this way.

Average collection period, or days' receivables The ratio of accounts receivables to sales, or the total amount of credit extended per dollar of daily sales (average AR/sales 365).

Average cost In the context of investing, refers to the average cost of shares or stock bought at different prices over time.

Average cost of capital A firm's required payout to bondholders and stockholders expressed as a percentage of capital contributed to the firm. Average cost of capital is computed by dividing the total required cost of capital by the total amount of contributed capital.

Average discount rate Purchasers tender their competitive bids on a discount rate basis. The weighted, or adjusted mean of all bids accepted in Treasury bill auctions.

Average down A strategy used by investors to reduce the average cost of shares, in which the investor purchases more shares with a fixed amount of capital as the price of the shares decreases. The investor receives more shares per dollar and decreases the average price per share.

Average life Also referred to as the weighted-average life (WAL). The average number of years that each dollar of unpaid principal due on the mortgage remains outstanding. Average life is computed as the weighted average time to the receipt of all future cash flows, using as the weights the dollar amounts of the principal paydowns.

Average maturity The average time to maturity of securities held by a mutual fund. Changes in interest rates have greater impact on funds with longer average maturity.

Average up A strategy used by investors to lower the overall cost of shares by buying as many shares with a given amount of capital in an increasing market. Buying Rs.1000 worth of shares at Rs.30, Rs.35, Rs.40, and Rs.45, for instance, will make the average cost of the shares Rs.36.65, lower than the average price of Rs.37.50.

Avoidable costs Costs that are not incurred if a particular course of action is taken or an alternative decision is made. For example, if a specific product is not produced, material and labour costs may not be incurred. In this instance material and labour costs are avoidable costs. Variable costs are often avoidable costs, whereas fixed costs, such as business rates, are not avoidable in the short term.

Away from us Used in context of general equities, to characterize role of a competing broker/dealer. Trading away from us signifies that stock is bought and/or sold with institutions using other trading firms.

Axe to grind Used in context of general equities. Involvement in a security, whether through a position, order, or inquiry.

B

Back away In the context of general equities, to withdraw from a previously declared interest, indication, or transaction; broker-dealer's failure, as a market maker in a given security, to make good on a bid/offer for the minimum quantity.

Back door One of the methods by which the Bank of England injects cash into the money market. The bank purchases Treasury bills at the market rate rather than by lending money directly to the discount houses (the front door method) when it acts as lender of last resort.

Back fee The fee paid on the extension date if the buyer wishes to continue the option.

Back freight The cost of shipping goods back to the port of destination after they have been overcarried. If the master overcarried the goods for reasons beyond his control, the shipowner may be responsible for paying the back freight. If delivery was not accepted in reasonable time at the port of destination and the master took the goods on or sent them back to the port of shipment, the back freight would be the responsibility of the cargo owner.

Back on the shelf In the context of general equities, permanently canceled order/interest in a stock by a customer. See: Take a powder.

Backdating In the context of mutual funds, a feature allowing fundholders to use an earlier date on a letter of intent to invest in a mutual fund in exchange for a reduced sales charge, e.g. Giving retroactive value to purchases from the earlier date.

Backed in In the context of general equities, to describe the result of unanticipated events that allow for a purchase at a discount or a sale at a premium.

Back-end load fund A mutual fund that charges investors a fee to sell (redeem) shares, often ranging from 4% to 6%. Some back-end load funds impose a full commission if the shares are redeemed within a designated length of time, such as one year. The commission decreases, the longer the investor holds the shares. The formal name for the back-end load is the contingent deferred sales charge, or CDSC

Backflush accounting A method of costing a product based on a management philosophy that includes having the minimum levels of stock available; in these circumstances, the valuation of stocks becomes less important, making the complex use of absorption costing techniques unnecessary. Backflush accounting works backwards; after the actual costs have been determined they are allocated between stocks and cost of sales to establish profitability. There is no separate accounting for work in progress.

Back-to-back loan A loan in which two companies in separate countries borrow each other's currency for a specific time period and repay the other's currency at an agreed-upon maturity.

Backup line A commercial paper issuer's bank line of credit covering maturing notes if, for some reason, selling new notes to cover the maturing notes is not possible.

Backward channel A channel of distribution for recycling, in which the customary flow from producer to ultimate user is reversed. For example, bottles, cans, and paper are now returned to the producer for recycling.

Backwardation A market condition in which futures prices are lower in the distant delivery months than in the nearest delivery month. This situation may occur in when the costs of storing the product until eventual delivery are effectively subtracted from the price today. The opposite of contango.

Bad debt An amount owed by a debtor that is unlikely to be paid for example, due to a company going into liquidation. The full amount should be written off to the profit and loss account of the period or to a provision for bad and doubtful debts as soon as it is foreseen, on the grounds of prudence. Bad debts subsequently recovered either in part or in full should be written back to the profit and loss account of the period (or to a provision for bad and doubtful debts).

Bailing out In the context of securities, refers to selling a security or commodity quickly, regardless of the price. May occur when an investor no longer wants to sustain further losses on a stock. Also refers to relieving an individual, corporation, or government entity in financial trouble.

Balance of payments A statistical compilation formulated by a sovereign nation of all economic transactions between residents of that nation and residents of all other nations during a stipulated period of time, usually a calendar year.

Balance of trade Net flow of goods (exports minus imports) between countries.

Balance on goods and services Netting of transaction balances, including the net amount of payments of interest and dividends to foreign investors and investments, as well as receipts and payments resulting from international tourism. Also known as Trade Balance.

Balance sheet A statement of the total assets and liabilities of an organization at a particular date, usually the last day of the accounting period. The first part of the statement lists the fixed and current assets and the liabilities, the second part shows how they have been financed; the totals for each part must be equal. Under the Companies Act the balance sheet is one of the primary statements to be included in the financial accounts of a company. The Companies Act requires that the balance sheet of a company must give a true and fair view of its state of affairs at the end of its financial year, and must comply with statute as to its form and content. There are two possible balance-sheet formats. In both formats, corresponding amounts for the preceding financial year should be shown. A balance sheet does not necessarily value a company, as some assets may be omitted, or given an unrealistic value.

Balanced budget A budget in which the income equals expenditure.

Balance-sheet asset value The book value of an asset as shown on the balance sheet. For tangible fixed assets this is the cost less accumulated depredation (although freehold land is generally not subject to depreciation). Intangible assets are shown at cost less amortization. Current assets are valued at the lower of cost and net realizable value. Under the alternative accounting rules given in the *Companies Act (1985)*, the historical cost of certain assets (for example, buildings and stocks) may be replaced by current cost.

Balancing charge The charge that may be assessable to corporation tax on the disposal of an asset when the proceeds realized on the sale of the asset exceed the written-down value, for tax purposes. The balancing charge amounts to the difference between the proceeds and the written-down value. For example, if the written-down value is £23,000 and the proceeds on disposal were £30,000, there would be a balancing charge of the difference of £7000. The balancing charge is deducted from the other allowances for the period. If the charge exceeds the allowances available, the net amount is added to the profit for the period and assessed to tax.

Balloon maturity Any large principal payment due at maturity for a bond or loan with or without a a sinking fund requirement.

Ballot The document distributed at the annual meeting to shareholders of record who wish to vote their shares in person.

Baltic Exchange A former commodity and freight-chartering exchange in the City of London. It took its name from the trade in grain with Baltic ports, which was the mainstay of the business in the 18th century. Most of its trade, including the Baltic International Freight Futures Exchange (BIFFEX), has been taken over by the London Commodity Exchange The building was badly damaged by an IRA bomb in 1992.

BAN (Bank anticipation notes) Notes issued by states and municipalities to obtain interim financing for projects that will eventually be funded long term through the sale of a bond issue.

Bank A commercial institution licensed as a taker of deposits. Banks are concerned mainly with making and receiving payments on behalf of their customers, accepting deposits, and making short-term loans to private individuals, companies, and other organizations. In the UK, the banking system comprises the Bank of England (the central bank), the commercial banks, merchant banks, branches of foreign

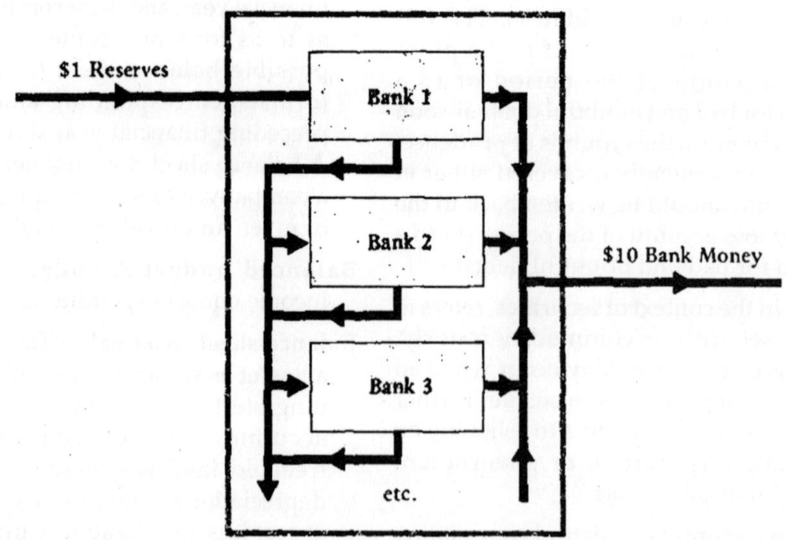

Fig. Bank Expansion of Money

and Commonwealth banks, the TSB Group, the National Savings Bank, and the National Girobank (*see*giro). The first (1990) building society to become a bank in the UK was the Abbey National after its public flotation; other building societies have now followed this precedent. In other countries banks are also usually supervised by a government controlled central bank.

Bank anticipation notes (BAN) Notes issued by states and municipalities to obtain interim financing for projects that will eventually be funded long term through the sale of a bond issue.

Bank bill A bill of exchange issued or guaranteed (accepted) by a bank. It is more acceptable than a trade bill as there is less risk of non-payment and hence it can be discounted at a more favourable rate, although to some extent this depends on the bank's credit rating.

Bank charge The amount charged to a customer by a bank, usually for a specific transaction, such as paying in a sum of money by means of a cheque or withdrawing a sum by means of fan automated teller machine. However, modern practice is to provide periods of commission-free banking by waiving most charges on personal current accounts. The commercial banks have largely been forced to take this step as a result of the free banking services offered by building societies. However, business customers invariably pay tariffs in one form or another.

Bank collection float The time that elapses between when a check is deposited into a bank account and when the funds are available to the depositor, during which period the bank is collecting payment from the payer's bank.

Bank deposit A sum of money placed by a customer with a bank. The deposit may or may not attract interest and may be instantly accessible or accessible at a time agreed by the two parties. Banks may use a percentage of their customers' deposits to lend on to other customers; thus most deposits may only exist on paper in the bank's books. Money on deposit at a bank is usually held in either a deposit account or a current account, although some banks now offer special high-interest accounts.

Bank for International Settlements (BIS) An international bank headquartered in Basel, Switzerland, which serves as a forum for monetary cooperation among several European central banks, the Bank of Japan, and the U.S. Federal Reserve System. Founded in 1930 to handle the German payment of World War I reparations, it now monitors and collects data on international banking activity and promulgates rules concerning international bank regulation.

Bank guarantee An undertaking given by a bank to settle a debt should the debtor fail to do so. A bank guarantee can be used as a security for a loan but the banks themselves will require good cover in cash or counter-indemnity before they issue a guarantee. A guarantee has to be in writing to be legally binding. Such guarantees often contain indemnity clauses, which place a direct onus on the guarantor. This onus leaves the guarantor liable in law in all eventualities.

Bank regulation The formulation and issuance by authorized agencies of specific rules or regulations, under governing law, for the conduct and structure of banking.

Bank run (bank panic) A series of unexpected cash withdrawals caused by a sudden decline in depositor confidence or fear that the bank will be closed by the chartering agency, i.e. many depositors withdraw cash almost simultaneously. Since the cash reserve a bank keeps on hand is only a small fraction of its deposits, a large number of withdrawals in a short period of time can deplete available cash and force the bank to close and possibly go out of business.

Bank statement A regular record, issued by a bank or building society, showing the credit and debit entries in a customer's cheque account, together with the current balance. The frequency of issue will vary with the customer's needs and the volume of transactions going through the account. Cash dispensers enable customers to ask for a statement whenever they are needed.

Banker's acceptance A short-term credit investment created by a non-financial firm and guaranteed by a bank as to payment. Acceptances are traded at discounts from face value in the secondary market. These instruments have been a popular investment for money market funds. They are commonly used in international transactions.

Bankers Automated Clearing System (BACS) A company owned by the UK banks that operates a computerized payments clearing service. It is commonly used by companies for paying employees. *See also* **Association for Payment Clearing Services.**

Banknote An item of paper currency issued by a central bank. Banknotes developed in England from the receipts issued by London goldsmiths in the 17th century for gold deposited with them for safe-keeping. These receipts came to be used as money and their popularity as a medium of exchange encouraged the goldsmiths to issue their own banknotes, largely to increase their involvement in banking and, particularly, moneylending. Now only the Bank of England and the Scottish and Irish banks in the UK have the right to issue notes. Originally all banknotes were fully backed by gold and could be exchanged on demand for gold; however, since 1931 the promise on a note to 'pay the bearer on demand' simply indicates that the note is legal tender.

Bankruptcy clerk This office receives and is responsible for the management and safe keeping of the documents filed under a bankruptcy case or adversary proceeding. The clerk is responsible for keeping a docket for each case, a claims register, correct copies of court documents filed in a case, as well as an index of all cases and adversary proceedings filed. Additionally, the clerk is often responsible for noticing creditors of deadlines within a case or proceeding.

Bankruptcy code The federal bankruptcy statute called the bankruptcy code.

Bankruptcy cost view The argument that expected indirect and direct bankruptcy costs offset the other benefits from leverage so that the optimal amount of leverage is less than 100% debt financing.

Bankruptcy court A unit of the United States District Court (federal court) specializing in the handling of bankruptcy cases.

Bankruptcy The state of individuals who are unable to pay their debts and against whom a **bankruptcy order** has been made by a court. The order deprives bankrupts of their property, which is then used to pay their debts. Bankruptcy proceedings are started by a **bankruptcy petition**, which may be presented to the court by (1) a creditor or creditors; (2) a person affected by a voluntary arrangement to pay debts set up by the debtor under the *Insolvency Act* (3) the Director of Public Prosecutions; or (4) the debtor. The grounds for a creditors' petition are that the debtors appear to be unable to pay their debts or to have reasonable prospects of doing so, i.e. that the debtor has failed to make arrangements to pay a debt for which a statutory demand has been made or that a judgement debt has not been satisified. The grounds for a petition by a person bound by a voluntary arrangement are that the debtor has not complied with the terms of the arrangement or has withheld material information. Debtors themselves may present a petition on the grounds that they are unable to pay their debts. Once a petition has been presented, debtors may not dispose of any of their property. The court may halt any other legal proceedings against the debtor. An interim receiver may be appointed. This will usually be the official receiver, who will take any necessary action to protect the debtor's estate. A special manager may be appointed if the nature of the debtor's business requires it. The court may make a bankruptcy order at its discretion. Once this has happened, the debtor is an undischarged bankrupt, who is deprived of the ownership of all property and must assist the official receiver in listing it, recovering it, protecting it, etc. The official receiver becomes manager and receiver of the estate until the appointment of a trustee in bankruptcy. The bankrupt must prepare a statement of affairs for the official receiver within 21 days of the bankruptcy order. A public examination of the bankrupt may be ordered on the application of the official receiver or the creditors, in which the bankrupt will be required to answer questions about his or her affairs in court. Within 12 weeks the official receiver must decide whether to call a meeting of creditors to appoint a trustee in bankruptcy. The trustee's duties are to collect, realize, and distribute the bankrupt's estate. The trustee may be appointed by the creditors, the court, or the Secretary of State and must be a qualified insolvency practitioner, or the official receiver. All the property of the bankrupt is available to pay the creditors, except for the following: equipment necessary for him or her to continue in employment or business, necessary domestic equipment; and income required for the reasonable domestic needs of the bankrupt and his or her family. The court has discretion whether to order sale of a house in which a spouse or children are living. All creditors must prove their claims to the trustees. Only unsecured claims can be proved in bankruptcy. When all expenses have been paid, the trustee will divide the estate.

Banque de France The central bank of France, which has approximately the same status as the Bank of England. Established in 1800 and nationalized in January 1946, it has some 214 branches.

Bar chart (bar diagram) A chart that presents statistical data by means of rectangles (i.e. bars) of differing heights. For example, the sales figures for a range of products for an accounting period may be presented in this way, the different sizes of the bars enabling the users to see at a glance how each product has performed during the period.

Bar code A code, consisting of an array of parallel rectangular bars and spaces, printed on a package for sale in a retail shop. When a laser scanner (bar-code reader) is passed over the bar code at the retailer's checkout till (*see*electronic point of sale), the price and description of the goods are displayed on the till screen and the computer-controlled stock record is simultaneously reduced.

Barriers to entry Factors that prevent competitors from entering a particular market. These factors may be innocent, e.g. an absolute cost advantage on the part of the firm that dominates the market, or deliberate, such as high spending on advertising to make it very expensive for new firms to enter the market and establish themselves. Other entry barriers may result from a firm's technological advantage, often protected by patents, or from a firm's existing access to end users as a result of its control of the distribution network. Barriers to entry reduce the level of competition in a market, i.e. they make it less contestable, thereby by enabling incumbents to charge higher-than-competitive prices. *See also*limit price. If a particular market for a product is dominated by a large multinational corporation, barriers to entry for smaller firms may be formidable: the multinational may be able to produce the product more cheaply because of economies of scale, the existing product may have built up a strong brand loyalty, the multinational may control the supply of raw material, or own the patent rights to all or part of the production process.

Barriers to exit Factors that make it difficult for a company to leave a market that is no longer profitable or that has ceased to provide an acceptable return on capital. For example, the workforce producing the product may not be redeployable, the plant and machinery producing it may be unsaleable, the product may have a niche in the company's product range that would affect sales of the whole range if it was withdrawn, etc.

Barron's confidence index Index measuring the ratio of the average yield on 10 top-grade bonds to the average yield on 10 intermediate-grade bonds. The discrepancy between high-rated top-grade bonds and low-rated bond yields establishes a measure that is indicative of investor confidence.

Barter A method of trading in which goods or services are exchanged without the use of money. It is a cumbersome system, which severely limits the scope for trade. Money, used as a medium of exchange, enables individuals to trade with each other at much greater distance and through whole chains of intermediaries, which are inconceivable in a barter system. The modern equivalent of barter is known as countertrading.

Barter The trading/exchange of goods or services without using currency.

Base currency Applies mainly to international equities. Currency in which gains or losses from operating an international portfolio are measured.

Base period A particular period of time used for comparative purposes when measuring economic data.

Base rate 1. The rate of interest used as a basis by banks for the rates they charge their customers. In practice most customers will pay a premium over base rate to take account of the bank's risk involved in lending, competitive market pressures, and to regulate the supply of credit.

2. An informal name for the rate at which the Bank of England lends to the discount houses, which effectively controls the lending rate throughout the banking system. The abolition of the minimum lending rate in 1981 heralded a loosening of government control over the banking system, but the need to increase interest rates in the late 1980s (to control inflation and the balance of payments deficit) led to the use of this term in this sense.

Base stock A certain volume of stock, assumed to be constant in that stock levels are not allowed to fall below this level. When the stock is valued, this proportion of the stock is valued at its original cost. This method is not normally acceptable under Statement of Standard Accounting Practice 9 for financial accounting purposes.

Base year (base date) The first of a series of years in an index. It is often denoted by the number 100, enabling percentage rises (or falls) to be seen at a glance. For example, if a price index indicates that the current value is 120, this will only be meaningful if it is compared to an earlier figure. This may be written: 120 (base year 1985 = 100), making it clear that there has been a 20% increase in prices since 1985.

Basel Accord Agreement concluded among country representatives in 1988 in Switzerland to develop standardized risk-based capital requirements for banks across countries.

Basic balance In a balance of payments, the basic balance is the net balance of the combination of the current account and the capital account.

Basic rate of income tax A rate of income tax between the lower and higher rates. In the UK it is 24%, which is applied to taxable income, for 1996-97, in the band £3901 to £25,500.

Basis of assessment The schedules under which personal income or business profits are assessed in the UK for each tax year. The individual rules for each schedule identify the profits or income to be assessed in that year. In many cases this is not the profits or income arising in the actual year. In the case of a partnership that has been trading for many years, the profits for the year to 30 April 1996, i.e. those arising during the period 1 May 1995 to 30 April 1996, will form the basis of the assessment for the tax year 1997-98. This is known as the prior-year basis of assessment. Other income received during the year, e.g. building society interest received, is assessed on an actual basis and so for 1996-97 the basis of assessment will be the tax year, i.e. the interest received during the year 6 April 1996 to 5 April 1997.

Basis point In the bond market, the smallest measure used for quoting yields is a basis point. Each percentage point of yield in bonds equals 100 basis points. Basis points also are used for interest rates. An interest rate of 5% is 50 basis points higher than an interest rate of 4.5%. Sometimes referred to as BPS, BIPS, and pronounced "Bips"

Basis risk Unexpected changes in the basis between the placing and the lifting of a hedge. Basis risk is in excess of convergence.

Basis The price an investor pays for a security plus any out-of-pocket expenses. It is used to determine capital gains or losses for tax purposes when the stock is sold. Also, for a futures contract, the difference between the cash price and the futures price observed in the market.

Basket Applies to derivative products. Group of stocks that is formed with the intention of either being bought or sold all at once, usually to perform index arbitrage or a hedging program.

Basket options Packages that involve the exchange of more than two currencies against a base currency at expiration. The basket option buyer purchases the right, but not the obligation, to receive designated currencies in exchange for a base currency, either at the prevailing spot market rate or at a prearranged rate of exchange. A basket option is generally used by multinational corporations with multicurrency cash flows since it is generally cheaper to buy an option on a basket of currencies than to buy individual options on each of the currencies that make up the basket.

Batch costing A form of costing in which the unit costs are expressed on the basis of a batch produced. This is particularly appropriate where the cost per unit of production would result in an infinitesimal unit cost and where homogeneous units of production can conveniently be collected together to form discrete batches.

Batch processing A method of processing data, using a computer, in which the programs to be executed are collected together into groups, or batches, for processing. All the information needed to execute the programs is loaded into the computer at the start so that it can work without further intervention. This contrasts with interactive processing, in which information is fed into the computer during processing. Batch processing is used for such tasks as preparing payrolls, maintaining inventory records, and producing reports.

Batch production A manufacturing process in which medium to high volumes of similar items are made in batches, rather than continuously, with the product moving from process to process in batches. The key to batch production is the careful scheduling of work to ensure good utilization of capacity and to minimize the capital locked up in work in progress. Examples in which batch production are used include small engineering works and book printers.

Budget manual A manual setting out the administrative procedures and operations that should be applied in the operation of the budgetary control system. It covers guidelines for the operation of the budget committee and the budget centres and includes such information as levels of responsibility, budget timetable, budget preparation, and budget-revision procedures.

BDS Statistic A statistic based upon the correlation integral which examines the probability that a purely random system could have the same scaling properties as the system under study. See: Correlation Integral.

Bear A dealer on a stock exchange, currency market, or commodity market who expects prices to fall. A bear market is one in which a dealer is more likely to sell securities, currency, or goods than to buy them. A bear may even sell securities, currency, or goods without having them. This is known as selling short or establishing a bear position. The bear hopes to close (or cover) such a short position by buying in at a lower price the securities, currency, or goods already sold. The difference between the purchase price and the original sale price represents the successful bear's profit. A concerted attempt to force prices down by one or more bears by sustained selling is called a bear raid. In a bear squeeze, sellers force prices up against someone known to have a bear position that has to be covered. *Compare*bull.

Bear hug An approach to the board of a company by another company indicating that an offer is about to be made for their shares. If the target company indicates that it is not against the merger, but wants a higher price, this is known as a teddy bear hug.

Bear market Any market in which prices exhibit a declining trend. For a prolonged period, usually falling by 20% or more.

Bear raid In the context of general equities, attempt by investors to move the price of a stock opportunistically by selling large numbers of shares short. The investors pocket the difference between the initial price and the new, lower price after this maneuver. This technique is illegal under SEC rules, which stipulate that every short sale must be on an uptick.

Bear spread Applies to derivative products. Strategy in the options or futures markets designed to take advantage of a fall in the price of a security or commodity. A bear spread with call options is created by buying a call option with a certain strike price and selling a call option on the same stock with a lower strike price (with the same expiration date). A bear spread with put options is where an investor buys a put with a high strike price and sells a put with a low strike price. With futures, the investor sells the nearby contract and purchases the next out contract. All of these strategies are designed to profit from a fall in the underlying asset's price.

Bearer bond bonds that are not registered on the books of the issuer. Such bonds are held in physical form by the owner, who receives interest payments by physically detaching coupons from the bond certificate and delivering them to the paying agent.

Bearer security (bearer bond) A security for which proof of ownership is possession of the security certificate; this enables such bonds to be transferred from one person to another without registration. This is unusual as most securities are registered, so that proof of ownership is the presence of the owner's name on the security register. Eurobonds are bearer securities, enabling their owners to preserve their anonymity, which can have taxation advantages. Bearer bonds are usually kept under lock and key, often deposited in a hank. Dividends are usually claimed by submitting coupons attached to the certificate.

Bearer share Security not registered on the books of the issuing corporation and thus payable to possessor of the shares. Negotiable without endorsement and transferred by delivery, thus avoiding some of the control associated with ordinary shares. Dividends are payable upon presentation of dividend coupons, which are dated or numbered. Applies mainly to international equities.

Benchmark interest rate Also called the base interest rate, it is the minimum interest rate investors will demand for investing in a non-Treasury security. It is also tied to the yield to maturity offered on a comparable-maturity Treasury security that was most recently issued ("on-the-run").

Benchmark issue Also called on-the-run or current-coupon issue or bellwether issue. In the secondary market, the benchmark issue is the most recently auctioned Treasury issues for each maturity.

Benchmark The performance of a predetermined set of securities, used for comparison purposes. Such sets may be based on published indexes or may be customized to suit an investment strategy.

Beneficial Owner As used for most purposes under the federal securities laws. A beneficial owner of stock is any person or entity with sole or shared power to vote or dispose of the stock. This SEC definition is intended to include a holder who enjoys the benefits of ownership although the shares may be held in another name.

Beneficiary Term used to refer to the person who receives the benefits of a trust or the recipient of the proceeds of a life insurance policy.

Benefit taxation A form of taxation in which taxpayers pay tax according to the amounts of benefit that they receive from the system. Such a system of taxation is, in practice, very difficult to apply unless specific charges are made for specific services, such as metered electricity charges.

Benefits in kind Benefits other than cash arising from employment. The UK tax legislation seeks to assess all earnings to tax, whether they be in the form of cash or in kind. The treatment of benefits depends on the level of total earnings, including the value of any benefits, and whether the employee is a director of a company. For employees earning less than £8500, the benefits are only assessable if they are capable of being turned into cash, such as credit tokens or vouchers, living accommodation, and payment by the employer of an employee's personal liability. For all directors and higher-paid employees, with total earnings (including benefits) in excess of £8500, the benefits must be reported on form P11D by the employer at the end of the fiscal year. This form will include details of company cars and associated fuel provided by the employer, beneficial loans, mobile telephones, medical insurance provided by the employer, subscriptions paid, and any costs paid on the employee's behalf. These benefits will be assessed to tax. This often takes the form of a restriction to the income tax code.

Benelux An association of countries in western Europe, consisting of Belgium, the Netherlands, and Luxembourg. Apart from geographical proximity these countries have particularly dose economic interests, recognized in their 1947 customs union. In 1958 the Benelux countries joined the European Economic Community.

Best-efforts sale A method of securities distribution/ underwriting in which the securities firm agrees to sell as much of the offering as possible and return any unsold shares to the issuer. As opposed to a guaranteed or fixed price sale, where the underwriter agrees to sell a specific number of shares (with the securities firm holding any unsold shares in its own account if necessary).

Beta (Mutual Funds) The measure of a fund's or stocks risk in relation to the market. A beta of 0.7 means the fund's total return is likely to move up or down 70% of the market change; 1.3 means total return is likely to move up or down 30% more than the market. Beta is referred to as an index of the systematic risk due to general market conditions that cannot be diversified away.

Beta The measure of an asset's risk in relation to the market (for example, the S&P500) or to an alternative benchmark or factors. Roughly speaking, a security with a beta of 1.5, will have move, on average, 1.5 times the market return. [More precisely, that stock's excess return (over and above a short-term money market rate) is expected to move 1.5 times the market excess return).] According to asset pricing theory, beta represents the type of risk, systematic risk, that cannot be diversified away. When using beta, there are a number of issues that you need to be aware of: (1) betas may change through time; (2) betas may be different depending on the direction of the market (i.e. betas may be greater for down moves in the market rather than up moves); (3) the estimated beta will be biased if the security does not frequently trade; (4) the beta is not necessarily a complete measure of risk (you may need multiple betas). Also, note that the beta is a measure of comovement, not volatility. It is possible for a security to have a zero beta and higher volatility than the market.

Bid bond A bid "performance" bond consisting of a small percentage (1-3%) of the tender contract price, refunded to losers once the contract is awarded.

Bid price This is the quoted bid, or the highest price an investor is willing to pay to buy a security. Practically speaking, this is the available price at which an investor can sell shares of stock.

Big Bang The upheaval on the London Stock Exchange (LSE) when major changes in operation were introduced on 27 October, 1986. The major changes enacted on that date were: (a) the abolition of LSE rules enforcing a dual-capacity system; (b) the abolition of fixed commission rates charged by stockbrokers to their clients. The measures were introduced by the LSE in return for an undertaking by the government (given in 1983) that they would not prosecute the LSE under the *Restrictive Practices Act*. Since 1986 the Big Bang has also been associated with the globalization and modernization of the London securities market.

Big Board A nickname for the New York Stock Exchange (NYSE). Also known as The Exchange. More than 2,000 common and preferred stocks are traded. Founded in 1792, the NYSE is the oldest exchange in the United States, and the largest. It is located on Wall Street in New York City.

Bilateral monopoly A situation in which a monopoly seller bargains with a monopoly buyer. The classic application of bilateral monopoly is to the negotiation of wages between a union and a firm. The most famous solution (1950) is that stated by John Nash, which suggests that both sides will settle for the wage that maximizes the benefits from cooperation.

Bilateral Netting Bilateral netting - the consolidation of all swap agreements between two counterparties into one master agreement. The result is that if one counterparty bankrupts, that counterparty cannot seek to collect on any swaps that are in-the-money to them while at the same time refusing to pay out on any that are out-of-the-money. Instead, the master agreement sets out that in this event all swaps between the two counterparties will be netted; only then will the bankrupt company receive money, and then only if they are net in-the-money.

Bilateral trade agreement An agreement on trade policy between two countries, usually concerning a reduction in tariffs or other protective barriers. Although bilateral deals tend to fragment worm trade, governments are attracted by their relative simplicity. The USA, in particular, has used such deals in free-trade agreements with Mexico and Canada and to hasten the resolution of problems with particular trading partners, notably Japan and other Asian countries.

Bill of entry A detailed statement of the nature and value of a consignment of goods prepared by the shipper of the consignment for customs entry.

Bill of exchange An unconditional order in writing, addressed by one person (the drawer) to another (the drawee) and signed by the person giving it, requiring the drawee to pay on demand or at a fixed or determinable future time a specified sum of money to or to the order of a specified person (the payee) or to the bearer. If the bill is payable at a future time the drawee signifies acceptance, which makes the drawee the party primarily liable upon the bill; the drawer and endorsers may also be liable upon a bill. The use of bills of exchange enables one person to transfer to another an enforceable right to a sum of money. A bill of exchange is not only transferable but also negotiable, since if a person without an enforceable right to the money transfers a bill to a holder in due course, the latter obtains a good title to it

Bill of lading A contract between an exporter and a transportation company in which the latter agrees to transport the goods under specified conditions that limit its liability. It is the exporter's receipt for the goods as well as proof that goods have been or will be received.

Bill of sale 1. A document by which a person transfers the ownership of goods to another. Commonly the goods are transferred conditionally, as security for a debt, and a conditional bill of sale is thus a mortgage of goods. The mortgagor has a right to redeem the goods on repayment of the debt and usually remains in possession of them; the mortgagor may thus obtain false credit by appearing to own them. An absolute bill of sale transfers ownership of the goods absolutely. **2.** A document recording the change of ownership when a ship is sold; it is regarded internationally as legal proof of ownership.

Billing cycle The time elapsed between billing periods for goods sold or services rendered.

Black economy Economic activity that is undisclosed, as to disclose it would render the earnings involved liable to taxation or even cause those engaged to be imprisoned (if they are claiming state benefits and have lied about their earnings). The black economy is thought to be significant in many countries, largely due to the benefits of evading tax. Earnings made in the black economy do not appear in national statistics.

Black Friday A precipitous drop in a financial market . The original Black Friday occurred on September 24, 1869, when prospectors attempted to corner the gold market. .

Black Monday Refers to October 19, 1987, when the Dow Jones Industrial Average fell 508 points on the heels of sharp drops the previous week. On Monday, October 27, 1997, the Dow dropped 554 points. While the point drop set a new record, the percentage decline was substantially less than in 1987.

Blank cheque A cheque in which the amount payable is not stated. The person writing the cheque (the drawer) may impose a maximum amount to be drawn by the cheque by writing on it, for example, 'Not more than £100'.

Blocked funds Cash flows generated by a foreign project that cannot be immediately repatriated to the parent firm because of capital flow restrictions imposed by the host government.

Blow-off top A steep and rapid increase in price followed by a steep and rapid drop. This is an indicator seen in charts and used in technical analysis of stock price and market trends.

Blue chip Colloquial name for any of the ordinary shares in the most highly regarded companies traded on a stock market. Originating in the USA, the name comes from the colour of the highest value chip used in poker. Bluechip companies have a well-known name, a good growth record, and large assets. The main part of an institution's equity portfolio will consist of blue chips.

Blue chip stocks Common stock of well-known companies with a history of growth and dividend payments.

Blue list Daily financial publication featuring bonds offered for sale by dealers and banks that represent billions of dollars in par value. Also available on-line at www.bluelist.com.

Blue-chip company Used in the context of general equities. Large and creditworthy company. Company renowned for the quality and wide acceptance of its products or services, and for its ability to make money and pay dividends. Gilt-edged security.

Board broker Employee of the Chicago Board Options Exchange who manages away from the market orders, which cannot be executed immediately.

Board of Directors Individuals elected by the shareholders of a corporation who carry out certain tasks established in the charter.

Board of Governors of the Federal Reserve System The managing body of the Federal Reserve System, which sets policies on bank practices and the money supply.

Board room A room at a brokerage firm where its clients can watch an electronic board displaying stock prices and transactions. Also refers to the room where Board of Directors meetings take place.

Bona fide In good faith, honestly, without collusion or fraud. A bona fide purchaser for value without notice is a person who has bought property in good faith, without being aware of prior claims to it (for example, that it is subject to a trust). The purchaser will not be bound by those claims, unless (if the property is land) they were registered.

Bond An IOU issued by a borrower to a lender. Bonds usually take the form of fixed-interest securities issued by governments, local authorities, or companies. However, bonds come in many forms: with fixed or variable rates of interest, redeemable or irredeemable, short- or long-term, secured or unsecured, and marketable or unmarketable. Fixed-interest payments are usually made twice a year but may alternatively be credited at the end of the agreement (typically 5 to 10 years). The borrower repays a specific sum of money plus the face value (PAR) of the bond. Most bonds are unsecured and do not grant shares in an organization Bonds are usually sold against loans, mortgages, credit-card income, etc., as marketable securities. A discount bond is one sold below its face value; a premium bond is one sold above par.

Bond value With respect to convertible bonds, the value the security would have if it were not convertible. That is, the market value of the bond minus the value of the conversion option.

Bonded goods Imported goods on which neither customs duty nor excise have been paid although the goods are dutiable. They are stored in a bonded warehouse until the duty has been paid or the goods re-exported.

Bonded warehouse A warehouse, usually close to a sea port or airport, in which goods that attract customs duty or excise are stored after being imported, pendingpayment of the duty or the re-export of the goods. The owners of the warehouse are held responsible for ensuring that the goods remain in bond until the duty is paid; they may only be released in the presence of a customs officer.

Bondpar A system that monitors and evaluates the performance of a fixed-income portfolio, as well as the individual securities held in the portfolio. BONDPAR decomposes the return into those elements beyond the manager's control—such as the interest rate environment and client-imposed duration policy constraints—and those that the management process contributes to, such as interest rate management, sector/quality allocations, and individual bond selection.

Bonus 1. An extra payment made to employees by management, usually as a reward for good work, to compensate for something (e.g. dangerous work) or to share out the profits of a good year's trading.

2. An extra amount of money additional to the proceeds, which is distributed to a policyholder by an insurer who has made a profit on the investment of a life-assurance fund. Only holders of with-profits policies are entitled to a share in these profits and the payment of this bonus is conditional on the life assurer having surplus funds after claims, costs, and expenses have been paid in a particular year.

3. Any extra or unexpected payment. *See also*no-claim bonus; reversionary bonus ; terminal bonus.

Book cash A firm's cash balance as reported in its financial statements. Also called ledger cash.

Book of prime entry A book or record in which certain types of transaction are recorded before becoming part of the double-entry book-keeping system. The most common books of prime entry are the day book, the cash book, and the journal.

Book to bill The book-to-bill ratio is the ratio of orders taken (booked) to products shipped and bills sent (billed). The ratio measures whether the company has more orders than it can deliver (>1), equal amounts (=1), or less (<1). This ratio is of significant interest to investors/ traders in the high-technology sector.

Book to market The ratio of book value to market value of equity. A high ratio is often interpreted as a value stock (the market is valuing equity relatively cheaply compared to book value). This is the same as a low price-to-book value ratio. Value managers often form portfolios of securities with high book to market values.

Book value A company's book value is its total assets minus intangible assets and liabilities, such as debt. A company's book value might be more or less than its market value.

Book value per share The ratio of stockholder equity to the average number of common shares. Book value per share should not be thought of as an indicator of economic worth, since it reflects accounting valuation (and not necessarily market valuation).

Book-Entry Registered ownership of stock without the issuance of a corresponding stock certificate, as is the case with dividend reinvestment and direct purchase plans, employee plans and Direct Registration System issuances. Periodic statements of ownership are issued instead of certificates.

Book-entry securities The Treasury and federal agencies are moving to a book-entry system in which securities are not represented by engraved pieces of paper but are maintained in computerized records at the Fed in the names of member banks, which in turn keep records of the securities they own as well as those they are holding for customers. In the case of other securities where a book-entry has developed, engraved securities do exist somewhere in quite a few cases. These securities do not move from holder to holder but are usually kept in a central clearinghouse or by another agent.

Bought deal A method of raising capital for acquisitions or other purposes, used by quoted companies as an alternative to a rights issue or placing. The company invites market makers or banks to bid for new shares, selling them to the highest bidder, who then sells them to the rest of the market in the expectation of making a profit. Bought deals originated in the USA and are becoming increasingly popular in the UK, although they remain controversial as they violate the principle of preemption rights.

Bought deal Security issue in which one or two underwriters buy the entire issue. Also known as a guaranteed or fixed-price sale; opposite of a best-efforts sale.

Bounce A check returned by a bank because it is not payable, usually because of insufficient funds. Also used in the context of securities to refer to the rejection and ensuing reclamation of a security; a stock price's abrupt decline and recovery.

Boutique A small, specialized brokerage firm that offers limited services and products to a limited number of clients. Antithesis of financial supermarket.

Box spread This strategy refers to a type of option arbitrage in which both a bull spread and a bear spread are implemented for an almost-riskless position. One spread is implemented using put options and the other is implemented with calls. The spreads may both be debit spreads (call bull spread vs. put bear spread) or both credit spreads (call bear spread vs. put bull spread).

Bracket indexation A change in the upper and lower limits of any particular taxation bracket in line with an index of inflation. This is needed in times of inflation to avoid fiscal drag (the tax system collecting unduly large amounts of tax).

Brady bonds Bonds issued by emerging countries under a debt reduction plan.

Branch accounting An accounting system in which each department or branch of a business is established as a separate cost centre or accounting centre. The net profit per branch may be added together to arrive at the profit for the whole business. Branch accounts may be prepared to show the performance of both a main trading centre (i.e. the head office) and subsidiary trading centres (i.e. branches) but with all the accounting records being maintained by head office. Alternatively, *separate entity* branch accounts are prepared in which branches maintain their own records, which are later combined with head-office records to prepare accounts for the whole business.

Brand extension Using a successful brand name to launch a new or modified product in a separate category. A successful brand helps a company enter new product categories more easily. For example, a car manufacturer may use a brand name to launch a lawn mower. Brand extensions do not always work, however, as one manufacturer of canned foods found with the unsuccessful launch of a pet food.

Brand loyalty Support by consumers for a particular brand or product. Brand loyalty is usually the result of continued satisfaction with a product or its price and is reinforced by effective and heavy advertising. Strong brand loyalty, which is often subjective or subconscious, reduces the impact of competitive brand promotions and brand switching unless it is for an improved product.

Brand manager (product manager) The executive responsible for the overall marketing and promotion of a particular branded product. Responsibilities range from setting objectives for the product to managing and coordinating its sale. Brand managers emerged in the 1930s at Proctor and Gamble but are now to be found employed by most producers of consumer goods.

Breach of contract A failure by a party to a contract to perform the obligations in that contract or an indication of an intention not to do so. An indication that a contract will be breached in the future is called repudiation or an anticipatory breach; it may be either expressed in words or implied from conduct. Such an implication arises as the only reasonable inference from a person's acts in this respect. For example, an anticipatory breach occurs if a person contracts to sell a car to A but sells and delivers it to B before the delivery date agreed with A. The repudiation of a contract entitles the injured party to treat the contract as discharged and to sue immediately for damages for the loss sustained. The same procedure only applies to an actual breach if it constitutes a fundamental breach, i.e. a breach of a major term of the contract. In either an anticipatory or an actual breach, the injured party may, however, decide to affirm the contract instead (*see*affirmation of contract). When an actual breach relates only to a minor term of the contract (a warranty) the injured party may sue for damages but has no right to treat the contract as discharged. The process of treating a contract as discharged by reason of repudiation or actual breach is sometimes referred to as rescission. Other remedies available under certain circumstances for breach of contract are an injunction and specific performance.

Break-even payment rate The prepayment rate of a MBS coupon that will produce the same CFY as that of a predetermined benchmark MBS coupon. Used to identify for coupons higher than the benchmark coupon the prepayment rate that will produce the same CFY as that of the benchmark coupon; and for coupons lower than the benchmark coupon the lowest prepayment rate that will do so.

Break-even point Refers to the price at which a transaction produces neither a gain nor a loss. In the context of options, the term has the additional definitions:

1. Long calls and short uncovered calls: strike price plus premium.

2. Long puts and short uncovered puts: strike price minus premium.

3. Short covered call: purchase price of underlying stock minus premium.

4. Short put covered by short stock: short sale price of underlying stock plus premium.

Break-even tax rate The tax rate at which a party to a prospective transaction is indifferent between entering into and not entering into the transaction.

Break-forward A contract on the money market that combines the features of a forward-exchange contract and a currency option. The forward contract can be undone at a previously agreed rate of exchange, enabling the consumer to be free if the market moves in his favour. There is no premium on the option; the cost is built into the fixed rate of exchange.

Breakpoint For mutual funds, the point at which the amount invested reduces the sales charge is called the "breakpoint." Each mutual fund may have several breakpoints; the larger the investment, the greater the discount. Note that the actual reduction in the sales charge is known as the "breakpoint discount". Also, the term "breakpointing" is sometimes used to refer to the offering of breakpoint discounts. The practice of soliciting mutual fund purchases just below the breakpoint (to earn more commissions) is considered unethical and in violation of NASD rules. See: right of accumulation.

Broad Market Usually refers to indices such as the Wilshire 5000 that track the performance of 5,000 securities, rather than the more narrow measures such as the Dow Jones Industrial Average and the S&P 500.

Broken up Used for listed equity securities. Prevented from executing a trade (committed to upstairs) due to exchange priority rules excluding one's order (e.g., higher bid/lower offer on floor, market order to satisfy).

Broker An agent who brings two parties together, enabling them to enter into a contract to which the broker is not a principal. The broker's remuneration consists of a brokerage, usually calculated as a percentage of the sum involved in the contract but sometimes fixed by a tariff. Brokers are used because they have specialized knowledge of certain markets or to conceal the identity of a principal, in addition to introducing buyers to sellers.

Broker-dealer Any person, other than a bank, engaged in the business of buying or selling securities on its own behalf or for others. See: Dealer.

Brokered CD A certificate of deposit issued by a bank or thrift institution bought by a brokerage firm in bulk for the purpose of reselling to brokerage customers. A broker CD features a higher interest rate, usually 1% higher, and are FDIC insured and do not usually have commissions.

Bubble theory A theory under which security prices sometimes move wildly above their true values, or the price falls sharply until the "bubble bursts". It is also possible for a bubble to deflate gradually.

Budapest Stock Exchange Established in 1864, the major securities market of Hungary.

Budget 1. A financial or quantitative statement, prepared prior to a specified accounting period, containing the plans and policies to be pursued during that period. It is used as the basis for budgetary control. Generally a functional budget is drawn up for each functional area within an organization, but in addition it is also usual to produce a capital budget, a cash-flow budget, stock budgets, and a master budget, which includes a budgeted profit and loss account and balance sheet.

2. In the UK, the government's annual budget, which is presented to parliament by the Chancellor of the Exchequer towards the end of the calendar year (usually in November or December). It contains estimates for the government's income and expenditure, together with the tax rates and the fiscal policies designed to meet the government's financial goals for the succeeding fiscal year.

Budget committee The committee responsible for the operation of the budgetary control process within an organization. The membership and responsibilities of the committee vary between organizations, but a typical committee might comprise a chief executive as chairman, the functional managers as members, and a financial manager as committee secretary or budget director. The committee is responsible for ensuring the formulation of the budgets according to the directives and policies communicated by the board of directors, scrutinizing the various budgets for coordination and acceptability, and ultimately submitting the budgets (or budget revisions) to the board of directors for approval.

Budget cost allowance The amount of budgeted expenditure that a cost centre or budget centre is allowed to spend according to its budget, having regard to the level of activity (or other basis of cost incurrence) actually achieved during the budget period. The budget cost allowance is usually based on the level of activity achieved and whether the cost item is classified as a fixed cost or a variable cost.

Budget In a financial planning context the word 'budget' (as a noun) strictly speaking means an amount of money that is planned to spend on a particularly activity or resource, usually over a trading year, although budgets apply to shorter and longer periods. An overall organizational plan therefore contains the budgets within it for all the different departments and costs held by them. The verb 'to budget' means to calculate and set a budget, although in a looser context it also means to be careful with money and find reductions (effectively by setting a lower budgeted level of expenditure). The word budget is also more loosely used by many people to mean the whole plan. In which context a budget means the same as a plan. For example in the UK the Government's annual plan is called 'The Budget'. A 'forecast' in certain contexts means the same as a budget - either a planned individual activity/resource cost, or a whole business/ corporate/ organizational plan. A 'forecast' more commonly (and precisely in my view) means a prediction of performance - costs and/or revenues, or other data such as headcount, % performance, etc., especially when the 'forecast' is made during the trading period, and normally after the plan or budget' has been approved. In simple terms: budget = plan or a cost element within a plan; forecast = updated budget or plan. The verb forms are also used, meaning the act of calculating the budget or forecast.

Budget surplus The amount by which government revenues exceed government spending.

Buffer stock A stock of a commodity owned by a government or trade organization and used to stabilize the price of the commodity. Usually the manager of the buffer stock is authorized to buy the commodity in question if its price falls below a certain level, which is itself reviewed periodically, to enable producers to find a ready market for their goods at a profitable level. If the price rises above another fixed level the buffer stock manager is authorized to sell the commodity on the open market. Thus, producers are encouraged to keep up a steady supply of the commodity and users are reassured that its price has a ceiling. This arrangement is often effective, but may collapse during a boom or slump. *See also* United Nations Conference on Trade and Development.

Builder buydown loan A mortgage loan on newly developed property that the builder subsidizes during the early years of the development. The builder uses cash to buy down the mortgage rate to a lower level than the prevailing market loan rate for some period of time. The typical buydown is 3% of the interest-rate amount for the first year, 2% for the second year, and 1% for the third year (also referred to as a 3-2-1 buydown).

Building society A financial istitution that accepts deposits, upon which it pays interest, and make loans for house purchase or house improvement secured by mortgages. The developed from the Friendly Society movement in the late 17th century and are non-profit making. They are regulated by the Building Societies Act (1986). The societies accept deposits into a variety of accounts, which offer different interest rates and different withdrawal terms, or into 'shares', which often require longer notice of withdrawal. Interest on all building-society accounts is paid net of income tax, the society paying the tax direct to the Inland Revenue. The societies attract both large and small savers, with average holdings being about £5000. • Loans made to persons wishing to purchase property are usually repaid by regular monthly instalments of capital and interest over a number of years. Another method, which is growing in popularity, is an endowment mortgage in which the capital remains unpaid until the maturity of an assurance policy taken out on the borrower's life; in these arrangements only the interest and the premiums on the assurance policy are paid during the period of the loan. • Since the 1986 Act, building societies have been able to widen the range of services they offer; this has enabled them to compete with the high-street banks in many areas. They offer cheque accounts, which pay interest on all credit balances, cash cards, credit cards, loans, money transmission, foreign exchange, personal financial planning services (shares, insurance, pensions, etc.), estate agency, and valuation and conveyancing services. The distinction between banks and building societies is fast disappearing, indeed some building societies have obtained the sanction of their members to become public limited companies. These changes have led to the merger of many building societies to provide a national network that can compete with the Big Four banks. Competition is well illustrated in the

Bull CD A bull CD pays its holder a specified percentage of the increase in return on a specified market index while guaranteeing a minimum rate of return.

Bull CD, Bear CD A bull CD pays its holder a specified percentage of the increase in return on a specified market index while guaranteeing a minimum rate of return. A bear CD pays the holder a fraction of any fall in a given market index.

Bull market Any market in which prices are in an upward trend.

Bull note A bond whose redemption value is linked to a price index (e.g. FTSE 100 Index; *see*Financial Times Share Indexes) or a commodity price (e.g. the price of gold). Thus, a holder of a bull note will receive on redemption an amount greater than the principal of the bond if the relevant index or price has risen (but less if it has fallen). With a bear note the reverse happens. Bull and bear notes are therefore akin to an ordinary bond plus an option, providing opportunities for hedging and speculating.

Bull-bear bond Bond whose principal repayment is linked to the price of another security. The bonds are issued in two tranches: In the first tranche repayment increases with the price of the other security, and in the second tranche repayment decreases with the price of the other security.

Bullet contract A guaranteed investment contract purchased with a single (one-shot) premium.

Bullet strategy A strategy in which a portfolio is constructed so that the maturities of its securities are highly concentrated at one point on the yield curve.

Bullion Gold, silver, or some other precious metal used in bulk, i.e. in the form of bars or ingots rather than in coin. Central banks use gold bullion in the settlement of international debts. In the London bullion market bullion brokers act as agents for both buyers and sellers and also trade as principals.

Bullish Word used to describe an investor's attitude. Bullish refers to an optimistic outlook, while bearish means a pessimistic outlook.

Bunching Describes the act of traders combining round-lot orders for execution at the same time. Bunching can also be used to combine odd-lot orders to save the odd-lot differential for customers. Also used to refer to the pattern on the ticker tape when a series of trades for a security appear consecutively.

Bundesbank The German central bank (literally, federal bank). It controlled the 11 regional banks in the former West Germany and now oversees all German banks.

Business cycle (trade cycle) The process by which investment, output, and employment in an economy tend to fluctuate up and down in a regular pattern causing booms and depressions, with recession and

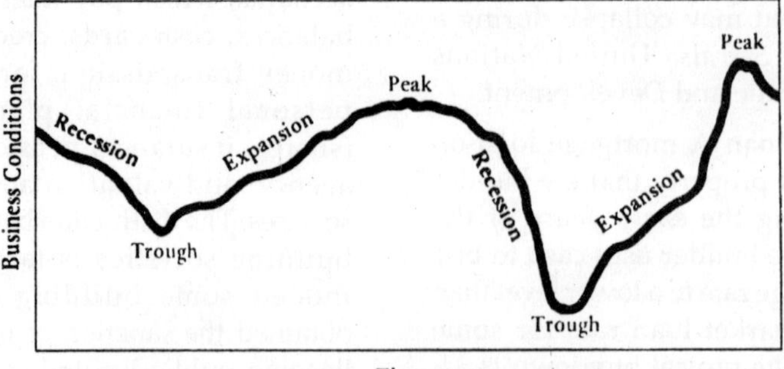

Fig. Business Cycle

recovery as intermediate stages. The business cycle is one of the major unsolved mysteries in economics. One explanation of business cycles is that they result from the desire of political parties to achieve re-election; it is easier for a party to impose its stringent decisions shortly after an election and then stimulate demand in the run-up to the next election, thus creating a business cycle.

Business judgement rule The rule that the courts will not generally interfere in the conduct of a business. For example, the courts will not substitute their judgement for that of the directors of a company unless the directors are acting improperly. The rule is often invoked when directors are accused of acting out of self-interest in takeover bids.

Business market All the organizations that buy goods and services to use in the production of other products and services or for the purpose of reselling or renting them to others at a profit.

Business plan A detailed plan setting out the objectives of a business over a stated period, often three, five, or ten years. A business plan is drawn up by many businesses, especially if the business has passed through a bad period or if it has had a major change of policy. For new businesses it is an essential document for raising capital or loans. The plan should quantify as many of the objectives as possible, providing monthly cash flows and production figures for at least the first two years, with diminishing detail in subsequent years; it must also outline its strategy and the tactics it intends to use in achieving its objectives. Anticipated profit and loss accounts should form part of the business plan on a quarterly basis for at least two years, and an annual basis thereafter. For a group of companies the business plan is often called a corporate plan.

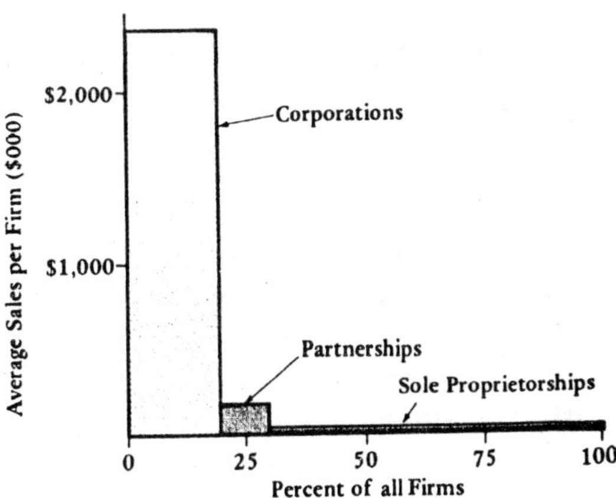

Fig. Business Organization

Business property relief An inheritance tax relief available on certain types of business property. For a business or interest in a business, including partnership share, the relief is 100%. Land or buildings owned and used in a company under the control of the donor or a partnership in which the donor was a partner, attract 50% relief.

Business segment reporting Reporting the results of the separate divisions or subsidiaries of a business.

Business strategy An overall strategy for a firm that coordinates the separate functional areas of a business. It defines the business objectives, analyses the internal and external environments, and determines the strategic direction of the firm. Each firm operates in a competitive environment and seeks to formulate a strategy that will provide it with an advantage over its rivals: design, quality, innovation, and branding are examples of ways in which competitive advantages may be established. Some firms may seek to diversify into new markets, either through internal growth (i.e. by expanding their existing products or introducing new ones) or by external growth through mergers, takeovers, joint ventures, or strategic partnerships (seediversification). Each of these methods carries different levels of risk.

Business-to-business marketing Marketing

aimed at selling a product for any use other than personal consumption. The buyer may be a manufacturer, a reseller, a government body, a nonprofit-making institution, or any organization other than an ultimate consumer.

Bust-up takeover A leveraged buyout in which the buyer sells off the assets of the target company to repay the debt that financed the takeover.

Butterfly spread Applies to derivative products. Complex option strategy that involves buying a call option with a relatively low strike price; buying a call option with a relatively high strike price; and selling two call options with an intermediate strike price. Essentially, this is a bear call spread stacked on top of a bull call spread. One can also do this with puts. The investor buys a put with a low strike, buys a put at high strike and sells two puts at intermediate strike price. The payoff diagram resembles the shape of a butterfly.

Buyback The covering of a short position by purchasing a long contract, usually resulting from the short sale of a commodity. See: Short covering, stock buyback. Also used in the context of bonds. The purchase of corporate bonds by the issuing company at a discount in the open market. Also used in the context of corporate finance. When a firm elects to repurchase some of the shares trading in the market.

Buyer 1. A person who makes a purchase of a product or service. Quite often, this person will be different from the 'influencer' and 'consumer' of the product or service. For example, a mother may — on the advice of a friend — make a purchase, which is consumed by her child.

2. A person in an organization's buying centre (or decision-making unit) with formal authority to select the supplier and arrange the terms of purchase. *See*business buyer behaviour.

Buying the index Purchasing the stocks in the S&P 500 in the same proportion as the index to achieve the same return.

Buyout Purchase of a controlling interest (or percent of shares) of a company's stock. A leveraged buy out is effected with borrowed money.

Bylaw Amendment Limitations These provisions limit shareholders' ability to amend the governing documents of the corporation. This might take the form of a supermajority vote requirement for charter or bylaw amendments, total elimination of the ability of shareholders to amend the bylaws, or the ability of directors beyond the provisions of state law to amend the bylaws without shareholder approval.

By-product pricing Setting the price for by-products in order to make the price of the main product more competitive. For example, in producing processed meats, chemicals, or oil there are often by-products, which — if they had to be disposed of — would make the main product uncompetitive. The producer therefore attempts to sell these by-products at the best possible price in order to keep the main product competitive.

C

Cage A section of a brokerage firm used for receiving and disbursing funds.

Calendar effect Describes the tendency of stocks to perform differently at different times. For example, a number of researchers have documented that historically, returns tend to be higher in January compared to other months (especially February). Others have documented returns patterns across days of the week and within the day. Some of these patterns are found in volume and volatility as well as returns.

Calendar spread Applies to derivative products. A strategy in which there is a simultaneous purchase and sale of options of the same class at the same strike prices, but with different expiration date.

Call option An option contract that gives its holder the right (but not the obligation) to purchase a specified number of shares of the underlying stock at the given strike price, on or before the expiration date of the contract.

Call provision An embedded option granting a bond issuer the right to buy back all or part of an issue prior to maturity.

Call swaption A swaption in which the buyer has the right to enter into a swap as a fixed-rate payer. The writer therefore becomes the fixed-rate receiver/floating-rate payer.

Callable Applies mainly to convertible securities. Redeemable by the issuer before the scheduled maturity under specific conditions and at a stated price, which usually begins at a premium to par and declines annually. Bonds are usually called when interest rates fall so significantly that the issuer can save money by issuing new bonds at lower rates.

Callable bonds Fixed-rate bonds, usually convertibles, in which the issuer has the right, but not the obligation, to redeem (call) the bond at par during the life of the bond. The call exercise price may be at par, although it is usually set at a premium. A grace period (i.e. a period during which the borrower is unable to call the bond) will usually be included in the terms of the agreement; conversion after the grace period will only be possible if certain conditions are met, usually related to the price of the underlying share.

Cancellation price The lowest price at which the manager of a unit trust may offer to redeem units on a particular day. The cancellation price is calculated on the basis of a formula laid down by the Securities and Investment Board.

Cannibalization A market situation in which increased sales of one brand results in decreased sales of another brand within the same product line, usually because there is little differentiation between them. For instance, two drinks marketed by the same manufacturer and packaged in almost the same colour could cause increased sales of one to be achieved at the expense of decreased sales of the other.

Cannot compete In the context of general equities, cannot accommodate customers at that price level (i.e., compete with other market makers), often because there is no natural opposite side of the trade.

Capital 1. The total value of the assets of a person less liabilities.

2. The amount of the proprietors' interests in the assets of an organization, less its liabilities.

3. The money contributed by the proprietors to an organization to enable it to function; thus share capital is the amount provided by way of shares and the loan capital is the amount provided by way of loans. However, the capital of the proprietors of companies not only consists of the share and loan capital, it also includes retained profit, which accrues to the holders of the ordinary shares. *See also*reserve capital.

4. In economic theory, a factor of production, usually either machinery and plant (physical capital) or money (financial capital). However, the concept can be applied to a variety of other assets (*see* human capital). Capital is generally used to enhance the productivity of other factors of production (e.g. combine harvesters enhance the productivity of land; tools enhance the value of labour) and its return is the reward following from this enhancement. In general, the rate of return on capital is called profit.

Capital account Net result of public and private international investment and lending activities.

Capital asset (fixed asset) An asset that is expected to be used for a considerable time in a trade or business (*compare*current assets). Examples of capital assets in most businesses are land and buildings, plant and machinery, investments in subsidiary companies, goodwill, and motor vehicles, although in the hands of dealers these assets would become current assets. The costs of these assets are normally written off against profits over their expected useful life spans by deducting an item for depreciation from their book value each year.

Capital asset A long-term asset, such as land or a building, not purchased or sold in the normal course of business.

Capital asset pricing model (CAPM) An economic theory that describes the relationship between risk and expected return, and serves as a model for the pricing of risky securities. The CAPM asserts that the only risk that is priced by rational investors is systematic risk, because that risk cannot be eliminated by diversification. The CAPM says that the expected return of a security or a portfolio is equal to the rate on a risk-free security plus a risk premium.

Capital charge rate. The capital charge is the cost of capital times the amount of invested capital. This capital charge is a dollar amount. By capital charge rate is just the cost of capital. In other words, the capital charge rate is the rate or return required on invested capital.

Capital expenditures Amount used during a particular period to acquire or improve long-term assets such as property, plant, or equipment.

Capital gain When a stock is sold for a profit, the capital gain is the difference between the net sales price of the securities and their net cost, or original basis. If a stock is sold below cost, the difference is a capital loss.

Capital gains tax The tax levied on profits from the sale of capital assets. A long-term capital gain, which is achieved once an asset is held for at least 12 months, is taxed at a maximum rate of 20% (taxpayers in 28% tax bracket) and 10% (taxpayers in 15% tax bracket). Assets held for less than 12 months are taxed at regular income tax levels, and, since January 1, 2000, assets held for at least five years are taxed at 18% and 8%.

Capital market Traditionally, this has referred to the market for trading long-term debt instruments (those that mature in more than one year). That is, the market where capital is raised. More recently, capital markets is used in a more general context to refer to the market for stocks, bonds, derivatives and other investments.

Capital requirements Financing required for the operation of a business, composed of long-term and working capital plus fixed assets.

Capital shares One of two types of shares in a dual-purpose investment company, which entitle the holder to the appreciation or depreciation in the value of a portfolio, as well as the gains from trading in the portfolio. Antithesis of income shares.

Capital stock Stock authorized by a firm's charter and having par value, stated value, or no par value. The number and the value of issued shares are usually shown, together with the number of shares authorized, in the capital accounts section of the balance sheet. See: Common stock.

Capital structure The makeup of the liabilities and stockholders' equity side of the balance sheet, especially the ratio of debt to equity and the mixture of short and long maturities.

Capitalization method A method of constructing a replicating portfolio in which the manager purchases a number of the largest-capitalized names in the index stock in proportion to their capitalization.

Capitalization ratios Also called financial leverage ratios, these ratios compare debt to total capitalization and thus reflect the extent to which a corporation is trading on its equity. Capitalization ratios can be interpreted only in the context of the stability of industry and company earnings and cash flow.

Capitalization table A table showing the capitalization of a firm, which typically includes the amount of capital obtained from each source - long-term debt and common equity - and the respective capitalization ratios.

Capitalized interest Interest that is not immediately expensed, but rather is considered as an asset and is then amortized through the income statement over time.

Capped-Style option A capped option is an option with an established profit cap or cap price. The cap price is equal to the option's strike price plus a cap interval for a call option or the strike price minus a cap interval for a put option. A capped option is automatically exercised when the underlying security closes at or above (for a call) or at or below (for a put) the Option's cap price.

Car A loose quantity term sometimes used to describe the amount of a commodity underlying one commodity contract; e.g., "a car of bellies." Derived from the fact that quantities of the product specified in a contract once corresponded closely to the capacity of a railroad car.

Care and maintenance (c & m) Denoting the status of a building, machinery, ship, etc., that is not currently in active use, usually as a result of a fall in demand, but which is being kept in a good state of repair so that it can be brought back into use quickly, if needed.

Cargo insurance An insurance covering cargoes carried by ships, aircraft, or other forms of transport. On an FOB contract the responsibility for insuring the goods for the voyage rests with the buyer. The seller's responsibility ends once the goods have been loaded onto the ship or aircraft (or train in the case of FOR contracts). On a c.i.f. contract, insurance is the responsibility of the seller up to the port of destination. On a c & f contract, the seller arranges the shipment and pays the freight but the buyer is responsible for the insurance during the voyage (as in an FOB contract).

Carriage and Insurance Paid To (CIP) Seller is responsible for the payment of freight to carry goods to a named overseas destination. The seller is also responsible for providing cargo insurance at minimum coverage against the buyer's risk of loss or damage to the goods during transport. The risk of loss or damage is transferred from the seller to the buyer once the goods are delivered into the carrier's custody. This term may be used for any mode of transport.

Carrier A person or firm that carries goods or people from place to place, usually under a contract and for a fee. A common carrier, who has to have an A licence in the UK, provides a public service and must carry any goods or people on regular routes; must charge a reasonable rate; and is liable for all loss or damage to goods in transit. A limited carrier, who requires a B licence, carries only certain types of goods (e.g. liquid chemicals in tankers), and may refuse to carry anything else. A private carrier, with a C licence, carries only his or her own goods.

Carry Trade A trade where you borrow and pay interest in order to buy something else that has higher interest. For example, with a positively sloped term structure (short rates lower than long rates), one might borrow at low short term rates and finance the purchase of long-term bonds. The carry return is the coupon on the bonds minus the interest costs of the short-term borrowing. Of course, if long-term interest rates unexpectedly rose(and long-term bond prices fell as a result), the carry trade could become unprofitable. Indeed, if this occured, there could be a number of investors trying to unwind the carry trade, which would involve selling the long-term bonds. It is possible that this could exacerbate the increase in long-term interest rates, i.e. push the rates even higher. Related: Currency Carry Trade.

Carrying charge The fee a broker charges for carrying securities on credit, such as on a margin account. Also, any component of a futures basis, such as storage costs, interest charges or insurance costs on the underlying interest.

Carry-over The quantity of a commodity that is carried over from one crop to the following one. The price of some commodities, such as grain, coffee, cocoa, and jute, which grow in annual or biannual crops, is determined by the supply and the demand. The supply consists of the quantity produced by the current crop added to the quantity in the hands of producers and traders that is carried over from the previous crop. Thus, in some circumstances the carryover can strongly influence the market price.

Cartel A group of businesses or nations that act together as a single producer to obtain market control and to influence prices in their favor by limiting production of a product. The United States has laws prohibiting cartels.

Case of need An endorsement written on a bill of exchange giving the name of someone to whom the holder may apply if the bill is not honoured at maturity.

Cash & carry Applies to derivative products. Combination of a long position in a stock/ index/commodity and short position in the underlying futures, which entails a cost of carry on the long position. Also known as cash and carry arbitrage.

Cash and carry Purchase of a security and simultaneous sale of a future, with the balance being financed with a loan or repo.

Cash and equivalents The value of assets that

can be converted into cash immediately, as reported by a company. Usually includes bank accounts and marketable securities, such as government bonds and Banker's Acceptances. Cash equivalents on balance sheets include securities (e.g., notes) that mature within 90 days.

Cash book The book of prime entry in which are recorded receipts into and payments out of the organization's bank account (*compare*petty cash). The Cash book, unlike most other books of prime entry, is also an account, as its balance shows the amount due to or from the bank.

Cash conversion cycle The length of time between a firm's purchase of inventory and the receipt of cash from accounts receivable.

Cash cycle In general, the time between cash disbursement and cash collection. In net working capital management, it can be thought of as the operating cycle less the accounts payable payment period.

cash deal A transaction (either buying or selling) on the London Stock Exchange in which settlement is made immediately, i.e. usually on the following day.

Cash discount An incentive offered to purchasers of a firm's product for payment within a specified time period, such as ten days.

Cash dividend A dividend paid in cash to a company's shareholders. The amount is normally based on profitability and is taxable as income. A cash distribution may include capital gains and return of capital in addition to the dividend.

Cash flow coverage ratio The number of times that financial obligations (for interest, principal payments, preferred stock dividends, and rental payments) are covered by earnings before interest, taxes, rental payments, and depreciation.

Cash flow from operations A firm's net cash inflow resulting directly from its regular operations (disregarding extraordinary items such as the sale of fixed assets or transaction costs associated with issuing securities), calculated as the sum of net income plus non-cash expenses that were deducted in calculating net income.

Cash flow In investments, it represents earnings before depreciation , amortization and non-cash charges. Sometimes called cash earnings. Cash flow from operations (called funds from operations) by real estate and other investment trusts is important because it indicates the ability to pay dividends.

Cash flow The amount of cash being received and expended by a business, which is often analysed into its various components. A cash-flow projection (or cash budget) sets out all the expected payments and receipts in a given period. This is different from the projected profit and loss account and, in times of cash shortage, may be more important. It is on the basis of the cash-flow projec- tion that managers arrange for employees and creditors to be paid at appropriate times. *See also*discounted cash flow.

Cash investments Short-term debt instruments—such as commercial paper, banker's acceptances, and Treasury bills—that mature in less than one year. Also known as money market instruments or cash reserves.

Cash Legal tender in the form of banknotes and coins that are readily acceptable for the settlement of debts.

Cash management Refers to the efficient management of cash in a business in order to put the cash to work more quickly and to keep the cash in applications that produce income, such as the use of lock boxes for payments.

Cash on delivery (COD) Terms of trade in which a supplier will post goods to a customer, provided the customer pays the postman or delivery man the full invoice amount when they are delivered. It was extensively used in mail order (seemail-order house), but the use of telephone ordering using credit cards has reduced the amount of COD business.

Cash ratio (liquidity ratio) The ratio of the cash reserve that a bank keeps in coin, banknotes, etc., to its total liabilities to its customers, i.e. the amount deposited with it in current accounts and deposit accounts. Because cash reserves earn no interest, bankers try to keep them to a minimum, consistent with being able to meet customers' demands. The usual figure for the cash ratio is 8%. The Bank of England may from time to time set specific levels of cash, which the banks must deposit with it. In the USA there have always been strict cash ratios set by the Federal Reserve Bank.

Cash transaction A transaction in which exchange is immediate in the form of cash, unlike a forward contract (which calls for future delivery of an asset at an agreed-upon price).

Cash-on-cash return A method used to find the return on investments when there is no active secondary market. The yield is determined by dividing the annual cash income by the total investment. See: Current yield or yield to maturity.

Cash-surrender value An amount the insurance company will pay if the policyholder ends a whole life insurance policy.

Catalogue marketing Direct marketing by means of catalogues mailed to a list of customers or made available in stores (*see also*catalogue store).

Catalogue store A retail operation that sells a wide selection of fast-moving goods at discount prices. The selling medium is a catalogue, with very little stock on display. The stores are not designed for the comfort of the customer, but the simple premises enable prices to be kept competitive.

Causal quantitative models Techniques used to forecast such future trends as demand for a product, based on establishing a causal relationship between two or more variables. For example, average temperature, time of year, and a local demographic profile can be used to predict ice-cream sales in an area. The reliability of the techniques is dependent on the correct identification of significant variables. Mathematical techniques, using regression analysis, are used to establish a functional relationship between variables.

Causal research Marketing research designed to determine whether a change in one variable has caused an observed change in another. For example, such research might determine whether a rise in temperature was the cause of an increase in sales of ice-cream.

Caveat A proviso or qualification that limits liability by putting another party on notice. For example, a retailer may sell goods subject to the caveat that no guarantee of their suitability for a particular purpose is given. The purchaser in these circumstances has no remedy against the retailer should the goods in fact turn out to be unsuitable for that particular purpose.

Cease-and-desist order An order issued after notice and opportunity for hearing, requiring a depository institution, a holding company or a depository institution official to terminate unlawful, unsafe or unsound banking practices. Cease-and-desist orders are issued by the appropriate federal regulatory agencies under the Financial Institutions Supervisory Act and can be enforced directly by the courts.

Ceiling The highest price, interest rate, or other numerical factor allowable in a financial transaction.

Cell-base manufacturing A process design that identifies the common manufacturing characteristics in batches of apparently different products and groups of machinery to enable their common features to be organized in a way that provides some of the benefits of line production.

Census An official count of a population for demographic, social, or economic purposes. In the UK, population censuses have been held every ten years since 1801 and since 1966 supplementary censuses have been held halfway through the ten-year period A census of distribution quantifying the wholesale and retail distribution, has been carried out approximately every five years since 1950. A census of production, recording the output of all manufacturing, mining, quarrying, and building industries, together with that of all public services, has been held every year since 1968.

Céntimo 1. A monetary unit of Costa Rica, worth one hundredth of a colón.

2. A monetary unit of Paraguay, worth one hundredth of a guaraní.

3. A monetary unit of Spain and Andorra, worth one hundredth of a peseta.]

4. A monetary unit of Venezuela, worth one hundredth of a bolívar.

Central bank A bank that provides financial and banking services for the government of a country and its commercial banking system as well as implementing the government's monetary policy. The main functions of a central bank are: to manage the government's accounts; to accept deposits and grant loans to the commercial banks; to control the issue of. the public debt; to help manage the exchange rate when necessary; to influence the interest rate structure and control the money supply; to hold the country's reserves of gold and foreign currency; to manage dealings with other central banks; and to act as lender of last resort to the banking system. Examples of major central banks include the Bank of England in the UK, the Federal Reserve Bank of the USA (*see* Federal Reserve System), the Deutsche Bundesbank in Germany, and France's Banque de France.

Central parity The maintenance of party between the currencies of members of the European Union in relation to the European Currency Unit, with a very narrow margin, up or down, for fluctuation. This parity of currency values is a major feature of the move towards a common monetary system.

Cents per share The amount of a mutual fund's dividend or capital gains distributions that a shareholder will receive for each share owned.

Certain annuity (terminable annuity) A form of investment contract that pays fixed sums at scheduled intervals to an individual after he or she attains a specified age; it runs for a specified number of years.

Certificate of Accrual on Treasury Securities (CATS) Refers to a zero-coupon US Treasury issue that is sold at a deep discount from the face value and pays no coupon interest during its lifetime, but returns the full face value at maturity.

Certificate of deposit (CD) Also called a time deposit, this is a certificate issued by a bank or thrift that indicates a specified sum of money has been deposited. A CD bears a maturity date and a specified interest rate, and can be issued in any denomination. The duration can be up to five years.

Certificate of incorporation The certificate that brings a company into existence; it is issued to the shareholders of a company by the Registrar of Companies. It is issued when the memorandum and articles of association have been submitted to the Registrar of Companies, together with other documents that disclose the proposed registered address of the company, details of the proposed directors and company secretary, the nominal and issued share capital, and the capital duty. The statutory registration fee must also be submitted. Until the certificate is issued, the company has no legal existence.

Certificate of insurance A certificate giving

abbreviated details of the cover provided by an insurance policy. In a motor-insurance policy or an employers'-liability insurance policy, the information that must be shown on the certificate of insurance is laid down by law and in both cases the policy cover does not come into force until the certificate has been delivered to the policyholder.

Certificate of Origin A document certifying the country of origin for goods sold internationally.

Certified Public Accountant (CPA) An accountant who has met certain standards, including experience, age, and licensing, and passed exams in a particular state.

Certified stock (Certificated stock) Stocks of a commodity that have been examined and passed as acceptable for delivery in fulfilment of contracts on a futures market.

Chairman The most senior officer in a company, who presides at the annual general meeting of the company and usually also at meetings of the board of directors. He may combine the roles of chairman and managing director, especially in a small company of which he is the majority shareholder, or he may be a figurehead, without executive participation in the day-to-day running of the company. He is often a retired managing director. In the USA the person who performs this function is often called the president. If this office is filled by a woman, she is known as a chairwoman or chairperson To avoid this complication the officer is now sometimes referred to as the chair.

Change agent A group or individual whose purpose is to bring about a change in existing practices of an organization that have become entrenched routines. In many instances this role is fulfilled by outside management consultants as they are regarded as more objective and therefore their advice is more acceptable to those affected by the changes.

Changes in financial position Sources of funds internally provided from operations that alter a company's cash flow position: depreciation, deferred taxes, other sources, and capital expenditures.

Channel cooperation Cooperation between members of a distribution channel as a result of harmonious marketing objectives and strategies. Problems can emerge when specific channel members have too much power or act negatively. Restriction of lines, limits on the number of new products, or the demand for promotional bonuses (typically a powerful retailer demanding cash for a promotion) by retailers and wholesalers can limit competition.

Charitable trust A trust set up for a charitable purpose that is registered with the Charity Commissioners. Charitable trusts do not have to pay income tax if they comply with the regulations of the Charity Commissioners.

Chartered bank A bank in the USA that is authorized to operate as a bank by a charter granted either by its home state (state-chartered bank) or by the federal government (nationally chartered bank). In Canada, a similar charter is granted to banks by the Comptroller of Currency. The use of charters dates back to the use of Royal Charters, which predated Acts of Parliament.

Cheap money (easy money) A monetary policy of keeping interest rates at a low level. This is normally done to encourage an expansion in the level of economic activity by reducing the costs of borrowing and investment. It was used in the 1930s to help recovery after the Depression and during World War II to reduce the cost of government borrowing. *Compare*dear money.

Cheapest to deliver issue The acceptable Treasury security with the highest implied repo rate; the rate that a seller of a futures contract can earn by buying an issue and then delivering it at the settlement date.

Checkwriting Free checkwriting privileges offered with nonretirement accounts for select mutual funds.

Cheque A preprinted form on which instructions are given to an account holder (a bank or building society) to pay a stated sum to a named recipient It is the most common form of payment of debts of all kinds. (*see also*cheque account; current account).

N a crossed cheque two parallel lines across the face of the cheque indicate that it must be paid into a bank account and not cashed over the counter (a general crossing). A special crossing may be used in order to further restrict the negotiability of the cheque, for example by adding the name of the payee's bank An open cheque is an uncrossed cheque that can be cashed at the bank of origin. An order cheque is one made payable to a named recipient "or order", enabling the payee to either deposit it in an account or endorse it to a third party, i.e. transfer the rights to the cheque by signing it on the reverse. In a blank cheque the amount is not stated; it is often used if the exact debt is not known and the payee is left to complete it. However, the drawer may impose a maximum by writing "under £. . ." on the cheque. *See also*marked cheque; returned cheque; stale cheque.

Chinese wall Communication barrier between financiers (investment bankers) and traders. This barrier is erected to prevent the sharing of inside information that bankers are likely to have.

Chose in action A right of proceeding in a court of law to obtain a sum of money or to recover damages. Examples include rights under an insurance policy, a debt, and rights under a contract. A chose in action is a form of property and can be assigned, sold, held in trust, etc.

Churning Excessive trading of a client's account in order to increase the broker's commissions.

Circle Underwriters, actual or potential, often seek out and "circle" investor interest in a new issue before final pricing. The customer circled has basically made a commitment to purchase the issue if it is available at an agreed-upon price. If the actual price is other than that stipulated, the customer supposedly has first offer at the actual price.

Circus swap A fixed-rate currency swap against floating US dollar LIBOR payments. An acronym that stands for Combined Interest Rate and CUrrency Swap.

Claim dilution A decrease in .he likelihood that one or more of a firm's claimants will be fully repaid, including time value of money considerations.

Classical system of corporation tax A system of taxing companies in which the company is treated as a taxable entity separate from its own shareholders. The profits of companies under this system are therefore taxed twice, first when made by the company and again when distributed to the shareholders as dividends.

Classified advertising The form of advertising that consists of small type- set or semi-display advertisements grouped together in such categories as cars for sale, furniture for sale, flats to let, etc., usually in a paper or magazine. They are usually inserted by direct contact between the advertiser and the advertising department of the publication.

Classified Board Also known as Staggered Board: is one in which the directors are placed into different classes and serve overlapping terms. Since only part of the board can be replaced each year, an outsider who gains control of a corporation may have to wait a few years before being able to gain control of the board. This slow replacement makes a classified board an effective delays of takeovers. Sometimes known as a delay provision.

Claused Bill of Lading A bill of lading with a notation that indicates damage or shortage.

Also called foul bill of lading and are the opposite of clean bills of lading.

Clawback Money that a government takes back from members of the public by taxation, especially by higher-rate tax, having given the money away in benefits, such as increased retirement pensions. Thus the money is clawed back from those who have no need of the extra benefit (because they are paying higher-rate taxes).

Clean floating A government policy allowing a country's currency to fluctuate without direct intervention in the foreign-exchange markets. In practice, clean floating is rare as governments are frequently tempted to manage exchange rates by direct intervention by means of the official reserves, a policy sometimes called managed floating. However, clean floating does not necessarily mean that there is no control of exchange rates, as they can still be influenced by the government's monetary policy.

Clean opinion An auditor's opinion reflecting an unqualified acceptance of a company's financial statements.

Clean price The price of a gilt-edged security excluding the accrued interest since the previous dividend payment. Interest on gilt-edged stocks accrues continuously although dividends are paid at fixed intervals (usually six months). Prices quoted in newspapers are usually clean prices, although a buyer will normally pay for and receive the accrued income as well as the stock itself.

Clean report of findings A report issued by an inspection firm, indicating that price has been verified, that the goods have been inspected prior to shipment, and that both conform to buyer specifications.

Clear title Title to ownership that is untainted by any claims on the property or disputed interests, and therefore available for sale. This is usually checked through a title search by a title company.

Clearing corporations Organizations that are affiliated with exchanges and are used to complete securities transactions by taking care of validation, delivery, and settlement.

Clearing cycle The process by which a payment made by cheque, etc., through the banking system is transferred from the payer's to the payee's account. Historically, in the UK the average time for the clearing cycle to be completed was three days, although by 1994 this had been reduced to two days in many cases. However, with this reduction in the cycle, the payee's bank may not have received confirmation from the payer's bank that funds are available to provide value, cleared for fate.

Clearing house A centralized and computerized system for settling indebtedness between members. The best-known in the UK is the Association for Payment Clearing Services (APACS), which enables the member banks to offset claims against one another for cheques and orders paid into banks other than those upon which they were drawn. Similar arrangements exist in some commodity exchanges, in which sales and purchases are registered with the clearing house for settlement at the end of the accounting period.

Client account A bank or building society account operated by a professional person (e.g. a solicitor, stockbroker, agent, etc.) on behalf of a client. A client account is legally required for any company handling investments on a client's behalf; it protects the client's money in case the company becomes insolvent and makes dishonest appropriation of the client's funds more difficult. For this reason money in client accounts should be quite separate from the business transactions of the company or the professional person.

Clientele effect The grouping of investors who have a preference that the firm follow a particular financing policy, such as the amount of leverage it uses.

Clone fund A new fund set up in a fund family to emulate another successful fund.

Close company A company resident in the UK that is under the control of five or fewer participators or any number of participators who are also directors. There is also an alternative asset-based test, which applies if five or fewer participators, or any number who are directors, would be entitled to more than 50% of the company's assets on a winding-up. The principal consequences of being a close company are that certain payments made to shareholders can be treated by the Inland Revenue as a distribution, as can loans or quasi-loans. Close investment companies do not qualify for the reduced rate of corporation tax. There are a number of other consequences. In the USA dose companies are known as closed companies.

Closed corporation A corporation whose shares are owned by just a few people, having no public market.

Closed economy A theoretical model of an economy that neither imports nor exports and is therefore independent of economic factors in the outside world. No such economy exists, although the foreign sector of the US economy is relatively small and most economic decisions are taken by the US government independently of the rest of the world.

Closed fund A mutual fund that is no longer issuing shares, mainly because it has grown too large.

Cluster analysis A statistical technique that identifies clusters of stocks whose returns are highly correlated within each cluster and relatively uncorrelated across clusters.

Co-agent An institution appointed by the issuer as co-transfer agent accepts and transfers certificates and sends daily activity journals to the primary record-keeping agent. A co-agent does not maintain security holder records, but is used to facilitate the transfer of stock in a geographic region not easily accessible to the issuer or its principal transfer agent.

Code of conduct A statement setting out the guidelines regarding the ethical principles and acceptable behaviour expected of a professional organization or company. For example, the Market Research Society has a professional code of conduct and utility companies have customer charters.

Coding notice A document issued by the Inland Revenue as part of the PAYE system showing the amount of income tax allowances to be set against an individual's taxable pay. The total of allowances is reduced to a code by dropping the last digit and adding a letter depending on the taxpayer's circumstances (e.g. H for higher personal allowance, L for lower single person's allowance, etc.).

Coefficient of determination A measure of the goodness of fit of the relationship between the dependent and independent variables in a regression analysis; for instance, the percentage of variation in the return of an asset explained by the market portfolio return. Also known as R-square.

Cofinancing agreements Joint participation of the World Bank and other agencies or lenders in providing funds to developing countries.

Cognitive process The broad range of mental activities, including perception. learning, thinking, information processing, and reasoning, that involve the interpretation of stimuli and the organization of thoughts and ideas. This process has relevance to the attitudes of potential customers and their evaluation of the flow of information provided for them by advertising and other communications seeking to persuade them to make purchases.

Cold calling A method of selling a product or service in which a sales representative makes calls, door-to-door, by post, or by telephone, to people who have not previously shown any interest in that product or service. In view of the high cost of maintaining a salesforce, many companies direct their sales representatives to those potential customers or who are favourably disposed to the supplier (either as an existing customer or by replying to a press advertisement). When companies wish to extend their market to new customers, the sales representative must be willing to call on a large number of people, many of whom will show no interest in the product. The primary objective in cold calling is to establish a favourable relationship quickly, enabling the representative to explain the benefits of their product or service before being dismissed.

Collar Refers to the ceiling and floor of the price fluctuation of an underlying asset. A collar is usually set up with options, swaps, or by other agreements. In corporate finance, the collar strategy of buying puts and selling calls is often used to mitigate the risk of a concentrated position in (sometimes) restricted stock. When the restricted owner can't sell the stock, but needs to diversify the risk, a collar transaction is one of the few tools available. Many corporate executives who receive chunks of their compensation in restricted stock need to employ this strategy to mitigate the diversification risk in their overall portfolio.

Collateral Property pledged to secure the payment of a debt.

When a borrower grants a lender a lien on the borrower's inventory (to secure repayment of the loan), the inventory is the collateral for the loan.

Collateral trust bonds A bond in which the issuer (often a holding company) grants investors a lien on stocks, notes, bonds, or other financial asset as security. Compare mortgage bond. Collateralized Loan Obligations, and Collateralized Mortgage Obligations,

Collateralized loan obligation (CLO) A security backed by a pool of commercial or personal loans , structured so that there are several classes of bondholders with varying maturities, called tranches. Similar in structure to Collateralized Mortgage Obligations.

Collateralized mortgage obligation (CMO) A security backed by a pool of pass-throughs , structured so that there are several classes of bondholders with varying maturities, called tranches. The principal payments from the underlying pool of pass-through securities are used to retire the bonds on a priority basis as specified in the prospectus.

Collectivism An economic system in which much of the planning is carried out by a central government and the means of production owned by the community. This system was formerly common in several eastern-bloc countries.

Collector of Taxes A civil servant responsible for the collection of taxes for which assessments have been raised by Inspectors of Taxes and for the collection of tax under PAYE.

Collusive duopoly A form of duopoly in which producers collude with each other in a price-fixing agreement, thus forming a virtual monopoly, and negotiate a profit-sharing arrangement.

Column inches (column centimetres) A measurement of area of typeset matter equal to the width of a column of type in a newspaper or multiplied by its depth. Column centimetres are now replacing column inches in practice, although most public relations consultancies still refer to column inches when measuring the coverage their activities have achieved.

Commercial loan selling A transaction involving two banks and one business customer. Bank A grants a loan to the customer and then sells that loan agreement to Bank B. Bank A makes a profit on the sale;

Bank B has a loan book it would not otherwise achieve; and the customer borrows at a favourable rate. Commercial loan selling is common in the USA.

Commercial mortgage backed securities Similar to MBS but backed by loans secured with commercial rather than residential property. Commercial property includes multi-family, retail, office, etc., They are not standardized so there are a lot of details associated with structure, credit enhancement, diversification, etc., that need to be understood when valuing these instruments.

Commercial paper Short-term promissory notes either unsecured or backed by assets such as loans or mortgages issued by a corporation. The maturity of commercial paper is typically less than 270 days; the most common maturity range is 30 to 50 days or less. They are usually sold, like Treasury bills, at a discount

Commercial risk The risk that a foreign debtor will be unable to pay its debts because of business events, such as bankruptcy.

Commercialization The stage in the development of a new product during which a decision is made to embark on its full-scale production and distribution.

Commetments for capital expenditure Expenditure on fixed assets to which a company is committed for the future. The aggregate amounts of contracts for capital expenditure not provided for in the accounts for the year and the aggregate amount of capital expenditure authorized by the directors but not yet accounted for should be disclosed in the notes to the accounts (according to the *Companies Act*, Schedule 4). These disclosures will usually be made in the directors' report.

Commingling In the context of securities, this involves mixing customer-owned securities with brokerage firm-owned securities. This process is referred to as rehypothecation, which is the use of customers' collateral to secure their loans. This is legal with customer consent, although some securities and collateral must be kept separately.

Commission The fee paid to a broker to execute a trade, based on number of shares, bonds, options, and/or their dollar value. In 1975, deregulation led to the establishment of discount brokers, who charge lower commissions than full service brokers. Full service brokers offer advice and usually have a staff of analysts who follow specific industries. Discount brokers simply execute a client's order and usually do not offer an opinion on a stock. Also known as a round-turn. Commissions are known as round-turn only in futures trading, since the commission is assessed only after liquidation of the position.

Committed facility An agreement between a bank and a customer to provide funds up to a specified maximum at a specified interest rate (usually based upon an agreed margin over the London Inter Bank Offered Rate) for a certain period. The total cost will be the interest rate plus the mandatory liquid asset cost. The agreement will include the conditions that must be adhered to by the borrower for the facility to remain in place.

Commodity broker A broker who deals in commodities, especially one who trades on behalf of principals in a commodity market (*see also*futures contract). The rules governing the procedure adopted in each market vary from commodity to commodity and the function of brokers may also vary. In some markets brokers pass on the names of their principals, in others they do not, and in yet others they are permitted to act as principals. Commodity brokers, other than those dealing in metals, are often called produce brokers.

commodity market A market in which commodities are traded. The main terminal markets in commodities are in London and New York, but in some commodities there are markets in the country of origin. Some commodities are dealt with at auctions (e.g. tea), each lot being sold having been examined by dealers, but most dealers deal with goods that have been classified according to established quality standards (*see*certified stock). In these commodities both actuals and futures (*see*futures contract) are traded on commodity exchanges, often with daily callovers, in which dealers are represented by commodity brokers. Many commodity exchanges offer option dealing in utures, and settlement of differences on futures through a clearing house. As commodity prices fluctuate widely, commodity exchanges provide users and producers with hedging facilities with outside speculators and investors helping to make an active market, although amateurs are advised not to gamble on commodity exchanges.

He fluctuations in commodity prices have caused considerable problems in developing countries, from which many commodities originate, as they are often important sources of foreign currency, upon which the economic welfare of the country depends. Various measures have been used to restrict price fluctuations but none have been completely successful. *See also* International Petroleum Exchange ;

Common shares In general, a public corporation has two types of shares, common and preferred. The common shares usually entitle the shareholders to vote at shareholders meetings. The common shares have a discretionary dividend.

Common stock equivalent A convertible security that is traded like an equity issue because the optioned common stock is trading above the conversion price.

Common stock ratios Ratios that are designed to measure the relative claims of stockholders to earnings (cash flow per share), and equity (book value per share) of a firm.

Common stock Securities that represent equity ownership in a company. Common shares let an investor vote on such matters as the election of directors. They also give the holder a share in a company's profits via dividend payments or the capital appreciation of the security. Units of ownership of a public corporation with junior status to the claims of secured/ unsecured creditors, bondholders and preferred shareholders in the event of liquidation.

Comnmunity Bank A smaller bank that is regulated by the Office of the Comptroller of Currency (OCC).Currently, there is no official definition of Community Bank, i.e. in terms of asset size.

Company A corporate enterprise that has a legal identity separate from that of its members; it operates as one single unit, in the success of which all the members participate. An incorporated company is a legal person in its own right, able to own property and to sue and be sued in its own name. A company may have limited liability (a limited company), so that the liability of the members for the company's debts is limited. An unlimited company is one in which the liability of the members is not limited in any way. There are various different types of company: a chartered company is one formed under Royal Charter. This was the earliest type of company to exist and they were influential in the development of both foreign trade and colonization; an example is the Hudson's Bay Company. Chartered companies, however, are now rare, unless a charter is required for prestige purposes, as it might be for a new university. A joint-stock company is a company in which the members pool their stock, trading on the basis of their joint stock. This differs from the earlier merchant corporations or regulated companies of the 14th century, in which each member traded with his own stock but agreed to obey the rules of the company.

Registered company, one registered under the *Companies Acts,* is the most common type of company. A company may be registered either as a public limited company or a private company. A public limited company must have a name ending with the initials 'plc' and have an authorized share capital of at least £50,000, of which at least £12,500 must be paid up. The company's memorandum must comply with the format in Table F of the Companies Regulations (1985). It may offer shares and securities to the public. The regulation of such companies is stricter than that of private companies. Most public companies are converted from private companies, under the re-registration procedure laid down in the *Companies Act.* A private company is any registered company that is not a public company. The shares of a private company may not be offered to the public for sale. The legal requirements for such a company are less strict; for example, there is no minimum issued or paid-up share capital requirement and small and medium-sized private companies need not file full accounts. A statutory company is a company formed by special Act of parliament. These are generally public utilities that were either not nationalized (for example, certain water authorities) or that have been privatized (such as British Gas and British Telecom). Their powers and privileges depend upon the Act under which they were formed.

Company formation The procedure to be adopted for forming a company in the UK. The subscribers to the company must send to the Registrar of Companies a statement giving details of the registered address of the new company together with the names and addresses of the first directors and secretary, with their written consent to act in these capacities. They must also give a declaration (declaration of compliance) that the provisions of the *Companies Acts* have been complied with and provide the memorandum of association and the articles of association Provided all these documents are in order the Registrar will issue a certificate of incorporation and a certificate enabling it to start business. In the case of a public limited company additional information is required.

Company limited by shares An incorporated organization in which the liability of members is limited by the memorandum of association to the amounts paid, or due to be paid for shares. In the UK this is the most popular form of company.

Company seal The common seal with the company's name engraved on it in legible characters. It is used to authenticate share certificates and other important documents issued by the company. The articles of association set out how and when the seal is to be affixed to contracts. Unless it is affixed to any contract required by English law to be made under seal, the company will not be bound by that contract.

Company secretary An officer of a company. The appointment is usually made by the directors. The secretary's duties are mainly administrative, including preparation of the agenda for directors' meetings. However, the modern company secretary has. an increasingly important role, which may include managing the office and entering into contracts on behalf of the company. Duties imposed by law include the.

Comparison advertising Advertising that compares one brand, either directly or indirectly, with other brands. This is quite often used in car advertising, by comparing the availability of a range of features on competitors' models in the same price bracket.

Competitive Market Refers to a market in which a very large number of small buyers and sellers trade independently, and as such no one trader is able to significantly influence price.

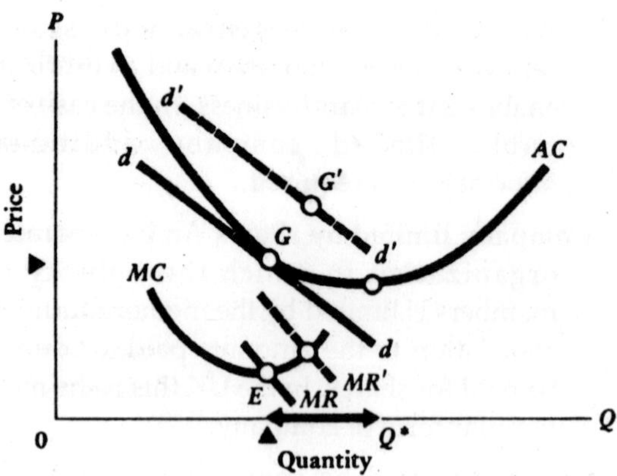

Fig. Competitive Market

Completion The conveyance of land in fulfillment of a contract of sale. The purchaser will have obtained an equitable interest in the land at the date of the exchange of contracts but will not become the full legal owner until completion. If the seller refuses to complete, the court may grant a decree of specific performance. The date on which completion must take place is stated in the contract of sale.

Completion undertaking An undertaking either .

(1) to complete a project so that it meets certain specified performance criteria on or before a certain specified date, or.

(2) to repay project debt if the completion test cannot be met.

Composite rate A special rate of tax that applied to banks and building societies from 1951 to 1991 Ms rate of tax was used to deduct tax from interest payments. The nuin disadvantage of the tax was that non-taxpayers were unable to reclaim the tax deducted from the interest paid to then. The change in 1991 resulted in tax being paid at basic rate, which could be reclaimed. A provision was also introduced enabling non-taxpayers to elect to have the interest paid gross.

Compound annual return (CAR) The total return available from an investment or deposit in which the interest is used to augment the investment. The more frequently the interest is credited, the higher the CAR The CAR is usually quoted on a gross basis. The return, taking into account the deduction of tax at the basic rate on the interest, is known as the compound not annual rate (CNAR) .

Compound interest Interest paid on previously earned interest as well as on the principal.

Compounding frequency The number of compounding periods in a year. For example, quarterly compounding has a compounding frequency of 4.

Compounding The process of accumulating the time value of money forward in time. For example, interest earned in one period earns additional interest during each subsequent time period.

Comprehensive due diligence investigation The investigation of a firm's business in conjunction with a securities offering to determine whether the firm's business and financial situation and its prospects are adequately disclosed in the prospectus for the offering.

Comptroller The corporate manager responsible for the firm's accounting activities. Sometimes referred to as the contoller (which means the same thing).

Compulsory liquidation,(compulsory winding-up) The winding-up of a company by a court. A petition must be presented both at the court and the registered office of the company. Those by whom it may be presented include: the company, the directors, a creditor, an official receiver, and the Secretary of State for Trade and Industry. The grounds on which a company may be wound up by the court include: a special resolution of the company that it be wound up by the court; that the company is unable to pay its debts; that the number of members is reduced below two; or that the court is of the opinion that it would be just and equitable for the company to be wound up. The court may appoint a provisional liquidator after the winding-up petition has been presented; it may also appoint a special manager to manage the company's property. On the grant, of the order for winding-up, the official receiver becomes the liquidator and continues in office until some other person is appointed, either by the creditors or the members.

Compulsory purchase The compulsory acquisition of land by the state when it is required for some purpose under the Town and Country Planning legislation. Compensation is paid on the basis of market value. *Compare*requisitioning

Concentrated segmentation (niche marketing) A comparatively small segment of a market that has been identified usually by a small company, in which to concentrate their efforts. For example, a company might decide to offer a service converting cine films to video tapes; as long as this activity does not threaten the profitability of the manufacturers of either video tapes or cine films, it can establish a profitable niche in the market.

Concentration account A single centralized account into which funds collected at regional locations (lockboxes) are transferred.

Concentration services Movement of cash from different lockbox locations into a single concentration account from which disbursements and investments are made.

Concept test A technique used in marketing research to assess the reactions of consumers to a new product or a proposed change to an existing product (*see*new product development). Before an organization invests in production facilities for a new product it writes a concept statement describing the proposed product and commissions a market researcher to interview a small number of potential consumers either in group discussion or depth interview. Respondents are shown the concept statement and their reactions are explored in detail The results of these interviews help the organization to understand what it should do with the proposed product if it is to be successful

Conditional call Applies mainly to convertible securities. Circumstances under which a company can effect an earlier call, usually stated as percentage of a stock's trading price during a particular period, such as 140% of the exercise price during a 40-day trading span.

Conditional call options A protective guarantee that, in the event a high yield bond is called, the issuing corporation will replace the bond with a noncallable bond of the same life and terms as the bond that is being called.

Conditional sales contracts Similar to equipment trust certificates except that the lender is either the equipment manufacturer or a bank or finance company to whom the manufacturer has sold the conditional sales contract.

Conditionality The terms under which the International Monetary Fund (IMF) provides balance-of-payments support to member states. The principle is that support will only be given on the condition that it is accompanied by steps to solve the underlying problem. Programmes of economic reform are agreed with the member; these emphasize the attainment of a sustainable balance-of-payments position and boosting the supply side of the economy. Lending by commecial banks is frequently linked to IMF conditionality.

Confidence indicator A measure of investors' faith in the economy and the securities market. A low or deteriorating level of confidence is considered by many technical analysts as a bearish sign.

Confirm me out Used for listed equity securities. "Go to the floor and check with the specialist or floor broker that my previously active order has been canceled and was not executed". One does not have to honor any trade reported after being given a "firm out".

Confirmation The written statement that follows any "trade" in the securities markets. Confirmation is issued immediately after a trade is executed. It spells out settlement date, terms, commission, etc.

Confirmed Letter of Credit A letter of credit which a bank other than the bank that opened it agrees to honor as though they had themselves issued it. This additional confirmation is in addition to the obligation of the bank which issued the letter of credit.

Confirming house An organization that purchases goods from local exporters on behalf of overseas buyers. It may act as a principal or an agent, invariably pays for the goods in the exporters' own currency, and purchases on a contract that is enforceable in the exporters' own country. The overseas buyer, who usually pays the confirming house a commission or its equivalent, regards the confirming house as a local buying agent, who will negotiate the best prices on its behalf, arrange for the shipment and insurance of the goods, and provide information regarding the goods being sold and the status of the various exporters.

Conflict between bondholders and stockholders These two groups may have interests in a corporation that conflict. Sources of conflict include dividends, distortion of investment, and underinvestment. Protective covenants work to resolve these conflicts.

Conflict of interests A situation that can arise if a person (or firm) acts in two or more separate capacities and the objectives in these capacities are not identical. The conflict may be between self-interest and the interest of a company for which a person works or it could arise when a person is a director of two companies, which find themselves competing. The proper course of action in the case of a conflict of interests is for the person concerned to declare any interests, to make known the way in which they conflict, and to abstain from voting or sharing in the decision-making procedure involving these interests.

Conforming loans Mortgage loans that meet the qualifications of Freddie Mac or Fannie Mae, which are bought from lenders and issued as pass-through securities.

Conglomerate A diverse group of companies, usually managed by a holding company. There is usually little integration and few transactions between each of the subsidiaries, often because their activities and products are unrelated. This form of business structure is sometimes not well regarded by investors because conglomerates may lack a clear strategic focus and the overall value of the group can undervalue the combined individual values of the operating subsidiaries. However, one of the main advantages of a conglomerate is that the risk is spread over a number of companies operating in different markets. Therefore, if one company is performing poorly it can be counterbalanced by others, which support the overall performance of the group.

Consignment Transfer of goods to a seller while title to the merchandise is retained by the owner.

Consistency concept A concept used accounting that ensures consistency of treatment of like items within each accounting period and from one period to the next. It also ensures that accounting policies are consistently applied. It is a principle contained in the Companies Act and in the Statement of Standard

Accounting Practice 2, 'Disclosure of Accounting Policies'.

Consolidated balance sheet The balance sheet of a group providing the financial information contained in the individual financial statements of the parent company of the group and its subsidiary undertakings, combined subject to any necessary consolidation adjustments. It must give a true and fair view of the state of affairs of a group as at the end of the financial year, and its form and content should comply with Schedule 4 of the Companies Act. If the balance sheet formats require the disclosure of the balances attributable to group undertakings (creditors, debtors, investments), the information should be analysed to show the amounts attributable to parent and fellow subsidiary undertakings of the parent company, and amounts attributable to unconsolidated subsidiaries.

Consolidated goodwill The difference between the fair value of the consideration given by an acquiring company when buying a business and the aggregate of the fair values of the separable net assets acquired. Goodwill is generally a positive amount. Goodwill should be eliminated by writing it off immediately to the reserves or alternatively by amortization to the profit and loss account over its useful life, to comply with Statement of Standard Accounting Practice 22, 'Accounting for Goodwill'.

Consolidated mortgage bond A bond that covers several units of property, sometimes refinancing mortgages on the properties.

Consolidated profit and loss account A combination of the individual profit and loss accounts of the members of a group of or subject to any consolidation adjustments. The consolidated profit and loss account must give a true and fair view of the profit and loss of the undertakings included in the consolidation A parent company may be exempted, under section 230 of the Companies Act, from publishing its own profit and loss account if it prepares group accounts. The individual profit and loss account must be approved by the directors, but may be omitted from the company's annual accounts. In such a case, the company must disclose its profit or loss for the financial year and also state in its notes that it has taken advantage of this exemption.

Consols Government securities that pay interest but have no redemption date. The present bonds, called consolidated annuities or consolidated stock, are the result of merging several loans at various different times going back to the 18th century. Their original interest rate was 3% on the nominal price of £100; most now pay 2½% and therefore stand at a price that makes their annual yield slightly above that of long-dated gilt-edged securities, e.g. at £27¾ they yield about 9%.

Constructive total loss A loss in which the item insured is not totally destroyed but is so severely damaged that it is not worth repairing. The *Marine Insurance Act (1906)* defines a constructive total loss as one in which "the subject matter insured is reasonably abandoned on account of its actual total loss appearing to be unavoidable, or because it could not be preserved from actual total loss without an expenditure which would exceed its value when the expenditure had been incurred."

Consular Invoice A document prepared by the shipper and certified in the country of origin by a consul of the country of importation. It shows the transaction details and origin of the goods.

Consumable materials Materials that are used in a production process although, unlike direct materials, they do not form part of the prime cost.

Examples are cooling fluid for production machinery, lubricating oil, and sanding discs. In circumstances in which direct materials of small value are used, such as cotton or nylon thread or nails and screws, they are sometimes treated in the same way as consumable materials.consumer advertising The advertising of goods or services specifically aimed at the potential end-user, rather than at an intermediary in the selling chain. *Compare*trade advertising.consumer buying behaviour The buying behaviour of individuals and households who buy goods and services for personal consumption. A number of different people, playing different roles, have been identified in the decision to make a specific purchase:

Initiator, the person who first suggests or thinks of the idea of buying a particular product or service;

Influencer, a person whose views or advice influences the buying decision;

Decider, the person who ultimately makes the decision to buy (or any part of it; this decision includes whether to buy, what to buy, how to buy, and where to buy;

Buyer, the person who makes the actual purchase;

User, the person who uses the produce;

Service. It is usual to regard a consumers decision to make a specific purchase as having five stages: problem recognition.

Information searchevaluation.

Consumer Durables are consumer goods, such as cars, refrigerators, and television sets, whose useful life extends over a relatively long period. The purchase of a consumer durable falls between consumption and investment. Consumer non-durables or disposables are goods that are used up within a short time after purchase. They include food, drink, newspapers, etc.

Consumer groups The groups to which consumers belong and which influence their behaviour. These groups can be small primary groups, such as the family or a work group, or they can be larger secondary groups, such as professional associations or trade unions. Reference groups serve as direct (faceto-face) or indirect points of reference, providing comparisons that help to form a person's attitudes or behaviour. Within these groups are aspirational groups, to which people would like to belong; quite often, the aspiring person assumes the values, attitudes, and even the dress of the group. Opinion leaders are important within reference groups, because their special skills, knowledge, or personality enables them to influence others.

Consumer market The market for consumer goods, as opposed to the industrial market, in which buyers of goods are not the end users.

Consumer preference The way in which consumers in a free market choose to divide their total expenditure in purchasing goods and services. Using a limited number of assumptions, an individual's preferences can be built up into a utility function (*see*consumer theory). Applying price theory to the utility functions of individuals enables a model to be constructed of the behaviour of markets in an economy.

Consumer price index The name for the Retail Price Index in the USA and some other countries.

Consumer research Any form of marketing research undertaken among the final consumers of a product or service. For example, a manufacturer supplying man-made fibres to a shirt factory might undertake consumer research, interviewing purchasers of shirts, in order to establish the merits, or otherwise, of the fibres. *Compare* industrial marketing research.

Consumer theory The theory explaining the choices made by individuals and households in terms of the concept of utility. Consumers are assumed to be able to order their preferences (seeconsumer preference) in such a way that they choose a basket of goods that maximize their utility, subject to the constraint that their income is limited. The application of price theory and demand theory to this problem forms the basis of microeconomics (together with the theory of the firm) and also of macroeconomics.

Contagion Excess correlation of delivering or bond returns. For example, under usual conditions we might observe a certain level of correlation of market returns. A period of contagion would be associated with much higher-than-expected correlation. Some examples are the conjectured contagion in East Asian markets beginning in July 1997 when the Thai currency devalued and the impact across many emerging markets of the Russian default. Contagion is difficult to identify because you need some sort of measure of the expected correlation. It is complicated because correlations are known to change through time, for example, see Erb, Harvey and Viskanta's article in the 1994 Financial Analysts Journal. In periods of negative returns, correlations (and volatility) are known to increase, so what might appear to be excessive may not be contagion.

Contango The former practice of carrying the purchase of stocks and shares over from one account day on the London Stock Exchange to the next.

Contingencies Potential gains and losses known to exist at the balancesheet date although the actual outcomes will only be known after one or more events have occurred (or not occurred). Depending on the nature of a particular contingency, it may be appropriate to include it in the financial statements or to show it as a note to the accounts; Statement of Standard Accounting Practice 18 provides guidance on this issue. Generally, accountants apply the prudence concept and will disclose information on contingent losses more readily than on contingent gains.

Contingency An additional amount or percentage added to any cash flow item (ie. Capex). Care is needed to ensure it is either to be spent or to remain as a cushion.

Contingency theory of management accounting The theory that there is no single management accounting system acceptable to all organizations or any system that is satisfactory in all circumstances in a single organization. Consequently, accounting systems are contingent upon the circumstances that prevail at any time; they must be capable of development in order to take into consideration such factors as changes in the environment, competition, organizational structures, and technology.

Contingent annuity (reversionary annuity) An annuity in which the payment is conditional on a specified event happening. The most common form is an annuity purchased jointly by a husband and wife that begins payment after the death of one of the parties (*see* joint-life and last-survivor annuities).

Contingent claim A claim that can be made only if one or more specified outcomes occur.

Contingent conversion trigger Used in the context of convertible instruments. The price of the stock must exceed the trigger price before the bond holder can convert to common stock at a pre-established conversion price. The trigger price exceeds the conversion price. In addition, after a certain number of years, the convertible instrument usually specifies that both the conversion price and the contingent conversion trigger will increase every year by, for example, a rate equal to LIBOR.

Contingent immunization An arrangement in which the money manager pursues an active bond portfolio strategy until an adverse investment experience drives the then-available potential return down to the safety-net level. When that point is reached, the money manager is obligated to pursue an immunization strategy to lock in the safety-net level return.

Contingent loss A loss that depends upon the

outcome of some contingency. For example, if there is a substantial legal claim for damages against a company, there is a contingent loss, if the claim has to be settled. A material contingent loss should be accrued in the financial statements if it is considered probable that a future event will confirm the loss and if it can be estimated with reasonable accuracy at the date on which the financial statements are approved by the board of directors. If the loss is not accrued, it should be disclosed, unless the possibility of making the loss is remote. Accounting for a contingent loss is therefore subjective; this can be reduced to some extent by seeking independent expert advice.

Contingent Voting Power Enables preferred stockholders to vote when the company fails to satisfy the agreement between itself and the preferred stockholders.

Continuous processing A process in which very high volumes are produced on dedicated plant continuously, e.g. cement production or electricity generation. The plant usually involves very high capital expenditure. Unplanned shut-downs are very costly, typically involving a major refurbishment, for example to replace cracked furnace linings. Most continuous processes are largely self-controlling or are monitored by computer with only minor operator involvement. The product is often made to stock and is therefore vulnerable to market fluctuations.

Continuous random variable A random value that can take any fractional value within specified ranges, as contrasted with a discrete variable.

Continuous-operation costing A system of costing applied to industries in which the method of production is continuous processing; examples include electricity generation and bottling. Because the product is homogeneous, this costing system is essentially a form of average costing in which the unit cost is obtained by dividing the total production cost by the number of units produced. Contra A book-keeping entry on the opposite side of an account to an earlier entry, with the object of cancelling the effect of the earlier entry.

Contra accounts Accounts that can be offset, one against the other. For example, if Company A owes money to Company B and Company B also owes money to Company A, the accounts can be offset against each other, enabling both debts to be settled by one payment.

Contra broker The broker on the buy side of a sell order or the sell side of a buy order.

Contra proferentem A rule of interpretation primarily applying to documents. If any doubt or ambiguity arises in the interpretation of a document, the rule requires that the doubt or ambiguity should be resolved against the party who drafted it or who uses it as a basis for a claim against another. For instance, a plaintiff who sues for breach of a written contract can expect that any ambiguity in the terms of the contract will be resolved against him. The expression derives from the Latin: *verba chartarum fortuis accipiuntur contra proferentum*, the words of a contract are construed more strictly against the person proclaiming them.

Contraband Illegally imported or exported goods, i.e. goods that have been smuggled into or out of a country.

Contract A legally binding agreement. Agreement arises as a result of an offer and acceptance, but a number of other requirements must be satisfied for an agreement to be legally binding. There must be consideration (unless the contract is by deed); the parties must have an intention to create legal relations; the parties must have capacity to contract (i.e. they must be competent to enter a legal obligation, by not being a minor, mentally disordered, or drunk); the agreement must comply with any formal legal requirements; the agreement must be legal (*see*illegal contract); and the agreement must not be rendered void either by some common-law or statutory rule or by some inherent defect.

N general, no particular formality is required for the creation of a valid contract. It may be oral, written, partly oral and partly written, or even implied from conduct. However, certain contracts are valid only if made by deed (e.g. transfers of shares in statutory companies, transfers of shares in British ships, legal mortgages, certain types of lease) or in writing (e.g. hire-purchase agreements, bills of exchange, promissory notes, contracts for the sale of land made after 21 September 1989), and certain others, though valid, can only be enforced if evidenced in writing (e.g. guarantees, contracts for the sale of land made before 21 September 1989). *See also* service contract. Certain contracts, though valid, may be liable to be set aside by one of the parties on such grounds as misrepresentation or the exercise of undue influence. *See also* affirmation of contract; breach of contract.

Contract costing A costing technique applied to long-term contracts, such as civil-engineering projects, in which the costs are collected by contract. A partic- ular problem of long-term projects is the determination of annual profits to be taken to the profit and loss account when the contract is incomplete. This requires the valuation of work in progress at the end of the financial year.

Contract for service A contract undertaken by a self-employed individual. The distinction between a contract for service (self-employed) and a service contract (employee) is fundamental in establishing the tax position. With a contract for service the person may hire and pay others to carry out the work, will be responsible for correcting unsatisfactory work at their own expense, and may make losses as well as profits.

Contract note A document sent by a stockbroker or commodity broker to a client as evidence of having bought (in which case it may be called a bought note) or sold (a sold note) securities or commodities in accordance with the client's instructions. It will state the quantity of securities or goods, the price, the date (and sometimes the time of day at which the bargain was struck), the rate of commission the cost of the transfer stamp and VAT (if any), and -finally — the amount due and the settlement date.

Contractual plan A plan in which fixed dollar amounts of mutual fund shares are purchased through periodic investments, usually featuring some sort of additional incentive for the fixed period payments.

Contrarian An investment style that leads one to buy assets that have performed poorly and sell assets that have performed well. There are two possible reasons this strategy might work. The first is a mean-reversion argument; that is, if the asset has deviated from its usual level, it should eventually return to that usual level. The second reason has to do with overreaction. Investors might have overreacted to bad news sending the asset price lower than it should be.

Contribution 1. The amount that, under marginal-costing principles, a given transaction produces to cover fixed overheads and to provide profit. The unit contribution is normally taken to be the selling price of a given unit of merchandise, less the variable costs of producing it. Once the total contributions exceed the fixed overheads, all further contribution represents pure profit. The total contribution is the product of the unit contribution and the number of units produced. This is based on the assumptions that the marginal cost and the sales value will be constant.

2. The sharing of claim payments between two or more insurers who find themselves insuring the same item, against the same risks, for the same person. As that person is not entitled to claim more than the full value

of the item once, each insurer pays a share. For example, if a coat was stolen from a car, it might be insured under both a personal-effects insurance and a motor policy. As the policyholder is only entitled to the value of the coat (and cannot profit from the theft), each insurer contributes half of the loss.

Contribution income statement The presentation of an income statement or profit and loss account using the marginal costing or variable costing layout in which the fixed costs are not charged to the individual

Control chart A graphical representation of the statistically derived performance limits (normally ±2 standard deviations), which is used in statistical process control. By recording samples on a control chart, it is possible to identify the trends and to take remedial action before performance goes outside the limits. *See also*concurrent control; feedforward control.

Control Limits The upper and lower limits on the acceptable level of cash that minimizes the sum of the opportunity cost of excessive cash and the cost of marketable security transactions.

Control parameters In a nonlinear dynamic system, the coefficient of the order parameter; the determinant of the influence of the order parameter on the total system. See: Order Parameter.

Control process A process that helps an organization to determine whether or not its objectives are being achieved and whether or not adjustments need to be made to meet them. A good control system should help managers to determine why goals and objectives are not being met; knowing the cause of failure will help to decide the corrective actions. Control activities should relate directly to tactical and operational plans; in turn, these will affect and influence the organization's long-term strategic plans.

Control stock The shares owned by the controlling shareholders of a corporation. Sometimes refers to stock that has voting rights rather than stock that carries no voting rights. In a situation where all stock has voting rights, it sometimes refers to the shareholdings of one investors or a group of investors that effectively control the firm.

Control The ability to direct the financial and operating policies of another undertaking with a view to gaining economic benefits from its activities. One company is said to control another company if it holds more than 50% of that company's share capital, voting power, distributable income, or net assets in a winding-up If one company has control in this sense over another, the two companies should produce consolidated financial statements.

Controlled disbursement A service that provides for a single presentation of checks each day (typically in the early part of the day).

Controller The corporate manager responsible for the firm's accounting activities. Sometimes referred to as the comptroller (which means the same thing).

Controlling interest An interest in a company that gives a person control of it. A shareholder having a controlling interest in a company would normally need to own or control more than half the voting shares. However, in practice, a shareholder might control the company with considerably less than half the shares, if the other shares are held by a large number of people For legal purposes, a director is said to have a controlling interest in a company if he or she alone, or together with a spouse, minor children, and the trustees of any settlement in which he or she has an interest, owns more than 20% of the voting shares in a company or in a company that controls that company.

Convenience yield The extra advantage that firms derive from holding the commodity rather than the future.

Convention statement An annual statement filed by a life insurance company in each state where it does business in compliance with that state's regulations. The statement and supporting documents show, among other things, the assets, liabilities, and surplus of the reporting company.

Convergence The movement of the price of a futures contract toward the price of the underlying cash commodity. At the start, the contract price is higher because of the time value. But as the contract nears expiration, the futures price and the cash price converge.

Conversion premium The extent by which the conversion price of a convertible security exceeds the prevailing common stock price at the time the convertible security is issued. In general usage, the conversion premium is the amount by which the convertible security trades above its conversted value. For example, if a Rs.1,000 par bond is trading at Rs.1,100, it is convertible into 50 shares, and the shares are trading at Rs.21, the converted value is 50 X 20.50 = Rs.1,025, and the conversion premium is Rs.75.

Conversion price Applies mainly to convertible securities. Dollar value at which convertible bonds, debentures, or preferred stock can be converted into common stock, as specified when the convertible is issued.

Conversion ratio Applies mainly to convertible securities. Relationship that determines how many shares of common stock will be received in exchange for each convertible bond or preferred stock when a conversion takes place. It is determined at the time of issue and is expressed either as a ratio or as a conversion price from which the ratio can be figured by dividing the par value of the convertible by the conversion price.

Convertibility The extent to which one currency can be freely exchanged for another. Since 1979 sterling has been freely convertible. The International Monetary Fund encourages free convertibility, although many governments try to maintain some direct control over foreign-exchange transactions involving their own currency, especially if there is a shortage of hard-currency foreignexchange reserves.

Convertible arbitrage A practice, usually of buying a convertible bond and shorting a percentage of the equivalent underlying common shares, to create a positive cash flow position (with expected returns above the riskless rate) in a static environment and benefit from capital appreciation should the convertible's premium rise. This form of investing is far from riskless and requires constant monitoring.

Convertible eurobond A eurobond that can be converted into another asset, often through exercise of attached warrants.

Convertible exchangeable preferred stock Convertible preferred stock that may be exchanged, at the issuer's option, into convertible bonds that have the same conversion features as the convertible preferred stock.

Convertible preferred stock Preferred stock that can be converted into common stock at the option of the holder.

Convertible price The contractually specified price per share at which a convertible security can be converted into shares of common stock.

Convertible revolving credit A revolving credit that can be converted by mutual agreement into a fixed-term loan.

Convertible security A security that can be converted into common stock at the option of the securityholder; includes convertible bonds and convertible preferred stock.

Convertible term assurance A term assurance that gives the policyholder the option to widen the policy to become a whole life policy or an endowment assurance policy, without having to provide any further evidence of good health. All that is required is the payment of the extra premium. The risks of AIDS has meant that policies of this kind are no longer available, as insurers are not now prepared to offer any widening of life cover without evidence of good health.

Core capital The capital required of a thrift institution, which must be at least 2% of assets to meet the rules of the Federal Home Loan Bank.

Corn skill The distinctive capabilities that endow a particular business with a competitive advantage over its rivals. The degree to which the business can exploit this unique competence varies according to the extent to which it can be imitated by rival firms. Some skills, such as technologies or processes, may be protected by patents; others may have been built up by reputation or by marketing expertise.

Corner A monopoly established by an organization that succeeds in controlling the total supply of a particular good or service. It will then force the price up until further supplies or substitutes can be found. This objective has often been attempted, but rarely achieved, in international commodity markets. Because it is undesirable and has antisocial effects, a corner can now rarely be attempted as government restrictions on monopolies and antitrust laws prevent it.

Corporate acquisition The acquisition of one firm by anther firm.

Corporate anorexia A malaise that affects businesses after a severe cost-cutting phase, when too many creative people have been made redundant. In many cases, design teams have been broken up, permanently endangering the company's ability to develop new products. The firm's ability to expand production to maintain its competitive position in the market may also have been compromised by a misguided pressure to reduce costs by downsizing, resulting in the elimination of personnel who made a long-term contribution to the company's viability.

Corporate bond A bond or security representing a normal commercial loan in sterling which, if conversion is possible, can only be converted into a similar bond. In the UK corporate bonds are exempt from capital gains tax.

Corporate financing committee A committee of the NASD that reviews underwriters' SEC-required documents to ensure that proposed markups are fair and in the public interest.

Corporate income fund (CIF) A unit investment trust featuring a fixed portfolio of high-grade securities and other investments, usually with monthly distribution of income.

Corporate modelling The use of simulation models to assist the management of an organization in carrying out planning and decision making. A budget is an example of a corporate model.

Corporate processing float The time that elapses between receipt of payment from a customer and the depositing of the customer's check in the firm's bank account; the time required to process customer payments.

Corporate raider A person or company that buys a substantial proportion of the equity of another company (the target company) with the object of either taking it over or of forcing the management of the target company to take certain steps to improve the image of the company sufficiently for the share price to rise enough for the raider to sell the holding at a profit.

Corporate repurchase Active buying by a corporation of its own stock in the marketplace. Reasons for repurchase include putting idle cash to use, raising EPS, creating support for a stock price, increasing internal control (shark repellant), or stock for ESOP or pension plans. Repurchase is subject to rules, such as that buying must be on a zero minus or a minus tick, after the opening and before 3:30 p.m.

Corporate restructuring A change in the business strategy of a company by diversification into new areas, which it considers more attractive. Alternatively, following poor performance, it may seek to withdraw from an area by divestment or closure of parts of its business, which it no longer considers to be part of its future strategic focus.

Corporate social reporting Issues, such as employee reports, ethical reporting, green reporting, and other matters that may be considered of interest to readers of the annual and report accounts of a company in addition to information that caters for the financial interests of the shareholders. From time to time companies do disclose such information in their annual report, although it remains largely a voluntary activity.

Corporate venturing The provision of venture capital by one company, either directly or by means of a venturecapital fund for another company; the objectives are usually either to obtain information about the activities of the company requiring the venture capital or its markets or as a preliminary step towards acquiring that company. It may often be a means of moving into a fresh market cheaply and without needing to acquire the necessary expertise and personnel required to do so on its own.

Corporation A succession of persons or body of persons authorized by law to act as one person and having rights and liabilities distinct from the individuals forming the corporation. The artificial personality may be created by royal charter, statute, or common law. The most important type is the registered company formed under the *Companies Act*. Corporations sole are those having only one individual forming them; for example, a bishop, the sovereign, the Treasury Solicitor. Corporations aggregate are composed of more than one individual, e.g. a limited company. They may be formed for special purposes by statute; the BBC is an example. Corporations can hold property, carry on business, bring legal actions, etc, in their own name. Their actions may, however, be limited by the doctrine of ultra vires. *See also*public corporation.

Corporation tax (CT) Tax charged on the total profits of a company resident in the UK arising in each accounting period. The rate of corporation tax depends on the level of profits of the company. The small companies rate of 24% applies to companies with total profits plus franked investment income of £300,000 or less. Full-rate corporation tax of 33% applies to companies with total profits and franked investment income of over £1.5 million, both limits being those for the financial year 1996-97. There is a marginal relief for those companies with total profits plus franked investment income between those two limits.

Correlation coefficient A standardized statistical measure of the dependence of two random variables, defined as the covariance divided by the product of the standard deviations of two variables.

Correspondent bank A bank in a foreign country that offers banking facilities to the customers of a bank in another country. These arrangements are usually the result of agreements, often reciprocal, between the two banks. The most frequent correspondent banking facilities used are those of money transmission.

Cost 1. An expenditure, usually of money, for the purchase of goods or services.

2. An expenditure, usually of money, incurred in achieving a goal, e.g. producing certain goods, building a factory, or closing down a brand! *See also* current cost ; economic cost ; fixed cost ; historical cost ; marginal cost ; opportunity cost ; replacement cost.

Cost behaviour The changes that occur to total costs as a result of changes in activity levels within an organization. The total of the fixed costs tends to remain unaltered by changes in activity levels in the short term, whereas the total of the variable costs tends to increase or decrease in proportion to activity. There are also some costs that demonstrate semi-variable cost behaviour, i.e. they have both fixed and variable elements. The study of cost behaviour is important for breakeven analysis and also when considering decision-making techniques.cost-benefit analysis A technique used in capital budgeting that takes into account the estimated costs to be incurred by a proposed investment and the estimated benefits likely to arise from it. In a financial appraisal the benefits may arise from an increase in the revenue from a product or service, from saved costs, or from other cash inflows, but in an economic appraisal the economic benefits, such as the value of time saved or of fewer accidents resulting from a road improvement, often require to be valued.cost centre The area of an organization for which costs are collected for the purposes of cost ascertainment, planning, decision making, and control. Cost centres are determined by individual organizations; they may be based on a function, department, section, individual or any group of these. Cost centres are of two main types: production cost centres in an organization are those concerned with making a product, while service cost centres provide a service (such as stores, boilerhouse, or canteen) to other parts of the organization.cost classification The process of grouping expenditure according to common characteristics. Costs are initially divided into capital expenditure and revenue expenditure. In a production organization the revenue expenditure is classified sequentially to enable a product cost to build up in the order inwhich these costs tend to be incurred. Revenue costs would normally include: direct material, direct labour, roduction overheads, administration overheads, selling overheads, distribution overheads, research overheads.

Cost driver Any activity or series of activities that takes place within an organization and causes costs to be incurred. Cost drivers are used in a system of activity-based costing to charge costs to products or services. Cost drivers are applied to cost pools, which relate to common activities. Cost drivers are not restricted to departments or sections, as more than one activity may be identified within a department.

Cost of capital The return, expressed in terms of an interest rate, that an organization is required to pay for the capital used in financing its activities. As the capital of an organization can be a mix of equity share capital, loan capital, and debt, there is considerable debate as to whether or not the cost of capital increases as the gearing increases. Another approach to establishing the cost of capital is to compute a unique weighted average cost of capital for each organization, based on its particular mix of capital sources. The cost of capital is often used as a hurdle rate in discounted cash flow calculations.

Cost of carry Out-of-pocket costs incurred while an investor has an investment position. Examples include interest on long positions in margin account, dividend lost on short margin positions, and incidental expenses. Related: Net financing cost.

Cost of goods sold (COGS) The directly attributable costs of products or services sold, (usually materials, labour, and direct production costs). Sales less COGS = gross profit. Effetively the same as cost of sales (COS) see below for fuller explanation.

Cost The opposite of revenue. An expense that reflects the price of purchasing goods, services and financial instruments. A cash cost means that cash is given up today to the purchase. Also, the purchase price of an investment, which is compared to the sale proceeds to determine capital gain or loss.

Cost unit A unit of production for which the management of an organization wishes to collect the costs incurred. In some cases the cost unit may be the final item produced, for example a chair or a light bulb, but in other more complex products the cost unit may be a sub-assembly, for example an aircraft wing or a gear box. Cost units may also be expressed as batches of items, particularly when the unit cost of the individual product would be very small; for example, the cost unit for a manufacturer of pens might be the cost per thousand pens.

Cost-of-carry market Applies to derivative products. Futures contracts trade in a "cost-of-carry market" where the underlying commodity can be stored, insured, and converted into the future easily and inexpensively. Arbitrageurs, because of the ease of switching from the spot commodity to futures, will keep these markets in line with prevailing interest rates.

Cost-plus 1. Contract terms in which a supplier provides goods and services at cost plus an agreed fee or percentage. These terms are sometimes used if it is difficult to estimate costs in advance. They are, however, unpopular because the supplier is motivated to increase the costs rather than to reduce them and the buyer has to enter into an openended commitment. There are, however, circumstances in which such a contract is inevitable, for example if the costs of materials cannot be ascertained in advance or if it is not possible to estimate the amount of work to be done. This form of contract is sometimes used in the building trade.

2. A method of pricing work in which the cost of doing the work is marked up by a fixed percentage to cover either overheads or profit or both.

Countertrading The practice in international trading of paying for goods in a form other than by hard currency. For example, a South American country wishing to buy aircraft may countertrade (usually through a third party) by paying in coffee beans.

Country economic risk Developments in a national economy that can affect the outcome of an international financial transaction.

Country financial risk The ability of the national economy to generate enough foreign exchange to meet payments of interest and principal on its foreign debt.

Country risk The general level of political, financial, and economic uncertainty in a country which impacts the value of the country's bonds and equities. See:Sovereign risk.

Coupon The contractual interest obligation a bond or debenture issuer covenants to pay to its debtholders.

Covenant An agreed action to be undertaken (Positive) or not done (Negative). A breach of a covenant is a default.

Coverage ratios Ratios used to test the adequacy of cash flows generated through earnings for purposes of meeting debt and lease obligations, including the interest coverage ratio and the fixed-charge coverage ratio.

Covered A written option is considered to be covered if the writer also has an opposing market position on a share-for-share basis in the underlying security. That is, a short call is covered if the underlying stock is owned, and a short put is covered (for margin purposes) if the underlying stock is also short in the account. In addition, a short call is covered if the account is also long another call on the same security, with a striking price equal to or less than the striking price of the short call. A short put is covered if there is also a long put in the account with a striking price equal to or greater than the striking price of the short put.

Covered call A short call option position in which the writer owns the number of shares of the underlying stock represented by the option contracts. Covered calls generally limit the risk the writer takes because the stock does not have to be bought at the market price, if the holder of that option decides to exercise it.

Covered Put A put option position in which the option writer also is short the corresponding stock or has deposited, in a cash account, cash or cash equivalents equal to the exercise of the option. This limits the option writer's risk because money or stock is already set aside. In the event that the holder of the put option decides to exercise the option, the writer's risk is more limited than it would be on an uncovered or naked put option.

Covered writer An investor who writes options only on stock that he or she owns, so that option premiums may be collected.

Capital loss (allowable capital loss) The excess of the cost of an asset over the proceeds received on its disposal. Indexation of the cost of the asset, or indexation of the 31 March 1982 value, for those assets acquired before 31 March 1982, could be deducted from the proceeds to establish the indexed capital loss on disposals prior to 30 November 1993. The Finance Act (1994) introduced a restriction on the use of indexation to create or increase a capital loss, with a degree of transitional relief for disposals made in 1993-94 and 1994-95.

Capital maintenance concept 1. The financial capital maintenance concept is that the capital of a company is only maintained if the financial or monetary amount of its net assets at the end of a financial period is equal to or exceeds the financial or monetary amount of its net assets at the beginning of the period, excluding any distributions to, or contributions from, the owners.

2. The physical capital maintenance concept is that the physical capital is only maintained if the physical productive or operating capacity, or the funds or resources required to achieve this capacity, is equal to or exceeds the physical productive capacity at the beginning of the period, after excluding any distributions to, or contributions from, owners during the financial period.

Capital market A market in which long-term capital is raised by industry and commerce, the government, and local authorities. The money comes from private investors, insurance companies, pension funds, and banks and is usually arranged by issuing houses and merchant banks. Stock exchanges are also part of the capital market in that they provide a market for the shares and loan stocks that represent the capital once it has been raised. It is the presence and sophistication of their capital markets that distinguishes the industrial countries from the developing countries, in that this facility for raising industrial and commercial capital is either absent or rudimentary in the latter.

Capital reserves Undistributed profits of a company that for various reasons are not regarded as distributable to shareholders as dividends. These include certain profits on the revaluation of capital assets and any sums received from share issues in excess of the nominal value of the shares, which are shown in a share premium account. Capital reserves are also known as undistributable reserves.

Crash 1. A rapid and serious fall in the level of prices in a market.

2. A breakdown of a computer system. A program is said to crash if it terminates

abnormally. A computer is said to crash either if it suffers a mechanical failure, or if one of the programs running on it misbehaves in a way that causes the computer to stop. Generally, the computer must then be switched off, any necessary repairs made, and restarted.

Credit analysis The process of analyzing information on companies and bond issues in order to estimate the ability of the issuer to live up to its future contractual obligations.

Credit quality A measure of a bond issuer's ability to repay interest and principal in a timely manner. Rating agencies assign letter designations such as AAA, AA, and so forth. The lower the rating, the higher the probability of default.

Credit rating An evaluation of an individual's or company's ability to repay obligations or its likelihood of not defaulting See: Creditworthiness.

Credit scoring A statistical technique wherein several financial characteristics are combined to form a single score to represent a customer's creditworthiness.

Credit squeeze A government measure, or set of measures, to reduce economic activity by restricting the money supply. Measures used include increasing the interest rate (to restrain borrowing), controlling moneylending by banks and others, and increasing down payments or making other changes to hire-purchase regulations.

Credit Standards The guidelines a company follows to determine whether a credit applicant is creditworthy.

Credit Terms The conditions under which credit will be extended to a customer. The components of credit terms are: cash discount, credit period, net period.

Creditor One to whom an organization or person owes money. The balance sheet of a company shows the total owed to creditors and a distinction has to be made between creditors who will be paid during the coming accounting period and those who will not be paid until later than this.

Creditworthiness An assessment of a person's or a business's ability to pay for goods purchased or services received. Creditworthiness may be presented in the form of a credit rating from a credit-reference agency.

Cross rates The exchange rate between two currencies expressed as the ratio of two foreign exchange rates that are both expressed in terms of a third currency. Foreign exchange rate between two currencies other than the US dollar, the currency in which most exchanges are usually quoted.

Cross-border factoring Concluding a transaction by a network of factors across borders. The exporter's factor can contact correspondent factors in other countries to handle the collection of accounts receivable.

Cross-border risk Describes the volatility of returns on international investments caused by events associated with a particular country as opposed to events associated solely with a particular economic or financial agent.

Cross-Collateral An agreement among project participants to pool collateral, to allow recourse to each other's collateral.

Cross-currency interest-rate swap An interest-rate swap that exchanges one currency for another; the currencies revert to their original form at maturity. One of the two currencies will bear interest at a fixed rate and the other at a floating rate. *Compare*currency interestrate swap.

Crossover rate The return at which two alternative projects have the same net present value.

Cross-sectional analysis Assessment of relationships among a cross-section of firms, countries, or some other variable at one particular time.

Cross-sectional approach A statistical methodology applied to a set of firms at a particular point in time.

Crown jewel A particularly profitable or otherwise particularly valuable corporate unit or asset of a firm. Often used in risk arbitrage. The most desirable entities within a diversified corporation as measured by asset value, earning power, and business prospects; in takeover attempts, these entities typically are the main objective of the acquirer and may be sold by a takeover target to make the rest of the company less attractive. See: Scorched earth policy.

Crown jewel option A form of poison pill in which a company, defending itself against an unwanted takeover bid, writes an option that would allow a partner or other friendly company to acquire one or more of its best businesses at an advantageous price if control of the defending company is lost to the unwelcome predator. The granting of such an option may not always be in the best interests of the shareholders of the defending company.

Cumulative abnormal return (CAR) Sum of the differences between the expected return on a stock and the actual return that comes from the release of news to the market.

Currency 1. Any kind of money that is in circulation in an economy.

2. Anything that functions as a medium of exchange, including coins, banknotes, cheques, bills of exchange, promissory notes. etc.

3. The money in use in a particular country.

4. The time that has to elapse before a bill of exchange matures.

Currency appreciation An increase in the value of one currency relative to another currency. Appreciation occurs when, because of a change in exchange rates, a unit of one currency buys more units of another currency.

Currency arbitrage Taking advantage of divergences in exchange rates in different money markets by buying a currency in one market and selling it in another market. value of a portfolio of specific amounts of individual currencies, used as the basis for setting the market value of another currency. It is also referred to as a currency cocktail.

Currency Board Entity charged with maintaining the value of a local currency with respect to some other specified currency.

Currency Carry Trade A carry trade where you borrow and pay interest in order to buy something else that has higher interest. For currencies, it might be that you borrow in Yen (where the interest rate might be low) and use the proceeds to purchase U.S. dollar long term debt. While the trade might produce a positive return, it is risky in two dimensions. First, U.S. rates could increase diminishing the value of the bond you purchased. Second, the exchange rate could take an unfavorable move effectively increasing your borrowing costs. Related: Carry Trade.

Currency interest-rate swap A transaction involving the exchange of two currencies, both bearing interest at fixed rates or both bearing interest at floating rates. Currency option A contract giving the right either to buy or to sell a specified currency at a fixed exchange rate within a given period. The price agreed is called the strike price or exercise price. This is the price at which the buyer has the right to buy or sell the currency.

Currency overvaluation Applies mainly to international equities:

(1) consideration that a currency is overvalued if private demand for the currency at the going exchange rate is less than total private supply (i.e., central banks are buying up the difference, supporting the value of the currency through foreign exchange intervention);

(2) currency value exceeding purchasing power parity.

Current account 1. An active account at a bank or building society into which deposits can be paid and from which withdrawals can be made by cheque (*see also*cheque account). The bank or building society issues cheque books free of charge and supplies regular statements listing all transactions and the current balance. Banks sometimes make charges for current accounts, based on the number of transactions undertaken, but modern practice is to waive these charges if a certain credit balance has been maintained for a given period. Building societies usually make no charges and pay interest on balances maintained in a current account. Banks, in order to remain competitive, follow this practice, although they have traditionally not paid interest on currentaccount balances.

2. The part of the balance of payments account that records non-capital transactions.

3. An account in which intercompany or interdepartmental balances are recorded.

4. An account recording the transactions of a partner in a partnership that do not relate directly to his or her capital in the partnership.

Current account Net flow of goods, services, and unilateral transactions (gifts) between countries.

Current assets Value of cash, accounts receivable, inventories, marketable securities and other assets that could be converted to cash in less than 1 year.

Current cost accounting (CCA) A form of accounting in which the approach to capital maintenance is based on maintaining the operating capability of a business. Assets are valued at their value to the business, also known as their deprival value. The deprival value of an asset is the loss that a business would suffer if it were to be deprived of the use of the asset. This may be the replacement cost of the asset, its net realizable value, or its economic value to the business. Current cost accounting ensures that a business maintains its operating capacity by separating holding gains from operating profits to prevent them being distributed to shareholders. The current cost accounting profit figure is derived by making a number of adjustments to the historical cost accounting profit and loss account: these are the cost of sales adjustment, depredation adjustment, monetary workingcapital adjustment, and the gearing adjustment. The current-cost reserve in the balance sheet is used to 'collect' the current cost adjustments. Statement of Standard Accounting Practice 16, 'Current Cost Accounting', was issued in March 1980 but withdrawn in April 1988 after much criticism.

Current issue In Treasury securities, the most recently auctioned issue. Trading is more active in current issues than in off-the-run issues. Also known as on-the-run issue.

Current liabilities Amounts owed by a business to other organizations and individuals that should be paid within one year from the balance-sheet date. These generally consist of trade creditors, bills of exchange payable, amounts owed to group and related companies, taxation, social-security creditors, proposed dividends, accruals, deferred income, payments received on account, bank overdrafts, and short-term loans. Any long-term loans repayable within one year from the balance-sheet date should also be included. Current liabilities are distinguished from long-term liabilities on the balance sheet.

Cushion bonds High-coupon bonds trading at a premium that tend to fall in price much less than comparable bonds when interest rates rise (hence the cushion effect), because of their high coupons.

Custodian bank Applies mainly to international equities. Bank or other financial institution that keeps custody of stock certificates and other assets of a mutual fund, individual, or corporate client. See: Depository Trust Company (DTC).

Custom of the trade A practice that has been used in a particular trade for a long time and is understood to apply by all engaged in that trade. Such customs or practices may influence the way in which a term in a contract is interpreted and courts will generally take into account established customs of a trade in settling a dispute over the interpretation of a contract.

Customer service The services an organization offers to its customers, especially of industrial goods and expensive consumer goods, such as computers or cars. Customer services cover a wide variety of forms, including after-sales servicing, such as a repair and replacement service, extended guarantees, regular mailings of information, and, more recently, free phone telephone calls in case of complaints. The appeal of a company's products are greatly influenced by the customer services it offers.

Customization The design and development of a product to meet the specific requirements of a single customer. Because of the high cost and time required to produce a unique product, most firms prefer to provide a product that consists of a set of standard features, some of which have been modified. Many cars, for example, are available in several different body styles, colours, and trim combinations. In addition, optional extras are specified to enable buyers to customize their purchases.

Customized benchmarks A benchmark that is designed to meet a client's requirements and long-term objectives.

Customs Broker An individual or firm licensed by customs authorities to enter and clear imported goods through customs. The broker represents the importer in dealings with the customs authorities.

Customs entry The record kept by the Customs of goods imported into or exported from a country. In the UK a bill of entry is used for either imports (entry in) or exports (entry out). If no duty is involved with a consignment it is given free entry.

Customs invoice An invoice for goods imported or exported, which is prepared especially for the customs authorities.

Customs tariff A listing of the goods on which a country's government requires customs duty to be paid on being imported into that country, together with the rate of customs duty applicable. For the UK the Customs Tariff is published by the Stationery Office.

Customs union A union of two or more states to form a region in which there are no import or export duties between members but goods imported into the region bear the same import duties. The European Union is an example.

Customs union An agreement by two or more countries to erect a common external tariff and to abolish restrictions on trade among members.

Cut and paste A technique used in wordprocessing (*see*wordprocessor). The term describes the way a section of text, such as a paragraph, can be moved within a document: first, the section is 'cut' from its original position; then, it is 'pasted' into the new position.

Cut Off Date The date prescribed in the unclaimed property law in most states for determining the items of property that must be turned over to the state. See: Escheat.

Cyclical stock Stock that tends to rise quickly when the economy turns up and fall quickly when the economy turns down. Examples are housing, automobiles, and paper.

Cyclical unemployment Unemployment caused by a low level of aggregate demand associated with recession in the business cycle.

D/A Abbreviation for documents against acceptance.

Daily price limit The level within many commodity, futures, and options markets are allowed to rise or fall in a day. Exchanges usually impose a daily price limit on each contract.

Daisy chain The buying and selling of the same items several times over, for example stocks and shares. This may be done to 'inflate' trading activity (that is, the sale of the same items are being included in the sales figure more than once).

Damages Compensation, in monetary form, for a loss or injury, breach of contract, tort, or infringement of a right. Damages refers to the compensation awarded, as opposed to damage, which refers to the actual injury or loss suffered. The legal principle is that the award of damages is an attempt, as far as money can, to restore injured parties to the position they were in before the event in question took place; i.e. the object is to provide restitution rather than profit. Damages are not assessed in an arbitrary fashion but are subject to various judicial guidelines. In general, damages capable of being quantified in monetary terms are known as liquidated damages. In particular, liquidated damages include instances in which a genuine pre-estimate can be given of the loss that will be caused to one party if a contract is broken by the other party. If the anticipated breach of contract occurs this will be the amount, no more and no less, that is recoverable for the breach. However, liquidated damages must be distinguished from a penalty. Another form of liquidated damages is that expressly made recoverable under a statute. These may also be known as statutory damages if they involve a breach of statutory duty or are regulated or limited by statute.

Dawn raid An attempt by one company or investor to acquire a significant holding in the equity of another company by instructing brokers to buy all the shares available in that company as soon as the stock exchange opens, usually before the target company knows that it is, in fact, a target. The dawn raid may provide a significant stake from which to launch a takeover bid. The conduct of dawn raids is now restricted by the City Code on Takeovers and Mergers.

Day around order A day order that supersedes (cancels and replaces) the previous order by altering its size or price limit.

Day book A specialized journal or book of prime entry recording specific transactions.

For example, the sales day book records invoices for sales, the purchase day book records invoices received from suppliers. Day-book entries are transferred to memorandum ledgers, such as the debtors' ledger and the creditors' ledger, while totals of entries are transferred to the nominal ledger control accounts, such as the debtors' ledger control account and the creditors' ledger control account.

Day loan A loan from a bank to a broker prior to the delivery of securities. Upon the delivery of the securities, a day loan becomes a regular broker call loan for which securities serve as collateral.

Day order In the context of general equities, request from a customer to either buy or sell stock, that, if not canceled or executed the day it is placed, expires automatically. All orders are day orders unless otherwise specified. Traders often make calls before the opening to check for renewals.

Days of grace The extra time allowed for payment of a bill of exchange or insurance premium after the actual due date. With bills of exchange the usual custom is to allow 3 days of grace (not including Sundays and Bank Holidays) and 14 days for insurance policies.

De facto (Latin: in fact) Denoting that something exists as a matter of fact rather than by right. For example, a plaintiff may have de facto control of a property. This compares to de jure (Latin: in law), which denotes that something exists as a matter of legal right. For example, a planning authority may have acted de jure in refusing a planning application. In international law, one government may recognize another de facto, i.e. acknowledge that it is in control of the country even though it has no legal right to be. *De jure* recognition acknowledges that it has a legal right to govern. The basis of the distinction, however, is more political than legal.

Deadweight debt A debt that is incurred to meet current needs without the security of an enduring asset. It is usually a debt incurred by a government; the national debt is a deadweight debt incurred by the UK government during the two World Wars.

Deal flow In investment banking, the rate at which new deals are referred to a brokerage firm.

Dealer An entity that stands ready and willing to buy a security for its own account (at its bid price) or sell from its own account (at its ask price). Individual or firm acting as a principal in a securities transaction. Principals are market makers in securities, and thus trade for their own account and risk. Antithesis of broker. See: Agency.

Dealer loan Overnight, collateralized loan made to a dealer financing his position by borrowing from a money market bank.

Dealer market Where traders specializing in particular commodities buy and sell assets for their own accounts.

Dealer options Over-the-counter options, such as those offered by government and mortgage-backed securities dealers.

Death Spiral Convertible Used by companies that are in such bad shape, that there is no other way to get financing. This instrument is similar to a convertible bond, but convertible at a discount to the share price at issuance and for a fixed dollar amount rather than a specific number of shares. The further the stock falls, the more shares you get. Popular in the mid to late 1990s. Also known as toxic convertibles or floorless convertibles.

Death-valley curve A curve on a graph showing how the venture capital invested in a new company falls as the company meets its start-up expenses before its income reaches predicted levels. This erosion of capital makes it difficult for the company to interest further investors in providing additional venture capital.

Debenture 1. The most common form of long-term loan taken by a company. It is usually a loan repayable at a fixed date, although some debentures are irredeemable securities; these are sometimes called perpetual debentures. Most debentures also pay a fixed rate of interest, and this interest must be paid before a dividend is paid to shareholders. Most debentures are also secured on the borrower's assets, although some, known as naked debentures or unsecured loan stock, are not. In the USA debentures are usually unsecured, relying only on the reputation of the borrower. In a secured debenture, the bond may have a fixed charge (i.e. a charge over a particular asset) or a floating charge. If debentures are issued to a large number of people (for example in the form of debenture stock or loan stock) trustees may be appointed to act on behalf of the debenture holders. There may be a premium on redemption and some debentures are convertible, i.e. they can be converted into ordinary shares on a specified date, usually at a specified price. The advantage of debentures to companies is that they carry lower interest rates than, say, overdrafts and are usually repayable a long time into the future. For an investor, they are usually saleable on a stock exchange and involve less risk than equities.

2. A deed under seal setting out the main terms of such a loan. **3.** A form of bank security covering corporate debt, either fixed or floating; in either case the bank ranks as a preferred creditor in the event of liquidation.

Debenture bond An unsecured bond whose holder has the claim of a general creditor on all assets of the issuer not pledged specifically to secure other debt. Compare subordinated debenture bond and collateral trust bonds.

Debenture stock A type of stock that makes fixed payments at scheduled intervals of time. Debenture stock differs from a debenture in that it has the status of equity, not debt, in liquidation.

Debit An expense, or money paid out from an account. A debit transaction is one which the net cost is greater than the net sale proceeds. See also Credit.

Debit balance The amount that is owed to a broker by a margin customer for loans the customer uses to buy securities.

Debit card A card that resembles a credit card but which debits a transaction account (checking account) with the transfers occurring contemporaneously with the customer's purchases. A debit card may be machine readable, allowing for the activation of an automated teller machine or other automated payments equipment.

Debit note A document sent by an organization to a person showing that the recipient is indebted to the organization for the amount shown in the debit note. Debit notes are rare as invoices are more regularly used; however, a debit note might be used when an invoice would not be appropriate, e.g. for some form of inter-company transfer other than a sale of goods or services.

Debit spread Applies to derivative products. Difference in the value of two options, when the value of the option bought exceeds the value of the one sold. One buys a "debit spread." Antithesis of a credit spread.

Debt A sum owed by one person to another. In commerce, it is usual for debts to be settled within one month of receiving an invoice, after which interest may be incurred. A long-term debt may be covered by a bill of exchange, which can be a negotiable instrument. *See also*debenture

Debt collection agency An organization that specializes in collecting the outstanding debts of its clients, charging a commission for doing so. Because of the historical stigma attached to the phrase 'debt collection', these agencies prefer to be called commercial collection agencies.

Debt instrument An asset requiring fixed dollar payments, such as a government or corporate bond.

Debt leverage Amplification of the return earned on equity when an investment or firm is financed partially with borrowed money.

Debt limitation A bond covenant that restricts in some way the firm's ability to incur additional indebtedness.

Debt relief Reducing the principal and/or interest payments on LDC loans.

Debt rescheduling A negotiation concerning outstanding loans in which the debtor has repayment difficulties. The rescheduling can take the form of an entirely new loan or an extension of the existing loan repayment period, deferring interest or principal repayments. When debt rescheduling concerns less developed countries, the new agreement may involve a lowering of interest rates or an offer of an aid package of foreign investment to the country to offset some of the existing debt. In the late 1980s the major, UK commercial banks wrote off or made provision for nearly £10 billion of debt to less developed countries — often Latin American countries who had simply stopped repaying their earlier loans. In the late 1980s and early 1990s a wide range of rescheduling schemes were adopted, including some in which the debt, or part of it, is converted into internal aid programmes.

Debt-equity ratio A ratio used to examine the financial structure or gearing of a business. The long-term debt, normally including preference shares, of a business is expressed as a percentage of its equity. A business may have entered into an agreement with a bank that it will maintain a certain debt-equity ratio; if it breaches this agreement the loan may have to be repaid. A highly geared company is one in which the debt is higher than the equity, compared to companies in a similar industry. A highly geared company offers higher returns to shareholders when it is performing well but should be regarded as a speculative investment. The debt—equity ratio is now sometimes expressed as the ratio of the debt to the sum of the debt and the equity.

Debtor One who owes money to another. In balance sheets, debtors are those who owe money to the organization and a distinction has to be made between those who are expected to pay their debts during the next accounting period and those who will not pay until later.

Decile rank Performance over time, rated on a scale of 1-10.1 indicates that a mutual fund's return was in the top 10% of funds being compared, while 3 means the return was in the top 30%. Objective Rank compares all funds in the same investment strategy category. All Rank compares all funds.

Decision making The act of deciding between two or more alternative courses of action. In the running of a business, accounting information and techniques are used to facilitate decision making, especially by the provision of decision models, such as discounted cash flow, critical-path analysis, marginal costing, and breakeven analysis.

Decision model A model that simulates the elements or variables inherent in a business decision, together with their relationships to each other and the constraints under which they operate; the purpose of the model is to enable a solution to be arrived at in keeping with the objectives of the organization. For example, a linear programming decision model may arrive at a particular production mix that, having regard to the constraints that exist, either minimizes costs or maximizes the contribution. Other decision models include decision trees, discounted cash flow, breakeven analysis, and budgets.

Decision trees Diagrams that illustrate the choices available to a decision maker and the estimated outcomes of each possible decision. Each possible decision is shown as a separate branch of the tree, together with each estimated outcome for each decision and the subjective probabilities of these outcomes actually occurring (*see*illustration overleaf). From this information the expected values for each outcome can be determined, which can provide valuable information in decision making.

Declaration date The date on which a firm's directors meet and announce the date and amount of the next dividend.

Declaration of dividend A statement in which the directors of a company announce that a dividend of a certain amount is recommended to be paid to the shareholders. The liability should be recognized as soon as the declaration has been made and the appropriate amount is included under current liabilities on the balance sheet. The company pays the dividend net of income tax to the shareholders and pays the tax due in the form of advance corporation tax.

Declaration of solvency A declaration made by the directors of a company seeking voluntary liquidation that it will be able to pay its debts within a specified period not exceeding 12 months from the date of the declaration. It must contain a statement of the company's assets and liabilities, and a copy must be sent to the Registrar of Companies. A director who participates in a declaration of solvency without reasonable grounds will be liable to a fine or imprisonment on conviction. *See* members voluntary liquidation.

Dedicated capital Total par value (number of shares issued, multiplied by the par value of each share). Also called dedicated value.

Deductive reasoning The use of general fact to provide accurate information about a specific situation.

Deed of covenant A legal document, which must be in a specified form, used to transfer income from one person to another with a view to making a saving in tax. It authorizes regular annual payments to be made, which must normally be at least six (except in the case of payments to charities, when it can be three). The person making the payment deducts income tax at the basic rate from the payment, in most cases obtaining tax relief on it. Any recipient who is exempt from tax (e.g. a charity) can reclaim the tax deducted. In certain cases, such as payments to charities, tax relief at higher rates may be available to the payer; this does not apply to student children.

Deed of partnership A partnership agreement drawn up in the form of a deed. It covers the respective capital contributions of the partners, their entitlement to interest on their capital, their profit-sharing percentages, agreed salary, etc.

Deed poll A deed having a straight edge at the top, as opposed to an indenture. A deed poll was used when only one party was involved in an action, e.g. when a person declared a wish to be known by a different name. Deeds commonly now have straight edges and are used for all purposes. .

Deep in the money A call option with an exercise price substantially below the underlying stock's market price. Also put option with an exercise price substantially above the underlying stock's market price. Often substantially below is defined as more than one strike price below (for calls)/above (for puts) the current value of the underlying security.

Deep-gain security A security issued at a deep discount or redeemed at a premium. The terms of the issue mean that the amount of discount or premium cannot be allocated evenly over the life of the bond. From 14 March 1989 the whole of the excess of the proceeds over the cost became assessable for income tax.

Default Failure to make timely payment of

interest or principal on a debt security or to otherwise comply with the provisions of a bond indenture.

Default interest A higher interest rate payable after default.

Default premium A differential in promised yield that compensates the investor for the risk inherent in purchasing a corporate bond that entails some risk of default. Often the premium is measured as the yield over and above a government bond yield of similar coupon and maturity.

Default risk The risk that an issuer of a bond may be unable to make timely principal and interest payments. Also referred to as credit risk (as gauged by commercial rating companies).

Default The failure to make timely payment of interest or principal on a debt security or to otherwise comply with the provisions of a bond indenture. A breach of a covenant. In context of project financing, a technical default signals a project parameter is outside defined or agreed limits or a legal matter is not yet resolved.

Defeasance The setting aside by a borrower of cash or bonds sufficient to service the borrower's debt. Both the borrower's debt and the offsetting cash or bonds are removed from the balance sheet. In securities trading, where a clearing house becomes counterparty to each side of a trade, after the trade has been agreed. This is necessary to facilitate netting, and reduce counterparty risk exposure. The term has become popular recently, because of the growth of central counterparty clearing services in European cash equities markets.

Defensive interval ratio A ratio that demonstrates the ability of a business to satisfy its current debts by calculating the time for which it can operate on current liquid assets, without needing revenue from the next period's sales. Current assets less stock is divided by the projected daily operational expenditure less non-cash charges. Projected daily operational expenditure is calculated by dividing by 365 the total of the cost of goods sold operating expenses, and other cash expenses.

Deferred compensation An amount that has been earned but is not actually paid until a later date, typically through a payment plan, pension, or stock option plan.

Deferred credit (deferred liability) Income received or recorded before it is earned, under the accruals concept. The income will not be included in the profit and loss account of the period but will be carried forward on the balance sheet until it is matched with the period in which it is earned. A common example of a deferred credit is a government grant. The grant is shown as a separate item or under creditors in the balance sheet, and an annual amount is transferred to the profit and loss account until the deferred credit balance is brought to nil.

Deferred debit (deferred asset; deferred expense) An item of expenditure incurred in an accounting period but, under the accruals concept, not matched with the income it will generate. Instead of being treated as an operating cost for that period, it is treated as an asset with the intention of treating it as an operating cost to be charged against the income it will generate in a future period.

Deferred equity A common term for convertible bonds because of their equity component and the expectation that the bond will ultimately be converted into shares of common stock.

Deferred futures The most distant months of a futures contract. A bond that sells at a discount and does not pay interest for an initial period, typically from three to seven years. Compare step-up bond and payment-in-kind bond.

Deferred income A payment received in advance of the period to which it relates. Under the accruals concept credit will not be taken for it in the profit and loss account of the period; instead it will be shown as a deferred credit balance in the balance sheet. Amounts are transferred from the deferred credit balance to the profit and loss account of future periods until the deferred credit balance in the balance sheet is brought to nil.

Deferred nominal life annuity A monthly fixed-dollar payment beginning at retirement age. It is nominal because the payment is fixed in dollar amount at any particular time, up to and including retirement.

Deflation A general fall in the price level; the opposite of inflation. As with inflation, a general change in the price level should, in theory, have no real effect. However, if traders are holding goods whose prices fall, they may suffer such large losses that they are forced into bankruptcy. However, agents holding money are simultaneously better off, although there may be a lag in increasing expenditure, during which a recession may occur. The only major deflation in this century occurred during the Great Depression in the 1920s and 1930s. Since them governments have avoided deflation wherever possible.

Delivery date Date by which a seller must fulfill the obligations of a forward or futures contract.

Delivery month The month in which commodities on a futures contract must be handed over to the buyer. A month is specified but usually any day within that month will fulfil the contract obligations.

Delivery note A document, usually made out in duplicate, that is given to the consignee of goods when they are delivered to him. The consignee, or his representative, signs one copy of the delivery note as evidence that the goods have been received.

Delivery notice The written notice given by the seller of its intention to make delivery against an open, short futures position on a particular date

Delivery options The options available to the seller of an interest rate futures contract, including the quality option, the timing option, and the wild card option. Delivery options make the buyer uncertain of which Treasury Bond will be delivered or when it will be delivered.

Delivery order A written document from the owner of goods to the holder of the goods (e.g. a warehouse company) instructing them to release the goods to the firm named on the delivery order or to the bearer (if made out to "bearer"). A delivery order backed by a dock or warehouse warrant may be accepted by a bank as security for a loan.

Delivery points Those points designated by futures exchanges at which the financial instrument or commodity covered by a futures contract may be delivered in fulfillment of such contract.

Delivery The tender and receipt of an actual commodity or financial instrument in settlement of a futures contract.

Delivery versus payment A transaction in which the buyer's payment for securities is due at the time of delivery (usually to a bank acting as agent for the buyer) upon receipt of the securities. The payment may be made by bank wire, check, or direct credit to an account.

Delphi technique A qualitative forecasting technique that seeks to eliminate the effects of personal relationships and domination by strong personalities. A coordinator sends a written question to each relevant expert, who then replies in writing. The replies are edited and summarized. On the basis of the summary, new questions are issued to the experts, who again reply in writing. The cycle is repeated until the coordinator is satisfied that the process has provided objective assessment of the issue in question (e.g. the likely sales of a new product).

Delta neutral Describes value of a portfolio not affected by changes in the value of the asset on which the options are written

Delta The ratio of the change in price of an option to the change in price of the underlying asset. Also called the hedge ratio. Applies to derivative products. For a call option on a stock, a delta of 0.50 means that for every Rs.1.00 that the stock goes up, the option price rises by Rs.0.50. As options near expiration, in-the-money call option contracts approach a delta of 1.0, while in-the-money put options approach a delta of -1. See: hedge ratio, neutral hedge. Call deltas range from 0.00 to +1.00; put deltas range from 0.00 to -1.00. If the call delta is 0.69, the put delta is -0.31 (call delta minus 1 equals put delta; 0.69 -1 =-0.31).

Demand curve A curve on a graph relating the quantity of a good demanded to its price. Economists usually expect the demand curve to slope downwards, i.e. an increase in the price of a good brings a lower level of demand.

Table Many factors, price and non-price after demand.

Factor Affecting Demand	Example for Automobiles
1. OWn price	Higher own price reduces quantity demanded
2. Average income	As incomes rise, people increase car purchases
3. Population	Higher population increases car purchases
4. Price of related goods	Lower gasolie price raise the demand for cars.
5. Tastes	Americans buy more cars than Europeans, other things equal
6. Particular factors	Special factors include availability of subways, TV viewing, dating patterns, etc.

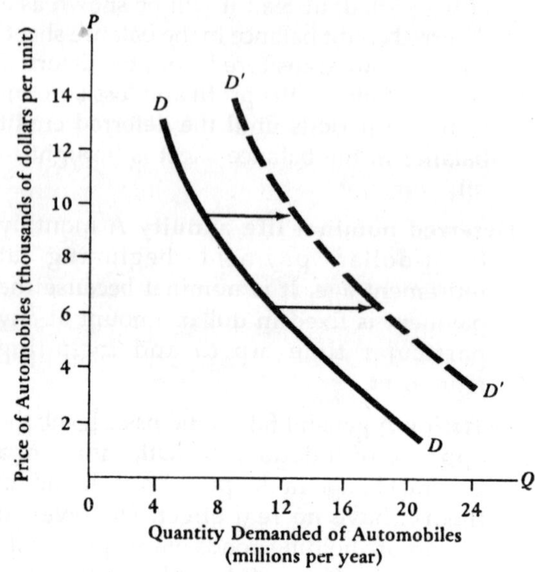

Fig. Demand

Demerger A business strategy in which a large company or group of companies splits up so that its activities are carried on by two or more independent companies. Alternatively, subsidiaries of a group are sold off. Demerging was popular in the late 1980s when large conglomerates became unfashionable. One of the main reasons for demerging is to improve the value of the company's shares, particularly if one part of a group's value can be better reflected by a separate share quotation.

Demurrage 1. Liquidated damages payable under a charterparty (see chartering), at a specified daily rate for any days (demurrage days) required for completing the loading or discharging of cargo after the lay days have expired.

2. Unliquidated damages to which a shipowner is entitled if, when no lay days are specified, the ship is detained for loading or unloading beyond a reasonable time.

3. Liquidated damages included in any contract to compensate one party if the other is late in fulfilling his obligations. This

occurs frequently in building contracts. Even if the loss caused by the delay is less than the demurrage, the demurrage must be paid in full.

Deposit 1. A sum of Money paid by a buyer as part of the sale price of something in order to reserve it. Depending on the terms agreed, the deposit may or may not be returned if the sale is not completed.

2. A sum of money left with an organization, such as a bank, for safekeeping or to earn interest or with a broker, dealer, etc., as a security to cover any trading losses incurred.
3. A sum of money paid as the first instalment on a hire-purchase agreement. It is usually paid when the buyer takes possession of the goods.

Deposit account An account with a bank from which money cannot be withdrawn by cheque (*compare*cheque account). The interest paid will depend on the current rate of interest and the notice required by the bank before money can be withdrawn, but it will always be higher than that on a current account.

Depreciation A non-cash expense (also known as non-cash charge) that provides a source of free cash flow. Amount allocated during the period to amortize the cost of acquiring long-term assets over the useful life of the assets. To be clear, this is an accounting expense not a real expense that demands cash. The sum of depreciation expenses of prior years leads to the balance sheet item Accumulated Depreciation.

Depreciation rate The percentage rate used in the straight-line method and diminishing-balance method of depreciation in order to determine the amount of depreciation that should be written off a capital asset and charged against income or the profit and loss account.

Depressed price In the context of stocks, stock whose market price is low in comparison to stocks in its sector.

Depression (slump) An extended or severe period of recession. Depressions occur infrequently and some economists believe they occur in long (about 50-year) cycles. The most recent Great Depression occurred in the 1930s; prior to that they occurred in the periods 1873-96, 1844-51, and 1810-17. Depressions are usually associated with falling prices (*see*deflation) and large-scale involuntary unemployment. They are often preceded by major financial crashes, e.g. the Wall Street crash of 1929.

Derivative A financial instrument, the price of which has a strong correlation with a related or underlying commodity, currency, or financial instrument. The most common derivatives are futures contracts and options. These standard products can be customized with regard to maturity, quantity, or pricing structure for a particular client. A derivative market is a futures or options market derived from a cash market. For example, the market for traded options in shares (the London International Financial Futures and Options Exchange) is a derivative market of the London Stock Exchange.

Derivative action A legal action brought by a shareholder on behalf of a company, when the company cannot itself decide to sue. A company will usually sue in its own name but if those against whom it has a cause of action are in control of the company (i.e. directors or majority shareholders) a shareholder may bring a derivative action. The company will appear as defendant so that it will be bound by, and able to benefit from, the decision The need to bring such an action must be proved to the court before it can proceed.

Descending tops A chart pattern which in which each successive peak in a security's price is lower than the preceding peak over a period of time. Antithesis of ascending tops.

Design risk The risk associated with the impact on project cash flow from deficiencies in design or engineering. Also known as engineering risk.

Designated order turnaround system (DOT) Computerized order entry system that allows orders to buy or sell large baskets of stock to be transmitted immediately to the specialist on the exchange, where execution will occur quickly, depending on the basket size. Also used for odd-lot transactions to occur at the prices and quantities available.

Designed capacity The maximum theoretical output of an operating system in units per time, taking no account of such operational limitations as staffing decisions, set-up times, unforeseeable failures, maintenance, etc. *Compare*effective capacity.

Desk research A marketing research study using mainly external published data and material but also including some internal reports, company records, etc.

Desk The New York Federal Reserve Bank's trading desk (or securities department) where all transactions of the Federal Reserve System are executed in the money market or the government securities market.

Desktop evaluation The process of deciding whether a computer system or program can perform a particular task by testing it in realistic, or desktop, circumstances. This contrasts with evaluation on a theoretical basis, using technical data supplied by the manufacturer.

Detachable warrant A warrant entitles the holder to buy a given number of shares of stock at a stipulated price. A detachable warrant is one that may be sold separately from the package it may have originally been issued with (usually a bond).

Detrend To remove the general drift, tendency, or bent of a set of statistical data as related to time. Often accomplished by regressing a variable or a time index and perhaps the square of the time index and capturing the residuals. A stochastic detrend would be to subtract a moving-average (say for five years) from the value of the variable.

Deutsche Börse AG (DBAG) Deutsche Börse AG (DBAG) is the operating company for the German cash and derivatives markets. It has four subsidiaries: Deutsche Börse Clearing AG, Deutsche Börse Systems AG, Frankfurter Wertpapierbörse (FWB), and the derivatives market, EUREX Deutschland (formerly the Deutsche Terminbörse).

Devaluation A decrease in the spot price of a currency. Often initiated by a government announcement.

Developing countries Countries that often have abundant natural resources but lack the capital and entrepreneurial and technical skills required to develop them. The average income per head and the standard of living in these countries is therefore far below that of the industrial nations.

The developing countries, in which some 70% of the world's population lives, are characterized by poverty, overpopulation and poor educational facilities. Many depend on a single product for their exports and are therefore vulnerable in world markets. Some have developed small low-technology companies, but market mechanisms (particularly distribution) do not exist.

Development cost The expenditure incurred organization in improving a product or process. The treatment of development costs in the accounts of the organization is determined by the UK Statement of Standard Accounting Practice 13, which deals with the treatment of research and development costs. If there is a reasonable expectation that the product or process on which the development expenditure is being spent is likely to create future income, the organization may capitalize the expenditure in the balance sheet, to be written off against the income when it arises. In all other circumstances development expenditure must be written off to the profit and loss account as soon as it is incurred.

Dialing for dollars A term used to describe the practice of cold calling, but which has negative implications as it is frequently applied to salespeople selling speculative or fraudulent investments.

Diamonds Units of interest in the diamonds trust, a unit investment trust that serves as an index to the Dow Jones Industrial Average in that its holdings consist of the 30 component stocks of the Dow.

Diary method A procedure for collecting purchase data and the media habits of consumers; it requires respondents to complete a written report (in a diary) of their behaviour. This is either collected by or mailed to a marketing research company weekly or monthly.

Diary panel A group of shops and shoppers who keep a regular record of all purchases or purchases of selected products, for the purpose of marketing research.

Difference check The difference in interest payments that is paid to a swap counterparty to close out a deal.

Difference from S&P A mutual fund's return minus the change in the Standard & Poors 500 Index for the same time period. A notation of -5.00 means the fund return was 5 percentage points less than the gain in the S&P, while 0.00 means that the fund and the S&P had the same return.

Differential disclosure The practice of reporting conflicting or markedly different information in official corporate statements including annual and quarterly reports and 10-Ks and 10-Qs.

Differential pricing A method of pricing a product in which the same product is supplied to different customers, or different market segments, at different price. This approach is based on the principle that to achieve maximum market penetration the price charged should be what a particular market will bear. For example, the prices charged for similar motor vehicles in the UK market have been higher than those charged in the rest of Europe.

Differential swap Swap between two LIBO rates of interest, e.g. yen LIBOR for dollar LIBOR. Payments are in one currency

Differentiated marketing (multiple market segmentation) 1. Marketing in which provision is made to meet the special needs of certain consumers. For example, weight watchers require diet drinks and left-handed people require left-handed scissors.

2. A marketing exercise in which the marketer selects more than one target market and then develops a separate marketing mix for each. The differentiation strategy for each of these target markets is a strategy in which the marketer offers a product that is unique in the industry, provides a distinct advantage, or is otherwise set apart from competitors' brands in some way other than price.

Diffusion of innovation The process by which the sale of new products and services spreads among customers. Initially, only those with confidence in the new product or who like taking risks will try it out. Once the innovators have accepted the product, a larger group of early adopters will come into the market. These opinion leaders will in turn bring about a wider acceptance by consumers. The diffusion process can be speeded up by making new products more attractive, for example by giving away free samples or by special introductory prices.

Diffusion process A conception of the way a stock's price changes that assumes that the price takes on all intermediate values. dirty price. Related: full price

Digits deleted Designation on securities exchange tape meaning that because the tape has been delayed, some digits have been dropped (e.g., 26 1/2 becomes 6 1/2).

Dilution Diminution in the proportion of income to which each share is entitled.

Dilution of equity An increase in the number of ordinary shares in a company without a corresponding increase in its assets or profitability. The result is a fall in the value of the shares as a result of this dilution.

Dilution protection Standard provision that changes the conversion ratio in the case of a stock dividend or extraordinary distribution to avoid dilution of a convertible bondholder's potential equity position. Adjustment usually requires a split or stock dividend in excess of 5% or issuance of stock below book value.

Dilutive effect Result of a transaction that decreases earnings per common share (EPS).

Diminishing-balance method (reducing-balance method) A method of computing the depreciation of a capital asset in an accounting period, in which the percentage to be charged against income is based on the depreciated value at the beginning of the period . This has the effect of reducing the annual depreciation charge against profits year by year. The annual percentage to be applied to the annual depreciated value is determined by the formula: rate of depreciation = 1 - (S/C)1/N, where N = estimated life in years, S = estimated scrap value at the end of its useful life, and C = original cost.

Dinar 1. (DA) The standard monetary unit of Algeria, divided into 100 centimes. 2. The standard monetary unit of Bahrain (BD), Iraq (ID), Jordan (JD), Kuwait (KD), and Yemen (YD), divided into 1000 fils.

3. (TD) The standard monetary unit of Tunisia, divided into 1000 millimes.

4. (Din.) The standard monetary unit of Yugoslavia, divided into 100 paras.

5. (LD) The standard monetary unit of Libya, divided into 1000 dirhams.

6. An Iranian monetary unit, worth one hundredth of a rial.

Dinkie Denoting an affluent married couple who may be expected to be extensive purchasers of consumer goods. The word is formed from double-income no-kids.

Dip Slight drop in securities prices after a sustained uptrend. Analysts often advise investors to buy on dips, meaning to buy when a price is momentarily weak. See: Correction, break, crash.

Direct access A method of extracting information from a computer memory. Information stored in a type of memory that supports direct access, such as random-access memory and magnetic disk, can be retrieved, or accessed, immediately regardless of its location within the memory. *Compare*sequential access.

Direct Claim A financial claim issued by a deficit unit to acquire funds for investment in real assets.

Direct cost 1. Product costs that can be directly traced to a product or cost unit. They are usually made up of direct materials (which can be charged directly to the product by means of materials requisitions), direct labour (charged by means of time sheets, time cards, or computer direct data entries), and direct expenses (which are subcontract costs charged by means of an invoice from the subcontractor). The total of direct materials, direct labour, and direct expenses is known as the prime cost.

2. Departmental or cost centre overhead costs that can be traced directly to the appropriate parts of an organization, without the necessity of cost apportionment. They can therefore be allocated to the cost centres. For example, the costs of a maintenance section that serves only one particular cost centre should be charged directly to that cost centre. *Compare*indirect costs.

Direct costs of financial distress Costs such as fees or penalties incurred as a result of bankruptcy or liquidation proceedings.

Direct debit A form of standing order given to a bank by an account holder to pay regular amounts from his cheque account to a third party Unlike a normal standing order, however, the amount to be paid is not specified; the account holder trusts the third party to claim from Ins bank an appropriate sum. In the USA this system is known as reverse wire transfer.

Direct deposit A method of payment which electronically credits your checking or savings account.

Direct deposit service A service that electronically transfers all or part of any recurring payment—including dividends, paychecks, pensions, and Social Security payments—directly to a shareholder's account.

Direct estimate method A method of cash budgeting based on detailed estimates of cash receipts and cash disbursements category by category.

Direct investment The purchase of a controlling interest in a company or at least enough interest to have enough influence to direct the course of the company.

Direct labour cost (direct wages) Expenditure on wages paid to those operators who are directly concerned with the production of a product, service, or cost unit. It is one of the cost classifications making up the prime cost of a cost unit; it is quantified as the product of the time spent on each activity (collected by means of time sheets or job cards) and the rate of pay of each operator concerned. A percentage of the direct labour cost is sometimes used as a basis for absorbing production overheads to the cost unit in absorption costing.

Direct labour Workers directly concerned with the production of a product, service, or cost unit, such as machine operators, assembly and finishing operators, etc *Compare*indirect labour.

Direct lease Lease in which the lessor purchases new equipment from the manufacturer and leases it to the lessee.

Direct marketing Selling by means of dealing directly with consumers rather than through retailers. Traditional methods include mail order (*see*mail-order house), direct-mail selling, cold calling, telephone selling, and door-to-door calling. More recently telemarketing, direct radio selling, magazine and TV advertising, and on-line computer shopping have been developed. *See also*direct-response TV advertising; home shopping channels.

Direct materials cost Expenditure on direct materials. It is one of the cost classifications that make up the prime cost of a cost unit and is ascertained by collecting together the quantities of each material used on each product by means of materials requisitions and multiplying the quantities by the cost per unit of each material. A percentage on direct materials cost is sometimes used as a basis for absorbing production overheads to the cost unit in absorption costing.

Direct placement Selling a new issue not by offering it for sale publicly, but by placing it with one of several institutional investors.

Direct Purchase Plan A plan that enables interested first-time individual investors to purchase a company's stock directly from the company or without the direct intervention of a broker. The administrator also ensures the safekeeping of the shares by registering them directly on the books of the company. Eliminates the need for shareholders to hold on to physical certificates.

Direct Registration System A system, sometimes referred to as DRS, that allows electronic direct registration of securities in an investor's name on the books for the transfer agent or issuer, and allows shares to be transferred between a transfer agent and broker electronically. DRS provides investors with a different way of holding their securities in certificate or street form. Under DRS, investors can elect to have their securities registered directly on the issuer's records in book-entry form. An investor electing to hold a security in a DRS book-entry position will receive a statement from the issuer or its transfer agent verifying ownership of the security. The investor can subsequently transfer electronically the DRS book-entry position to their bank or broker/ dealer.

Direct search market Buyers and sellers seek each other directly and transact directly.

Direct stock-purchase programs Investors purchase securities directly from the issuer.

Direct taxation Taxation, the effect of which is intended to be borne by the person or organization that pays it. Economists distinguish between direct taxation and indirect taxation. The former is best illustrated by income tax, in which the person who receives the income pays the tax and his income is thereby reduced. The latter is illustrated by VAT, in which the tax is paid by traders but the effects are borne by the consumers who buy the trader's goods. In practice these distinctions are rarely clear-cut. Corporation tax is a direct tax but there is evidence that its incidence can be shifted to consumers by higher prices or to employees by lower wages. Inheritance tax could also be thought of as a direct tax on the deceased, although its incidence falls on the heirs of the estate.

Direct write-off method In the USA. the procedure of writing off bad debts as they occur instead of creating a provision for them. Although this practice is unacceptable for financial reporting purposes, as it ignores the matching concept and prudence concept, it is the only method allowed for tax purposes.

Directive 1. An instruction to carry out certain money-market operations, particularly instructions given by the US Federal Reserve System

2. A legislative decision by the European Union's Council of Ministers and Parliament, which is binding on member states but allows them to decide how to enact the required legislation.

Direct-mail selling (direct-mail marketing) A form of direct marketing in which sales literature or other promotional material is mailed directly to selected potential purchasers. The seller may be the producer of the products or a business that specializes in this form of marketing. The seller may build up his own list of potential customers or he may buy or rent a list (*see*list renting).

Directors, usually act together, although power may be conferred (by the articles of association) on one or more directors to exercise executive powers; in particular there is often a managing director with considerable executive power.

The first directors of a company are usually named in its articles of association or are appointed by the subscribers; they are required to give a signed undertaking to act in that capacity, which must be sent to the Registrar of Companies. Subsequent directors are appointed by the company at a general meeting, although in practice they may be appointed by the other directors for ratification by the general meeting. Directors may be discharged from office by an ordinary resolution with special notice at a general meeting, whether or not they have a service contract in force They may be disqualified for fraudulent trading or wrongful trading or for any conduct that makes them unfit to manage the company.

Directors owe duties of honesty and loyalty to the company (fiduciary duties) and a duty of care; their liability in negligence depends upon their personal qualifications (e.g. a chartered accountant must exercise more skill than an unqualified man). Directors need no formal qualifications. Directors may not put their own interests before those of the company, may not make contracts (other than service contracts) with the company, and must declare any personal interest in work undertaken by the company. Their formal responsibilities include: presenting to members of the company, at least annually, the accounts of the company and a directors' report; keeping a register of directors, a register of directors' shareholdings, and a register of shares; calling an annual general meeting; sending all relevant documents to the Registrar of Companies; and submitting a statement of affairs if the company is wound up

Directors' remuneration consists of a salary and in some cases directors' fees, paid to them for being a director, and an expense allowance to cover their expenses incurred in the service of the company. Directors' remuneration must be disclosed in the company's accounts and shown separately from any pension payments or compensation for loss of office.

Directors' Duties In the context of corporate governance, Directors' Duties refers to stated responsibilities of the company's Board of Directors. These provisions allow directors to consider constituencies other than shareholders when considering a merger. These constituencies may include, for example, employees, host communities, or suppliers. This provision provides boards of directors with legal basis for rejecting a takeover that would have been beneficial to shareholders. A majority of states have Directors Duties Laws.

Directors' interests The interests held by directors in the shares and debentures of the company of which they are a director. The directors' interests can also include options on shares and debentures of the company. These interests must be disclosed to comply with the Companies Acts.

Directors' remuneration (directors emoluments) The amounts received by directors from their office or employment, including all salaries, fees, wages, perquisites, and other profit as well as certain expenses and benefits paid or provided by the employer, which are deemed to be remuneration or emoluments.

Directors' report An annual report by the directors of a company to its shareholders, which forms parts of the companys accounts required to be filed with the Registrar of Companies under the *Companies Act* (1985). The information that must be given includes the principal activities of the company, a fair review of the developments and position of the business with likely future developments, details of research and development, significant issues on the sale, purchase, or valuation of assets, recommended dividends, transfers to reserves, names of the directors and their interests in the company during the period employee statistics, and any political or charitable gifts made during the period. *See also*medium-size company; small company.

Directorship Used in the context of general equities. Stock status whereby a trader may not maintain positions in the security, due to an investment bank employee serving as a director on the corporation's Board of Directors done to avoid conflicts of interest; signified by a flashing "D" on Quotron. Contrast to restricted.

Direct-response TV advertising (direct-action advertising) An advertisement designed to stimulate immediate purchase or encourage some other direct response. Television spots, typically 60 to 120 seconds long, attempt to describe a product or service, persuade viewers of its merits, and provide a free phone number for ordering.

Dirty bill of lading (foul or claused bill of lading) A bill of lading carrying a clause or endorsement by the master or mate of the ship on which goods are carried to the effect that the goods (or their packing) arrived for loading in a damaged condition.

Dirty float A system of floating exchange rates in which the government occasionally intervenes to change the direction of the value of the country's currency.

Dirty price Bond price including accrued interest, i.e., the price paid by the bond buyer.

Disability income insurance An insurance policy that insures a worker in the event of an occupational mishap resulting in disability. Insurance benefits compensate the injured worker for lost pay.

Disbursement A payment made by a professional person, such as a solicitor or banker, on behalf of a client. This is claimed back when the client receives an account for the professional services.

Disbursement float A decrease in book cash but no immediate change in bank cash, generated by checks written by the firm.

Discharge Discharge of debts is the goal of the debtor in a bankruptcy filing. The type of discharge depends on whether the debtor is a corporation or an individual. It also depends on the particular chapter under which the discharge is granted. Unless a specific debt is determined to be non-dischargeable, all of the debtor's debts become unenforceable. If a creditor attempts to enforce payment of a discharged debt, it violates the order of discharge.

Discharge of lien An order terminating a lien on property.

Discharge To release a person from a binding legal obligation by agreement, by the performance of an obligation, or by law. For example, the payment of a debt discharges the debt; similarly, a judicial decision that a contract is frustrated discharges the parties from performing it

Disclaimer of opinion An auditor's statement that does not express any opinion regarding the company's financial condition.

Disclosure A company's release of all information pertaining to the company's business activity, regardless of how that information may influence investors.

Disclosure statement An explanatory document which accompanies the plan and the solicitation of votes in a Chapter 11 case. No one is allowed to solicit votes for a plan without there being an approved disclosure statement. After notice to all parties, the court holds a hearing to determine if the disclosure statement contains adequate information to allow a typical creditor to make an informed decision about the plan.

Discontinued operations Divisions of a business that have been sold or written off and that no longer are maintained by the business.

Discontinuous innovation An entirely new product introduced to the market to perform a function that no previous product has performed. Such a product requires new consumption or usage patterns to be developed. The introduction of the home personal computer, for example, illustrates how new marketing systems had to be established. In addition to new retailers, consumers had to be encouraged to buy and use PCs.

Discount Arbitrage A riskless arbitrage in which a discount option is purchased and an opposite position is taken in the underlying security. The arbitrageur may either buy a call at a discount and simultaneously sell the underlying security (basic call arbitrage) or may buy a put at a discount and simultaneously buy the underlying security (basic put arbitrage). See also Discount.

Discount bond Debt sold for less than its principal value. If a discount bond pays no coupon, it is called a zero coupon bond.

Discount broker A brokerage house featuring relatively low commission rates in comparison to a full-service broker.

Discount Convertible: Difference between gross parity and a given convertible price.

Most often invoked when a redemption is expected before the next coupon payment, making it liable for accrued interest. Antithesis of premium.

Discount factor (present-value factor) A factor that, when multiplied by a particular year's predicted cash flow, brings the cash flow to a present value. The factor takes into consideration the number of years from the inception of the project and the hurdle rate that the project is expected to earn before it can be regarded as feasible. The factor is computed using the formula: discount factor = $1/(1+r)^t$, where r = hurdle rate required and t = the number of years from project inception.

n practice there is little necessity to compute discount factors when carrying out appraisal calculations as they are readily available in discount tables. Most computer spreadsheet programs now include a discounted cash flow routine, which also obviates the need for using discount factors.

Discount Interest Interest at a beginning of the loan. For example if you take out a one-year loan of Rs.100 at a discount interest rate of 10%, you would receive Rs.90 at the outset.

Discount market The part of the money market consisting of banks, discount houses, and bill brokers. By borrowing money at short notice from commercial banks or discount houses, bill brokers are able to discount bills of exchange, especially Treasury bills, and make a profit. The loans are secured on unmatured bills.

Discount payment The difference between the face value and the price paid for a security.

Discount rate The interest rate that the Federal Reserve charges a bank to borrow funds when a bank is temporarily short of funds. Collateral is necessary to borrow, and such borrowing is quite limited because the Fed views it as a privilege to be used to meet short-term liquidity needs, and not a device to increase earnings. In context of NPV or PV calculations, the discount rate is the annual percentage applied. In the context of project financing, the discount rate is often the all-in interest rate or the interest rate plus margin.

Discount securities Non-interest-bearing money market instruments that are issued at a discount and redeemed at maturity for full face value, e.g., US Treasury bills.

Discount store A retail organization that regularly sells standard merchandise (national brands) at lower prices than supermarkets, etc, by accepting lower margins and selling a higher volume. In the UK, several major grocery retailers operate in this way.

Discount yield The yield or annual interest rate on a security sold to an investor at a discount. A bond that is sold at Rs.4875 that matures to Rs.5000 has a discount of Rs.125. To calculate the discount yield: (discount divided by the face value of the security) multiplied by the (number of days in the year divided by the number of days to maturity).

Discounted basis Selling something on a discounted basis is selling below what its value will be at maturity, so that the difference makes up all or part of the interest.

Discounted cash flow (DCF) A method of capital budgeting or capital expenditure appraisal that predicts the stream of cash flows, both inflows and outflows, over the estimated life of a project and discounts them, using a cost of capital or hurdle rate, to present values or discounted values in order to determine whether the project is likely to be feasible. A number of appraisal approaches use the DCF principle, namely the net present value, the internal rate of return, and the profitability index Most computer spreadsheet programs now include a DCF appraisal routine.

Discounting back *Reducing a future payment or receipt to its present equiva-* lent by taking account of the interest, which when added to the present equivalent for the relevant number of years would equate to the future payment or receipt.

Discrete random variable A random variable that can take only a certain specified set of individual possible values-for example, the positive integers 1, 2, 3, . . . For example, stock prices are discrete random variables, because they can only take on certain values, such as Rs.10.00, Rs.10.01 and Rs.10.02 and not Rs.10.005, since stocks have a minimum tick size of Rs.0.01. By way of contrast, stock returns are continuous not discrete random variables, since a stock's return could be any number.

Discretion Freedom given to the floor broker by an investor to use his judgment regarding the execution of an order. Discretion can be limited, as in the case of a limit order that gives the floor broker some distance from the stated limit price to use his judgment in executing the order. Discretion can also be unlimited, as in the case of a market-not-held order. See also: Market Not Held Order.

Discretionary account An account placed with a stockbroker, securities house, or commodity broker in which the broker is empowered to carry out transactions on this account without referring back for approval to the principal. Principals normally set parameters for their accounts, but with a discretionary account the broker has more discretion, only reporting back on purchases, sales, profits and losses, and the value of the portfolio.

Discretionary cash flow Cash flow that is available after the funding of all positive net present value (NPV) capital investment projects; it is available for paying cash dividends, repurchasing common stock, retiring debt, and so on.

Discretionary costs (managed costs) Costs incurred as a result of a managerial decision; the extent of these costs is consequently subject to managerial discretion. A characteristic of such costs is that they are often for a specified amount or subject to a specific formula, such as a percentage of sales revenue. Examples include advertising and research expenditure.

Discretionary income The amount of income a consumer has available after purchasing essentials such as food and shelter.

Discretionary order 1. An order given to a stockbroker, commodity broker, etc, to buy or sell a stated quantity of specified securities or commodities, leaving the broker discretion to deal at the best price.

2. A similar order given to a stockbroker in which the sum of money is specified but the broker has discretion as to which security to buy for his client.

Discretionary order A type of buy order or sell order that gives the broker the freedom and power to make the execution at any time and price that is seen fit and reasonable, given the investor's goals.

Discretionary Proposition A proposal on a proxy card that brokers can cast in favor of management if they have not yet heard from the beneficial holder ten days before the annual meeting. See: Ten-Day Rule

Discretionary reserves Balance sheet accounts representing temporary accumulations of earnings from the current year or the recent past.

Discretionary trust A trust in which the shares of each beneficiary are not fixed by the settlor in the trust deed but may be varied at the discretion of some person or persons (often the trustees). In an exhaustive discretionary trust all the income arising in any year must be paid out during that year, although no beneficiary has a right to any specific sum. In a nonexhaustive discretionary trust, income may be carried forward to subsequent years and no beneficiary need receive anything. Such trusts are useful when the needs of the beneficiaries are likely to change, for example when they are children.

Discretionary trust In the context of mutual funds, refers to a mutual fund or unit trust whose management decides on the best way to use the assets without restriction to a specific type of security. In the context of trusts, refers to a personal trust in which a trustee has the power of decision as to how much income or principal each beneficiary receives.

Discriminant analysis A statistical process that links the probability of default to a specified set of financial ratios.

Discriminate analysis A statistical process that links the probability of default to a specified set of financial ratios.

Discriminating monopoly A monopoly in which the supplier's products or services are sold to consumers at different prices; by dividing the market into segments and charging each market segment the price it will bear, the monopolist increases profits. An example is the different domestic and industrial tariffs operated by suppliers of electricity in many countries.

Discriminating tariff A tariff that is not imposed equally by a country or group of countries on all its trading partners. The abolition of discriminating tariffs is one of the purposes of the General Agreement on Tariffs and Trade and other tradings blocs.

Disinflation A decrease in the rate of inflation.

Disintermediation Withdrawal of funds from a financial institution in order to invest them directly.

Disintermediation Withdrawal of funds from a financial_institution in order to invest them directly.

Disinvestment A reduction in capital investment reflected by a decrease in capital goods and a company's decision not to replace depleted capital goods.

Dismissal An order terminating the case prior to its normal end.

Where a debtor in possession fails to cooperate with the court, the court may dismiss the case and terminate the proceedings leaving creditors free to resume collection efforts.

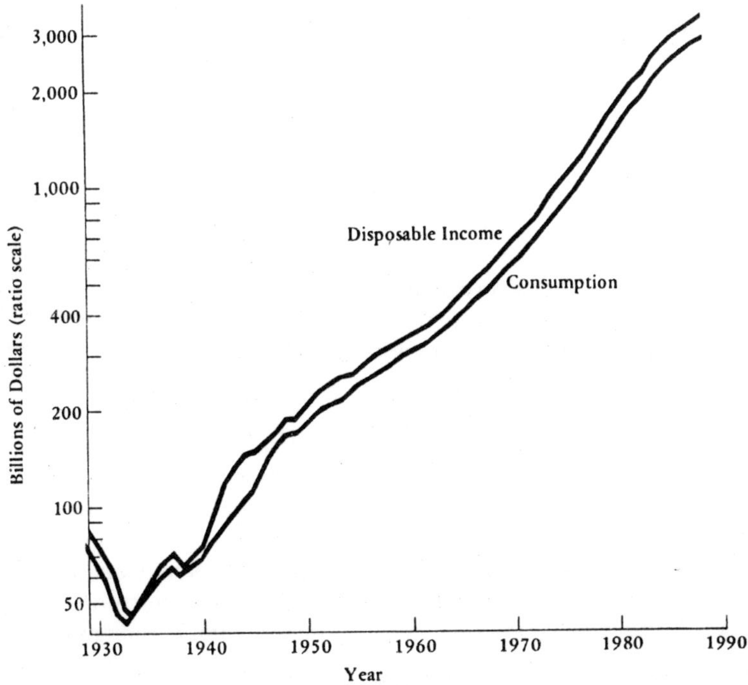

Fig. Disposable income

Dismissal with prejudice Case dismissed and barring the right to bring or maintain an action on the same claim.

Dismissal without prejudice Case dismissed with no rights or privileges to be considered as waived or lost, such as the ability to file again.

Disorderly Market A characterization of market conditions whereby there is excessive volatility at a time when there is no news. The volatility is often caused by order imbalances. In some markets, shorts trying to cover can cause disorderly conditions. If disorderly conditions arise, sometimes trading is halted.

Display advertising Advertising, such as a full-page or quarter-page advertisement, often containing a logo or illustration, rather than simple classified advertising.

Disposable income The amount of personal income an individual has after taxes and government fees, which can be spent on necessities, or non-essentials, or be saved.

Disposals account An account used to record the disposal of a fixed asset. The original cost (a debit entry), accumulated depreciation (a credit entry), and the amount received (credit entry) are transferred to the account, the balancing figure being any profit (debit entry) or loss (credit entry) on disposal.

Disproportionate stratified sampling A probability sampling method in which the sample size of a particular stratum is not proportional to the population strata. For example, a stratum could be large supermarkets, which may only account for 20% of all grocery stores - although they account for 80% of grocery sales. In this case, a disproportionate sample would be used to represent the large supermarkets to reflect their sales (i.e. 80%) rather than the number of stores.

Distrain To seize goods as a security for the performance of an obligation, especially the seizure of goods by a landlord because a tenant is in arrears with his rent.

Distress sale The selling of assets under adverse conditions, e.g., an investor may have to sell securities to cover a margin call.

Distressed securities A security of a firm that has declared or is about to declare bankruptcy. In the context of hedge funds, a style of management that focuses on securities of companies that have declared bankruptcy and may be undergoing reorganization. Investment holdings can include bonds as well as stock in these firms.

Distributable profits The profits of a company that are legally available for distribution as dividends. They consist of a company's accumulated realized profits after deducting all realized losses, except for any part of these net realized profits that have been previously distributed or capitalized. Public companies, however, may not distribute profits to such an extent that their net assets are reduced to less than the sum of their called-up capital (*see*share capital) and their undistributable reserves (*see*capital reserves).

Distributed After a Treasury auction, there will be many new issues in dealer's hands. As those issues are sold, it is said that they are distributed.

Distributed logic A computer system that supplements the main computer with remote terminals capable of doing some of the computing, or with electronic devices capable of making simple decisions, distributed throughout the system.

Distributed processing A system of processing data in which several computers are used at various locations within an organization instead of using one central computer. The computers may be linked to each other in a local area network, allowing them to cooperate, or they may be linked to a larger central computer, although a significant amount of the processing is done without reference to the central computer.

Distribution 1. A payment by a company from its distributable profits, usually by means of a dividend.

2. A dividend or quasi-dividend on which advance corporation tax is payable. For UK tax purposes a distribution includes, in addition to dividends, any payment, whether in cash or kind, out of the assets of a company in respect of shares of that company, except repayments of capital

3. The allocation of goods to consumers by means of wholesalers and retailers.

4. The division of property and assets according to law, e.g. of a bankrupt person or a deceased person *See also* qualifying distribution.

Distribution area An established price range in which a stock has been trading for a significant amount of time. See: Accumulation area.

Distribution by coupon Classification of a portfolio's securities according to coupon rate – the interest rate that an issuer promises to pay, expressed as an annual percentage of face value.

Distribution by credit quality Classification of a portfolio's securities according to credit rating.

Distribution by issuer Classification of a portfolio's holdings by type of issuer or type of instrument.

Distribution by maturity An indicator of interest rate risk. In general, the higher the concentration of longer-maturity issues, the more a portfolio's share price will fluctuate in response to changes in interest rates.

Distribution Cost Advantage A source of competitive advantage that depends on the efficient delivery of a product or service to customers.

Distribution period The few days between the Board of Directors' declaration of a stock dividend (declaration date) and the date of record, or the date an individual must own shares to be entitled to a dividend.

Distribution plan A mutual fund's plan to charge distribution costs such as advertising to the investors of the fund.

Distribution schedule The frequency (monthly, quarterly, semiannually, or annually) of a mutual fund's scheduled distributions of dividends or capital gains.

Distribution stock A small amount of a specific stock that forms part of a larger block of stock that is sold small amount by small amount so as not to disrupt the stock's market price.

Distributions Payments from fund or corporate cash flow. May include dividends from earnings, capital gains from sale of portfolio holdings and return of capital. Fund distributions can be made by check or by investing in additional shares. Funds are required to distribute realized capital gains (if any) to shareholders at least once per year if they are not to be taxed by the fund itself. Some corporations offer Dividend Reinvestment Plans (DRP).

Diversifiable risk Related: Unsystematic risk

Diversification 1. Movement by a manufacturer or trader into a wider field of products. This may be achieved by buying firms already serving the target markets or by expanding existing facilities. It is often undertaken to reduce reliance on one market, which may be diminishing (e.g. tobacco), to balance a seasonal market (e.g. ice cream), or to provide scope for general growth. This is the major means that enables a firm to change its strategic direction and may be accompanied by the divestment of certain activities.

2. The spreading of an investment portfolio over a wide range of companies to avoid serious losses if a recession is localized to one sector of the market. .

Dividend A dividend is a payment made per share, to a company's shareholders by a company, based on the profits of the year, but not necessarily all of the profits, arrived at by the directors and voted at the company's annual general meeting. A company can choose to pay a dividend from reserves following a loss-making year, and conversely a company can choose to pay no dividend after a profit-making year, depending on what is believed to be in the best interests of the company. Keeping shareholders happy and committed to their investment is always an issue in deciding dividend payments. Along with the increase in value of a stock or share, the annual dividend provides the shareholder with a return on the shareholding investment.

Dividend clawback An arrangement under which sponsors of a project agree to contribute as equity any prior dividends received from the project to the extent necessary to cover any cash deficiencies.

Dividend clientele A group of shareholders who prefer that the firm follow a particular dividend policy. Such a preference may be based on comparable tax situations.

Dividend Disbursing Agent A commercial bank or financial_institution that disburses dividend to the securityholders. Usually a Transfer Agent is also the Dividend Disbursing Agent.

Dividend discount model (DDM) A model for valuing the common stock of a company, based on the present value of the expected cash flows.

Dividend growth model An approach that assumes dividends grow at a constant rate in perpetuity. The value of the stock equals next year's dividends divided by the difference between the required rate of return and the assumed constant growth rate in dividends.

Dividend limitation A bond convenant that restricts in some way the firm's ability to pay cash dividends.

Dividend Order A letter or form signed by the shareholder instructing a corporation to issue and forward dividend and/or interest payments to a specific person or entity other than the registered owner, such as a bank or broker.

Dividend payout ratio Percentage of earnings paid out as dividends.

Dividend reinvestment plan (DRP) Automatic reinvestment of shareholder dividends in more shares of a company's stock, often without commissions. Some plans provide for the purchase of additional shares at a discount to market price. Dividend reinvestment plans allow shareholders to accumulate stock over the Long term using dollar cost averaging. The DRP is usually administered by the company without charges to the holder.

Dividend requirement The annual earnings minimum required for payment of dividends on a preferred stock.

Dividend rollover plan An investment strategy that entails the purchasing before and selling after of a stock right before its ex-dividend date in order to collect the dividends paid out by the stock and capture a trade profit.

Dividend trade roll/play Used for listed equity securities. Method of buying and selling stocks around their ex-dividend dates so as to collect the dividend (which is 80% tax-exempt) offset by a fully-taxable capital loss. Predicated on the 80% current exemption that some corporations receive on dividend income.

Dividend waiver A decision by a major shareholder in a company not to take a dividend, usually because the company cannot afford to pay it.

Dividend warrant The cheque issued by a company to its shareholders when paying dividends. It states the tax deducted and the net amount paid This document must be sent by non-taxpayers to the Inland Revenue when claiming back the tax.

Dividend yield (Funds) Indicated yield

represents return on a share of a mutual fund held over the past 12 months. Assumes fund was purchased 1 year ago. Reflects effect of sales charges (at current rates), but not redemption charges.

Dividend yield (Stocks) Indicated yield represents annual dividends divided by current stock price.

Dividends per share Dividends paid for the past 12 months divided by the number of common shares outstanding, as reported by a company. The number of shares often is determined by a weighted average of shares outstanding over the reporting term.

Dividends-received deduction A corporate tax deduction on income allowed by company A that is in ownership of shares of company B and receives dividends on the shares of company B.

Divisionalization The splitting up of a large organization into separate divisions, based on the products or services provided or on the geographical location of a particular plant. Although some management functions are repeated separately in each division, and economies of scale are not maximized, many large organizations that encompass a diversity of activities are structured in this way. Because decision making is decentralized to the divisional managers, a swift response to a change in markets is possible. Coordination between divisions is achieved by a central management.

Divisor Used in construction of stock indices. Suppose there 10 stocks in an index, each worth Rs.10 and the index is at 100. Now suppose that one of the stocks must be replaced with another stock that is worth Rs.20. If no adjustment is made to the divisor, the total value of the index would be110 after the swapping. yet there should be no increase in value because nothing has happened other than switching the two constituents. The solution is to change the divisor; in this case from 1.00 to 1.10. Note that the value of the index, 110/1.1, is now exactly 100 - which is where it was prior to the swap.

Docket The list of documents and actions within a bankruptcy case or adversary proceeding that is kept by clerk of the Bankruptcy Court. The docket also generally lists the time by which responses to pleadings are required by the Court, as well as the time and place for hearings.

Doctrine of sovereign immunity Principle that a nation may not be tried in another country without its consent.

Document merge The process of combining two or more documents to produce a single document. This is a common operation in wordprocessing (*see*wordprocessor). An example is the combining of data produced by a spreadsheet program into a document produced by a wordprocessing program.

Documentary bill A bill of exchange attached to the shipping documents of a consignment of goods. These documents include the bill of lading, insurance policy, dock warrant, invoice, etc

Documentary Collection A service provided by banks to sellers in obtaining payments. This service is usually transacted by the seller's bank through the buyer's bank, with the latter presenting the shipping documents to the buyer in exchange for payment or for signing a promissory note like instrument called a time draft.

Documented discount notes Commercial paper backed by normal bank lines plus a letter of credit from a bank stating that it will pay off the paper at maturity if the borrower does not. Such paper is also referred to as LOC (letter of credit) paper.

Documents against acceptance (D/A) A method of payment for goods that have been exported in which the exporter sends the shipping documents with a bill of exchange to a bank or agent at the port of destination.

The bank or agent releases the goods when the bill has been accepted by the consignee. *Compare*cash against documents.

Documents against acceptance Shipping documents held by the buyer's bank until the buyer has accepted (signed) the draft.

Dollar bears Traders who capitalize on a falling dollar by buying other foreign currencies directly.

Dollar bonds Municipal revenue bonds for which quotes are given in dollar prices. Not to be confused with "US Dollar" bonds, a common term of reference in the Eurobond market.

Dollar return The return realized on a portfolio for any evaluation period, including (1) the change in market value of the portfolio and (2) any distributions made from the portfolio during that period.

Dollar roll Similar to the reverse repurchase agreement-a simultaneous agreement to sell a security held in a portfolio with purchase of a similar security at a future date at an agreed-upon price.

Dollar shortage Results when a nation importing US goods cannot pay for them without the aid of the United States.

Dollar The standard monetary unit of Antigua and Barbuda (ECRs.), Australia (Rs.A), Bahamas (BRs.), Barbados (BDSRs.), Belize (BZRs.), Bermuda (BdRs.), the British Virgin Islands (USRs.), Brunei (BRs.), Canada (CanRs.), Cayman Islands (CIRs.), Dominica (ECRs.). Fiji (FRs.), Grenada (ECRs.), Guam (USRs.), Guyana (GRs.), Hong Kong (HKRs.), Jamaica (JRs.), Kiribati (Rs.A), Liberia (LRs.), Micronesia (USRs.), New Zealand (Rs.NZ), Puerto Rico (USRs.), Saint Kitts and Nevis (ECRs.), Singapore (SRs.), the Solomon Islands (SIRs.), Taiwan (NTRs.), Trinidad and Tobago (TTRs.) Tuvalu (Rs.A), the USA (USRs.), the Virgin Islands (USRs.), and Zimbabwe (ZRs.), in all cases divided into 100 cents.

Dollar-weighted rate of return Also called the internal rate of return, the interest rate that will make the present value of the cash flows from all the subperiods in the evaluation period plus the terminal market value of the portfolio equal to the initial market value of the portfolio.

Domestic bonds Bonds issued and traded within the internal market of a country and denominated in the currency of that country.

Domestic International Sales Corporation (DISC) A U.S. corporation that receives a tax incentive for export activities.

Domestic market Part of a nation's internal market representing the mechanisms for issuing and trading securities of entities domiciled within that nation. Compare external market and foreign market.

Domestic series Nonmarketable interest and noninterest-bearing securities issued periodically by the Treasury to the Resolution Funding Corporation (RFC) for investment of funds authorized under section 21B of the Federal Home Loan Bank Act.

Domicile (domicil) 1. The country or place of a person's permanent home, which may differ from that person's nationality or place where they are a resident. Domicile is determined by both the physical fact of residence and the continued intention of remaining there. For example, a citizen of a foreign country who is resident in the UK is not necessarily domiciled there unless there is a clear intention to make the UK a permanent home. Under the common law, it is domicile and not residence or nationality that determines a person's civil status, including the capacity to marry. A corporation may also have a domicile, which is determined by its place of registration. Under English law a child normally takes the domicile of its father unless the child is illegitimate or the parents divorce, when the child takes its mother's domicile. The domicile of choice is the domicile an individual chooses to take up. It involves taking up permanent residence in a country other than the domicile of origin with the intention of never returning to live in the original country. The domicile of origin is acquired at birth and is usually that of the father except in the case of illegitimate children or those of divorced parents, when the child takes the mother's domicile. In order to prove that the domicile of origin has been relinquished in favour of the domicile of choice, tangible changes have to be made. Links with the country of origin must be severed and active steps taken to become involved with the country that is the domicile of choice, e.g. by making a will under the laws of the domicile of choice. The status of domicile of choice will be kept under review and if actions are subsequently taken to re-establish links with the domicile of origin, the individual's domicile could revert to the domicile of origin. The domicile of dependency was a concept that applied to women whose marriage took place prior to 1 January 1974. In these circumstances the woman acquired the domicile of her husband For marriages since 31 December 1973 the position is determined under the provisions of the *Domicile and Matrimonial Proceedings Act* (1973), which allows a woman to retain her own domicile of origin or domicile of choice on marriage, if it differs from that of her husband. 2. In banking, an account is said to be domiciled at a particular branch and the customer treats that branch as his or her main banking contact. Customers may be charged for using other branches as if they were their own. Computer technology, however, now allows customers to use many branches as if their account was domiciled there.

Donor One who gives property or assets to someone else through the vehicle of a trust.

Door-to-door retailing Selling doorto-door, office-to-office, or at home-sales parties. Originally started in the USA with the Fuller Brush Company, this is now a popular method of marketing. In the UK cosmetics and plastic containers are widely marketed at home-sale parties. The Japanese market is the largest direct-sales market with some Rs.25 billion sales, mainly because most Japanese cars are sold door-to-door.

Dormant company A company that has had no significant accounting transactions for the accounting period in question. Such a company need not appoint auditors.

Double auction market Systems by which listed securities are bought and sold through brokers on the securities exchanges, as distinguished from the OTC market, where trades are negotiated. Unlike the conventional auction with one auctioneer and many buyers, double auction markets consist of many sellers and many buyers.

Double auction system A market consisting of many sellers and many buyers, as opposed to a conventional auction with one market maker and many buyers.

Double dip Used for listed equity securities. Dividend roll in which the "dividend capturer" already owns the stock cum dividend. Also used when tax depreciation is accessed in two countries concurrently.

Double-declining-balance depreciation method (DDB) An accounting methodology in which the depreciation rate used is double the rate used under the straight-line method. In addition, the rate is applied to the full purchase cost of the asset, whereas under the straight-line method the rate is applied to the cost net of salvage value.

Double-entry book-keeping A method of recording the transactions of a business in a set of accounts, such that every transaction has a dual aspect and therefore needs to be recorded in at least two accounts. For example, when a person (debtor) pays cash to a business for goods purchased, the cash held by the business is increased and the amount due from the debtor is decreased by the same amount; similarly, when a purchase is made on credit, the stock is decreased and the amount owing to creditors is increased by the same amount. This double aspect enables the business to be controlled because all the books of accounts must balance.

Double-tax agreement Agreement between two countries that taxes paid abroad can be offset against domestic taxes levied on foreign dividends.

Doubling option A sinking fund provision that may allow repurchase of twice the required number of bonds at the sinking fund call price.

Doubtful debts Money owed to an organization, which it is unlikely to receive. A provision for doubtful debts may be created, which may be based on specific debts or on the general assumption that a certain percentage of debtors' amounts are doubtful. As the doubtful debt becomes a bad debt, it may be written off to the provision for doubtful debts or alternatively charged to the profit and loss account if there is no provision.

Dovish Refers to the tone of language used to describe a situation and the associated implications for actions. For example, if the Federal Reserve bank refers to inflation in a dovish tone, it is unlikely that they would take agressive actions. Similarly, a CEO might use dovish language to describe an important event facing the firm. This indicates that the firm is unlikely to take strong actions. Dovish sometimes means conciliatory. Opposite of hawkish.

Dow Jones Industrial Average The best known U.S. index of stocks. A price-weighted average of 30 actively traded blue-chip stocks, primarily industrials including stocks that trade on the New York Stock Exchange. The Dow, as it is called, is a barometer of how shares of the largest US companies are performing. There are hundreds of investment indexes around the world for stocks, bonds, currencies, and commodities.

Dow Theory Used in the context of general equities. Technical theory that a major trend in the stock market must be confirmed by simultaneous movement of the Dow Jones Industrial Average and the Dow Jones Transportation Average to new highs or lows.

Down round Refers to a round of venture capital financing that is raised at a lower firm valuation than the previous round.

Down volume When a stock decreases in value on a particular day, the volume in that stock is considered down volume. Related: Up volume.

Downgrade A classic negative change in ratings for a stock, and or other rated security.

Downside Protection Generally used in connection with covered call writing, this is the cushion against loss, in case of a price decline by the underlying security, that is afforded by the written call option. Alternatively, it may be expressed in terms of the distance the stock could fall before the total position becomes a loss (an amount equal to the option premium), or it can be expressed as percentage of the current stock price.

Downsizing A company's reduction in the number of employees, number of bureaucratic levels, and overall size in an attempt to increase efficiency and profitability.

Downtick A trade in a particular stock at a price lower than the trade immediately preceding it. On U.S. stock exchanges, you cannot sell a stock short on a downtick.

Drachma (Dr) The standard monetary unit of Greece, divided into 100 lepta.

Draft An unconventional order in writing - signed by a person, usually the exporter, and addressed to the importer - ordering the importer or the importer's agent to pay, on demand (sight draft) or at a fixed future date (time draft), the amount specified on its face.

Dragon markets A colloquial name for the emerging markets and ecoromies in the Pacific basin, including Indonesia, Malaysia, the Philippines, and Thailand. They are characterized by dynamic growth and high savings ratios.

Draining reserves Federal Reserve System's course of action to tighten the money supply by (1) raising a bank's minimum reserve requirements, (2) selling bonds in the open market, (3) raising the rate at which banks borrow from the Fed, or (4) through draw-downs.

Dram The standard monetary unit of Armenia, divided into 100 couma.

Drawback The refund of import duty by the Customs and Excise when imported goods are re-exported. Payment of the import duty and claiming the drawback can be avoided if the goods are stored in a bonded warehouse immediately after unloading from the incoming ship or aircraft until reexport.

Drawer 1. A person who signs a bill of exchange ordering the drawee to pay the specified sum at the specified time.

2. A person who signs a cheque ordering the drawee bank to pay a specified sum of money on demand.

Drawings Assets (cash or goods) with- drawn from an unincorporated business by its owner. If a business is incorporated, drawings are usually in the form of dividends or scrip dividends. Drawings are shown in a drawings account, which is used, for example, by the partners in a partnership.

Drip feed The continual investment of capital in a small and growing company as the company needs it, rather than investing a lump sum at the company's inception.

Drop lock An issue in the bond market that combines the benefits of a bank loan with the benefits of a bond. The borrower arranges a variable-rate bank loan on the understanding that if longterm interest rates fall to a specified level, the bank loan will be automatically refinanced by a placing of fixedrate long-term bonds with a group of institutions. They are most commonly used on the international market.

Drop, the With the dollar roll transaction the difference between the sale price of a mortgage-backed pass-through, and its re-purchase price on a future date at a predetermined price.

Drop-dead day The date on which a deadline is final, with no exceptions.

Drop-dead fee A term of British origin referring to fee that must be paid if a deal falls through because of financing issues.

Dual pricing A method of pricing transfers of goods and services when trading takes place between the divisions of an organization. The dual prices method charges a low price, say a price based on the marginal cost, to the buying division, while at the same time crediting a high price, say a price based on full cost pricing, to the selling division. The benefit is said to arise by encouraging a buying division to buy within the organization without penalizing a selling division. However, this is only likely to be beneficial to the organization as a whole if the selling division has sufficient spare capacity to supply the buying division's needs. A compensating entry to eliminate unrealized profits is required in the books of the head office when consolidation of the divisional results takes place.

Dual trading The custom of a trader on the commodities market to deal for its own account and the investor's account at the same time.

Dual-capacity system A system of trading on

a stock exchange in which the functions of stockjobber and stockbroker are carried out by separate firms. In a single-capacity system the two functions can be combined by firms known as market makers. Dual capacity existed on the London Stock Exchange prior to October, 1986, since when a single-capacity system has been introduced.

Dual-currency issues Eurobonds that pay coupon interest in one currency but pay the principal in a different currency.

Dual-purpose fund A closed-end fund consisting of two classes of shares. The two classes are preferred shares, on which shareholders receive all the dividends and interest from the portfolio, and common shares, on which shareholders receive all the capital gains.

Due bill An instrument evidencing the obligation of a seller to deliver securities sold to the buyer. Occasionally used in the bill market.

Due diligence An internal audit of a target firm by an acquiring firm. Offers are often made contingent upon resolution of the due diligence process.

Due diligence meeting Meeting legally required to be held by an underwriter to enable brokers to question a new issuer about an upcoming issue.

Dumping The selling of goods abroad at prices below their marginal cost, which implies that the seller is making a loss. It has often been argued that developing countries wishing to establish an industry should use tariffs and quotas to ensure a monopoly for producers at home, while selling goods cheaply overseas. However, this practice inevitably leads to claims of dumping by other countries. The General Agreement on Tariffs and Trade allows countries to prevent dumping by imposing tariffs on goods that are being dumped, although it is always difficult to establish conclusively that a particular price level constitutes dumping. Dumping is prohibited in the EU.

Duopoly A market in which there are only two producers or sellers of a particular product or service and many buyers. The profits in such an imperfect form of competition are in practice usually less than could be achieved if the two suppliers merged to form a monopoly but more than if the two allowed competition to force them into marginal costing. *See also*collusive duopoly.

Duplicate Proxy A second proxy received on an account. If the second proxy bears a more recent date than the first proxy, and has a different voting pattern, the second proxy will override the first.

Duplicative portfolio Mainly applies to derivative products. Basket of stocks that imitates the price movement of another set of securities (e.g., S&P 500 index).

Dupont system of financial control Highlights the fact that return on assets (ROA) can be expressed in terms of the profit margin and asset turnover.

Durable merchandise Goods that have a relatively lengthy life (television sets, radios, etc.).

Duration A common gauge of the price sensitivity of an asset or portfolio to a change in interest rates.

Duration matching strategy An immunization technique that matches asset duration with the duration of the liabilities.

During Some official strikes unions pay strike pay to their members, while funds for doing so last.

Dutch auction Auction in which the lowest price necessary to sell the entire offering becomes the price at which all securities offered are sold. This technique has been used in Treasury auctions. Often used in risk arbitrage. Auction system in which the price of an item (stock) is gradually lowered until it meets a responsive bid (government T-bills) or offer (corporate repurchase) and is sold. In a corporate repurchase, a range of prices is set by the company within which shareholders are invited to tender their shares. The tender offer is open for a specific period of time (i.e., 20 days), and the quantity of stock to be purchased is stated as well, subject to proration if more shares are tendered than can be legally purchased under the stated terms (often an additional amount equal to 20% of outstanding shares can be purchased). The price paid is that at which the amount stated to be purchased can be sold. Compare to double auction system.

Dutch disease The deindustrialization of an economy as a result of the discovery of a natural resource. So named because, it occurred in Holland after the discovery of North Sea gas; it has also been applied to the UK since the discovery of North Sea oil. The discovery of such a resource lifts the value of the country's currency, making manufactured goods less competitive; exports therefore decline and imports rise.

Duty A tax on imports, exports, or consumption goods.

Dynamic asset allocation An asset allocation strategy in which the asset mix is quantitatively shifted in response to changing market conditions, as in a portfolio insurance strategy, for example.

Dynamic For option strategies, describing analyses made during the course of changing security prices and during the passage of time. This is as opposed to an analysis made at expiration of the options used in the strategy. A dynamic break-even point is one that changes as time passes. A dynamic follow-up action is one that will change as either the security price changes or the option price changes or time passes.

Dynamic Cobweb Tools of supply and demand are not restricted to unchanging situations but can also be used to analyze dynamic situations of change.

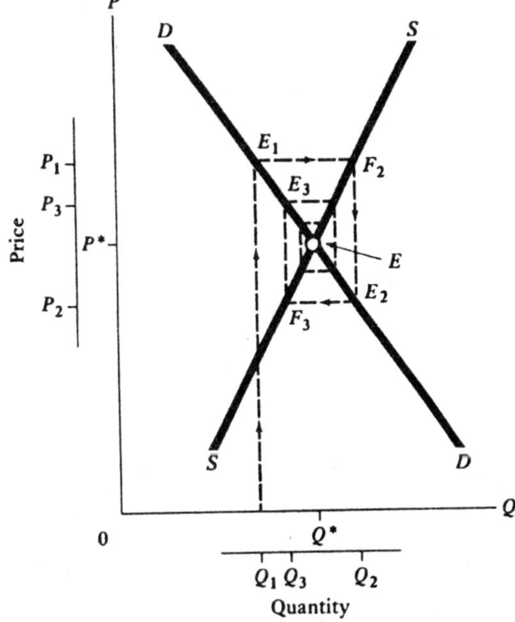

Fig. Dynamic Cobweb

Dynamic hedging A strategy that involves rebalancing hedge positions as market conditions change; a strategy that seeks to insure the value of a portfolio using a synthetic put option.

Dynamical Noise When the output of a dynamical system becomes corrupted with noise, and the noisy value is used as input during the next iteration. Also called System Noise. See: Observational Noise.

Dynamical Systems A system of equations where the output of one equation is part of the input for another. A simple version of a dynamical system is linear simultaneous equations. Non-linear simultaneous equations are nonlinear dynamical systems.

E

Each way A broker's commission from his or her involvement on both the purchase and the sale side of a security.

Early adopter A member of a group of consumers who purchases a new product soon after it has been introduced, but after the innovators. *See* adoption of innovations.

Early Exercise (assignment) The exercise or assignment of an option contract before its expiration date.

Early withdrawal penalty Penalty paid by the holder of a fixed-term investment penalizing an investor who withdraws money before the agreed-upon maturity date.

Earned income Compensation earned from employment, which includes wages, salary, tips, and compensation.

Earned income credit A tax credit for taxpayers with children.

Earnest money Money given to a seller by a buyer to demonstrate the buyer's good faith. If the deal falls through, the deposit is usually forfeited.

Earning asset An asset that generates income, e.g., income from rental property.

Earning power Earnings before interest and taxes (EBIT) divided by total assets.

Earnings before interest after taxes (EBIAT) A financial measure defined as revenues less cost of goods sold and selling, general and administrative expenses. In other words, operating and nonoperating profit before the deduction of interest plus cash income taxes. Equivalent to EBIT minus cash taxes.

Earnings before interest and, taxes (EBIT) A financial measure defined as revenues less cost of goods sold and selling, general, and administrative expenses. In other words, operating and nonoperating profit before the deduction of interest and income taxes.

Earnings before interest, taxes, and depreciation (EBITD) A financial measure defined as revenues less cost of goods sold and selling, general, and administrative expenses. In other words, operating and nonoperating profit before the deduction of interest and income taxes. Depreciation expenses are not included in the costs.

Earnings before taxes (EBT) A financial measure defined as revenues less cost of goods sold and selling, general, and administrative expenses. In other words, operating and nonoperating profit before the deduction of income taxes.

Earnings per share (eps) The profit in pence attributable to each ordinary share in a company, based on the consolidated profit for the period, after tax and after deducting minority interests and preference dividends. This profit figure is divided by the number of equity shares in issue that rank for dividend in respect of the period. The eps may be calculated on a net basis or a nil basis. Using the net basis, the tax charge includes any irrecoverable advance corporation tax and any unrelieved overseas tax arising from the payment or proposed payment of dividends. The nil basis excludes both these items from the tax charge. Statement of Standard Accounting Practice 3, 'Earnings per Share', requires the use of the net basis; the nil basis should also be disclosed if it produces a figure that is materially different from the net basis. The eps should be shown on the face of the profit and loss account on the net basis, both for the period under review and for the corresponding previous period. The basis of calculating the earnings per share should be disclosed on the face of the profit and loss account or in the notes to the accounts.

Earnings surprises Positive or negative differences from the consensus forecast of earnings by institutions such as First Call or IBES. Negative earnings surprises generally have a greater adverse affect on stock prices than the reciprocal positive earnings surprise on stock prices.

Earnings yield The ratio of earnings per share, after allowing for tax and interest payments on fixed interest debt, to the current share price. The inverse of the price-earnings ratio. It is the total twelve months earnings divided by number of outstanding shares, divided by the recent price, multiplied by 100. The end result is shown in percentage terms. We often look at earnings yield because this avoids the problem of zero earnings in the denominator of the price-earning ratio.

Earn-out agreement (contingent contract) An agreement to purchase a company in which the purchaser pays a lump sum at the time of the acquisition, with a promise to pay more (a contingent consideration) if certain criteria, usually specified earnings levels, are met for a specified number of years. This method of acquisition has been popular in 'people' businesses, in particular advertising agencies. The agreement provides a means of retaining the previous owners in the business, with a motivation to maintain the company's profitability.

Easement A right, such as a right of way, right of water, or right of support, that one owner of one piece of land (the dominant tenement) may have over the land of another (the servient tenement). The right must benefit the dominant tenement and the two pieces of land must be reasonably near each other. The right must not involve expenditure by the owner of the servient tenement and must be analogous to those rights accepted in the past as easements. An easement may be granted by deed or it may be acquired by prescription (lapse of time, during which it is exercised without challenge); it may also be acquired of necessity (for example, if A sells B a piece of land that B cannot reach without crossing A's land) or when 'continuous and apparent' rights have been enjoyed with the part of the land sold before it was divided. Existing easements over the land of third parties pass with a conveyance of the dominant tenement.

Eclectic paradigm A theory that posits three types of advantages benefiting a multinational corporation: ownership-specific, location-specific, and market internalization advantages.

Ecocentric management Management that treats environmental issues as a core business concern, rather than an externality. Recognizing the relationship that the firm has with the natural environment, it regards taking full account of the environmental impact it produces as part of management's responsibility.

Econometrics The quantitative science of modelling the economy. Econometric models help explain and predict variables of interest.

Economic appraisal A method of capital budgeting that makes use of discounted cash flow techniques to determine a preferred investment. However, instead of using annual projected cash flows in the analysis, the technique discounts over the project's life the expected annual economic costs and economic benefits. It is mainly used in the assessment of governmental or quasi-governmental projects, such as road, railway, and port developments.

Economic batch quantity A refinement of the economic order quantity to take into account circumstances in which the goods are produced in batches. The formula is: $Q = [2cdr/h(r - d)]^{1/2}$, where Q = quantity to be purchased or manufactured, c = cost of processing an order for delivery, d = demand in the period for that stock item, h = cost of holding a unit of stock, and r = rate of production.

Economic benefits The projected benefits revealed by an economic appraisal. Economic benefits are usually gains that can be expressed in financial terms as the result of an improvement in facilities provided by a government, local authority, etc. For example, the economic benefits arising from the construction of a new or improved road might include lower vehicle operating costs, time savings for the road users, and lower accident costs as a result of fewer accidents. In each case the savings would be in economic terms, that is, excluding the effect of taxes and subsidies within the economy. *See also*economic costs.

Economic costs The projected costs revealed by an economic appraisal. Economic costs differ from financial costs in that they exclude the transfer payments within the economy, which arise when an investment is made. In the construction of a road, for example, the economic costs exclude taxes and import duties on the materials and plant used in its construction, while any subsidies made are added back to the costs.

Economic earnings The real flow of cash that a firm could pay out forever in the absence of any change in the firm's productive capacity.

Economic environment Factors that affect the buying power and spending patterns of consumers. These factors include income distribution, changes in purchasing power as a result of inflation, the state of the country's economy, etc. While the economic environment can be influenced extensively by government activity, companies need to monitor these changes in order to predict their sales estimates.

Economic growth An increase in the nation's capacity to produce goods and services. Usually refers to real GDP growth.

Economic Life The time period over which an asset's NPV is maximized. Economic life can be less than absolute physical life for reasons of technological obsolescence, physical deterioration, or product life cycle.

Economic of scale (scale effect) Reductions in the average cost of production, and hence in the unit costs, when output is increased. If the average costs of production rise with output, this is known as diseconomies of scale. Economies of scale can enable a producer to offer his product at more competitive prices and thus to capture a larger share of the market. Internal economies of scale occur when better use is made of the factors of production and by using the increased output to pay for a higher proportion of the costs of marketing, financing, and development, etc. Internal diseconomies can occur when a plant exceeds its optimum size, requiring a disproportionate unwieldy administrative staff. External economies and diseconomies arise from the effects of a firm's expansion on market conditions and on technological advance.

Economic order quantity (EOQ) A decision model, based on differential calculus, that determines the optimum order size for purchasing (sometimes called the economic purchase quantity) or manufacturing (economic manufacturing quantity) an item of stock. The optimum order quantity is that which equates the total ordering and total holding costs. The formula used is: Q = v(2cd/h), where *Q* = quantity to be purchased or manufactured, *c* = cost of processing an order for delivery, d = demand in the period for that stock item, and *h* = cost of holding a unit of stock. Holding costs include the costs of capital tied up in stock, storage costs (e.g. insurance), special environmental conditions, and the possible costs of obsolescence or deterioration. Ordering costs include the paperwork, transport costs, and volume-related discount variations.

Economic risk In project financing, the risk that the project's output will not be salable at a price that will cover the project's operating and maintenance costs and its debt service requirements.

Economic sanctions Action taken by one country or group of countries to harm the economic interest of another country or group of countries, usually to bring about pressure for social or political change. Sanctions normally take the form of restrictions on imports or exports, or on financial transactions. They may be applied to specific items or they may be comprehensive trade bans. There is considerable disagreement over their effectiveness. Critics point out that they are easily evaded and often inflict more pain on those they are designed to help than on the governments they are meant to influence. They can also harm the country that imposes sanctions, through the loss of export markets or raw material supplies. In addition the target country may impose retaliatory sanctions. *See also*embargo.

Economic surplus For any entity, the difference between the market value of all its assets and the market value of its liabilities.

Economic union An agreement between two or more countries that allows the free movement of capital, labor, all goods and services, and involves the harmonization and unification of social, fiscal, and monetary policies.

Economic value added (EVA) A method of performance evaluation that adjusts accounting performance for investors' required return on investment. Suppose a division produces a 12% return on capital invested. Given the risk of the division's business line, if investors would usually require 14% on capital invested for this level of risk, the division destroyed shareholder value by the EVA metric. This Stern-Stewart has a trade mark on this term.

Economic value The present value of expected future cash flows. For example, the economic value of a fixed asset would be the present value of any future revenues it is expected to generate, less the present value of any future costs related to it.

Economics The study of the economy. See also: Macroeconomics; microeconomics; Keynesian economics, monetarism, and supply-side economics.

Economies of scale The decrease in the marginal cost of production as a firm's extent of operations expands.

Economies of scope Scope economies exist whenever the same investment can support multiple profitable activities less expensively in combination than separately.

Economy, efficiency, and effectiveness The primary performance measures for an operating system. In order for a system to be effectively managed, it is necessary to have feedback. Economy requires feedback on the cost of the inputs to a system. Efficiency measures how successfully the inputs have been transformed into outputs. Effectiveness measures how successfully the system achieves its desired outputs. Because effectiveness involves the subjective reaction of the customer, it is the most difficult to measure.

ECU Abbreviation for European Currency Unit.

Edge corporations Specialized banking institutions, authorized and chartered by the Federal Reserve Board of Governors in the U.S., that are allowed to engage in transactions of a foreign or international character. They are not subject to restrictions on interstate banking. Foreign banks operating in the U.S. are permitted to organize and own an edge corporation.

Effective capacity The capacity of a system in units per time, taking into account such factors as staffing decisions (e.g. whether to run 24 hours a day, 7 days a week, or to operate one 8-hour shift, 5 days a week), set-up times, maintenance, and an allowance for unforeseeable failures. *Compare*designed capacity.

Effective convexity The convexity of a bond calculated with cash flows that change with yields.

Effective duration The duration calculated using the approximate duration formula for a bond with an embedded option, reflecting the expected change in the cash flow caused by the option. Measures the responsiveness of a bond's price taking into account the expected cash flows will change as interest rates change due to the embedded option.

Effective Interest Rate The annual rate at which an investment grows in value when interest is credited more often than once a year.

Effective margin (EM) Used with SAT performance measures, the amount equal to the net earned spread, or margin of income, on assets in excess of financing costs for a given interest rate and prepayment rate scenario.

Effective rate A measure of the time value of money that fully reflects the effects of compounding.

Effective sale A sale based on the most recent round-lot price, which determines the price of the next odd lot. The difference created between the last round-lot price and the odd-lot price is referred to as the odd-lot differential.

Effective tax rate The average tax rate that is applicable in a given circumstance. In many cases the actual rate of tax applying to an amount of income or to a gift may not, for various reasons, be the published rate of the tax; these reasons include the necessity to gross up, the complex effects of some reliefs, and peculiarities in scales of rates. The effective rate is therefore found by dividing the additional tax payable as a result of the transaction by the amount of the income, gift, or whatever else is involved in the transaction.

Effective yield Yield or return on a short-term investment after adjustment for the change in exchange rates over the period of concern.

Efficiency The degree and speed with which a market accurately incorporates information into prices.

Efficient capital market A market in which new information is very quickly reflected accurately in share prices.

Efficient diversification The organizing principle of modern portfolio theory, which maintains that any risk-averse investor will search for the highest expected return for any level of portfolio risk.

Efficient frontier The combinations of securities portfolios that maximize expected return for any level of expected risk, or that minimizes expected risk for any level of expected return.

Efficient Market Hypothesis In general the hypothesis states that all relevant information is fully and immediately reflected in a security's market price thereby assuming that an investor will obtain an equilibrium rate of return. In other words, an investor should not expect to earn an abnormal return (above the market return) through either technical analysis or fundamental analysis. Three forms of efficient market hypothesis exist: weak form (stock prices reflect all information of past prices), semi-strong form (stock prices reflect all publicly available information) and strong form (stock prices reflect all relevant information including insider information).

Efficient portfolio A portfolio that provides the greatest expected return for a given level of risk (i.e., standard deviation), or, equivalently, the lowest risk for a given expected return.

Elasticity of demand The degree of buyers' responsiveness to price changes. Elasticity is measured as the percent change in quantity divided by the percent change in price. A large value (greater than 1) of elasticity indicates sensitivity of demand to price, e.g., luxury goods, where a rise in price causes a decrease in demand. Goods with a small value of elasticity (less than 1) have a demand that is insensitive to price, e.g., food, where a rise in price has little or no effect on the quantity demanded by buyers.

Elasticity of supply The degree of producers' responsiveness to price changes. Elasticity is measured as the percent change in quantity divided by the percent change in price. A large value (greater than 1) of elasticity indicates sensitivity of supply to price, e.g., luxury goods, where a rise in price causes an increase in supply. Goods with a small value of elasticity (less than 1) have a supply that is insensitive to price, e.g., food, where a rise in price has little or no effect on the amount that producers supply.

Electronic funds transfer (EFT) Transfer of funds electronically rather than by check or cash. The Federal Reserve's Fedwire and automated clearninghouse services are EFT systems.

Electronic funds transfer at point of sale (EFTPOS) The automatic debiting of a purchase price from the customer's bank or credit-card account by a computer link between the checkout till and the bank or credit-card company. This system is still experimental in the UK, and can only work when the customer has a debit card or credit card recognized by the retailer. The system is in use in some petrol filling stations and large stores. In addition to a debit card or credit card, the user may also be required to have a personal identification number (PIN) with which to identify himself to the computer.

Electronic Funds Transfer Systems A variety of systems and technologies for transferring funds (money) electronically rather than by check. Includes Fedwire, automated clearringhouses (ACHs) and other automated systems.

Electronic office (paperless office) An office that has been computerized; i.e. traditional methods have been replaced by electronic ones. For example, wordprocessing packages have replaced typewriters, spreadsheet packages have replaced manual accounts, and e-mail has replaced paper mail.

Electronic point of sale (EPOS) A computerized method of recording sales in retail outlets, using a laser scanner at the checkout till to read bar codes printed on the items' packages. Other advantages over conventional checkout points include a more efficient use of checkout staff time, the provision of a more detailed receipt to the customer, and real-time stock control, enabling rapid replenishment of stock and minimizing the stock that has to be held. In consumer goods sales, the data can also be used to produce consumer profiles for marketing purposes.

Electronic shopping A form of direct marketing in which a two-way system links consumers to a seller's computerized catalogue by cable or telephone lines using a computer or television screen.

Electronic transfer of funds (ETF) The transfer of money from one bank account to another by means of computers and communications links. Banks routinely transfer funds between accounts using computers; another variety of ETF is the telebanking service enabling the Viewdata network to be used for banking in a customer's home.

Eleven bond index An index based on the average yield of 11 municipal bonds that mature in 20 years and carry an average AA rating. The eleven bonds used to calculate the index are also found in the 20 bond index, which serves as a benchmark in tracking municipal bond yields.

Eligible bankers' acceptances In the BA market, an acceptance may be referred to as eligible because it is acceptable by the Fed as collateral at the discount window and/or because the accepting bank can sell it without incurring a reserve requirement.

Elliott Wave Theory Technical market timing strategy that predicts price movements on the basis of historical price wave patterns and their underlying psychological motives. Robert Prechter is a famous Elliott Wave theorist.

Elves A term the host uses to refer to guests on the PBS television show, "Wall Street Week", who are technical analysts attempting to predict the direction of stock prices over the next six months.

Embargo A ban on some or all of the trade with one or more countries. A trade embargo is a form of economic sanction. Prominent examples include the international ban on trade in arms with South Africa and on certain hightechnology products with the Eastern bloc. Full embargos are rare and difficult to apply in practice.

Embedded option An option that is part of the structure of a bond that provides either the bondholder or issuer the right to take some action against the other party, as opposed to a bare option, which trades separately from any underlying security.

Embezzlement A form of theft in which an employee dishonestly appropriates money or property given to him or her on behalf of an employer. The special offence of embezzlement ceased to exist in 1969.

EMCF Abbreviation for European Monetary Cooperation Fund.

Emerging Company Marketplace (ECM) A service once offered by the American Stock Exchange to help small growth companies fulfill special listing requirements. The service is no longer available.

Emerging Markets Free index (EMF) A Morgan Stanley Capital International index created to track stock markets in selected emerging markets that are open to foreign investment like Argentina, Chile, Jordan, Malaysia, Mexico, Philippines, and Thailand.

Emerging markets fund A mutual fund that invests primarily in countries with developing economies (that is, those that are becoming industrialized). Emerging markets funds tend to be more volatile than domestic stock funds due to currency fluctuation and political instability. Consequently, fund prices can fluctuate dramatically.

Emoluments Amounts received from an office or employment including all salaries, fees, wages, perquisites, and other profits as well as certain expenses and benefits paid or provided by the employer, which are deemed to be emoluments.

Employee contribution An employee's own deposit to a company retirement plan.

Employee participation 1. The encouragement of motivation in a workforce by giving shares in the company to employees. Employee shareholding (*see* employee share ownership plan) is now an important factor in improving industrial relations.

2. The appointment to a board of directors of a representative of the employees of a company, to enable the employees to take part in the direction of the company.

Employee stock fund A firm-sponsored program that enables employees to purchase shares of the firm's common stock on a preferential basis.

Employee stock ownership plan (ESOP) A company contributes to a trust fund that buys stock on behalf of employees.

Employee Stock Purchase Plan (ESPP) A plan usually linked to a corporation's payroll deduction system allowing employees to purchase shares at a discount from current market value.

Employers'-liability insurance An insurance policy covering an employer's legal liability to pay compensation to any of his employees suffering or contracting injury or disease during the course of their work This type of insurance is compulsory by law for anyone who employs another person (other than members of their own family) under a contract of service. A certificate of insurance must be displayed at each place of work, confirming that employer's liability insurance is in force and giving details of the policy number and the insurer's name and address.

Employment agency An organization that introduces suitable potential employees to employers, charging the employer a fee, usually related to the initial salary, for the service. Employment agencies also provide temporary staff, in which they are the employers, the temporary staff member being charged out at an hourly rate. Employment agencies that specialize in finding suitable managers and executives for a firm, or finding suitable jobs for executives who want a change, are often known as head hunters. Such agencies will often provide a short list of candidates in order to save their client's time in personnel selection. Employment agencies run by the government are called job centres.

Employment rate The percentage of the labor force that is employed. The employment rate is one of the economic indicators that economists examine to help understand the state of the economy. See also: Unemployment rate.

Empowerment In human resource management, the giving of increased responsibility and a measure of control to employees in their working lives. The concept is based on the view that people need personal satisfaction and fulfilment in their work and that responsibility and control increase satisfaction. Employees like empowerment because an increase in responsibility usually leads to greater rewards and enhanced prospects. On the other hand, it has been criticized because staff are asked to become more accountable without being given more authority. Empowerment also has the added advantage that it enables potential talents to be identified and developed, either by the employing organization or the individual.

N general, empowerment is a motivational strategy, or part of a process of re-engineering organizational structures to remove layers of management to make the system respond more rapidly to customers' requirements.

EMS Abbreviation for European Monetary System.

EMU Abbreviation for European Monetary Union. *See*European Monetary System.

EMV Abbreviation for expected monetary value.

Encryption The encoding of data to be transmitted by the Internet, for instance by e-mail The fear of unauthorized access being gained to confidential information, e.g. credit cards and company financial details, has placed a restriction on the commercial development of the Internet. Systems are being developed to enable this data to be transmitted in an encrypted form, decipherable only by those in possession of the appropriate key.

Encumbered A property owned by one party on which a second party reserves the right to make a valid claim, e.g., a bank's holding of a home mortgage encumbers property.

End-of-day sweep An automatic transfer of funds from one bank account held by a company to another of its bank accounts, usually one that pays interest on deposits. The sweep takes place at the end of every day, or at the end of the day when certain conditions are met, for instance that the cleared value exceeds £50,000 in one of the accounts.

End-of-year convention Treating cash flows as if they occur at the end of a year as opposed to the date convention. Under the end-of-year convention, the present is time 0, the end of year 1 occurs one year hence; and so on.

Endorsement (indorsement) 1. A signature on the back of a bill of exchange or cheque, making it payable to the person who signed it. A bill can be endorsed any number of times, the presumption being that the endorsements were made in the order in which they appear, the last named being the holder to receive payment. If the bill is blank endorsed, i.e. no endorsee is named, it is payable to the bearer. In the case of a restrictive endorsement of the form "Pay X only", it ceases to be a negotiable instrument. A special endorsement, when the endorsee is specified, becomes payable to order, which is short for 'in obedience to the order of'. 2. A signature required on a document to make it valid in law. 3. An amendment to an insurance policy or cover note, recording a change in the conditions of the insurance.

Endowment funds Investment funds established for the support of institutions such as colleges, private schools, museums, hospitals, and foundations. The investment income may be used for the operation of the institution and for capital expenditures.

Enhanced indexing Also called indexing-plus, an indexing strategy whose objective is to exceed or replicate the total return performance of some predetermined index.

Enhancement An innovation that has a positive impact on one or more of a firm's existing products.

Enterprise Value The market capitalization of a firm's equity plus the market value of the firm's debt. Often the value of assets that are non-core are excluded from the final calculation.

Enterprise zone An area, designated as such by the government, in which its aim is to restore private-sector activity by removing certain tax burdens and by relaxing certain statutory controls. Benefits, which are available for a 10-year period, include: exemption from rates on industrial and commercial property; 100% allowances for corporation- and income-tax purposes for capital expenditure on industrial and commercial buildings; exemption from industrial training levies; and a simplified planning regime.

Entireties property Typically husband and wife, a form of indivisible co-ownership of realty or personalty estate that upon death of one, survivor takes title to the whole estate.

Entrepôt trade Trade that passes through a port, district, airport, etc., before being shipped on to some other country. The entrepôt trade may make use of such a port because it is conveniently situated on shipping lanes and has the warehouses and customs facilities required for re-export or because that port is the centre of the particular trade concerned and facilities are available there for sampling, testing, auctioning, breaking bulk, etc. Much entrepôt trade used to pass through London, but since the decline of the London docks other European ports, such as Rotterdam, have taken its place. Singapore and Hong Kong are also centres of the entrepôt trade.

Entrepreneur A person starting a new company who takes on the risks associated with starting the enterprise, which may require venture capital to cover start-up costs.

Entry A record made in a book of account, register, or computer file of a financial transaction, event, proceeding, etc.

Equillibrium A term used to describe a situation of economic agents or of a ggregates of economic agents such as markets. Applied to an individual agent, such as a consumer or firm, it is used to describe a situation in which the agent is under no pressures or incentives to alter current levels or states of economic action, because given his aspirations and the constraints he faiths he to improve his position i terms of any used to denote a situation in which, in the aggregate, buyers and sellers get satisfied with the current combination of prices and quantities bought or sold, and so are under no incentive to change their present

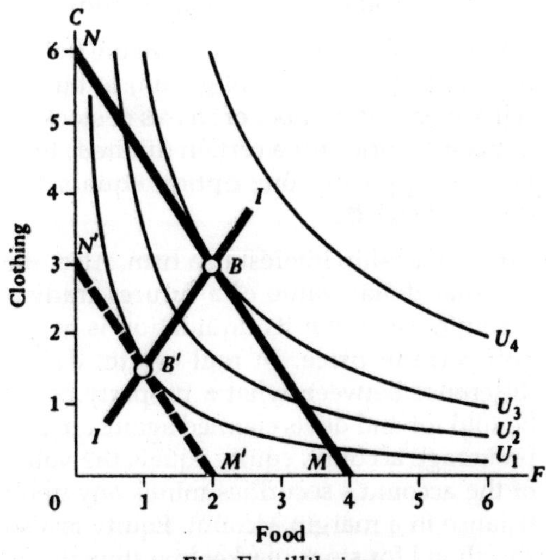

Fig. Equillibrium

Equipment trust certificates Certificates issued by a trust that was formed to purchase an asset and lease it to a lessee. When the last of the certificates has been repaid, title of ownership of the asset reverts to the lessee.

Equitable apportionment The process of sharing common costs between cost centres in a fair manner, using a basis of apportionment that reflects the way in which the costs are incurred by the cost centres.

Equitable interest An interest in, or ownership of, property that is recognized by equity but not by the common law. A beneficiary under a trust has an equitable interest. Any disposal of an equitable interest (e.g. a sale) must be in writing. Some equitable interests in land must be registered or they will be lost if the legal title to the land is sold. Similarly, equitable interests in other property will be lost if the legal title is sold to a bona fide purchaser for value who has no notice of the equitable interest. In such circumstances the owner of the equitable interest may claim damages from the person who sold the legal title.

Equity 1. A beneficial interest in asset. For example, a person having a house worth £100,000 with a mortgage of £20,000 may be said to have an equity of £80,000 in the house.

2. The net assets of a company after all creditors (including the holders of preference shares) have been paid off.

3. The amount of money returned to a borrower in a mortgage or hire-purchase agreement, after the sale of the specified asset and the full repayment of the lender of the money.

4. The ordinary share capital of a company

5. The system of law developed by the medieval chancellors and later by the Court of Chancery. It is distinguished from the common law, which was developed by the king's courts, having originated from the residual jurisdiction delegated by the king to the chancellor. A citizen dissatisfied by the common law could petition the chancellor, who might grant relief on an ad-hoc basis. In time this developed into a complementary but separate system of law providing remedies unavailable at common law, such as specific performance of a contract rather than damages. Until 1873 equity was applied and administered by the Court of Chancery, and equitable remedies were not available in the common law courts.

Equity accounting The practice of showing in a company's accounts a share of the undistributed profits of another company in which it holds a share of the equity (usually a share of between 20% and 50%). The share of profit shown by the equity-holding company is usually equal to its share of the equity in the other company. Although none of the profit may actually be paid over, the company has a right to this share of the undistributed profit.

Equity cap An agreement in which one party, for an upfront premium, agrees to compensate the other at specific time periods if a designated stock market benchmark is greater than a predetermined level.

Equity capital The part of the share capital of a company owned by ordinary shareholders, although for certain purposes, such as pre-emption rights, other classes of shareholders may be deemed to share in the equity capital and therefore be entitled to share in the profits of the company or any surplus assets on winding up.

Equity carve out Usually occurs when a company decides to IPO one of their subsidiaries or divisions. The company usually only offers a minority share to the equity market. Also known as carve out.

Equity claim Also called a residual claim; a claim to a share of earnings after debt obligations have been satisfied.

Equity contribution agreement An agreement to contribute equity to a project under certain specified conditions.

Equity dilution A reduction in the percentage of the equity owned by a shareholder as a result of a new issue of shares in the company, which rank equally with the existing voting shares.

Equity dividend cover A ratio that shows how many times the dividend to ordinary shareholders can be paid out of the profits of a company available for distribution. The higher the cover, the greater the certainty that dividends will be paid in the future.

Equity finance Finance raised from shareholders in the form of ordinary shares and reserves, as opposed to nonequity shares and to borrowing.

Equity floor An agreement in which one party agrees to pay the other at specific time periods if a specific stock market benchmark falls below a predetermined level.

Equity funding An investment consisting of a life insurance policy and a mutual fund. The insurance policy is paid by the collateral value of fund shares, giving the investor the advantages of insurance protection with the growth potential of a mutual fund.

Equity multiplier Total assets divided by total common stockholders' equity; the total assets per dollar of stockholders' equity.

Equity options Securities that give the holder the right (but not the obligation) to buy or sell a specified number of shares of stock, at a specified price for a certain (limited) time period. Typically one option equals 100 shares of stock. .

Equity Ownership interest in a firm. Also, the residual dollar value of a futures trading account, assuming its liquidation is at the going trade price. In real estate, dollar difference between what a property could be sold for and debts claimed against it. In a brokerage account, equity equals the value of the account's securities minus any debit balance in a margin account. Equity is also shorthand for stock market investments. he share capital of a company that consists of its equity shares as opposed to its nonequity shares.

Equity swap A swap in which the cash flows exchanged are based on the total return on some stock market index and an interest rate (either a fixed rate or floating rate). Related: Interest rate swap.

Equity-linked policy An insurance or assurance policy in which a proportion of the premiums paid are invested in equities. The surrender value of the policy is therefore the selling price of the equities purchased; as more premiums are paid the portfolio gets larger. Although investment returns may be considerably better on this type of policy than on a traditional endowment policy, the risk is greater, as the price of equities can fall dramatically reducing the value of the policy. With unitlinked policies, a much wider range of investments can be achieved and the risk is correspondingly reduced

Equivalent annual cash flow Annuity with the same net present value as the company's proposed investment.

Equivalent loan Given the after-tax stream associated with a lease, the maximum amount of conventional debt that the same period-by-period after-tax debt service stream is capable of supporting.

Equivalent units (effective units) Unfinished units of production that remain in a process at the end of a period as work in progress (or process). Degrees of completion are assigned to each cost classification, which, when applied to the number of units in work in progress, give an equivalent number of completed units. For example, if the number of units of work in progress is, say, 3000 units, the table below shows how the equivalent units are given in terms of the degrees of completion. The equivalent units have an impact on the valuation of opening and closing work in process.

Ergonomics The study of the physical and psychological aspects of work; it is primarily concerned with designing jobs and work environments that are safe and efficient. For example, such human characteristics as size, reach, and flexibility must match the machines and tools that people have to use.

Erosion An innovation that has a negative impact on one or more of a firm's existing assets.

Escalation clause A clause in a contract authorizing the contractor to increase the price in specified conditions of all or part of the services or goods he has contracted to supply. Escalation clauses are common in contracts involving work over a long period in times of high inflation.

Escalation of commitment Increasing the resources available to an unsuccessful venture in the hope of recovering past losses. This is sometimes called a creeping commitment. This policy is often adopted in new product development when the company continues to pour in funds, irrespective of the likelihood of success, in an attempt to recover some of its investment. It is known colloquially as throwing good money after bad.

Escheat Period The period of elapsed time required by applicable state law for property to be presumed abandoned.

Escheatment The process of turning over unclaimed or abandoned property to a state authority. Escheatment laws require mutual funds to turn over uncashed or returned check dollars and/or client account fund shares if the owner cannot be located within a length of time determined by each state.

Escrow A deed that has been signed and sealed but is delivered on the condition that it will not become operative until some stated event happens. It will become effective as soon as that event occurs and it cannot be revoked in the meantime. Banks often hold escrow accounts, in which funds accumulate to pay taxes, insurance on mortgaged property, etc.

Estate planning The preparation of a plan to carry out an individual's wishes as to the administration and disposition of his/her property before or after his/her death.

Estimated tax Tax to be paid quarterly on income that is not subject to withholding tax, including self-employed income, investment income, alimony, rent, and capital gains.

Ethical behaviour Behaviour judged to be good, just, right, and honourable, based on principles or guides from a specific ethical theory. However, ethical theories may vary from person to person, country to country, or company to company. Ethical realism accepts that although morality does not apply internationally, the ethical values of a trading partner should be respected.

Ethical investment (socially responsible investment) An investment made in a company not engaged in an activity that the investor considers to be unethical, such as armaments or tobacco, or investment in a company of which the investor approves on ethical grounds, e.g. one having a good environmental employment record.

Euclidean Geometry The Plane geometry learned in high school, based upon a few

Euro The currency unit to be used by the European Monetary Union (*see* European Monetary System).

Euro-ad An advertisement designed to be used in all countries of the European Union. Certain products, particularly those that have been in use for long enough for national traditions to build up, are not suitable for multinational advertisements. Food is an example: there would be no point in advertising tinned baked beans in France. On the other hand, advertisements for cars, a more recent product, can have an equal appeal in all European countries.

Eurobank A bank that regularly accepts foreign currency-denominated deposits and makes foreign currency loans.

Eurobond A bond issued in a eurocurrency, which is now one of the largest markets for raising money (it is much larger than the UK stock exchange). The reason for the popularity of the eurobond market is that secondary market investors can remain anonymous, usually for the purpose of avoiding tax. For this reason it is difficult to ascertain the exact size and scope of operation of the market. Issues of new eurobonds normally take place in London, largely through syndicates of US and Japanese investment banks; they are bearer securities, unlike the shares registered in most stock exchanges and interest payments are free of any withholding taxes. There are various kinds of eurobonds. An ordinary bond, called a straight, is a fixed-interest loan of 3 to 8 years duration; others include floating-rate notes, which carry a variable interest rate based on the LIBOR; and perpetuals, which are never redeemed. Some carry warrants and some are convertible.

Eurobond A bond that is.

(1) underwritten by an international syndicate,

(2) issued simultaneously to investors in a number of countries, and.

(3) issued outside the jurisdiction of any single country. Eurobonds are often bearer bonds.

Eurocheque A cheque drawn on a European bank, which can be cashed at any bank or bureau de change in the world that displays the European Union sign (of which there are some 200 000). It can also be used to pay for goods and services in shops, hotels, restaurants, garages, etc., that display the EU sign (over 4 million). The cheques are blank and are made out for any amount as required usually in the local currency. They have to be used with a Eurocheque Card, which guarantees cheques for up to about £100. In most cases, a commission of 125% is added to the foreign currency value of the cheque before it is converted to sterling and there is a cheque charge (for cheques drawn on UK banks).

Euroclear One of two principal clearing systems in the Eurobond market. It began operations in 1968, is located in Brussels, and is managed by Morgan Guaranty Bank.

Euroclear The Euroclear group is the world's largest settlement system for domestic and international securities transactions, covering both bonds and equities for financial institutions located in over 80 countries.

Eurocredit market Comprises banks that accept deposits and provide loans in large denominations and in a variety of currencies. The banks that constitute this market are the same banks that constitute the Eurocurrency market; the difference is that Eurocredit loans are longer-term than so-called Eurocurrency loans.

Eurocurrency A currency held in a European country other than its country of origin. For example, dollars deposited in a bank in Switzerland are eurodollars, yen deposited in Germany are euroyen, etc. Eurocurrency is used for lending and borrowing; the eurocurrency market often provides a cheap and convenient form of liquidity for the financing of international trade and investment. The main borrowers and lenders are the commercial banks, large companies, and the central banks. By raising funds in eurocurrencies it is possible to secure more favourable terms and rates of interest, and sometimes to avoid domestic regulations and taxation. Most of the deposits and loans are on a short-term basis but increasing use is being made of medium-term loans, particularly through the raising of eurobonds. This has to some extent replaced the syndicated loan market, in which banks lent money as a group in order to share the risk. Euromarkets emerged in the 1950s.

Eurocurrency deposit A short-term fixed-rate time deposit denominated in a currency other than the local currency (e.g., U.S. dollars deposited in a London bank).

Eurocurrency market The money market for borrowing and lending currencies that are held in the form of deposits in banks located outside the countries of the currencies issued as legal tender.

Eurodollar certificate of deposit A certificate of deposit paying interest and principal in dollars, but issued by a bank outside the United States, usually in Europe.

Eurodollar obligations Certificates of deposit issued in U.S. dollars by foreign banks and foreign branches of U.S. banks.

Eurodollars Dollars deposited in financial institutions outside the USA. The eurodollar market evolved in London in the late 1950s when the growing demand for dollars to finance international trade and investment coincided with a greater supply of dollars. The prefix 'euro' indicates the origin of the practice but it now refers to all dollar deposits made anywhere outside the USA. *See also* eurocurrency.

Euroequity issues Securities sold in the Euromarket. That is, securities initially sold to investors simultaneously in several national markets by an international syndicate.

Euroequity issues Securities sold in the Euromarket. That is, securities initially sold to investors simultaneously in several national markets by an international syndicate.

Euromarket 1. A market that emerged in the 1950s for financing international trade. Its principal participants are commercial banks, large companies, and the central banks of members of the EU. Its main business is in eurobonds issued in eurocurrencies. The largest euromarket is in London, but there are smaller ones in Paris and Brussels.

2. The European Union, regarded as one large market for goods.

Euro-medium term note (Euro-MTN) A non-underwritten Euronote issued directly to the market. Euro-MTNs are offered continuously rather than all at once as a bond issue is. Most Euro-MTN maturities are under five years.

European Association of Securities Dealers Automated Quotation (EASDAQ) European equivalent of Nasdaq.

European Bank for Reconstruction and Development Bank targeted at Eastern Europe and the former Soviet Union.

European Commission The single executive body formed in 1967 (as the Commission of the European Communities) from the three separate executive bodies of the European Coal and Steel Community, the European Atomic Energy Community, and the European Economic Community. It now consists of 20 Commissioners: two each from the UK, France, Germany, Spain, and Italy; and one each from Belgium, the Netherlands, Luxembourg, Ireland, Denmark Greece, Portugal, Austria, Finland, and Sweden. The Commissioners accept joint responsibility for their decisions, which are taken on the basis of a majority vote. The Commission initiates and implements EU legislation and mediates, between member governments. The Commissioners are backed by a staff of some 20 000 civil servants.

European Currency Unit (ECU) A currency medium and unit of account created in 1979 to act as the reserve asset and accounting unit of the European Monetary System (EMS). The value of the ECU is calculated as a weighted average of a basket of specified amounts of European Union currencies; its value is reviewed periodically as currencies change in importance and membership of the EU expands. Fluctuations in the value of the ECU in terms of the currencies of the member states are controlled by the EMS. The ECU also acts as the unit of account for all EU transactions. It has some similarities with the Special Drawing Rights of the International Monetary Fund; however, ECU reserves are not allocated to individual countries but are held in the European Monetary Cooperation Fund. Private transactions using the ECU as the denomination for borrowing and lending have proved popular.

European Currency Unit (ECU) An index of foreign exchange consisting of about 10 European currencies, originally devised in 1979.

European exchange rate mechanism (ERM) The system that countries in the European Union once used to pay exchange rates within bands around an ERM central value.

European Exercise A feature of an option that stipulates that the option may only be exercised at its expiration. Therefore, there can be no early assignment with this type of option. Most index options are European-style exercise.

European Investment Bank (EIB) A bank set up under the Treaty of Rome in 1958 to finance capital-investment projects in the European Economic Community (EEC). It grants long-term loans to private companies and public institutions for projects that further the aims of the Community. The members of the EIB are the 15 member states of the European Union, all of whom have subscribed to the Bank's capital of 62,013M ECU but most of the funds lent by the bank are borrowed on the EU and international capital markets. The bank is non-profit-making and charges interest at a rate that reflects the rate at which it borrows. Its headquarters are in Luxembourg.

European Monetary Institute (EMI) An organization set up by the Maastricht Treaty in 1991 in order to coodinate the economic and monetary policy of members of the European Union before the introduction of a single European currency in accordance with the objectives of European Monetary Union (EMU; *see*European System). The members of the EMI council are the governors of the central banks of the 15 EU countries.

European option Option that may be exercised only at the expiration date. Related: american option.

European Union (EU) An economic association of European countries founded by the Treaty of Rome in 1957 as a common market for six nations. It was known as the European Community before 1993 and is comprised of 15 European countries. Its goals are a single market for goods and services without any economic barriers and a common currency with one monetary authority. The EU was known as the European Community until January 1, 1994.

Evaluation period The time interval over which a money manager's performance is evaluated.

Event of default A critical clause in a loan agreement, the breaching of which will make the loan repayable immediately. The breaching of any covenant clause will be an event of default. Events of default also include failure to pay, failure to perform other duties and obligations, false representation and warranty, material adverse change, bankruptcy, and alienation of assets.

Event risk The risk that the ability of an issuer to make interest and principal payments will change because of rare, discontinuous, and very large, unanticipated changes in the market environment such as (1) a natural or industrial accident or some regulatory change or (2) a takeover, or corporate restructuring.

Evergreen fund A fund that provides capital for new companies and supports their development for a period with regular injections of capital.

Evergreen funding A British term referring to the gradual injection of capital into a new or existing enterprise.

Ex ante (Latin) Before the event. The phrase is used, for example, of a budget that is prepared as an estimate and sub- sequently compared with actual figures. *Compare*ex post

Ex ante return The expected return or anticipation return of an asset or portfolio.

Ex gratia (Latin: as of grace) Denoting a payment made out of gratitude, moral obligation, kindness, etc., rather than to fulfil a legal obligation. When an ex gratia payment is made, no legal liability is admitted by the payer.

Ex gratia pension A pension paid by an employer although there is no legal contractual or implied commitment to provide it.

Ex growth (of a share or a company) Having had substantial growth in the past but now not holding out prospects for immediate growth of earnings or value.

Ex post (Latin) Short for *ex post facto*: after the event. This abbreviation is used, for example, to refer to the collection of financial data for transactions after they have been effected.

Ex post return Related: Holding-period return

Ex quay Delivery terms for goods in which the seller pays all freight charges up to the port of destination, unloading onto the quay, and loading onto road or rail vehicles. Thereafter all transport charges must be paid by the buyer.

Ex warehouse Delivery terms for goods that are available for immediate delivery, in which the buyer pays for the delivery of the goods but the seller pays for loading them onto road or rail transport. *Compare*at warehouse.

Ex Works (EXW) A transaction in which the seller's only responsibility is to make the ordered goods available to the buyer at the seller's premises. The buyer bears the cost and risk in transporting the goods from the seller's premises to destination. Since this includes pre- carriage and export clearance in the seller's country, EXW is not a very practical Incoterm for U.S. exports.

Exact interest Interest paid based on the basis of a 365-day/year schedule by a bank or other financial_institution as opposed to a 360-day basis (ordinary interest). Difference can be material when large principal sums of money are involved.

Exact matching A bond portfolio management strategy that involves finding the lowest cost portfolio generating cash inflows exactly equal to cash outflows that are being financed by investment.

Ex-all The sale of a security without the privileges associated with the security such as dividends, voting rights, or warrants.

Examiner An examiner is an officer appointed by the bankruptcy court to review certain aspects of the operation of a Chapter 11 debtor. Aside from the investigative function, examiners have been used as mediators to assist the court in resolving disputes between and among various factions in a bankruptcy court.

Exante return The expected return of a portfolio based on the expected returns of its component assets and their weights.

Except for opinion An auditor's opinion reflecting the fact that the auditor was unable to audit certain areas of the company's operations because of restrictions imposed by management or other conditions beyond the auditor's control.

Exception A proxy which does not authorize the proxy committee to act on its behalf concerning any other business, adjournments or substitutions.

Exceptional items Costs or income affecting a company's profit and loss account that arise from the normal activities of the company but are of exceptional magnitude, either large or small These should be disclosed separately in arriving at the trading profit or loss but not after the normal trading profit or loss has been shown

Exceptional Return Residual return plus benchmark timing return. For a given asset with beta equal to one, if its residual return is 2%, and the benchmark portfolio exceeds its consensus expected returns by 1%, then the asset's exceptional return is 3%.

Excess accumulation The amount of a required minimum distribution that an IRA holder fails to remove from an IRA in a timely manner. Excess accumulations are subject to a 50% IRS penalty tax.

Excess contribution The amount by which an IRA contribution exceeds the allowable limits. If an excess contribution is not properly corrected, a 6% IRS penalty applies.

Excess kurtosis Kurtosis measures the "fatness" of the tails of a distribution. Positive excess kurtosis means that distribution has fatter tails than a normal distribution. Fat tails means there is a higher than normal probability of big positive and negative returns realizations. When calculating kurtosis, a result of +3.00 indicates the absence of kurtosis (distribution is mesokurtic). For simplicity in its interpretation, some statisticians adjust this result to zero (i.e. kurtosis minus 3 equals zero), and then any reading other than zero is referred to as excess kurtosis. Negative numbers indicate a platykurtic distribution; positive numbers indicate a leptokurtic distribution.

Excess policy An insurance policy in which the insured is responsible for paying a specified sum (the excess) of each claim and cannot make claims of a lower value than this excess. For example, a £100 excess on a motor-insurance policy means that the insured has to pay the first £100 of any claim and cannot make a claim on the policy for less than £100. This arrangement enables the insurer to offer the insurance at a lower premium than would otherwise be the case as he avoids the administrative cost of small claims and also makes a saving on claims paid out.

Excess profits tax Additional federal taxes

placed on the earnings of a business, used only in time of national emergency such as war.

Excess reserves Amount of reserves held by an institution in excess of its reserve requirement and required clearing balance. Also see reserves.

Excess return on the market portfolio The difference between the return on the market portfolio and the riskless rate.

Excess returns Difference between an asset's return and the riskless rate. Sometimes confused with abnormal returns, returns in excess of those required by some asset pricing model.

Exchange A marketplace in which shares, options and futures on stocks, bonds, commodities, and indexes are traded. Principal U.S. stock exchanges are: New York Stock Exchange (NYSE), American Stock Exchange (AMEX), and National Association of Securities Dealers Automatic Quotation System (Nasdaq).

Exchange control Restrictions on the purchase and sale of foreign exchange. It is operated in various forms by many countries, in particular those who experience shortages of hard currencies; sometimes different regulations apply to transactions that would come under the capital account of the balance of payments. There has been a gradual movement toward dismantling exchange controls by many countries in recent years. The UK abolished all form of exchange control in 1979.

Exchange controls Governmental restrictions on the purchase of foreign currencies by domestic citizens or on the purchase of the local domestic currency by foreigners.

Exchange distribution A sale on an exchange floor of a large block of stock in a single transaction. A broker bunches a large number of buy orders and sells the block all at once. The broker receives a special commission from the seller.

Exchange equalization account An account set up in 1932 and managed by the Bank of England on behalf of the government. It contains the official gold and foreign-exchange reserves (including Special Drawing Rights) of the UK and is used as a buyer of foreign exchange to support the value of sterling. Although all exchange controls were abolished in 1979, the Bank of England still makes use of this account to help to stabilize rates of exchange.

Exchange of assets Acquisition of another company by purchase of its assets in exchange for cash or stock.

Exchange of contracts A procedure adopted in the sale and purchase of land in which both parties sign their copies of the contract, having satisfied themselves as to the state of the property, etc., and agreed that they wish to be bound. There need be no physical exchange of documents; the parties or their advisers can exchange contracts by agreeing to do so orally (for example, by telephone). From that moment the contract is binding and can normally be enforced by specific performance. Contracts for the sale of land must be in writing. Acquisition of another company by purchase of its stock in exchange for cash or shares.

Exchange offer An offer by a firm to give one security, such as a bond or preferred stock, in exchange for another security, such as shares of common stock.

Exchange offer An offer by the firm to give one security, such as a bond or preferred stock, in exchange for another security, such as shares of common stock.

Exchange Rate Mechanism (ERM) The methodology by which members of the EMS maintain their currency exchange rates within an agreed upon range with respect to other member countries.

Exchange rate risk Also called currency risk, the risk of an investment's value changing because of currency exchange rates.

Exchange risk The variability of a firm's value that results from unexpected exchange rate changes or the extent to which the present value of a firm is expected to change as a result of a given currency's appreciation or depreciation.

Exchange The marketplace in which shares, options and futures on stocks, bonds, commodities and indices are traded. Principal US stock exchanges are: New York Stock Exchange (NYSE), American Stock Exchange (AMEX) and the National Association of Securities Dealers (NASDAQ)

Exchange Traded Funds (ETF) Also known as ETF. A basket of stocks similar to an index mutual fund. However, there are a number of important differences between ETFs and mutual funds. The ETF can be traded within the day, they can be shorted, purchased on margin and there even exists options on some ETFs.

Exchangeable Applies mainly to convertible securities. Means the issuer, if so stated, may substitute a convertible debenture for an existing convertible preferred with identical terms. Most often used when a corporation has an immediate need for equity capital and a low tax rate, and expects either or both conditions to change. This would make the debenture less attractive if the interest tax-deductibility is lost.

Excise duty A duty or tax levied on certain goods, such as alcoholic drinks and tobacco products, produced and sold within the UK, unlike customs duty (*see* customs tariff), which is levied on imports. Both excise and customs duties are collected by the Board of Customs and Excise.

Excise tax Federal or state tax placed on the sale or manufacture of a commodity, typically a luxury item e.g., alcohol.

Exclusive distribution A situation in which a distributor carries the products of one manufacturer and not those of competing manufacturers; this may give the distributor the exclusive right to distribute the company's products in other territories. *See*exclusive territory.

Ex-dividend date The first day of trading when the buyer of a stock is no longer entitled to the most recently announced dividend payment (i.e. the trade will settle the day after the record date, too late for the buyer to appear on the shareholder record and receive the dividend.) The date set by the NYSE (and generally followed on other U.S. exchanges) is currently two business days before the record date. A stock that has gone ex-dividend is denoted by an x in newspaper listings on that date.

Ex-dividend This literally means "without dividend." The buyer of shares when they are quoted ex-dividend is not entitled to receive a declared dividend. It is the interval between the record date and the payment date during which the stock trades without its dividend-the buyer of a stock selling ex-dividend does not receive the recently declared dividend. Antithesis of cum dividend (with dividend).

Execution The process of completing an order to buy or sell securities. Once a trade is executed, it is reported by a Confirmation Report; settlement (payment and transfer of ownership) occurs in the U.S. between 1 (mutual funds) and 5 (stocks) days after an order is executed. Settlement times for exchange listed stocks are in the process of being reduced to three days in the U. S.

Executive director (working director) A director of a company who is also an employee (usually full-time) of that company. An executive director will often have a specified role in the management of the company, such as finance director, marketing director, production director, etc. A non-executive director is a member of the board of directors and is therefore involved in planning and policy-making, but not in the day-today management of the company A nonexecutive director is often employed for prestige (if he is well known), for his experience or contacts, or for his specialist knowledge, which may only be required occasionally.

Executive pension plan (EPP) A pension for a senior executive or director of a company in which the company provides a tax-deductible contribution to the premium. An executive pension plan may be additional to any group pension scheme provided by the company, as long as the pension limit of two-thirds of the working salary is not exceeded.

Executive share option scheme An approved share option scheme that entitles a specified class of directors or employees to purchase shares in the company in which they are employed. Strict conditions need to be met in order to receive Inland Revenue approval for such a scheme. Once achieved there are no income tax charges on the grant or exercise of the option or on the growth in value of the shares. The only charge will be to capital gains tax when the shares are sold.

Executor An individual or trust institution nominated in a will and appointed by a court to settle the estate of a deceased person.

Exempt List Sophisticated investors, usually institutional investors, who are considered informed enough that new issues can be marketed to them without a prospectus. This exemption reduces the cost of private placements.

Exemptions Property which may be kept by an individual debtor and is not used for distribution to creditors. The bankruptcy code and state laws control the exemptions to which the debtor is entitled. • Examples of exemptions include: up to Rs.17,425 in equity in a personal residence; up to Rs.2,775 equity in an automobile; and certain personal property where the value of each item does not exceed Rs.450 and the cumulative value does not exceed Rs.9,300.

Exercise limit Cap on the number of option contracts of any one class of contract that can be exercised within a five-day period contract. There are no restrictions on exercise for the last 10 trading days before expiry. A stock option's exercise limit varies with the volume of the underlying stock.

Exercise price The price at which the underlying future or options contract may be bought or sold.

Exercising the option The act buying or selling the underlying asset via the option contract.

Exit value The net realizable value of an asset, i.e. its market price at the date of a balance sheet less the selling expenses. Exit values are effectively break-up values and are not consistent with the going-concern concept, which assumes that a business is continuing to trade.

Ex-legal A municipal bond offered without a law firm's legal opinion. A majority of bonds are issued with legal opinions.

Exogenous Describes facts outside the control of the firm. Converse of endogenous.

Exotic option Refers to options that are more complex than simple put or call options. For example, a Caput is a call option on a put option. Exotic options trade over-the-counter.

Expectations hypothesis theories Theories of the term structure of interest rates which include the pure expectations theory, the liquidity theory of the term structure, and the preferred habitat theory. These theories hold that each forward rate equals the expected future interest rate for the relevant period. These three theories differ, however, on whether other factors also affect forward rates, and how.

Expected dividend yield Total amount of dividends received during the life of a futures contract or total dividends received for holding a particular stock one year. See: Current yield.

Expected return The expected return on a risky

asset, given a probability distribution for the possible rates of return. Expected return equals some risk-free rate (generally the prevailing U.S. Treasury note or bond rate) plus a risk premium (the difference between the historic market return, based upon a well diversified index such as the S&P 500 and the historic U.S. Treasury bond) multiplied by the asset's beta. The conditional expected return varies through time as a function of current market information.

Expected Spot Rate The exchange rate between two currencies that is anticipated to prevail in the spot market on a given future date. It differs from the current spot rate primarily by the extent to which inflation expectations in the two currencies differ.

Expected value of perfect information The expected value if the future uncertain outcomes could be known minus the expected value with no additional information.

Expenditure tax (outlay tax) A tax on the expenditure of individuals or households. This form of taxation, of which VAT is an example, is often preferred by tax theorists to income taxes, as it does not distort the incentive to work.

Expenditure The costs or expenses incurred by an organization. They may be capital expenditure or revenue expenditure. Although expenditure is usually incurred by an outlay of money, expenditure may also arise in accounting by the acknowledgement of a liability, for example rent accrued due, which is regarded as expenditure in the period accrued although it will not be paid until a later date.

Expense ratio The percentage of the assets that were spent to run a mutual fund (as of the last annual statement). This includes expenses such as management and advisory fees, overhead costs and 12b-1 (distribution and advertising) fees. The expense ratio does not include brokerage costs for trading the portfolio, although these are reported as a percentage of assets to the SEC by the funds in a Statement of Additional Information (SAI). the SAI is available to shareholders on request. Neither the expense ratio or the SAI includes the transaction costs of spreads, normally incurred in unlisted securities and foreign stocks. These two costs can add significantly to the reported expenses of a fund. The expense ratio is often termed an Operating Expense Ratio (OER).

Experience rating A technique insurance companies use to determine the correct price of a policy premium.

Expiration cycle The recurring cycle of expiry months for which options on a particular security can be available. Basic options are placed in one of three cycles; Cycle 1 (the January/April/July/October, or the first month of each quarter); Cycle 2 (the second month of each quarter); or Cycle 3 (the third month of each quarter). At any one time, a basic option has contracts with three expiration dates outstanding. For example, in mid-February, options trading on cycle 3 will have March, June and September expiries available. Late in March, after the March options expire, a December contract will be added, thus offering June, September and December expiries. Higher-volume equity options, index options, and LEAPS can trade on other cycles, such as Cycle 4, Cycle 5 or Cycle 6. Cycle 4, for example, offers options in the two nearest months plus two months from Cycle 3. For example, in mid-April, there would be April, May, June and September expires available. A month later, there would be May, June, September and December expiries available for trading.

Expiration date The last day (in the case of American-style) or the only day (in the case of European-style) on which an option may be exercised. For stock options, this date is the Saturday immediately following the third Friday of the expiration month; brokerage firms may set an earlier deadline for notification of an option holder's intention to exercise. If Friday is a holiday, the last trading day will be the preceding Thursday.

Expiration time The time of day by which all exercise notices must be received on the expiration date. Technically, the expiration time is currently 11:59AM on the expiration date, but public holders of option contracts must indicate their desire to exercise no later than 5:30PM on the business day preceding the expiration date. The times are Eastern

Ex-pit transaction The closing out of a futures position off the exchange floor. Effected when two hedgers, one long and one short, make a private deal in the cash market, and no longer need their (equal and opposite) futures contracts to hedge. The hedgers contact the exchange and request the contracts be nullified without making a trade on the floor. This must be done (1) to ensure neither contract results in delivery/the requirement to deliver; (2) to properly reflect open interest; and (3) to eliminate the uncertainty of the fill price should a trade actually be done to offset the positions. Extremely rare. Also known as an EFP transaction, an exchange-for-physicals transaction or an against-actuals transaction.

Exploratory research Preliminary marketing research to clarify the exact nature of a problem to be solved. Quite often sales managers recognize that there is a problem but do not understand the dimensions of the problem. In these circumstances preliminary research is conducted with small numbers of potential customers to define the nature of the problem.

Export licence A licence required before goods can be exported from a country. Export licences are only required in the UK for certain works of art, antiques, etc., and cErtain types of arms and armaments.

Export-Import Bank (Eximbank) A US bank established by the US government to foster trade with the USA. It provides export credit guarantees and guarantees loans made by commercial banks to US exporters.

Exports Goods or services sold to foreign countries. Export selling may be achieved by using international marketing middlemen (indirect exporting) or by a company's own overseas branch or sales representatives or by a company's agents abroad (direct exporting). In terms of the balance of payments, goods are classified as visibles, while such services as banking, insurance, and tourism are treated as invisibles. The UK has traditionally relied on its invisibles to achieve its trade balance as it tends to spend more on imports than it receives in exports.

Exports Goods or services sold to parties in foreign countries.

Expropriation The official seizure by a government of private property. Any government has the right to seize such property, according to international law, if prompt and adequate compensation is given.

Extendable bond Bond whose maturity can be extended at the option of the lender or issuer.

Extendable notes Note with maturity that can be extended by mutual agreement between the issuer and investors.

Extended guarantee A servicing and maintenance cover that a manufacturer offers to a customer for a specified additional amount of time beyond the original guarantee. Extended guarantees usually have to be paid for.

Extended trial balance A trial balance that gives a vertical listing of all the ledger account balances with three additional columns for adjustments, accruals, and payments in advance, and a final two columns (each containing a debit and a credit side) that show the entries in the profit and loss account and the balance sheet.

Extension Voluntary arrangements to restructure a firm's debt, under which the payment date is postponed.

External audit An audit of an organization carried out by an auditor who is external to, and independent of, the organization. An example would be a statutory audit carried out on behalf of the shareholders of a limited company. *Compare*internal audit.

External finance Funding that is not generated by a firm's operations: new borrowing or a stock issue.

External funds Funds originating from a source outside the corporation to increase cash flow and to aid in expansion efforts, e.g., bank loan or bond offering.

External growth The means by which a business can grow by merger, takeover, or joint ventures, rather than by growing organically through its own internal development (*see*business strategy). External growth is widely used by companies as it can offer greater speed in achieving its corporate objectives than internal development. Typically firms can increase their market share by merging with or taking over a competitor in the same field (horizontal diversification). In mature or declining markets, restructuring of combined operations can lead to cost savings. Alternatively, a company may seek to gain greater control of its supply chain by expanding through merging or taking over a supplier or distributor (vertical integration). Unrelated external expansion may take place when a firm buys into a market in which it has no existing expertise; often the time taken to develop a significant presence on its own account is considered to be too long and risky, compared to buying an established business. Although great benefits may accrue from external growth, it also carries high risks, since the expected gains may often be slow to appear as a result of the restructuring and organizational changes involved as well as the high costs of financing the merger or takeover. Joint ventures, often between rival firms, sometimes between companies from different countries, are becoming a more common form of external growth. In such cases, the parties bring complementary strengths (technology, operations, marketing) and share the risks of the venture, although differences in company and country culture may cause difficulties.

External market Also referred to as the international market, the offshore market, or, more popularly, the Euromarket. A mechanism for trading securities that at issuance.

(1) are offered simultaneously to investors in a number of countries and,

(2) are issued outside the jurisdiction of any single country.

Externalities Those economic effects of a business that are not recorded in its accounts as they do not arise from individual transactions of the business. For example, local overcrowding may arise because a large number of employees have been attracted to the neighbourhood, thus incurring extra costs for roads, schools, health care, etc. More generally, an externality results from an economic choice that is not reflected in market prices. For example, siting a railway station close to a housing estate represents an externality to householders on that estate if they are not asked to pay for it. It is an external economy if the householders benefit from greater freedom to travel and an external diseconomy if the noise of trains keeps them awake at night. It is usually argued that governments should internalize external diseconomies, such as pollution, by means of taxation.

Extraordinary call Early redemption of a revenue bond because the revenue source paying the interest on the bond has been eliminated or has disappeared.

Extraordinary general meeting (EGM) Any general meeting of a company other than the annual general meeting. Most company's articles give the directors the right to call an EGM whenever they wish. Members have the right to requisition an EGM if they hold not less than 10% of the paid-up share capital; a resigning auditor may also requisition a meeting. Directors must call an EGM when there has been a serious loss of capital. The court may call an EGM if it is impracticable to call it in any other way. Those entitled to attend must be given 14 days' notice of the meeting (21 days if a special resolution is to be proposed).

Extraordinary item An unusual and unexpected one-time event that must be explained to shareholders in an annual or quarterly report, e.g., write down for a discontinued operation, employee fraud, a lawsuit, or other one-time events. Results are often presented with and without these items. The logic of excluding these items is that investors have a better notion of future performance if one-time events are excluded. Differs from an unusual item in that extraordinary items are (1) material; (2) non-recurring; and (3) outside the ordinary nature of the business.

Extraordinary positive value A positive net present value.

Extraordinary resolution A resolution submitted to a general meeting of a company, 14 days' notice of such a resolution is required, and the notice should state that it is an extraordinary resolution. 75% of those voting must approve the resolution for it to be passed.

F

Fact book A file containing information on the history of a product, including data on sales, distribution, competition, customers, and relevant marketing research undertaken, as well as a detailed record of the product's performance in relation to the marketing effort made on its behalf.

Factoring The buying of the trade debts of a manufacturer, assuming the task of debt collection and accepting the credit risk, thus providing the manufacturer with working capital. With service factoring involves collecting the debts, assuming the credit risk, and passing on the funds as they are paid by the buyer. With service plus finance factoring involves paying the manufacturer up to 90% of the invoice value immediately after delivery of the goods, with the balance paid after the money has been collected. This form of factoring is clearly more expensive than with service factoring. In either case the factor, which may be a bank or finance house, has the right to select its debtors. *See also*undisclosed factoring.

Factory costs (Factory expenses) The expenditure incurred by the manufacturing section of an organization. Factory costs include direct materials, direct labour, direct costs, and production overheads but not mark-up or profit.

Factory overhead (indirect manufacturing costs) Those manufacturing costs of an organization that cannot be traced directly to the product. Examples are factory rent, maintenance wages, and depreciation of general production machinery. In the USA factory overhead is known as factory burden.

Factory shop A retailing operation owned and operated by a manufacturer to sell the manufacturer's surplus, discontinued, or irregular goods.

Facultative reinsurance A form of reinsurance in which the terms, conditions, and reinsurance premium is individually negotiated between the insurer and the reinsurer. There is no obligation on the reinsurer to accept the risk or on the insurer to reinsure it if it is not considered necessary. The main differences between facultative reinsurance and coinsurance is that the policyholder has no indication that reinsurance has been arranged. In coinsurance, the coinsurers and the proportion of the risk they are covering are shown on the policy schedule. Also, coinsurance involves the splitting of the premium charged to the policyholder between the coinsurers, whereas the reinsurers charge entirely separate reinsurance premiums.

Fad A fashion that enters the market quickly, is adopted with great enthusiasm, peaks early, and declines rapidly. Fashion clothing and many novelty products fall into this category.

Failure mode and effect analysis (FMEA) A technique for analysing how systems might fail and what the consequences of that failure would be. Primarily used to calculate and predict risks for safety and insurance purposes, FMEA identifies all the ways in which it is possible for each component to fail and then follows through the consequence of each type of failure. From this data it is possible to develop preventive maintenance and contingency plans.

Fallen angel A security in the US market that has dropped below its original value; it may be sold for its increased yield.

Falsification of accounts A dishonest entry in a firm's books of account, made by an employee with the object of covering up the theft of goods or money from the firm.

Family brand A group of brand names for the products of a company, all of which contain the same word to establish their relationship in the minds of the consumers. *See also*product line.

Family life cycle The six stages of family life based on demographic data:

(1) young single people;

(2) young couples with no children;

(3) young couples with youngest child under six years;

(4) couples with dependent children;

(5) older couples with no children at home;

(6) older single people. These groups have been useful in marketing and advertising for defining the markets for certain goods and services, as each group has its own specific and distinguishable needs and interests.

Fast-moving consumer goods (FMCG) Products that move off the shelves of retail shops quickly, which therefore require constant replenishing. Fast-moving consumer goods include standard groceries, etc., sold in supermarkets as well as records and tapes sold in music shops.

Fate Whether or not a cheque or bill has been paid or dishonoured. A bank requested by another bank to advise fate of a cheque or bill is being asked if it has been paid or not.

Fault trees An analytical technique used to provide a model of the way components interact during the course of the failure of a system.

Fax A widely used method of communication between businesses, linked to the international telephone system. Firms purchase or hire a Fax machine and usually install a separate telephone line although small firms can use a joint Fax/telephone line. This enables messages, copies of documents, diagrams, plans, etc, to be sent for the cost of a telephone call of equal duration (approximately 30-60 seconds per A4 page). The Fax machine itself has the appearance of a desk-top copier. The Fax system operates worldwide.

Feedforward control An approach to controlling the performance of an operating system that attempts to forecast a future state and to take the necessary action before problems arise. For example, in a statistical process control system the operation may remain in control, but knowledge of the system enables the operator to extrapolate trends in control data and to take the action necessary to avoid the system becoming out of control. The success of this approach is dependent on the ability to predict the consequences of a particular event. *Compare*concurrent control; feedback control

Fee simple The most usual freehold estate in land. Although all land in England is theoretically held by the Crown, the owner of a fee simple (or his or her heirs) will own the land forever and may dispose of it as they wish both during the lifetime of the owner and by will. The land will revert to the Crown only if the owner dies without leaving a will and with no surviving relatives. This is the only type of freehold estate that can now exist in common law, as opposed to in equity.

Fiat money Money that a government has declared to be legal tender, although it has no intrinsic value and is not backed by reserves. Most of the. world's paper money is now flat money. *Compare* fiduciary issue.

Fictitious asset An asset, usually as shown in a balance sheet, that does not exist. The reasons for showing it may be fraudulent, the asset may have ceased to exist but has not been taken out of the accounts, or it may be an asset, such as goodwill, that no longer has any value. *Compare*intangible asset.

fidelity guarantee An insurance policy covering employers for any financial losses they may sustain as a result of the dishonesty of employees. Policies can be arranged to cover all employees or specific named persons. Because of the nature of the cover, insurers require full details of the procedure adopted by the organization in recruiting and vetting new employees and they usually reserve the right to refuse to cover a particular person without giving a reason.

Fiduciary 1. Denoting a person who holds property in trust or as an executor. Persons acting in a fiduciary capacity do so not for their own profit but to safeguard the interests of some other person or persons.

2. Denoting a loan that is made on trust, rather than against some security.

Fiduciary issue 1. In the UK, the part of the issue of banknotes by the Bank of England that is backed by government securities, rather than by gold. Nearly the whole of the note issue is now fiduciary. *Compare*fiat money.

2. Formerly, in the UK, banknotes issued by a bank without backing in gold, the value of the issues relying entirely on the reputation of the issuing bank.

Field experiments Marketing tests conducted outside the laboratory in a real market environment. *See*test marketing.

Field sales manager A district or regional sales manager (so called because his or her main concern is the control of salespeople in the field). The primary task of the field sales manager is to motivate and supervise sales personnel Depending on the organization, the field sales manager may also be involved in setting salesforce objectives, designing the strategy, and recruiting, selecting, training, and evaluating the field sales personnel.

Field selling Non-retail selling that takes place outside the employer's place of business, usually on the prospective customer's premises. The sales personnel involved are the company's personal link with customers and can bring back to the company much needed information about the customers.

Fifth-generation computer A very advanced type of computer, still under development These computers will use VLSI (very large-scale integration) silicon chips, containing millions of transistors on a single chip. Their input devices will be less cumbersome and more direct than keyboards; for example, they might recognize speech; they will also incorporate artificial intelligence.

File In computer technology, a collection of related information held on backing store that is treated as a unit. A file may contain a program, which can be copied into main store memory and executed; it may contain modules from which programs are built; or it may contain data of any kind. A data file is often subdivided into records. For example, the subscription information of a magazine might be kept in a subscriptions file, with a record for each customer.

Fillér A monetary unit of Hungary, worth one hundredth of a forint.

Fils 1. A monetary unit of the United Arab Emirates, worth one hundredth of a dirham **2.** A monetary unit of Bahrain, Iraq, Jordan, and Kuwait, worth one thousandth of a dinar. **3.** A monetary unit of Yemen, worth one hundredth of a riyal and one thousandth of a dinar.

Final accounts The accounts for a company produced at the end of its financial year (see annual accounts), as opposed to any interim accounts produced during the year, often after six months. Interim accounts are for the guidance of management and are often not audited, whereas final accounts must be audited and are open for inspection by the public.

final invoice An invoice that replaces a proforma invoice for goods. The proforma invoice is sent before all the details of the goods are known. The final invoice contains any missing information and states the full amount still owing for the goods.

Finance 1. The practice of manipulating and managing money.

2. The capital involved in a project, especially the capital that has to be raised to start a new business. **3.** A loan of money for a particular purpose, especially by a finance house.

Finance Act The annual *UK Act of Parliament* that changes the law relating to taxation, giving the rates of income tax, corporation tax, etc, proposed in the preceding Budget.

Finance bill A bill of exchange used for short-term credit. It cannot be sold on to another party in the same way as a banker's acceptance.

Finance company A company that provides finance, normally in the form of loans. As it tends to finance ventures with a high risk factor, the cost of borrowing is likely to be higher than that made by a clearing bank.

Finance house An organization, many of which are owned by commercial banks, that provides finance for hirepurchase agreements. A consumer, who buys an expensive item (such as a car) from a trader and does not wish to pay cash, enters into a hire-purchase contract with the finance house, who collects the deposit and instalments. The finance house pays the trader the cash price in full, borrowing from the commercial banks in order to do so. The finance house's profit is the difference between the low rate of interest it pays to the commercial banks to borrow and the high rate it charges the consumer.

Finance lease A lease that transfers substantially all the risks and rewards of ownership of an asset to the lessee. Under Statement of Standard Accounting Practice 21, 'Accounting for Leases and Hire Purchase Contracts', the lessee should record the finance lease as an asset in the balance sheet.

Financial accounting The branch of accounting concerned with classifying, measuring, and recording the transactions of a business. At the end of a period, usually a year but sometimes less, a profit and loss account and a balance sheet are prepared to show the performance and position of the business. Financial accounting is primarily concerned with providing a true and fair view of the activities of a business to parties external to it. To ensure that this is done correctly considerable attention will be paid to accounting concepts and to any requirements of legislation, accounting standards, and (where appropriate) the regulations of the stock exchange. Financial accounting can be separated into a number of specific activities, such as conducting audits, taxation, book-keeping, and insolvency.

Financial accountants need not be qualified, in that they need not belong to an accountancy body, although the majority of those working in public practice will be. Compare management accounting.

Financial adviser 1. Anyone who offers financial advice to someone else, especially one who advises on investments. *See also* financial adviser.

2. An organization, usually a merchant bank, who advises the board of a company during a takeover

Financial analysis The use of financial statements and the calculation of ratios, to monitor and evaluate the financial performance and position of a business.

financial appraisal The use of financial evaluation techniques to determine which of a range of possible alternatives is preferred. Financial appraisal usually refers to the use of discounted cash flow techniques but it may also be applied to any other approaches used to assess a business problem in financial terms, such as ratio analysis or profitability index.

financial control (financial management) The actions of the management of an organization taken to ensure that the costs incurred and revenue generated are at acceptable levels. Financial control is assisted by the provision of financial information to management by the accountant and by the use of such techniques as budgetary control and standard costing, which highlight and analyse any variances.

Financial futures A futures contract in currencies or interest rates (*see*interest-rate futures). Unlike simple forward contracts, futures contracts themselves can be bought and sold on specialized markets. Until about 1970 trading in financial futures did not exist, although futures and options were dealt in widely on commodity markets. However, instantaneous trading across the world coupled with accelerated international capital flows combined to produce great volatility in interest rates, stock-market prices, and currency exchanges. The result has been an environment in which organizations and individuals responsible for managing large sums need a financial futures and options market both to manage risks effectively and as a source of additional profit. In the UK financial futures and options are traded on the London International Financial Futures and Options Exchange (LIFFE). *See also*hedging; index futures; portfolio insurance.

Financial institution Any organization, such as a bank, building society, or finance house, that collects funds from individuals, other organizations, or government agencies and invests these funds or lends them on to borrowers. Some financial institutions are nondeposit-taking, e.g. brokers and life insurance companies, who fund their activities and derive their income from selling securities or insurance policies, or undertaking brokerage services. At one time there was a clear distinction and regulatory division between deposittaking and non-deposit-taking financial institutions. This is no longer the case; brokers and other companies now often invest funds for their clients with banks and in the money markets.

Financial Intermediaries, Managers and Brokers Regulatory Association Ltd (FIMBRA)

Financial intermediary 1. A bank, building society, finance house, insurance company, investment trust, etc., that holds funds borrowed from lenders in order to make loans to borrowers.

2. In the Financial Services Act (1986), a person or organization that sells insurance but is not directly employed by an insurance company (e.g. a broker, insurance agent, bank).

Financial modelling The construction and use of planning and decision models based on financial data to simulate actual circumstances in order to facilitate decision making within an organization. The financial models used include discounted cash flow, economic order quantity, decision trees, learning curves, and budgetary control.

Financial Ombudsmen 1. **(Banking Ombudsman)** An official in charge of a service set up in 1986 by 19 banks, which fund the service, to investigate complaints from bank customers against the service provided by member banks. In 1993 the service was extended to deal with complaints from small businesses (turnover less than £1M).

2. (**Building Societies Ombudsman**) An official in charge of a service set up in 1987 by all the UK building societies, who fund the service, to investigate complaints from building-society customers against the services provided by the building societies. 3. **(Insurance Ombudsman**) An official in charge of a service set up in 1981 by some 200 insurance companies, who fund the service, to settle disputes between insurance policyholders and member insurance companies. The Insurance Ombudsman's Bureau also runs the Unit Trusts Ombudsman Scheme, to settle disputes between unit-trust holders and unit-trust management companies. It was set up in 1988.

4. (Investment Ombudsman) An official for resolving disputes between members of IMRO (*see* Self-Regulating Organization) and their customers. The service was set up in 1989. The Personal Investment Authority (PIA) Ombudsman is responsible for resolving complaints against PIA members about personal investments.

5. (Pensions Ombudsman) An official set up by the Pension Schemes Act (1993) to resolve grievances between individuals and their pension schemes (but not National Insurance benefits).

Financial planning The formulation of short-term and long-term plans in financial terms for the purposes of establishing goals for an organization to achieve, against which its actual performance can be measured.

Financial report The financial statements of a company.

Financial Reporting Council (FRC) A UK company limited by guarantee, set up in 1989 to oversee and support the work of the Accounting Standards Board and the *Financial Reporting Review Panel* to encourage good financial reporting generally. Its chairman and three deputies are appointed jointly by the Secretary of State for Trade and Industry and the Governor of the Bank of England. It first met in May 1990.

Financial Reporting Review Panel, (FRRP) A UK company limited by guarantee; it is a subsidiary of the Financial Reporting Council, which acts as its sole director. The panel investigates departures from the accounting requirements of the Companies Act (1985) and is empowered to take legal action to remedy any such departures. Its remit covers the financial reports of public companies and large private companies. All other companies are subject to scrutiny by the Department of Trade and Industry. The panel does not look at all company accounts but has doubtful cases drawn to its attention.

Financial Services Act (1986) A UK act of parliament that came into force in April 1988. Its purpose was to regulate investment business in the UK by means of the Securities and Investment Board and its Self-Regulating Organizations. It provided legislation for many of the recommendations of the Gower Report.

Financial Statement and Budget Report (FSBR) The document published by the Chancellor of the Exchequer on Budget Day. It summarizes the provisions of the Budget as given in the Chancellor's speech to the House of Commons.

Financial statements The annual statements's summarizing a company's activities over the last year. They consist of the Profit and loss account, balance sheet, statement of total recognized gains and losses, and if required, the cash-flow statement, together with supporting notes.

Financial Statistics A monthly publication of the UKCentral Statistical Office giving a full account of financial statistics.

Financial Times Share indexes A number of share indexes published by the *Financial Times*, daily except Sundays and Mondays, as a barometer of share prices on the London Stock Exchange. The Financial Times Actuaries Share Indexes are calculated by the Institute of Actuaries and the Faculty of Actuaries as weighted arithmetic averages for 54 sectors of the market (capital goods, consumer goods, etc.) and divided into various industries. They are widely used by investors and portfolio managers. The widest measure of the market comes from the FTA All-Share Index of some 800 shares and fixed-interest stocks (increased from 657 in October 1992), which includes a selection from the financial sector. Calculated after the end of daily business, it covers 98% of the market and 90% of turnover by value. The FTA World Share Index was introduced in 1987 and is based on 2400 share prices from 24 countries. The Financial Times Industrial Ordinary Share Index (dr FT-30) represents the movements of shares in 30 leading industrial and commercial shares, chosen to be representative of British industry rather than of the Stock Exchange as a whole; it therefore excludes banks, insurance companies, and government stocks. The index is an unweighted geometric average, calculated hourly during the day and closing at 4.30 pm. The index started from a base of 100 in 1935 and for many years was the main day-to-day market barometer. It continues to be published but it has been superseded as the index by the Financial, Times-Stock Exchange 100 Share Index (FT-SE 100 or FOOTSIE), a weighted arithmetic index representing the price of 100 securities with a base of 1000 on 3 January 1984. This index is calculated minute-byminute and its constituents, whose membership is by market capitalization, above £1 billion, are reviewed quarterly. The index was created to help to support a UK equity-market base for a futures contract. In 1992 the index series was extended to create two further realtime indexes, the FT-SE Mid 250, comprising companies capitalized between £150 million and £1 billion, and the FTSE Actuaries 350, both based on 31 December 1985. These indexes are further broken down into Industry Baskets, comprising all the shares of the industrial sectors, to provide an instant view of industry performance across the market, and corresponding roughly to sectors defined by markets in New York and Tokyo. A FT-SE Small Cap Index covers 500-600 companies capitalized between £20 million and £150 million, calculated at the end of the day's business, both including and excluding investment trusts. The Financial Times Government Securities Index measures the movements of Government stocks (gilts). The newest indexes measure the performance of securities throughout the European market. The Financial Times-Stock Exchange Eurotrack 100 Index (FT-SE Eurotrack 100) is a weighted average of 100 stocks in Europe, which started on 29 October 1990, with a base of 1000 at the close of business on 26 October 1990. Quoted in Deutschmarks, the index combines prices from SEAQ and SEAQ International with up-to-date currency exchange rates. On 25 February 1991 the Financial Times-Stock Exchange Eurotrack 200 Index was first quoted, with the same base as the Eurotrack 100 to combine the constituents of the FT-SE 100 and the Eurotrack 100

Financial year 1. Any year connected with finance, such as a company's accounting period or a year for which budgets are made up.

2. A specific period relating to corporation tax. i.e. the year beginning 1st April (the year beginning 1st April 1988 is the financial year 1988). Corporation-tax rates are fixed for specific financial years by the Chancellor in his budget; if a company's accounting period falls into two financial years the profits have to be apportioned to the relevant financial years to find the rates of tax applicable. *Compare*fiscal year.

Financier A person who uses his own money to finance a business deal or venture or who makes arrangements for such a deal or venture to be financed by a merchant bank or other financial institution.

Fine trade bill A bill of exchange that is acceptable to the Bank of England as security, when acting as lender of last resort. It will be backed by a first-class bank or finance house.

Finished goods stock (finished goods inventory) The value of goods that have completed the manufacturing process and are available for distribution to customers. In any accounting period there will be opening stock of finished goods at the beginning of the period and closing stock of finished goods at the end of the period. Methods of valuing finished goods stock are covered by Statement of Standard Accounting Practice 9 and may include first-in-firstout cost or average cost methods.

Finite capacity loading Loading based on the assumption that capacity is finite, i.e. cannot be increased. This parameter aligns load to the available capacity, enabling realistic lead times to be calculated. *Compare*infinite capacity loading

Fire insurance An insurance policy covering financial losses caused by damage to property by fires. Most fire insurance policies also include cover for damage caused by lightning and explosions of boilers or gas used for domestic purposes. On payment of an additional premium, fire insurers are usually prepared to widen the cover to include such special perils as those arising from various weather-related or man-made causes.

Firewall A system usually inserted between a local area network and the Internet to filter incoming traffic, to try to eliminate viral infection, and to restrict the access of hackers to the system. It may create a bottleneck in the transmission of data to and from the system.

Firm 1. Any business organization.

2. A business partnership.

Firm commitment 1. An undertaking by a bank to lend up to a maximum sum over a period at a specified rate; a commitment fee usually has to be paid by the borrower, which is not returned if the loan is not taken up. **2**. An agreement in which an underwriter of a flotation in the USA assumes the risk that not all the securities issued will be sold, by guaranteeing to buy all the excess securities at the offer price.

Firm offer An offer to sell goods that remains in force for a stated period. For example, an 'offer firm for 24 hours' binds the seller to sell if the buyer accepts the offer within 24 hours. If the buyer makes a lower bid during the period that the offer is firm, the offer ceases to be valid. An offer that is not firm is usually called a quotation in commercial terms.

Firm order An order to a broker (for securities, commodities, currencies, etc.) that remains firm for a stated period or until cancelled. A broker who has a firm order from a principal does not have to refer back if the terms of the order can be executed in the stated period.

Firmware Computer programs or data that are stored in a memory chip. These are often built into a computer to make it unnecessary to load the programs or data from disk or from the keyboard. Wordprocessing programs, for example, on read-only memory (ROM) chips are built into some business computer systems. Firmware is also used where absolute reliability is required, for example in air-traffic control.

First-class paper (fine paper) A bill of exchange, cheque, etc., drawn on or accepted or endorsed by a first-class bank, finance house, etc.

First-in-first-out cost (FIFO cost) A method of valuing units of raw material or finished goods issued from stock based on using the earliest unit value for pricing the issues until all the stock received at that price has been used up. The next latest price is then used for pricing the issues, and so on. Because the issues are based on a FIFO cost, the valuation of closing stocks is described as being on the same FIFO basis. The method may also be used in process costing to value the work in process at the end of an accounting period.

First-loss policy A property insurance policy in which the policyholder arranges cover for an amount below the full value of the items insured and the insurer agrees not to penalize him for under-insurance. The main use of these policies is in circumstances in which a total loss is virtually impossible For example, a large warehouse may contain £2.5M worth of wines and spirits but the owner may feel that no more than £500,000 worth could be stolen at any one time. The solution is a first-loss policy that deals with all claim up to £500,000 but pays no more than this figure if more is stolen. First-loss policies differ from coinsurance agreements with the policyholders because the insured is not involved in claims below the first-loss level and the premiums are not calculated proportionately. In the above example, the premium might be as much as 80-90% of the premium on the full value.

First mortgage debenture A debenture with the first charge over property owned by a company. Such debentures are most commonly issued by property companies.

First notice day 1. The date on which a seller in a futures market informs the clearing house of an intention to deliver according to the terms of a particular futures contract.

2. The date on which the clearing house notifies the buyer of such an arrangement.

First-year allowance A capital allowance, at a rate of 40%, available during the period 1 November 1992 to 31 October 1993 in place of the writingdown allowance. It was also available before 1986 at varying rates. The allowance was given for the basis period provided the expenditure was incurred between 1 November 1992 and 31 October 1993.

Fiscal agent 1. A third party who acts on behalf of a bond issuer to pay subscribers to the issue and generally assist the issuer. **2.** An agent for one of the national financial institutions in the USA who acts as an adviser, for example, to the National Mortgage Association for its debt securities. **3.** In the USA, an agent empowered to collect taxes, revenues, and duties on behalf of the government. For example, the Federal Reserve Banks perform this function for the US Treasury.

Fiscal year In the UK, the year beginning on 6. April in one year and ending on 5 April in the next (the fiscal year 1992/93 runs from 6 April 1992 to 5 April 1993). This will change in 1994 so that it coincides with the calendar year, running from 1 January to 31 December. Income tax, capital-gains tax, and annual allowances for inheritance tax are calculated for fiscal years, and the UK Budget estimates refer to the fiscal year. In the USA it runs from 1 July to the following 30 June. The fiscal year is sometimes called the tax year or the year of assessment. *Compare* financial year.

Fixed capital The amount of capital tied up in the capital assets of an organization. *Compare*circulating capital.

Fixed capital formation An investment over a given period, as used in the national income accounts. It consists primarily of investment in manufacturing and housing. **Gross fixed capital formation** is the total amount of expenditure on investment, while net fixed capital formation includes a deduction for the depreciation of existing capital.

Fixed charge 1. (specific charge) A charge in which a creditor has the right to have a specific asset sold and applied to the repayment of a debt if the debtor defaults on any payments. The debtor is not at liberty to deal with the asset without the charge-holder's consent. *Compare*floating charge

2. The part of an expense that remains unchanged, irrespective of the amount of the commodity or service used or consumed. For example, in the UK both the electricity and gas industries operate tariffs consisting of a fixed charge, which remains unchanged irrespective of the consumption of energy, and a variable charge, based on the energy consumed.

fixed cost (fixed expense) An item of expenditure that remains unchanged, in total, irrespective of changes in the levels of production or sales. Examples are business rates, rent, and some salaries. *Compare*fixed production overhead; overhead costs; semi-variable cost; variable cost.

Fixed debenture A debenture that has a fixed charge as security. *Compare* floating debenture.

fixed exchange rate A rate of exchange between one currency and another that is fixed by government and maintained by that government buying or selling its currency to support or depress its currency. *Compare*floating exchange rate.

Fixed-interest security A type of security that gives a fixed stated interest payment once or twice per annum. They include gilt-edged securities, bonds, Preference shares, and debentures; as they entail less risk than equities they offer less scope for capital appreciation. They do, however, often give a better yield than equities.

The prices of fixed-interest securities tends to move inversely with the general level of interest rates, reflecting changes in the value of their fixed yield relative to the market. Fixed-interest securities tend to be particularly poor investments at times of high inflation as their value does not adjust to changes in the price level. To overcome this problem some gilts now give index-linked interest payments.

Fixed overhead cost The elements of the indirect costs of an organization's product that, in total, remain unchanged irrespective of changes in the levels of production or sales. Examples include administrative salaries, sales personnel salaries, and factory rent.

Fixed production overhead The elements of an organization's factory overheads that, in total, remain unchanged irrespective of changes in the level of production or sales. Examples include factory rent, depreciation of machinery using the straight-line method, and the factory manager's salary.

Fixed-rate loan A loan in which the interest rate is fixed at the start of the loan. It is standard for bond issues, but unusual for bank borrowing.

Fixed-rate mortgage A mortgage in which the rate of interest paid by the borrower is fixed, usually for the first few years of the loan.

Fixed spot A television advertising spot for which a premium is paid (normally 15%) to ensure that it is transmitted in a preselected commercial break during a particular programme.

Fixtures and fittings Items normally forming part of the setting in which an organization conducts its business, as distinct from the plant and machinery it uses in conducting the business.

Flag of convenience The national flag of a small country, such as Panama, Liberia, Honduras, or Costa Rica, flown by a ship that is registered in one of these countries, although the ship is owned by a national of another country. The practice of registering ships in these countries, and sailing them under flags of convenience, grew up in the post-war years and still continues. The object for the shipowners is to avoid taxation and the more stringent safety and humanitarian conditions imposed on ships and their crews by the larger sea-going nations.

Flash report In the USA, a management report that highlights key data for corrective action.

Fleet rating A single special premium rate quoted by an insurer for covering the insurance on a number of ships or vehicles owned by one person or company, rather than considering each one individually. The fleet need not consist of identical vehicles or vessels but common ownership is essential A common method of fleet rating is to examine the claims history of the fleet against the total premium. In this way one fleet member may have several claims (which would individually merit a premium adjustment) but if the rest of the fleet is claim-free no adjustment need be made. Insurers vary on the minimum number constituting a fleet.

Flexibility The ability to adapt an operating system to respond to changes in the environment. Increasingly seen as a source of competitive advantage in a rapidly changing market, it is an area of operations management in which Japanese practices have had a major impact. Flexibility has two dimensions: how quickly and how far an organization can change. Changes may be called for in terms of the product or service, i.e. the flexibility to introduce new products; mix flexibility is the ability to provide a wide range or mix of products; volume flexibility is the ability to produce different volumes to suit demand; delivery flexibility is the ability to change (usually advance) the delivery date.

Flexible budget A budget that takes into account circumstances resulting in the actual levels of activity achieved being different from those on which the original budget was based Consequently, in a flexible budget the budget cost allowances for each cost item are changed for the variable items to allow for the actual levels of activity achieved. This flexible budget compares with a fixed budget, in which no flexibility is included.

Flexible drawdown The drawdown in stages of funds available under a credit agreement.

Flexible manufacturing systems (FMS) Systems that integrate computer-aided manufacturing, numerically controlled machines, robotics, and automated guided vehicles. In this context 'flexible' is a relative term. The ability to reprogramme machines by means of software rather than the physical set-up makes FMS particularly applicable to handling batches of a fairly small range of products.

Flexitime (flexihours) A system of working, especially in offices, in which employees are given a degree of flexibility in the hours they work. Provided they work an agreed number of hours per day, they may start or finish work at different times. The object is usually to reduce time spent in rush-hour travelling.

Flip-flop FRN A perpetual floatingrate note (i.e. one without redemption), which can be converted into a shortterm note of up to four years' maturity and back again into a perpetual FRN.

Float 1. In the USA, the proportion of a corporation's stocks that are held by the public rather than the corporation.

2. Money created as a result of a delay in processing cheques, e.g. when one account is credited before the paying bank's account has been debited.

3. Money set aside as a contingency fund or an advance to be reimbursed.

4. *See* slack time.

Floating charge A charge over the assets of a company; it is not a legal charge over its fixed assets but floats over the charged assets until crystallized by some predetermined event. For example, a floating charge may be created over all the assets of a company, including its trading stock. The assets may be freely dealt with until a crystallizing event occurs, such as the company going into liquidation. Thereafter no further dealing may take place but the debt may be satisfied from the charged assets. Such a charge ranks in priority after legal charges and after preferred creditors in the event of a winding-up. It must be registered.

Floating exchange rate A rate of exchange between one currency and others that is permitted to float according to market forces. Most major currencies and countries now have floating exchange rates but governments and central banks intervene, buying or selling currencies when rates become too high or too low.

Floating policy An insurance policy that has only one sum insured although it may cover many items. No division of the total is shown on the policy and the policyholder is often able to add or remove items from the cover without reference to the insurers, provided that the total sum insured is not exceeded. Cover for contractors' plant and machinery is often arranged on this basis, because it enables them to purchase specialist equipment for a particular contract, without having to contact the insurers on every occasion.

Floating-rate certificate of deposit (FRCD) A certificate of deposit with a variable interest rate, normally for interbank lending purposes. Commonly bearing a one-year maturity, these certificates have interest rates that may be adjusted (e.g. every 90 days); they are usually expressed in eurodollars at a rate linked to LIBOR.

Floating-rate interest An interest rate on certain bonds, certificates of deposit, etc, that changes with the market rate in a predetermined manner, usually in relation to the base rate.

Floating-rate loan A loan that does not have a fixed interest rate throughout its life. Floating-rate loans can take various forms but they are all tied to a short-term market indicator; in the UK this is usually the London Inter Bank Offered Rate.

Floating-rate note (FRN) A eurobond with a floating-rate interest, usually based on the London Inter Bank Offered Rate. They first appeared in the 1970s and have a maturity of between 7 and 15 years. They are usually issued as negotiable bearer bonds. A perpetual FRN has no redemption.

Floating warranty A guarantee given by one person to another that induces this other person to enter into a contract with a third party. For example, a car dealer may induce a customer to enter into a hire-purchase contract with a finance company. If the car does not comply with the dealer's guarantee, the customer may recover damages from the dealer, on the basis of the hire-purchase contract, even though the dealer is not a party to that contract.

Floor trader A member of a stock exchange, commodity market, Lloyd's, etc., who is permitted to enter the dealing room of these institutions and deal with other traders, brokers, underwriters, etc. Each institution has its own rules of exclusivity, but in many, computer dealing is replacing face-to-face floor trading.

Florin (Af) The standard monetary unit of Aruba, divided into 100 cents.

Flotation The process of launching a public company for the first time by inviting the public to subscribe in its shares (also known as 'going public'). It applies both to private and nationalized share issues, and can be carried out by means of an introduction, issue by tender, offer for sale, placing, or public issue. After flotation the shares can be traded on a stock exchange. Flotation allows the owners of the business to raise new capital or to realize their investments. In the UK, flotation can be either on the main market through a full listing or on the alternative investment market, where less stringent regulations apply.

Flotsam 1. Items from a cargo or ship that is itself floating on the sea after a shipwreck. In British waters, it can be claimed by the ship's owners for 366 days after the shipwreck; thereafter it belongs to the Crown. *Compare*jetsam.

2. In marine insurance, items or rights that are lost as a result of breach of warranty. In such cases the policyholder has no right to make an insurance claim.

Flow chart A chart used to provide a model for the flow of material, information, or people through a system. The primary purpose of a flow chart is to identify the activities taking place within a system and to enable this information to be presented graphically. Flow charts can be used in the initial design 'layout of a process, to diagnose problems, or to optimize existing processes by, for example, identifying duplication of effort or elements in which no value is added. The principles of flow charting now form the basis for a range of computer packages enabling production and service environments to be simulated with 'what-if' scenarios. There are a number of conventional symbols used in flow charts. The important ones are the process box, which indicates a process taking place, and the decision lozenge, which indicates where a decision is needed.

Focus group An exploratory research group of 8 to 12 participants, led by a **moderator**, who meet for an in-depth discussion on one particular topic or concept. The participants are generally chosen for their relevance to the particular topic or concept By skilful use of probes and other interviewing devices, the group members are encouraged to respond in depth to the moderator. These group discussions largely depend for their success on the moderator, who must establish the right atmosphere in order to gain the maximum involvement and cooperation of all the group members. Focus groups are useful in the early stages of developing methods to understand the nature of business or organizational problems and for suggesting issues that should be covered in a questionnaire survey. Discussions are usually tape-recorded and (increasingly) videoed and observed through seethrough mirrors (with the full knowledge of the group members). Normally, these groups last for between one and three months.

Focus-group facility A facility belonging to a marketing research agency consisting of a conference room with a separate observation room. The facility could have audiovisual recording equipment, enabling observers to see and hear what is happening in the conference room, and sometimes seethrough mirrors. *See*focus group.

Focus-group moderator A person appointed by a marketing research agency on behalf of a client to lead a focus group. This person may need a background in psychology or sociology or, at least, in marketing.

Follow-the-leader strategy A pricing strategy in which an organization sets prices at the level established by the market leader. This strategy is especially common in organizations in weak competitive positions. It applies to both price increases and reductions.

Forced sale A sale that has to take place because it has been ordered by a court or because it is necessary to raise funds to avoid bankruptcy or liquidation.

Forced saving A government measure imposed

on an economy with a view to increasing savings and reducing expenditure on consumer goods. It is usually implemented by raising taxes, increasing interest rates, or raising prices.

Force majeure (French: superior force) An event outside the control of either party to a contract (such as a strike, riot, war, act of God) that may excuse either party from fulfilling his contractual obligations in certain circumstances, provided that the contract contains a force majeure clause. If one party invokes the force majeure clause the other may either accept that it is applicable or challenge the interpretation. In the latter case an arbitration would be involved.

Foreclosure The legal right of a lender of money if the borrower fails to repay the money or part of it on the due date. The lender must apply to a court to be permitted to sell the property that has been held as security for the debt. The court will order a new date for payment in an order called a foreclosure nisi. If the borrower again fails to pay, the lender may sell the property. This procedure can occur when a mortgagor fails to pay the mortgagee (bank, building society, etc.) the mortgage instalments, in which the security is the house in which the mortgagor lives. The bank, etc., then forecloses the mortgage, dispossessing the mortgagor.

Foreign-exchange broker A broker who specializes in arranging deals in foreign currencies on the foreignexchange markets. Most transactions are between commercial banks and governments. Foreign-exchange brokers do not normally deal direct with the public or with firms who require foreign currencies for buying goods abroad (who buy from commercial banks). Foreignexchange brokers earn their living from the brokerage paid on each deal.

Foreign-exchange dealer A person who buys and sells foreign exchange on a foreign-exchange market, usually as an employee of a commercial bank. Banks charge fees or commissions for buying and selling foreign exchange on behalf of their customers; dealers may also be authorized to speculate in forward exchange rates.

Foreign-exchange market An international market in which foreign currencies are traded. It consists primarily of foreign-exchange dealers employed by commercial banks (acting as principals) and foreign-exchange brokers (acting as intermediaries). Although tight exchange controls have been abandoned by many governments, including the UK government, the market is not entirely free in that the market is to some extent manipulated by the Bank of England on behalf of the government, usually by means of the exchange equalization account. Currency dealing has a spot market for delivery of foreign exchange immediately and a forward-exchange market. (*see* **Forward-exchange contract**) in which transactions are made for foreign currencies to be delivered at agreed dates in the future. This enables dealers, and their customers who require foreign exchange in the future, to hedge their purchases and sales. Options on future exchange rates can also be traded in London through the London International Financial Futures and Options Exchange.

Foreign investment Investment in the domestic economy by foreign individuals or companies. Foreign investment takes the form of either direct investment in productive enterprises or investment in financial instruments, such as a portfolio of shares. Countries receiving foreign investment tend to have a mixed attitude towards it. While the creation of jobs and wealth is welcome, there is frequently antagonism on the grounds that the country is being 'bought by foreigners'. Similarly, there is frequently resentment in the country whose nationals invest overseas, especially if there is domestic unemployment. Nevertheless, foreign investment is increasingly important in the economy of the modern world.

Foreign trade multiplier The effect that an increase in home demand has on a country's foreign trade. The primary effect is to increase that country's imports of raw materials. A secondary effect may be an increase in exports because the increased home demand enables manufacturers to be more competitive and also because the countries supplying the increased quantities of imports have more foreign exchange available to increase *their* imports.

Forensic accounting Accounting that involves litigation. In such circumstances accountants may be called on to provide expert investigations and evidence.

Forwardation A situation in a commodity market in which spot goods can be bought more cheaply than goods for forward delivery, enabling a dealer to buy spot goods and carry them forward to deliver them against a forward contract.

Forward dealing Dealing in commodities, securities, currencies, freight, etc., for delivery at some future date at a price agreed at the time the contract (called a forward contract) is made. This form of trading enables dealers and manufacturers to cover their future requirements by hedging their more immediate purchases. *See*futures contract.

Forward delivery Terms of a contract in which goods are purchased for delivery at some time in the future (*compare* spot goods). Commodities may be sold for forward delivery up to one year or more ahead, often involving shipment from their port of origin.

Forward-exchange contract An agreement to purchase foreign exchange at a specified date in the future at an agreed exchange rate. In international trade, with floating rates of exchange, the forward-exchange market provides an important way of eliminating risk on future transactions that will require foreign exchange. The buyer on the forward market gains by the certainty such a contract can bring; the seller, by buying and selling exchanges for future delivery, makes a market and earns his living partly from the profit he makes by selling at a higher price than that at which he buys and partly by speculation. There is also an active option market in forward foreignexchange rates. *See*foreign-exchange market.

Forward points (forward differential; forward margin) The amount to be added to or deducted from the spot foreign-exchange rate to calculate the forward exchange rate.

Forward price The fixed price at which a given amount of a commodity, currency, or a financial instrument is to be delivered on a fixed date in the future. A forward contract differs from a futures contract in that each forward deal stands alone.

Forward-rate agreement (FRA) 1. A contract between two parties that determines the rate of interest that will apply to a future loan or a deposit, which may or may not materialize. 2. A specified amount of a specified currency to be exchanged on an agreed future date at a specified rate of exchange.

Fractional banking A banking practice that some governments impose on their banks, calling for a fixed fraction between cash reserves and total liabilities. If governments increase the ratio of reserves to deposits, this indicates a tighter credit policy. In the USA large banks have to keep up to 12% of their deposits with their regional Federal Reserve Bank.

Franked investment income Dividends and other distributions from UK companies that are received by other companies. The principle of the imputation system of taxation is that once one company has paid corporation tax, any dividends it pays can pass through any number of other companies without carrying a further corporation-tax charge, hence the term 'franked'. The amount of tax credit included in the franked investment income can reduce the amount of advance corporation tax that the recipient company has to pay on its own dividends. Where franked investment income exceeds franked payments, the excess is carried forward to future accounting periods and can be set off against future franked payments (e.g. dividends). If the company has unused trading losses that could be carried back, a terminal loss, unused capital allowances, or a deduction for charges on income (e.g. debenture interest) then a claim can be made for the repayment of any unused tax credit.

Franked payment A dividend or other distribution from a UK company together with the amount of advance corporation tax (ACT) attributable to the dividend. Thus, if the basic rate of income tax is 25%, a franked payment is the dividend actually paid plus ? of it, i.e. it is a grossed-up dividend. In any accounting period ACT is actually paid on franked payments less franked investment income at the basic rate of income tax.

Franking machine A machine supplied by the Post Office to businesses who wish to frank their own mail. Instead of taking mail to a post office and using gummed stamps, the business uses the machine to produce slips of paper that state the postage and the date. The machine records the amount of money spent, which has to be remitted to the Post Office.

Fraud A false representation by means of a statement or conduct, in order to gain a material advantage. A contract obtained by fraud is voidable on the grounds of fraudulent misrepresentation. If a person uses fraud to induce someone to part with money he or she would not otherwise have parted with, this may amount to theft. *See also*fraudulent conveyance; fraudulent preference; fraudulent trading.

Fraudulent conveyance The transfer of property to another person for the purposes of putting it beyond the reach of creditors. For example, if A transfers his house into the name of his wife because he realizes that his business is about to become insolvent, the transaction may be set aside by the court under the provisions of the *Insolvency Act* (1986).

Fraudulent preference Putting a creditor of a company into a better position than he would have been in, at a time when the company was unable to pay its debts. If this occurs because of an act of the company within six months of winding-up (or two years if the preference is given to a person connected with the company), an application to the court may be made to cancel the transaction. The court may make any order that it thinks fit, but no order may prejudice the rights of a third party who has acquired property for value without notice of the preference.

Fraudulent trading The carrying on of the business of a company with intent to defraud creditors or for any other fraudulent purpose. This includes accepting money from customers when the company is unable to pay its debts and cannot meet its obligations under the contract. The liquidator of a company may apply to the court for an order against any person who has been a party to fraudulent trading to make such contributions to the assets of the company as the court thinks fit. Thus an officer of the company may be made personally liable for some of its debts. 'Fraudulent' in this context implies actual dishonesty or real moral blame; this definition has limited the usefulness of the remedy as fraud is notoriously difficult to prove. *See*wrongful trading.

Free alongside ship (fas) The basis of an export contract in which the seller pays for sending the goods to the port of shipment but not for loading them onto the ship. The seller must cover all insurance up to this point and any lighterage required. This may be the same as free on quay (FOQ) or free alongside quay (faq) as long as the ship can reach the dock If it is unable to, the buyer has to pay for lighterage.

Free asset ratio The ratio of the market value of an insurance company's assets to its liabilities.

Free competition (free economy) An economy in which the market forces of supply and demand control prices, incomes, etc, without government interference. In such an economy private enterprise is dominant, the public sector being very small.

Free depreciation A method of granting tax relief to organizations by allowing them to charge the cost of fixed assets against taxable profits in whatever proportions and over whatever period they choose. This gives businesses considerable flexibility, enabling them to choose the best method of depreciation depending on their anticipated cash flow, profit estimates, and taxation expectations.

Free docks The basis of an export contract in which the exporter pays to deliver the goods to the shipping dock, but the buyer arranges and pays for loading and shipping.

Freefone A telephone service provided to businesses by British Telecom in which the business pays for incoming calls. A customer dials the operator and asks to be connected to a specified Freefone number knowing that he will not have to pay for the call. This service is much used by mail-order houses.

Freehold An estate in land that is now usually held in fee simple. Land that is not freehold will be leasehold land.

Free in and out Denoting a selling price that includes all costs of loading goods (into a container, road vehicle, ship, etc) and unloading them (out of the transport).

Free list A list, issued by the Customs and Excise, of the goods that can be imported into the UK without paying duty.

Free lunch Economists' jargon for a nonexistent benefit. It derives from a 19th-century tavern that advertised free food, it being clearly understood that anyone attempting to exploit this offer without buying a drink would be thrown out. The phrase is now used to reflect the economist's belief that wherever there appears to be a free benefit, someone, somewhere, always pays for it.

Free market 1. A market that is free from government interference, prices rising and falling in accordance with supply and demand. **2.** A security that is widely traded on a stock exchange, there being sufficient stock on offer for the price to be uninfluenced by availability. **3.** A foreign-exchange market that is free from pegging of rates by governments, rates being free to rise and fall in accordance with supply and demand.

Free of all averages (faa) Denoting a marine insurance that covers only a total loss, general average and particular average losses being excluded.

Free of capture Denoting a marine insurance in which capture, seizure, and mutiny are excluded. This usually applies to policies taken out in wartime.

Free on board and trimmed Denoting an FOB contract in the coal trade in which the seller is responsible for delivering the coal to the ship, paying for it to be loaded, and ensuring that it is correctly stowed.

Free port A port, such as Bremerhaven, Gdansk. Rotterdam, or Singapore, that is free of customs duties. The area around the port, known as a foreign trade zone or free zone, specializes in entrepôt trade, as goods can be landed and warehoused before re-export without the payment of customs duties.

Freepost A UK business reply service provided by the Post office. Members of the public reply to advertisements and the advertiser pays the cost of postage after the reply is delivered.

Freesheets Local newspapers and magazines that are published and distributed without charge. As they are entirely dependent upon advertising revenue they must ensure high circulation and low costs to survive. For this reason, the advertising to editorial ratio is exceptionally high.

Free trade The flow of goods and services across national frontiers without the interference of laws, tariffs, quotas, or other restrictions. Since World War II, under the auspices of GATT, trade barriers have continuously, if slowly, fallen.

Freeze-out Pressure applied to minority shareholders of a company that has been taken over, to sell their stock to the new owners.

Freight 1. The transport of goods by sea (sea freight) or air (air freight). 2. The goods so transported. 3. The cost of shipping goods for a particular voyage by sea or by air. Freight is charged on the basis of weight (*see*deadweight cargo) or the volume occupied, described as weight or measurement. Usually the rate is quoted per tonne or per cubic metre, the freight paid being whichever is the greater amount. Air freights are usually quoted on the basis of per kilo or per 7000 cubic centimetres, whichever is the greater. Certain cargoes are charged on an ad valorem basis, expressed as a percentage of the FOB value. Freight is normally paid when the goods are delivered for shipment but in some cases it is paid at the destination (freigth forward).

Freight forward Denoting a shipment of goods in which the freight is payable by the buyer at the port of destination.

Freight insurance A form of marine or aviation insurance in which a consignor or a consignee covers loss of sums paid in freight for the transport of goods.

Freight note An invoice from a shipowner to a shipper showing the amount of freight due on a shipment of goods.

Freight release An endorsement on a bill of lading made by a shipowner or his agent, stating that freight has been paid and the goods may be released on arrival. Sometimes the freight release is a separate document providing the same authority.

Frequency 1. The number of times the average person in a target market is exposed to an advertising message during a given period. 2. The number of times an advertisement is repeated in a given medium (television, radio, etc.) within a given period, usually a month.

Frictional unemployment The amount of unemployment consistent with the efficient operation of an economy. Frictional unemployment exists because of lags between workers leaving one job and taking up another and because there are times of the year when many new workers (such as school leavers) enter the labour market; in these circumstances there is some delay in finding them all jobs.

Friendly Society A UK non-profitmaking association registered as such under the *Friendly Society Acts* (1896- 1992). Mutual insurance societies, dating back to the 17th century, they were widespread in the 19th and 20th centuries, until many closed in 1946 after the introduction of National Insurance. Some developed into trade unions and some large insurance companies are still registered as Friendly Societies. The *1992 Act* created a new framework for Friendly Societies, enabling them to extend their services and to compete with other organizations. The Act also established the Friendly Society Commission, to regulate their activities. In 1995 the Commission and the Treasury issued a document proposing changes to the *1992 Act*.

Fringe benefits 1. Non-monetary benefits offered to the employees of a company in addition to their wages or salaries. They include company cars, expense accounts, the opportunity to buy company products at reduced prices, private health plans, canteens with subsidized meals, luncheon vouchers, cheap loans, social clubs, etc. Some of these benefits, such as company cars, do not escape the tax net.

2. Benefits, other than dividends, provided by a company for its shareholders. They include reduced prices for the company's products or services, Christmas gifts, and special travel facilities.

Front-end fee A fee payable by a company that has borrowed a sum of money. It is paid shortly after signing the loan agreement and relates to the full amount of the loan, regardless of usage, subsequent or early repayment. It is usually made up of four components: a lead management fee; general management fee; underwriting fee; and the participation fee.

Front-end loading The initial charge to cover administrative expenses and commission included in the first payment of a loan instalment, unit trust investment, or insurance premium. The effect of this is to increase the first payment in relation to subsequent payments. Therefore, if an investment is being made on behalf of a client, the investment is the initial payment made by the client less the front-end load.

Front office-back office The division between those elements of an operating system that are transparent to the customer and with which they can interact (front office), and those that are hidden (back office). In the service sector there is a trade-off between the benefits that accrue from close and open relationships with the customer and the greater ease with which operations can be controlled when they are out of sight. Application of information technology to many service operations has allowed the customer to experience what appears to be a customized service, while at the same time allowing standardization of procedures.

Front running The practice by market makers on the London Stock Exchange of dealing on advance information provided by their brokers and investment analysis department, before their clients have been given the information. *See also* Chinese wall.

Frozen assets Assets that for one reason or another cannot be used or realized. This may happen when a government refuses to allow certain assets to be exported.

Frustration of contract The termination of a contract as a result of an unforeseen event that makes its performance impossible or illegal. A contract to sell an aircraft could be frustrated if it crashed before the contract was due to be implemented. Similarly an export contract could be frustrated if the importer was in a country that declared war on the country of the exporter.

FSBR Abbreviation for Financial Statement and Budget Report.

FTC Abbreviation for Federal Trade Commission.

FT Cityline A telephone service giving information on the Financial Times Ordinary Share Index (*see*Financial Times Share Indexes). The information is updated seven times each day.

FTII Abbreviation for Fellow of the Chartered Institute of Taxation. In order to achieve this qualification a thesis on an aspect of UK taxation has to be submitted and accepted as being of the appropriate standard

FTP (file transfer protocol) 1. The protocol that controls the way that files move across the Internet from one computer to another.

2. An application program that moves files using file transfer protocol.

FT-SE Eurotrack Indexes *See* Financial Times Share Indexes.

FT Share Indexes *See* **Financial Times Share Indexes.**

Full consolidation A method of accounting that allows a parent company to show in its balance sheet the assets and liabilities of subsidiaries at full market value and to allow for such items as goodwill.

Full cost pricing A method of setting the selling prices of a product or service that ensures the price is based on all the costs likely to be incurred in its supply.

Full employment The state in which all the economic resources of a country (and more specifically, its labour force) are fully utilized.

Full-line strategy (deep-line strategy) A strategy that involves a company in offering a large number of variations of its products. Companies employing a full-line strategy are seeking both a high market share and market growth. Leading detergent manufacturers offer up to 15 brands of their detergents.

Full listing A description of a company whose shares appear on the Official List of the main market of the London Stock Exchange. *See*listing requirements.

Full-text retrieval The ability to obtain a full copy of an article or document from a newspaper, journal or organizational source by means of the Internet. Many electronic information systems allow users to search for references and citations but for financial and copyright reasons it is often necessary to obtain hard copy from conventional sources. With the advent of easy worldwide communication and the increased use of CD-ROM as a storage medium, full-text retrieval is becoming more practical and necessary. fully diluted earnings per share The earnings per share for a company that takes into account the number of shares in actual issue as well as those that may be issued as a result of such factors as convertible loans and options or warranties.

Fully paid share A share in a company in which all calls for payment have been paid. The total paid will be the par value plus any premium.

Function A section or department of an organization that carries out a discrete activity, under the control of a manager or director. It is the section of the business for which functional budgets are produced. Examples of separate functions are production, sales, finance, and personnel.

Functional budget A financial or quantitative statement prepared for a function of an organization; it summarizes the policies and the level of performance expected to be achieved by that function for a budget period.

Functional strategy The contribution to the overall organizational objectives made by individual functions. Functional strategy is particularly concerned with the management of functions to achieve the optimum outputs and to maximize systems synergy.

Function costing The technique of collecting the costs of an organization by function and presenting them to the functional management in operating statements on a regular basis.

Functions of money In economics, money fulfils the functions of acting as a medium of exchange, a unit of account, as well as a store of value. None of the functions occur in an economy based on barter.

Fund 1. A resource managed on behalf of a client by a financial institution.

2. A separate pool of monetary and other resources used to support designated activities.

Fundamental term A term in a contract that is of such importance to the contract that to omit it would make the contract useless. Fundamental terms cannot be exempted from a contract, whereas lesser terms, known as conditions and warranties, can be exempted by mutual consent.

Funded debt The part of the national debt that the government is under no obligation to repay by a specified date. This consists mostly of Consols. *See also* funding operations.

Funded pension scheme A pension scheme that pays benefits to retired people from a pension fund invested in securities. The profits produced by such a fund are paid out as pensions to the members of the scheme.

Funding Paying short-term debt by arranging long-term borrowing. *See* funding operations.

Funding operations 1. The replacement of short-term fixed-interest debt (floating debt) by long-term fixed-interest debt (funded debt). This is normally associated with the government's handling of the National Debt through the operations of the Bank of England. The bank buys Treasury bills and replaces them with an equal amount of longerterm government bonds, thus lengthening the average maturity of government debt. This has the effect of tightening the monetary system, as Treasury bills are regarded by the commercial banks as liquid assets while bonds are not.

2. A change in the gearing of a company, in which short-term debts, such as overdrafts, are replaced by longer-term debts, such as debentures.

Fund manager (investment manager) An employee of one of the larger institutions, such as an insurance company, investment trust, or pension fund, who manages its investment fund. The fund manager decides which investments the fund shall hold, in accordance with the specified aims of the fund, e.g. high income, maximum growth, etc.

Fund of funds A unit trust belonging to an institution in which most of its funds are invested in a selection of other unit trusts owned by that institution. It is designed to give maximum security to the small investor by spreading the investment across a wide range.

Fungible issue A bond issued on the same terms and conditions as a bond previously issued by the same company. It has the advantage of having paperwork consistent with the previous bond and of increasing the depth of the market of that particular bond (*see*deep market; thin market). The gross redemption yield on the fungible issue will probably be different from that of the original issue, which is achieved by issuing the bond at a discount or a premium.

Fungibles 1. interchangeable goods, securities, etc., that allow one to be replaced by another without loss of value. Bearer bonds and banknotes are examples. 2. Perishable goods the quantity of which can be estimated by number or weight.

Futures contract An agreement to buy or sell a fixed quantity of a particular commodity currency, or security for delivery at a fixed date in the future at a fixed price. Unlike an option, a futures contract involves a definite purchase or sale and not an option to buy or sell; it therefore may entail a potentially unlimited loss. However, futures provide an opportunity for those who must purchase goods regularly to hedge against changes in price. For hedging to be possible there must be speculators willing to offer these contracts; in fact trade between speculators usually exceeds the amount of hedging taking place by a considerable amount. In London, futures are traded in a variety of markets. Financial futures are traded on the London International Financial Futures and Options Exchange; the London Commodity Exchange, which incorporates the Baltic International Freight Futures Exchange, deals with shipping and cocoa, coffee, and other foodstuffs; the London Metal Exchange with metals; and the International Petroleum Exchange with oil. In these futures markets, in many cases actual goods (*see*actuals) do not pass between dealers, a bought contract being cancelled out by an equivalent sale contract, and vice versa;

money differences arising as a result are usually settled through a clearing house (*see also*London Clearing House). In some futures markets only brokers are allowed to trade; in others, both dealers and brokers are permitted to do so. *See also*forward-exchange contract.

Future value The value that a sum of money (the present value) invested at compound interest will have in the future. If the future value is F, and the present value is P, at an annual rate of interest r, compounded annually for n years, $F = P(1 + r)^n$. Thus a sum with a present value of £1000 will have a future value of £1973.82 at 12% pa, after 6 years.

G

Game theory A mathematical theory, developed by J von Neumann (1903-57) and O Morgenstern (1902-77) in 1944, concerned with predicting the outcome of games of strategy (rather than games of chance) in which the participants have incomplete information about the others' intentions. Under perfect competition there is no scope for game theory, as individual actions are assumed not to influence others significantly; under oligopoly, however, this is not the case. Game theory has been increasingly applied to economics in recent years, particularly in the theory of industrial organizations.

Gaming contract A contract involving the playing of a game of chance by any number of people for money. A wagering contract involves only two people. In general both gaming contracts and wagering contracts are solid and no action can be brought to recover money paid or won under them.

Gap analysis A methodical tabulation of all the known requirements of consumers in a particular category of products, together with a cross-listing of all the features provided by existing products to satisfy these requirements. Such a chart shows up any gaps that exist and therefore provides a pointer to any new

Garnishee order An order made by a judge on behalf of a judgement creditor restraining a third party (often a bank), called a garnishee, from paying money to the judgement debtor until sanctioned to do so by the court. The order may also specify that the garnishee must pay a stated sum to the judgement creditor, or to the court, from the funds belonging to the judgement debtor.

Gatekeeper A manager in a large company who controls the flow of information. It is the gatekeeper who decides what information shall be passed upwards to a parent company and downwards to a subsidiary.

GATT Abbreviation for General Agreement on Tariffs and Trade.

Gazump To raise the price of, or accept a higher offer for, land, buildings, etc., on which a sale price has been verbally agreed but before the exchange of contracts has taken place The intending purchaser with whom the sale price had been verbally agreed has no remedy, even though he may have incurred expenditure on legal fees, surveys, etc. Proposals to reform the law in this respect have been put forward but not enacted.

Gazunder To reduce an offer on a house, flat, etc, immediately before exchanging contracts, having previously agreed a higher price. In a market in which house prices are falling, the unscrupulous buyer is aware that the seller will be extremely anxious to sell having incurred legal expenses and perhaps having bought another property. Gazundering is the dishonest buyer's opportunity in a falling market, as gazumping was the dishonest seller's opportunity in a rising market.

Gearing (capital gearing; equity gearing; financial gearing; leverage) The ratio of the long-term funds with a fixed interest charge, such as debentures and preference shares, making up a company's capital to its ordinary share capital A company is said to be high geared when its fixed interest capital is dominant and low geared when its capital is predominantly in ordinary shares, especially in relation to other similar companies. A high-geared company is considered to be a speculative investment for the ordinary shareholder and will be expected to show good returns when the company is doing well The US word leverage is increasingly used in the UK.

Gearing adjustment In current cost accounting, an adjustment that reduces the charge to the owners for the effect of price changes on depreciation, stock, and working capital It is justified on the grounds that a proportion of the extra financing is supplied by the loan capital of the business.

Gearing ratios (leverage ratios) Ratios that express a company's capital gearing. There are a number of different ratios that can be calculated from either the balance sheet or the Profit and loss account. Ratios based on the balance sheet usually express debt as a percentage of equity, or as a percentage of debt plus equity. Income gearing is normally calculated by dividing the profit before interest and tax by the gross interest payable to give the interest cover.

Generally accepted accounting principles (GAAP) In the USA, the rules, accounting standards, and accounting concepts followed by accountants in measuring, recording, and reporting transactions. There is also a requirement to state whether financial statements conform with GAAP. In the UK the concept is more loosely used but is normally taken to mean accounting standards and the requirements of company legislation and the stock exchange. There is an argument that compliance with these requirements places an undue burden on small companies, which should be exempt from certain accounting standards.

Generally accepted auditing standards (GAAS) In the USA, the broad rules and guidelines set down by the Auditing Standards Board of the American Institute of Certified Public Accountants (AICPA). In carrying out work for a client, a certified public accountant would apply the generally accepted accounting principles; if they fail to do so, they can be held to be in violation of the AICPA's code of professional ethics. general meeting A meeting that all the members of an association may attend.

General obligation bond In the USA, a security in which the government department with the authority to levy taxes has unconditionally promised payment.

General offer An offer of sale made to the general public rather than to a restricted number of people. For example, an object displayed in a shop window with a price tag is a general offer; the shopowner must sell to anyone willing to pay the price.

General price level An index that gives a measure of the purchasing power of money. In the UK, the bestknown measure is the Retail Price Index; in the USA it is the consumer price index.

Generic name The name of a class or category of products, such as videos or pens. Sometimes the name of an extensively promoted and successful brand comes to be used as a generic name; examples include Biro, Hoover, Levi, and Walkman, which are then often written without an initial capital letter.

Generic product (generic brand) A product that carries neither a manufacturer's name nor a distributor's brand. The goods are plainly packaged with stark lettering that simply lists the contents. These no-name brands offer no guarantees of quality and are produced and sold inexpensively.

Gensaki The Japanese money market for the resale and repurchase of medium- and long-term government securities.

Geodemographic segmentation Market segmentation in which consumers are grouped according to demographic variables, such as income and age, and identified by a geographic variable, such as post code or zip code. The base data is obtained from the census data. Two principles are involved: (1) people who live in the same neighbourhood, defined by a census enumeration district, are likely to share similar buying habits; (2) neighbourhoods can be categorized in terms of their populations: two or more neighbourhoods with similar populations can be placed in the same category.

Geographic information system A business tool for interpreting data based on geodemographic segmentation. It consists of a demographic database, digitized maps, and computer software. The main systems available are: *ACORN* – the first to be marketed, it has 39 neighbourhood types classified into 12 groups.

Mosaic – identifies 58 types of neighbourhood with 54 variables classified into 10 groups.

Pinpoint – has many of the features of both *ACORN* and Mosaic but claims to offer greater discrimination by focusing on individual postal sectors and so targets potential customers more closely. This system has 60 neighbourhood types, which are aggregated to 25, and finally reduced to 12 main types.

Geography-based organization An organizational structure in which a market area is divided into territories; marketing efforts are specialized and carried out by geographic area.

Geometric mean An average obtained by calculating the *n*th root of a set of n numbers. For example the geometric mean of 7, 100, and 107 is $\sqrt[3]{74900} = 42.15$, which is considerably less than the arithmetic mean of 71.3.

Gilt-edged security (gilt) A fixedinterest security or stock issued by the British government in the form of Exchequer stocks or Treasury stocks. Gilts are among the safest of all investments, as the government is unlikely to default on interest or on principal repayments. They may be irredeemable (*see*Consols) or redeemable. Redeemable gilts are classified as: long-dated gilts or longs (not redeemable for 15 years or more), medium-dated gilts or mediums (redeemable in 5 to 15 years), or shortdated gilts or shorts (redeemable in less than 5 years). Like most fixed-interest securities, gilts are sensitive not only to interest rates but also inflation rates. This led the government to introduce index-linked gilts in the 1970s, with interest payments moving in a specified way relative to inflation.

Most gilts are issued in units of £100. If they pay a high rate of interest (i.e. higher than the current rate) a £100 unit may be worth more than £100 for a period of its life, even though it will only pay £100 on redemption. Gilts bought through a stockbroker or bank are entered on the Bank of England Stock Register and income tax is deducted at source. Gilts can, however, be bought direct by post through the National Savings Stock Register, in which case tax is not deducted.

Globalization I. The process that has enabled investment in financial markets to be carried out on an international basis. It has come about as a result of improvements in technology and deregulation, with globalization, for example, investors in London can buy shares or bonds directly from Japanese brokers in Tokyo rather than passing through intermediaries. *See* also disintermedia- tion.

2. The internationalization of products and services by large firms (*see* globalization strategy). *See* also multinational enterprise.

Globalization strategy (standardization strategy) A marketing strategy that is standardized for use throughout the world. The strategy assumes that the behaviour of many consumers throughout the world has become very similar. It assumes that members of market segments require the same features and buy for the same reasons. It also anticipates that the cost savings obtained from production and marketing on a global scale will more than offset the lower sales achieved because the product has not been adapted for local markets. Part of the globalization strategy is to make use of these economies of scale to support the development of new products.

Global product A product that is marketed throughout the world with the same brand name, such as Coca Cola, Guinness, Levi, and McDonalds. The advantage of a global product is that it usually enables an advertisement or image to be used worldwide. However, one drawback is that some advertising slogans do not travel well For example, 'things come alive with Pepsi', when translated into Chinese, led the populace to believe that their ancestors would be brought back from the dead; the Vauxhall Nova encountered similar problems in Spain. where Nova means 'no go'!

Godfather offer A tender offer pitched so high that the management of the target company is unable to discourage shareholders from accepting it.

Going-concern concept A principle of accounting practice that assumes businesses to be going concerns, unless circumstances indicate otherwise. It assumes that an enterprise will continue in operation for the foreseeable future, i.e. that the accounts assume no, intention or necessity to liquidate or significantly curtail the scale of the enterprise's operation. The implication of this principle is that assets are shown at cost, or at cost less depredation, and not at their break-up values; it also assumes that liabilities applicable only on liquidation are not show. The **going-concern value** of a business is higher than the value that would be achieved by disposing of its individual assets, since it is assumed that the business has a continuing potential to earn profits. The concept is assumed in the preparation of financial statements. If an auditor thinks otherwise the auditors' report should be qualified.

Going-rate pricing Setting the price of products largely by following competitors' prices rather than by basing them on company costs or demand.

Gold card A credit card that entitles its holder to various benefits (e.g. an unsecured overdraft, some insurance cover, a higher limit, lower interest rates) in addition to those offered to standard card holders. The cards are available only to those on higher-thanaverage incomes.

Gold clause A clause in a loan agreement between governments stipulating that repayments must be made in the gold equivalent of the currency involved at the time either the agreement or the loan was made. The purpose is to protect the lender against a fall in the borrower's currency, especially in countries suffering high rates of inflation.

Gold coins Coins made of gold ceased to circulate after World War I. At one time it was illegal to hold more than four post-1837 gold coins and there have been various restrictions on dealing in and exporting gold coins at various times. Since 1979 (Exchange Control, Gold Coins Exemption, Order) gold coins may be imported and exported without restriction, except that gold coins more than 50 years old with a value in excess of £8000 cannot be exported without authorization from the Department of Trade and Industry. *See also*Britania coins, Krugerrand.

Golden hello A payment made to induce an employee to take up employment. The tax treatment depends on the nature of the payment; in some cases the taxpayer has successfully argued that the payment should be tax-free. However, in 1991 the House of Lords ruled that a payment made to a wellknown footballer by a football club, as an inducement to join a new club, was taxable.

Golden parachute A clause in the employment contract of a senior executive in a company that provides for financial and other benefits if the executive is sacked or decides to leave as the result of a takeover or change of ownership.

Golden share A share in a company that controls at least 51% of the voting rights. A golden share has been retained by the UK government in some privatization issues to ensure that the company does not fall into foreign or other unacceptable hands.

Gold standard A former monetary system in which a country's currency unit was fixed in terms of gold. In this system currency was freely convertible into gold and free import and export of gold was permitted. The UK was on the gold standard from the early 19th century until it finally withdrew in 1931. Most other countries withdrew soon after. *See also* International Monetary Fund

Good A commodity or service that is regarded by economists as satisfying a human need, An economic good is one that is both needed and sufficiently scarce to command a price. A *free good* is also needed but it is in abundant supply and therefore does not need to be purchased; air is an example. However, a commodity or service that is free but has required an effort to produce or obtain is not a free good in this sense. *See also*consumer goods.

Goods-in-transit insurance An insurance policy covering property against loss or damage while it is in transit from one place to another or being stored during a journey. Policies often specify the means of transport to be used which may include the postal service. Goods shipped solely by sea are not covered by this type of policy, in which case a marine insurance policy would be used.

Goodwill The difference between the value of the separable net assets of a business and the total value of the business. Purchased goodwill is the difference between the fair value of the price paid for a business and the aggregate of the fair values of its separable net assets. It may be written off to reserves or recognized as an intangible asset in the balance sheet and written off by amortization to the profit and loss account over its useful economic life. Internally generated goodwill should not be recognized in the Financial statements of an organization. The treatment of goodwill is governed by Statement of Standard Accounting Practice 22, 'Accounting for Goodwill'.

Government Accounting Standards Board In the USA, the organization responsible for accounting standards for government units. It is under the control of the Financial Accounting Foundation.

Government actuary A government officer with a small team of actuaries, with the responsibility of estimating future expenditure on the basis of observed population trends. He provides a consulting service to government departments and to

Commonwealth governments, advising on social security schemes and superannuation arrangements and on government supervision of insurance companies and Friendly Societies.

Government broker The stockbroker appointed by the government to sell government securities on the London Stock Exchange, under the instructions of the Bank of England. He is also the broker to the National Debt Commissioners. Until October 1986 (*see*Big Bang) the government broker was traditionally the senior partner of Mullins & Co. Since then, when the Bank of England started its own gilt-edged dealing room, the government broker has been appointed from the gilt-edged division of the Bank of England, although some of his functions are now undertaken by the gilts primary dealers.

Government grant An amount paid to an organization to assist it to pursue activities considered socially or economically desirable Grants may be revenuebased, i.e. made by reference to a specified category of revenue expenditure. Revenue-based grants should be credited to the profit and loss account in the same period as the revenue expenditure to which they relate. Capitalbased grants are made by reference to specified categories of capital expenditure and should be credited to the profit and loss account over the useful economic life of the asset to which they relate. Statement of Standard Accounting Practice 4, 'Accounting for Government Grants', gives guidance with respect to the treatment of grants.

Govornment market The market consisting of national and local government units that purchase or hire goods and services for carrying out their main functions. This includes defence procurement, education requirements (school buildings, books, teachers), medical services, road building, etc.

Greenmail (greymail) The purchase of a large block of shares in a company, which are then sold back to the company at a premium over the market price in return for a promise not to launch a bid for the company. This practice is not uncommon in the USA, where companies are much freer than in the UK to buy their own shares. Although the morality of greenmail is dubious, it can be extremely profitable.

Green marketing Marketing products that benefit the environment. The ecological properties of products are important in order that companies produce ecologically safer products, including recyclable and biodegradable packing. Better pollution controls and more energy-efficient production processes and product performance also form a part of green marketing. *See*product life cycle

Green product A product that its manufacturers claim will not cause damage to the environment, especially a new product or formulation that is being launched to replace one that is said to cause environmental damage.

Green reporting (environmental accounting) A report by the directors of a company that attempts to quantify the costs and benefits of that company's operations in relation to the environment. Although there are a number of advocates of the practice, few companies disclose in their annual accounts and report information on the impact their activities have had on the environment.

Gray knight In a takeover battle, a counter bidder whose ultimate intentions are undeclared. The original unwelcome bidder is the black knight, the welcome counter bidder for the target company is the white knight. The grey knight is an ambiguous intervener whose appearance is unwelcome to all.

Grey market 1. Any market for goods that are

in short supply. It differs from a black market in being legal; a black market is usually not. **2.** A market in shares that have not been issued, although they are due to be issued in a short time. Market makers will often deal with investors or speculators who are willing to trade in anticipation of receiving an allotment of these shares or are willing to cover their deals after flotation. This type of grey market provides an indication of the market price (and premium, if any) after flotation.

Grey wave A company that is thought to be potentially profitable and ultimately a good investment, but that is unlikely to fulfil expectations in the near future. The fruits of an investment in the present should be available when the investor has grey hair.

Gross domestic product (GDP) The monetary value of all the goods and services produced by an economy over a specified period. It is measured in three ways: (1) on the basis of expenditure, i.e. the value of all goods and services bought, including consumption, capital expenditure, increase in, the value of stocks, government expenditure, and exports less imports; (2) on the basis of income, i.e. income arising from employment, self-employment, rent, company profits (public and private), and stock appreciation; (3) on the basis of the value added by industry, i.e. the value of sales less the costs of raw materials. In the UK, statistics for GDP are published monthly by the government on all three bases, there are large discrepancies between each measure. Economists are usually interested in the real rate of change of GDP to measure the performance of an economy, rather than the absolute level of GDP. *See* also GDP deflator, gross national product (GNP); net national product (NNP) .

Gross income 1. The income of a person or an organization before the deduction of the expenses incurred in earning it

2. Income that is liable to tax but from which the tax has not been deducted. For many types of income, tax may be deducted at source (*see*deductions at source) leaving the taxpayer with a net amount.

Grossing up Converting a net return into the equivalent gross amount. For example, if a net dividend (*d*) has been paid after deduction of r% tax, the gross equivalent (g) is given by. g = 100d/(100 - r).

Gross interest The amount of interest applicable to a particular loan or deposit before tax is deducted. Interest rates may be quoted gross (as they are on government securities) or net (as in most building society deposits or bank deposits). The gross interest less the tax deducted at the basic rate of income tax gives the net interest. Any tax suffered is usually, but not always, available as a credit against tax liabilities.

Gross margin (gross profit gross profit margin) The difference between sales revenue of a business and the cost of goods sold. It does not include finance costs, administration, or the cost of distributing the goods. *Compare*net margin.

Gross margin ratio (gross profit percentage) A ratio of financial performance calculated by expressing the gross margin as a percentage of sales. With retailing companies in particular, it is regarded as a prime measure of their trading success. The only ways in which a company can improve its gross margin ratio are to increase selling prices and/or reduce its cost of sales.

Grow National Product(GNP) The gross domestic product (GDP) with the addition of interest, profits, and dividends received from abroad by UK residents. The GNP better reflects the welfare of the population in monetary terms, although it is not as accurate

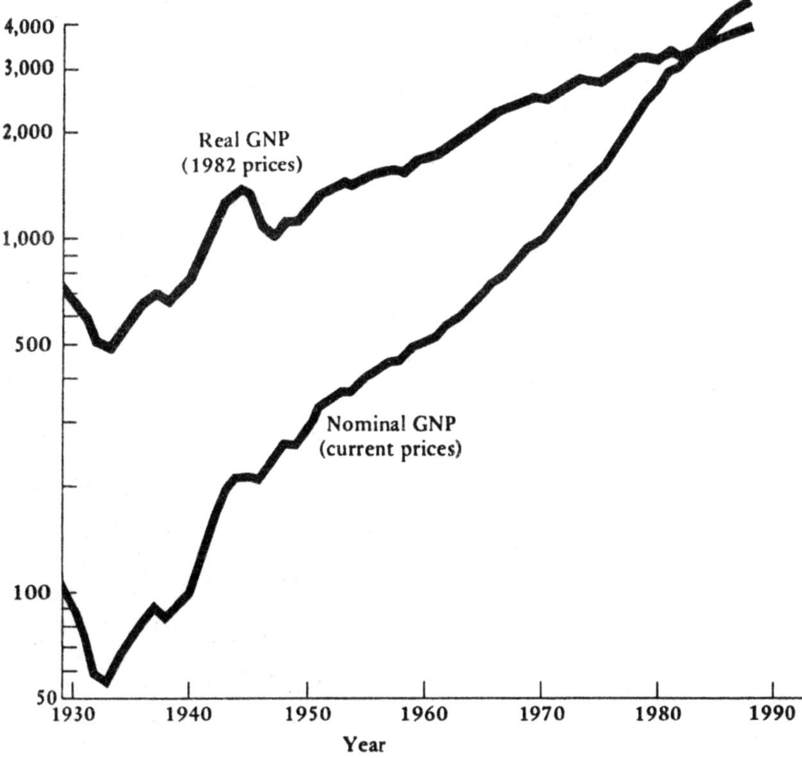

Fig. GNP

a guide as to the productive performance of the economy as the GDP. *See also*net national product (NNP) .

Gross receipts The total amount of money received by a business in a specified period before any deductions for costs, raw materials, taxation, etc. *Compare*net receipts.

Gross redemption yield (effective yield; yield to maturity) The internal rate of return of a bond bought at a specified price and held until maturity; it therefore includes all the income and all the capital payments due on the bond. The tax payable on the interest and the capital repayments is ignored.

Gross tonnage (register tonnage) *See*tonnage.

Gross weight 1. The total weight of a package including its contents and the packing. The net weight is the weight of the contents only, i.e. after deducting the weight of the packing (often called the tare). 2. The total weight of a vehicle (road or rail) and the goods it is carrying, as shown by a weighbridge. The net weight is obtained by deducting the tare of the vehicle.

Ground rent Rent payable under a lease that has been granted or assigned for a capital sum (premium). Normally. long leases on offices, flats, etc, are granted for such a premium, payable when the lease is first granted; in addition the leaseholder pays the landlord a relatively small annual ground rent *See also*rent.

Group A parent undertaking and its subsidiary or subsidiaries. *See***consolidation; consolidated financial statements.**

Group discussion A marketing research technique that brings between six and eight respondents together for at least an hour to discuss a marketing issue under the guidance of an interviewer. It has been found that, once such a group is relaxed, it will fully explore issues in a manner that shows what is important to it, using its own rather than marketers' terminology. The interviewer, who is a specialist acting on behalf of a client, first draws up a topic guide that identifies the points to be explored. A sample of respondents -who must not know one another — is recruited to match certain criteria (e.g. if the problem concerned tea all respondents would be tea drinkers). The group meets in the interviewer's home or in a hotel; with the discussion being taperecorded, the interviewer describes the problem and then adopts a passive role, allowing the group to discuss their views and interjecting further questions only if some aspect of the problem is not being explored adequately. After a minimum of four group discussions, the interviewer will have sufficient material to write a report and recommend a particular action. Group discussions are a qualitative procedure (see qualitative frequently used to determine attitudes to particular products or advertisements.

Group income A dividend paid by one group company to another, winch is exempt from advance corporation tax. The dividends received are not franked investment income of the receiving company and are not subject to corporation tax.

Group life assurance A life-assurance policy that covers a number of people, usually a group of employees or the members of a particular dub or association. Often a single policy is issued and premiums are deducted from salaries or club-membership fees. In return for agreement that all employees or members join the scheme, insurers are prepared to ask only a few basic questions about the health of a person joining. However, with the advent of AIDS insurers are no, longer prepared to waive all health enquiries, as they were a few years ago.

Group of Seven (G7) The seven leading industrial nations outside the communist bloc: USA, Japan, Germany, France, UK, Italy, and Canada. This group evolved from the first economic summit held in 1976 and is now an annual meeting attended by heads of state. The original aim was to discuss economic coordination but the agenda has since broadened to include political issues. However, increasing enthusiasm for international economic cooperation in the 1980s led to collective action, for example on exchange rates as a result of meetings of G7 finance ministers.

Growth curve A curve on a graph in which a variable is plotted as a function of time. The curve thus illustrates the growth of the variable. This may be used to show the growth of a population, sales of a product, price of a security, etc.

Growth industry Any industry that is expected to grow faster than others.

Growth rate The amount of change over a period in some of the financial characteristics of a company, such as sales revenue or profits. It is normally measured in percentage terms and can be compared to the Retail Price Index, or some other measure of inflation, to assess the real performance of the company.

Growth stage The stage in a product life cycle during which a product's sales start climbing rapidly. growth stocks Securities that are expected to offer the investor sustained capital growth. investors and investment managers often distinguish between growth stocks and income stocks. The former are expected to provide capital gains; the latter, high income. The investor will usually expect a growth stock to be an ordinary share in a company whose products are selling well and whose sales are expected to expand, whose capital expenditure on. new plant and equipment is high, whose earnings are growing, and whose management is strong, resourceful and investing in product development and long-term research.

Guaranteed minimum pension The earnings-related component of a state pension that a person would have been entitled to as an employee of a company, had he or she not contracted out of the State Earnings-Related Pension Scheme (SERPS). Any private pension contract must pay at least the guaranteed minimum pension if it is to be an acceptable replacement of a SERPS pension.

Guarantor A person who guarantees to pay a debt incurred by someone else if that person fails to repay it. A person who acts as a guarantor for a bank loan, for example, must repay the loan if the borrower fails to repay it when it becomes due.

Guilder The standard monetary unit of the Netherlands (f), the Antilles (NAf), and Surinam (Sf), divided into 100 cents.

Guinea A former British coin, originally made from gold from Guinea. First issued in 1663, it had various values; during most of the 18th century it stood at 21 shillings. Although it was replaced in 1817 by the sovereign, it persisted, not as a coin but as a unit of 21 shillings for charging for some professional services, until decimalization (1971).

habitual buying behaviour Consumer behaviour in the buying of cheap frequently purchased items, characterized by low consumer involvement and few significant brand differences. Purchases are made out of habit, rather than a strong commitment to a brand. Generally, price and sales promotions are used to stimulate sales, as consumers tend to base their decisions on past behaviour and have no need for further information.

Hang Seng Index An arithmetically weighted index based on the capital value of 33 stocks on the Hong Kong Stock Exchange. It was first quoted in 1964 and takes its name from the Hang Seng Bank. The number of stocks, 33, was chosen because the bank was founded in 1933 and 33 is a lucky number in Chinese astrology.

Hard currency A currency that is commonly accepted throughout the world; they are usually those of the western industrialized countries although other currencies have achieved this status, especially within regional trading blocs. Holdings of hard currency are valued because of their universal purchasing power. Countries with soft currencies go to great lengths to obtain and maintain stocks of hard currencies, often imposing strict restrictions on their use by the private citizen.

Hard dollars A fee paid to a US stockbroker, investment adviser, etc., for research, analysis, or advice, as opposed to soft dollars, which refers to the commission earned on purchases made.

Hard sell The use of selling methods that emphasize the supposed virtues of a product, repeat its name, and attempt to force it upon the consumer by forceful and unsubtle means. *Compare*soft sell.

Hard systems Systems that consist entirely of non-human elements, e.g. machines. The behaviour of all the elements in the system should be predictable and amenable to mathematical or other scientific modelling methods. *Compare*soft systems methodology.

Hardware The electronic and mechanical parts of a computer system; for example, the central processing unit, disk drive, screen, and printer. *Compare software*. Harvesting strategy Making a shortterm profit from a particular product shortly before withdrawing it from the market. This is often achieved by reducing the marketing support it enjoys, such as advertising, on the assumption that the effects of earlier advertising will still be felt and the product will continue to sell.

Haulage The charge made by a haulier (haulage contractor) for transporting goods, especially by road. If the goods consist of a large number of packages (e.g. 100 tonnes of cattlefood packed in 2000 bags each weighing 50 kilograms) there will be a separate charge for loading and unloading the vehicle.

Hedged funds Funding in which the managers invest in liquid instruments, such as currency and interest rate derivatives, with the aim of making profits from movements in foreignexchange or bond markets.

Hedging An operation undertaken by a trader or dealer who wishes to protect an open position, especially a sale or a purchase of a commodity, currency, security, etc., that is likely to fluctuate in price over the period that the position remains open. For example, a manufacturer may contract to sell a large quantity of a product for delivery over the next six months. If the product depends on a raw material that fluctuates in price, and if the manufacturer does not have sufficient raw material in stock, an open position will result. This open position can be hedged by buying the raw material required on a futures contract; if it has to be paid for in a foreign currency the manufacturer's currency needs can be hedged by buying that foreign currency forward or on an option. Operations of this type do not offer total protection because the prices of spot goods and futures do not always move together, but it is possible to reduce the vulnerability of an open position substantially by hedging.

Buying futures or options as a hedge is only one kind of hedging; it is known as long hedging. In short hedging, something is sold to cover a risk. For example, a fund manager may have a large holding of long-term fixed income investments and is worried that an anticipated rise in interest rates will reduce the value of the portfolio. This risk can be hedged by selling interestrate futures on a financial futures market. If interest rates rise the loss in the value of the portfolio will be offset by the profit made in covering the futures sale at a lower price.

Hedging against inflation Protecting one's capital against the ravages of inflation by buying equities or making other investments that are likely to rise with the general level of prices.

Hereditament Real property. Originally, it was property that would be inherited by the heir on intestacy. Corporeal hereditaments are physical real property, such as land, buildings, trees, and minerals. Incorporeal hereditaments are intangible rights, such as easements or profits à prendre, attached to land.

Hidden reserve Funds held in reserve but not disclosed on the balance sheet (they are also known as off balance sheet reserves or secret reserves). They arise when an asset is deliberately either undisclosed or undervalued. Such hidden reserves are permitted for some banking institutions but are not permitted for limited companies as they reduce profits and therefore the corporation tax liability of the company.

Hidden tax A tax, the incidence of which may be hidden from the person who is suffering it. An example could be a tax levied on goods at the whole level which increases the retail price in such a way that the final customer cannot detect either that it has happened or the amount of the extra cost. Again, a government might introduce a hidden tax by artificially causing the price of a certain utility to be raised above its usual commercial value.

Hidden unemployment (concealed unemployment) Unemployment that does not appear in government statistics, which are usually based on figures for those drawing unemployment benefit. In addition to these people there will be some who are doing very little productive work, although they are not claiming benefit. These include workers on short time for whom their employers are expecting to be able to provide work shortly; workers who are not usefully employed, although their employers think they are; and others not claiming benefit for a variety of reasons.

Hierarchy of effects The steps in the persuasion process that lead a consumer to purchase a particular product. They are awareness, knowledge, liking, preference, conviction, and finally a decision to purchase.

Higher-rate tax A higher rate of income tax than the basic rate of income tax. For 1996-97 higher-rate tax is payable on taxable income, after personal allowances and other allowances, of £25,501 and over. The rate of tax is 40%.

High-involvement product A product that involves the consumer in taking time and trouble before deciding on a purchase. This will include looking in several catalogues, shops, etc., to compare prices and the products themselves. High-involvement products include such major purchases as cars and houses, as well as hi-fi equipment and many other domestic products. *Compare*low-involvement product.

High yielder A stock or share that gives a high yield but is more speculative than most, i.e. its price may fluctuate.

HIP Abbreviation for human-information processing.

Hire purchase (HP) A method of buying goods in which the purchaser takes possession of them as soon as an initial instalment of the price (a deposit) has been paid and obtains ownership of the goods when all the agreed number of subsequent instalments have been paid. A hire-purchase agreement differs from a credit-sale agreement and sale by instalments (or a deferred payment agreement) because in these transactions ownership passes when the contract is signed. It also differs from a contract of hire, because in this case ownership never passes. Hire-purchase agreements were formerly controlled by government regulations stipulating the minimum deposit and the length of the repayment period. These controls were removed in 1982. Hire-purchase agreements were also formerly controlled by the *Hire Purchase Act (1965)*, but most are now regulated by the *Consumer Credit Act (1974)*. In this Act a hire-purchase agreement is regarded as one in which goods are bailed in return for periodical payments by the bailee; ownership passes to the bailee if the terms of the agreement are complied with and the option to purchase is exercised.

A hire-purchase agreement often involves a finance company as a third party. The seller of the goods sells them outright to the finance company, which enters into, a hire-purchase agreement with the hirer.

Histogram A graph of a frequency distribution constructed with rectangles, in which the areas of the rectangles are proportional to the frequencies.

Historical cost A method of valuing units of stock or other assets based on the original cost incurred by the organization. For example, the issue of stock using first-in-first-out cost or average cost charge the original cost against profits. Similarly, the charging of depreciation to the profit and loss account, based on the original cost of an asset, is writing off the historical cost of the asset against profits. An alternative approach is the use of current cost accounting.

Historical cost accounting A system of accounting based primarily on the original costs incurred in a transaction. It is relaxed to some extent by such practices as the valuation of stock at the lower of cost and net realizable value and, in the UK, revaluation of capital assets. The advantages of historical cost accounting are that it is relatively objective, easy to apply, difficult to falsely manipulate, and suitable for audit verification. In times of high inflation, however, the results of historical cost accounting can be misleading as profit can be overstated, assets understated in terms of current values, and capital maintenance is only concerned with the nominal amount of the capital invested rather than its purchasing power. Because of these defects it is argued that historical cost accounting is of little use for decision making, but attempts to replace it with such other methods as current cost accounting have failed.

Company legislation sets out the rules for the application of historical cost accounting to financial statements. Companies may also choose to use alternative accounting rules based on current cost accounting.

Historical summary A voluntary statement appearing in the annual accounts and report of some companies in which the main financial results are given for the previous five to ten years.

Holding company (parent company) A company in a group of companies that holds shares in other companies (usually, but not necessarily, its subsidiaries).

Holiday and travel insurance An insurance policy covering a variety of risks for the duration of a person's holiday. Although policies vary, an average policy covers the policyholder's baggage and personal effects for all risks (*see*allrisk policy); compensation for delays while travelling to or from the holiday, refund of deposits lost if the holiday has to be cancelled for any of a number of specific causes; theft or loss of money, traveller's cheques, or credit cards; payment of medical expenses or costs of flying home in the event of illness or injury; and payment of legal compensation for injuring other people or damaging their property because of negligence.

Holistic evaluation An evaluation of an advertising or marketing campaign as a whole, as distinct from an analysis of the results of its constituent parts.

Home audit Marketing research conducted in the home by means of panels of householders who keep a diary on a regular basis recording the products they buy and when they buy them.

Home banking Carrying out normal banking transactions by means of a home computer linked to a bank's computer. Payment of telephone bills was the first major use of the service in the USA, where it is largely restricted to New York and other major cities. In the UK limited experiments have been conducted but only a small minority of High-Street bank customers have access to the service; it is more common among business customers who make use of electronic banking services between business premises and the bank.

Home page A document, usually in graphical form that acts as the entry point to a World Wide Web server. Set out in a page format it provides hypertext access, to related menus and files.

Home shopping channels A form of television direct marketing in which television programmes or entire channels are dedicated to selling goods and services. While still relatively small in money terms in the UK and Europe, in the USA it is estimated that more than $4 billion is spent on goods and services in this manner. *Compare*direct-response TV advertising.

Honour policy A marine insurance policy that covers the risk of a person who has an interest in a voyage in circumstances in which the interest may be difficult to establish.

Horizontal marketing systems A channel arrangement in which two or more companies, at one level join together to follow a new marketing opportunity. Such joint ventures have become more commonplace as companies combine their capital production facilities, and marketing resources to compete in global markets.

Horizontal spread A combination of options with different expiry dates, such as a long put option combined with a short put option.

Horns and halo effect An effect on a person's judgement of another that is unduly influenced by a first impression; it may be either unfavourable (horns) or favourable (halo). The effects can be misleading when interviewing job applicants. For example, a well-qualified candidate who arrives late for an interview, for a good reason, may be passed over as a bad timekeeper, while a poorly qualified but punctual and well-groomed candidate may be offered the job.

Hostile bid A takeover bid that is unwelcome either to the board of directors of the target company or to its shareholders.

Hot money 1. Money that moves at short notice from one financial centre to another in search of the highest shortterm interest rates, for the purposes of arbitrage, or because its owners are apprehensive of some political intervention in the money market, such as a devaluation. Hot money can influence a country's balance of payments.

2. Money that has been acquired dishonestly and must therefore be untraceable.

Household insurance 1. An insurance policy that covers the structure of a home (often called buildings insurance). 2. An insurance policy covering the personal goods and effects kept inside a home (contents policy). A comprehensive householder's policy will cover both the buildings and the contents.

House journal A journal produced by a large organization for the benefit of its employees. It usually seeks to keep all levels of employees informed of the general policy of the company, to function as a forum for employees, and often to provide social news. It often helps to create a corporate identity, especilly if employees are widely distributed geographically.

Hull insurance Marine insurance or aircraft insurance covering the structure of a ship, boat, hovercraft, or aeroplane and the equipment maintained permanently on board.

Human capital The skills, general or specific, acquired by an individual in the course of training and work experience. The concept was introduced by Gary Becker in the 1960s in order to point out that wages reflect in part a return on human capital. This theory has been used to explain large variations in wages for apparently similar jobs and why even in a recession a firm may retain its workers on relatively high wages, in spite of involuntary unemployment. *Compare* physical capital.

Human-information processing (HIP) A study of the processes involved in decision making. Its importance in a business context is that an understanding of the way in which an individual uses information in the decision-making process should make it possible to determine the most appropriate information to be provided and its most suitable form.

Human-resource accounting (human-asset accounting) An attempt to recognize the human resources of an organization, quantify them in monetary terms, and show them on the balance sheet. A value is placed on such factors as the age and experience of employees as well as their future earning power for the company. Although this approach has aroused some interest, in practice considerable difficulty has been met in quantifying the value of human resources.

Human-resource management (HRM) The management of people to achieve individual behaviour and performance that will enhance an organization's effectiveness. HRM encourages individuals to set personal goals and rewards, guiding them to shape their behaviour in accordance with the objectives of the organization that employs them.

HRM was traditionally called personnel management ; however, current trends place a greater emphasis on the morale of employees and the ways of achieving consistent job satisfaction, by using more sophisticated psychological tests in selecting employees, by training employees to do more than one job, and by encouraging all the members of a workforce to accept responsibility (*see* empowerment ; self-actualization ; total quality management).

Hurdle rate The rate of interest in a capital budgeting study that a proposed project must exceed before it can be regarded worthy of consideration. The hurdle rate is often based on the cost of capital or the weighted average cost of capital adjusted by a factor to represent the risk characteristics of the projects under consideration.

Hybrid A synthetic financial instrument formed by combining two or more individual financial instruments, such as bonds with warrants.

Hyperinflation (galloping inflation) A very high rate of increase in the general price level (e.g. 50% per month).

Hypermarket A very large shop, usually having a selling area of at least 50 000 square feet (4645 square metres). that sells a wide range of products. Able to buy in bulk, hypermarkets are often located adjacent to towns and base their attraction on low prices and to car-owning consumers, who can make many of their purchases in, one place where parking facilities are provided. Hypermarkets are now widespread, especially in grocery and do-it-yourself retailing.

Hypothecation 1. An authority given to a banker, usually as a letter of hypothecation, enable the banker to sell goods that have been pledged (*see* pledge) as security for a loan. The goods have often been pledged as security in relation to a documentary bill, the banker being entitled to sell the goods if the bill is dishonoured by nonacceptance or non-payment.

2. A mortgage granted by a ship's master to secure the repayment with interest, on the safe arrival of the ship at her destination, of money borrowed during a voyage as a matter of necessity (e.g. to pay for urgent repairs). The hypothecation of a ship itself, with or without cargo, is called bottomry and is effected by a bottomry bond ; that of its cargo alone is respondentia and requires a respondentia bond. The bondholder is entitled to a maritime lien.

I

Idle capacity The part of the budgeted capacity within an organization that is unused. It is measured in hours using the same measure as production. Idle capacity can arise as a result of a number of causes in all of which the actual hours worked is less than the budgeted hours available. The reasons can include non-delivery of raw materials, shortage of skilled labour, or lack of sales demand.

Illegal contract A contract prohibited by statute (e.g. one between traders providing for minimum resale prices) or illegal at common law on the grounds of public policy. An illegal contract is totally void, but neither party (unless innocent of the illegality) can recover any money paid or property transferred under it, according to the maxim *ex turpi causa non oritur actio* (a right of action does not arise out of an evil cause). Related transactions may also be affected. A related transaction between the same parties (e.g. if X gives Y a promissory note for money due from him under an illegal contract) is equally tainted with the illegality and is therefore void. The same is true of a related transaction with a third party (e.g. if Z lends X the money to pay Y) if the original illegality is known to that party.

Illegal partnership A partnership formed for an illegal purpose and therefore disallowed by law. A partnership of more than 20 partners is illegal, except in the case of certain professionals, e.g. accountants, solicitors, and stockbrokers.

Image A composite mental picture formed by people about an organization or its products. One of the functions of advertising and public relations is to create a favourable image or to improve one that is unfavourable.

Immediate annuity An annuity contract that begins to make payments as soon as the contract has come into force.

Immediate holding company A company that has a controlling interest in another company, even though it is itself controlled by a third company, which is the holding company of both companies.

Immigrant remittances Money sent by immigrants from the country in which they work to their families in their native countries. These sums can be a valuable source of foreign exchange for the native countries.

Impact day The day on which the terms of a new issue of shares are made public.

Imperial units A system of units formerly widely used in the UK and the rest of the English-speaking world. It includes the pound (lb), quarter (qt), hundredweight (cwt), and ton (ton); the foot (ft), yard (yd), and mile (mi); and the gallon (gal), British thermal unit (btu), etc. These units have been largely replaced by metric units, although Imperial units persist in some contexts, e.g. beer is sold in pints, loose fruit and vegetables are sold in pounds, etc. *Compare*metric system ; US Customary units.

Impersonal account A ledger account that does not bear the name of a person. These accounts normally comprise the nominal accounts, having such names as motor vehicles, heat and light, and stock in trade.

Implied term A provision of a contract not agreed to by the parties in words but either regarded by the courts as necessary to give effect to their presumed intentions or introduced into the contract by statute (as in the case of contracts for the sale of goods). An implied term may constitute either a condition of the contract or a warranty; if it is introduced by statute it often cannot be expressly excluded.

Import deposit A method of restricting importation to a country, in which the importer is required to deposit a sum of money with the Customs and Excise authorities when goods are unported. The UK used this method in 1968-70, as an alternative to raising unport duties, when it had an agreement with GATT not to do so.

Import duty A tax or tariff on import goods. Import duties can either be a fixed amount or a percentage of the value of the goods. They have been a major type of barrier used to protect domestic production against foreign competition and they have also been an important source of government revenue, especially in developing countries.

Import entry form A form completed by a UK importer of goods and submitted to the Customs and Excise for assessment of the import duty, if any. When passed by the Customs the form functions as a warrant to permit the goods to be removed from the port of entry.

Import licence A permit allowing an importer to bring a stated quantity of certain goods into a country. Import licences are needed when import restrictions include import quotas, currency restrictions, and prohibition. They also function as a means of exchange control the licence both permitting importation and allowing the importer to purchase the required foreign currency.

Import quota An import restriction imposed on imported goods, to reduce the quantity of certain goods allowed into a country from a particular exporting country, in a stated period. The purpose may be to conserve foreign currency, if there is an unfavourable balance of payments, or to protect the home market against foreign competition (*see*protective duty). Quotas are usually enforced by means of import licences.

Import restrictions Restrictions imposed on goods and services imported into a country, which usually need to be paid for in the currency of the exporting country. This can cause a serious problem to the importing country's balance of payments, hence the need for restrictions, which include tariffs, import quotas, currency restrictions, and prohibition. Prohibition will also apply to preventing the importation of illegal goods (e.g. drugs, arms.) The restrictions may also be imposed to protect the home industry against foreign competition (*see*protective duty) or during the course of political bargaining.

Imports Goods or services purchased from another country. *See*import duty ; import quota ; import restrictions.

Import surcharge An extra tax applied by a government on certain imports in addition to the normal tariff. It is usually applied as a temporary measure by a government that is in difficulty with its balance of payments, but does not wish to violate its international tariff agreements.

Imprest account A means of controlling petty-cash expenditure in which a person is given a certain sum of money (float or imprest).

When he has spent some of it he provides appropriate vouchers for the amounts spent and is then reimbursed so that the float is restored. Thus at any given time the person should have either vouchers or cash to a total of the amount of his float.

Impulse buying The buying of a product by a consumer without previous intention and almost always without evaluation of competing brands. Impulse buying is encouraged by such factors as prominent shelf position; manufacturers are therefore keen to persuade retailers to give these advantages to their products.

Imputation system The system in which the advance corporation tax paid by a company making qualifying distributions is available to set against the gross corporation tax for the company. The shareholder receiving the dividend is treated as having suffered tax on the dividend and the tax credit is available to set against his or her own liability to tax.

Imputed cost A cost that is not actually incurred by an organization but is introduced into the accounting records in order to ensure that the costs incurred by dissimilar operations are comparable. For example, if rent is not payable by an operation it will be introduced as an imputed cost so that the costs may be compared with an operation that does pay rent.

IMRO Abbreviation for Investment Management Regulatory Organization. *See*Self-Regulating Organization.

In bond Delivery terms for goods that are available for immediate delivery but are held in a bonded warehouse. The buyer has to pay the cost of any Customs duty due, the cost of loading from the warehouse, and any further on-carriage costs.

Incentive stock option In the USA, the right given to employees to purchase a specified number of company shares at a specified price during a specified period. Only when the stock is sold by employees is it subject to tax.

Incestuous share dealing The buying and selling of shares in companies that belong to the same group, in order to obtain an advantage of some kind, usually a tax advantage. The legality of the transaction will depend on its nature.

Inchoaete instrument A negotiable instrument in which not all the particulars are given. The drawer of an inchoate instrument can authorize a third party to fill in a specified missing particular.

Incidence of taxation The impact of a tax on those who bear its burden. rather than those who pay it. For example, VAT is paid by traders, but the ultimate burden of it falls on the consumer of the trader's goods or services. Again, a company may pay corporation tax but if it then raises its prices or reduces its employees' wages to recoup the tax it may be said to have shifted the incidence.

Income 1. Any sum that a person or organization receives either as a reward for effort (e.g. salary or trading profit) or as a return on investments (e.g. rents or interest). From the point of view of taxation, income has to be distinguished from capital

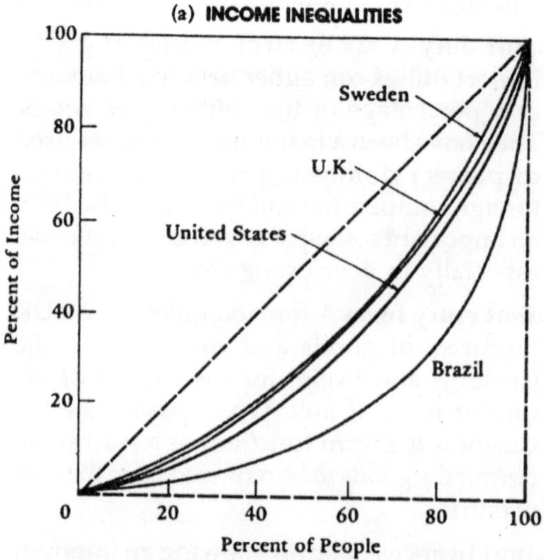

Fig. An Example of Income Inequalities

2. In economics, the flow of economic value attributed to an individual or group of individuals over a particular period. The total of income throughout an individual's life, minus expenditure, is equal to that person's wealth. Note that both income and wealth are generally considered to include more than simply money, for example, benefits in kind, services rendered by governments, and human capital should all be included.

Income and expenditure account An account, similar to a profit and loss account, prepared by an organization whose main purpose is not the generation of profit (e.g. a charity). It records the income and expenditure of the organization and results in either a surplus of income over expenditure or of expenditure over income.

Income bond 1. (National Savings Income Bond) A type of bond introduced by the Department for National Savings in 1982. They offer monthly interest payments on investments between £2000 and £250,000. Interest is taxable but not deducted at source. The bonds have a guaranteed life of 10 years.

2. See guaranteed-income bond.

Income distribution The payment by a unit trust of its half-yearly income to unit holders, in proportion to their holdings. The income distributed is the total income less the manager's service charge and income tax at the standard rate.

Income redistribution The use of taxation to ensure that the incomes of the richer members of the community are reduced and those of the poorer members are increased. Tax theorists tend to favour ability-to-pay taxation and progressive taxes for this purpose, combined with various forms of income support for the poorer members of the community.

Income smoothing The manipulation by companies of certain items in their financial statements so that they eliminate large movements in profit and are able to report a smooth trend over a number of years. The practice is pursued because of the belief that investors have greater confidence in companies that are reporting a steady increase in profits year by year. It is doubtful if any regulations can totally prevent this form of creative accounting.

Incomes policy A government policy aimed at controlling inflation and maintaining full employment by holding down increases in wages and prices by statutory or other means.

Income statement In the USA, the equivalent of a UKprofit and loss account.

Income stock A stock or share bought primarily for the steady and relatively high income it can be expected to produce. This may be a fixed-interest gilt-edged security or an ordinary share with a good yield record.

Income tax (IT) A direct tax on income. Income is not defined in UK tax legislation; amounts received are classified under various headings or schedules and these schedules are subdivided into cases. In order to be classed as income an amount received must fall into one of these schedules. There are some specific occasions when the legislation requires capital receipts to be treated as income for taxation purposes, e.g. when a landlord receives a lump sum on the granting of a lease. In the UK the importance of the distinction between income and capital has diminished since income and capital have been charged at the same rate. Prior to 6 April 1988 capital was charged at 30%, whereas the top rate of income tax was 60%. The tax is calculated on the taxpayer's taxable income, i.e. gross income less any income tax allowances and deductions. If the allowances and deductions exceed the gross income in a fiscal year, no income tax is payable. In the UK, from 6 April 1996 the first £3900 of taxable income is charged at a reduced rate of 20%, after which the basic rate of income tax applies. For those on high incomes, higher-rate tax is also charged.

Income tax allowances Allowances that may be deducted from a taxpayer's gross income before calculating the liability to income tax. Every individual who is a UK resident is entitled to a personal allowance. The level of allowance will depend on the age of the individual For 1996-97 married couples are entitled to a married couple's allowance of £1790 if the elder spouse is under 65. This increases to £3115 when the elder spouse is aged between 65 and 74, and to £3155 when the elder spouse is aged 75 and over. There is an additional personal allowance for single parents equal to the married couple's allowance. A widow's bereavement allowance, equal to the married couple's allowance, is available to a widow for the year of her husband's death and the year following, if she does not remarry during that time A blind person's relief is available to a blind person, which amounted to £1200 for 1996-97.

Income tax code A code number issued by the Board of Inland Revenue that takes account of the personal allowance available to the taxpayer together with any other additional allowances to which he or she is entitled, e.g. married couple's allowance. The code is used by the employer through the pay-as-you-earn scheme to calculate the taxable pay using tables supplied by the Inland Revenue. The income tax code can also be used to tax benefits in kind, such as company cars, by reducing the code number and so collecting more tax each tax week or month. The code provides a means of ensuring that the tax due for the fiscal year is deducted from the employee's earnings in equal weekly or monthly amounts.

Incompatibility The inability of one computer to handle data and programs produced for a different type of computer.

Inconvertible paper money Paper money that is not convertible into gold. Most paper money now falls into this category, although until 1931 in the UK, the Bank of England had an obligation to supply any holder of a bank note with the appropriate quantity of gold.

Incorporated Society of British Advertisers Ltd (ISBA) A society founded in 1900 as the Advertisers Protection Society. It is open to any advertising company and is concerned with all matters relating to advertising.

Incoterms A glossary of terms used in international trade published by the International Chamber of Commerce in Paris. It gives precise definitions to eliminate misunderstandings between traders in different countries.

Increasing capital An increase in the number or value of the shares in a company to augment its authorized share capital If the articles of association of the company do not permit this to be done (with the agreement of the members), the articles will need to be changed. A company cannot increase its share capital unless authorized to do so by its articles of association.

Incremental costs Costs incurred purely because of the occurrence of a spedcific event. An incremental cost is similar to a marginal cost; it tends to be used when the principal is the exclusion of all costs except those that would not have been incurred but for the specific project in view.

Indemnity 1. An agreement by one party to make good the losses suffered by another, usually by payment of money, repair, replacement or reinstatement This is the function of indemnity insurance.

2. An undertaking by a bank's client, who has lost a document (such as a share certificate or bill of lading), that the bank will be held harmless against any consequences of the document's absence if it proceeds to service the documents that have not been mislaid. The bank usually requires a letter

of indemnity to make sure that it suffers no loss.

Indemnity insurance Any insurance designed to compensate a policyholder for a loss suffered, by the payment of money, repair, replacement, or reinstatement. In every case the policyholder is entitled to be put back in the same financial position as he or she was immediately before the event insured against occurred. There must be no element of profit to the policyholder nor any element of loss. Most – but not all -insurance policies are indemnity contracts. For example, personal accident and life-assurance policies are not contracts of indemnity as it is impossible to calculate value of a lost life or limb (as the value of a car or other property can be calculated).

Indenture 1. A deed, especially one creating or transferring an estate in land. It derives its name from the former practice of writing the two parts of a two-part deed on one piece of parchment and separating the two parts by an irregular wavy line. The two parts of the indenture were known to belong together if the indented edges fitted together. A deed with a straight edge was known as a deed poll

2. *See* apprentice.

Independent demand The situation in which the demand for a product is independent of the production process, i.e. it is subject to market forces and therefore demand has to be forecast. When a car manufacturer produces a car, the number of wheels required is known before each unit can be sent out of the factory. However, the manufacturer does not know how many transmissions will fail or whether individuals will buy their replacements from other suppliers. They cannot therefore predict accurately how many spares to produce. *Compare*dependent demand.

Independent financial adviser (IFA) A person defined under the *Financial Services Act (1986)* as an adviser who is not committed to the products of any one company or organization. Such a person is licensed to operate by one of the Self-Regulating Organizations or Recognized Professional Bodies. With no loyalties except to the customer, the IFA must offer best advice from the whole market place. Eight categories of IFA exist, grouped into four main areas: advising on investments; arranging and transacting life assurance, pensions, and unit trusts; arranging and transacting other types of investments; and management of investments. All licensed independent financial advisers contribute to a compensation fund for the protection of their customers.

Independent intermediary A person who acts as a representative of a prospective policyholder in the arrangement of an insurance or assurance policy. In life assurance and pensions it is a person who represents more than one insurer and is legally bound to offer advice to clients on the type of assurance or investment contracts best suited to their needs. In general insurance the independent intermediary can represent more than six insurers and is responsible for advising clients on policies that best suit their needs. They must, themselves, have professional-indemnity insurance to cover any errors that they may make. Although, in both cases, intermediaries are the servants of the policyholder (and the insurer is therefore not responsible for their errors), they are paid by the insurer in the form of a commission, being an agreed percentage of the first or renewal premium paid by the policyholder.

Independent retailer A small retailer, generally owning between one and nine shops. They are now less important than multiple shops in many areas of retailing.

Indexation 1. The policy of connecting such economic variables as wages, taxes, social-security payments, annuities, or pensions to rises in the general price level (*see*inflation) This policy is often advocated by economists in the belief that it mitigates the effects of inflation. In practice, complete indexation is rarely possible, so that inflation usually leaves somebody worse off (e.g. lenders, savers) and somebody better off (borrowers). *See* Retail Price Index

2. An adjustment to take account of the rise in the Retail Price Index over the period of ownership of an asset. In the UK, indexation is applied to the cost, or 31 March 1982 value, of an asset. The indexed cost, or indexed 31 March 1982 value, I is deducted from. the proceeds of sale on disposal of the asset, in order to establish the. chargeable gain for capital gains tax purposes. Indexation was introduced to eliminate the part of the gain arising from inflation.

Index futures A futures contract on a financial futures mark et, such as, the London International Financial Futures and Options Exchange, which offers facilities for trading in futures and options on the FT-SE 100 Index and the FT-SE Eurotrack 100 Index (see Financial Times Share Indexes). On the FT-SE 100 Index the trading unit is £25 per index point; thus if a futures contract is purchased when the index stands at 2400, say, the buyer is covering an equivalent purchase of equities of £60,000 (£25 × 2400). If the index rises 100 points the. purchaser can sell a matching futures contract at this level, making a profit of £2500 (£25 × 100). On the FT-SE Eurotrack 100 Index the trading unit is DM100 per index point.

Index number A number used to represent the changes in a set of values between a base year and the present. If the index reflects fluctuations in a single variable, such as the price of a commodity, the index number can be calculated using the formula:

Index of Industrial Production A UK index produced by the Central Statistical Office to show changes in the volume of production of the main British industries and of the economy as a whole.

Indirect costs (indirect expenses) Expenses that cannot be traced directly to a product or cost unit and are therefore overheads (*compare*direct cost). As some indirect product costs may, however, be regarded as cost centre direct costs, the indirect cost centre costs are usually those costs requiring apportionment to cost centres in an absorption costing system.

indirect labour 1. Employees of a contractor who carry out work for another organization. This organization employs the contractor to supply the labour for a particular task For example, some councils use a contractor, and indirect labour, to collect refuse. **2.** The part of a workforce in an organization that does not directly produce goods or provide the service that the organization sells. Indirect labour includes office staff, cleaners, canteen workers, etc. *Compare* direct labour.

Indirect materials Those materials that do not feature in the final product but are necessary to carry out the production, such as machine oil, cleaning materials, and consumable materials. *Compare*direct materials.

Indirect method The method used for a cash-flow statement in which the operating profit is adjusted for non-cash charges and credits to reconcile it with the net cash flow from operating activities.

Indirect taxation Taxation that is intended to be borne by persons or organizations other than those who pay the tax (*compare*direct taxation). The principal indirect tax in the UK is VAT, which is paid by traders as goods or services enter into the chain of production, but which is ultimately borne by the consumer of the goods or services. One of the advantages of indirect taxes is that they can be collected from comparatively few sources while their economic effects can be widespread.

Individual retirement account (IRA) A US pension plan that allows annual sums to be set aside from earnings free of tax and accumulated in a fund, which pays interest. Basic-rate tax is payable once the saver starts to withdraw from the account, which must be done no later than the participant's 70th birthday. The concession is not available to employees in a company-pension scheme or profit-sharing scheme.

Indorsement See endorsement.

Industrial action Any form of coordinated action in an industrial dispute by employees, with or without the support of a trade union, that seeks to force employers to agree to their demands relating to wages, terms of employment, working conditions, etc. It may take the form of a go slow, overtime ban, work-to-rule, sit-down strike, or strike.

Industrial bank A relatively small finance house that specializes in hire purchase, obtaining its own funds by accepting long-term deposits, largely from the general public.

Industrial buildings Factories and ancillary premises used for manufacturing a product or for carrying on a trade in which goods are subjected to any process. For qualifying buildings there is a special category of capital allowance, known as industrial-buildings allowance. From 1 November 1993 this was restricted to a writing-down allowance of 4% on the straight-line method.

iIndustrial democracy Any system in which workers participate in the management of an organization or in sharing its profits. Workers' participation in decision making can improve industrial relations and add to workers' job satisfaction and motivation. Various schemes have been tried, including worker directors on the boards of some nationalized industries, workers' councils, and profit-sharing schemes.

Industrial development certificate (IDC) A certificate required by an industrial organization wishing to build a new factory in the UK or to extend an existing one. An IDC has to accompany any application for planning permission for industrial property. The certificate is issued by the Department of the Environment.

Industrial disability benefit Benefits paid by the Department of Social Security in the UK to compensate for disablement resulting from an industrial accident or from certain diseases due to the nature of a person's employment. A weekly payment is made on a scale of disablement ranging from 14% to 100%. No benefit is paid for less than 14% disablement. Claimants have to submit to a medical examination. The payment is made if the claimant is still suffering disability 15 weeks or more after the accident or onset of the disease.

Industrial engineering The application of work study techniques. It was traditionally regarded as a specialist function within organizations, with its own professional bodies. Increasingly, it is being taught as techniques to operations managers in organizations that take the more holistic approach associated with total quality management.

Industrial espionage Spying on a competitor to obtain trade secrets by dishonest means. The information sought often refers to the development of new products, innovative manufacturing techniques, commissioned market surveys, forthcoming advertising campaigns, research plans, etc. The means used include a wide range of telephoneand computer-tapping devices, infiltration of the competitor's workforce, etc.

Industrial life assurance A life-assurance policy, usually for a small amount, the premiums for which are paid on a regular basis (weekly or monthly) and collected by an agent of the assurance company, who calls at the policyholder's home. Records of the premium payments are kept in a book which -together with the policy document — has to be produced to make a claim. This type of assurance began in industrial areas (hence its name), where small weekly policies were purchased to help pay the funeral expenses of the policyholder. The company official who calls to collect the premium is called an agent or, in certain areas, a tally man. This type of insurance is now more widely known as home service assurance.

Industrial marketing research (business marketing research) Any form of marketing research undertaken among organizations that buy and add value to products and services but are not the final consumers. *Compare* consumer research.

Industrial medicine The health care provided by employers for their employees. Apart from the provision of first aid and nursing care in larger organizations, its main aims are prophylactic: the prevention of accidents by providing adequate guards for machinery, the prevention of disease by controlling noxious fumes, dusts, etc., and the prevention of stress by providing the best possible working environment. In addition, regular check-ups are often provided to monitor the health of members of staff.

Industrial relations The relationship between the management of an organization and its workforce. If industrial relations are good, the whole workforce will be well motivated to work hard for the benefit of the organization and its customers., The job satisfaction in such an environment will itself provide some of the rewards for achieving good industrial relations. If the industrial relations are bad, both management and workers will find the workplace an uncongenial environment, causing discontent, poor motivation, and a marked tendency to take self-destructive industrial action.

industrial tribunal Any of the bodies established under the UK employment protection legislation to hear disputes between employers and employees or trade unions relating to statutory terms and conditions of employment. For example, the tribunals hear complaints concerning unfair dismissal redundancy, equal pay, and maternity rights. Tribunals may also hear complaints from members of trade unions concerning unjustifiable disciplining by their union. The tribunal usually consists of a legally qualified chairman and two independent laymen; they are appointed by the Secretary of State for Trade and Industry. It differs from a civil court in that it cannot enforce its awards (this must be done by separate application to a court) and it can conduct its proceedings informally. Strict rules of evidence need not apply and the parties can present their own case or be represented by anyone they wish at their own expense; legal aid is not available. The tribunal has wide powers to declare a party in breach of a contract of employment, to award compensation, and to order the reinstatement or reengagement of a dismissed employee.

industry 1. An organized activity in which capital and labour are utilized to produce goods.

2. The sector of an economy that is concerned with manufacture.

Inertia selling A form of selling in which goods are sent to a potential customer on a sale-or-return basis, without his prior consent or knowledge. This method is especially prevalent with book record, and video clubs.

Infant industry A new industry that may merit some protection against foreign competition in the short term. The argument in favour of providing protection, say in the form of a tariff on imported competitors' goods, is that it would provide a period during which the new industry could streamline its process and make the necessary economies in order for it to become truly competitive with all market producers.

Infinite capacity loading Loading based only on the due dates for deliveries, taking no notice of whether capacity is available. This parameter encourages a critical examination of the options available for expanding capacity, e.g. overtime, subcontracting. *Compare*finite capacity loading.

Inflation A general increase in prices in an economy and consequent fall in the purchasing value of money. *See also* Retail Price Index.

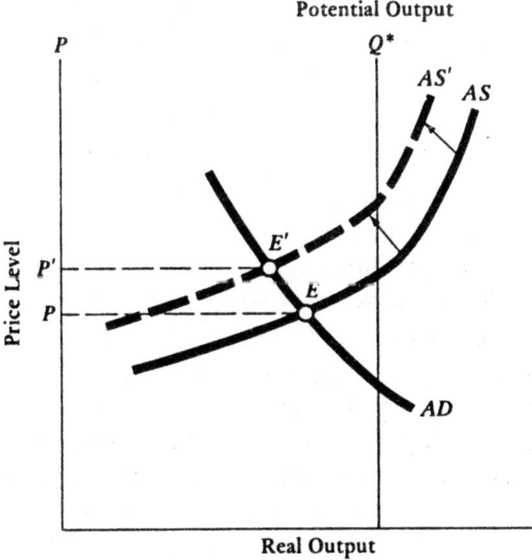

Fig. Inflation

Inflation accounting A method of accounting that, unlike historical cost accounting, attempts to take account of the fact that a monetary unit (e.g. the pound sterling) does not have a constant value; because of the effects of inflation, successive accounts expressed in that unit do not necessarily give a fair view of the trend of profits. The principal methods of dealing with inflation have been current cost accounting and current purchasing power accounting.

Inflationary gap 1. The difference between the total spending in an economy (both private and public) and the total spending that would be needed to full employment.

2. Government expenditure in excess of taxation income and borrowing from the public This excess will be financed by increasing the money supply, by printing paper money, or by borrowing from banks.

Infomercial A television commercial, usually 30 minutes long, that has the appearance of being a programme. The product advertised is repeatedly demonstrated, together with a telephone number to enable the viewer to order. Infomercials are widely used in the USA.

Infopreneural industry The manufacture and sale of electronic office and factory equipment for the distribution of information. *See*information technology.

Information technology (IT) The use of computers and other electronic means to process and distribute information. Information can be transferred between computers using cables, satellite links, or telephone lines. Networks of connected computers can be used to send e-mail or to interrogate databases, using such systems as Viewdata and Teletext. These systems also enable electronic transfer of funds to be made between banks, as well as telebanking and teleshopping from the home The same technology is used in the entertainment industry to provide cable and satellite television and videotape and laser disk films. *See also*Internet.

Infrastructure The goods and services, usually requiring substantial investment, considered essential to the proper functioning of an economy. For example, roads, railways, sewerage, and electricity supply constitute essential elements of a community's infrastructure. Since the infrastructure benefits the whole of economy, it is often argued that it should be funded, partly if not wholly, by the government by means of taxation.

Inherent vice A defect or weakness of an item, especially of a cargo, that causes it to suffer some form of damage or destruction without the intervention of an outside cause. For example, certain substances, such as jute, when shipped in bales, can warm up spontaneously, causing damage to the fibre. Damage by this cause is excluded from most cargo insurance policies as an excepted peril.

Inheritance tax A form of wealth tax on amounts inherited, usually in excess of a specified amount that is free of tax (£200,000 in the UK, from the tax year 1996-97). A true inheritance tax would be based on the amounts inherited by an individual or possibly by a household. The inheritance tax introduced in 1986 in the UK is not strictly such a tax since it taxes the total estate left by a deceased person together with a tapered charge on gifts inter vivos that that person may have made up to seven years before his or her death. This charge is added to the cumulative total used to calculate the amount of inheritance tax due. For the purposes of working out any inheritance tax due, the estate is said to be comprised of chargeable assets only, such as houses and stocks and shares.

Initial charge The charge paid to the managers of a unit trust by an investor when units are first purchased. For most trusts the initial charge ranges between 5% and 6%, as laid down in the trust deed. Money-market unit trusts and exempt unit trusts have lower initial charges and in some cases no initial charge.

Initial yield The gross initial annual income from an asset divided by the initial cost of that asset. *Compare*gross redemption yield.

Injunction An order by a court that a person shall do, or refrain from doing, a particular act. This is an equitable remedy that may be granted by the High Court wherever 'just and convenient'. The county court also has a limited jurisdiction to grant injunctions. An interlocutory injunction lasts only until the main action is heard. Interim injunctions, lasting a short time only, may be granted on the application of one party without the other being present (an ex parte interim injunction) if there is great urgency. Prohibitory Injunctions forbid the doing of a particular act; mandatory injunctions order a person to do some act. Failure to obey an injunction is contempt of court and punishable by a fine or imprisonment.

Inland bill A bill of exchange that is both drawn and payable in the UK. Any other bill is classed as a foreign bill.

Innovation Any new approach to designing, producing, or marketing goods that gives the innovator or his company an advantage over competitors. By means of patents, a successful innovator can enjoy a temporary monopoly, although eventually competitors will find ways of supplying a valuable market. Some companies rely on bringing out new products based on established demand, while others develop technological innovations that open up new markets.

Input tax Value added tax paid by a taxable person on purchasing goods or services from a VAT-registered trader. The input tax, excluding irrecoverable input VAT, is set against the output tax in order to establish the amount of VAT to be paid to the tax authorities.

Inquiry test A method of testing the response to an advertisement or a particular medium by comparing the number of inquiries received as a result of it.

Inscribed stock (registered stock) Shares in loan stock, for which the names of the holders are kept in a register rather than by the issue of a certificate of ownership. On a transfer, a new name has to be registered, which makes them cumbersome and unpopular in practice.

Inside director in the USA, an employee of a company who has been appointed to the board of directors.

Insider dealing (insider trading) Dealing in

company securities with a view to making a profit or avoiding a loss while in possession of information that if generally known, would affect their price. Under the *Companies Securities (Insider Dealing) Act (1985)* those who are or have been connected with a company (e.g. the directors, the company secretary, employees, and professional advisers) are prohibited from such dealing on or, in certain circumstances, off the stock exchange if they acquired the information by virtue of their connection and in confidence. The prohibition extends to certain unconnected persons to whom the information has been conveyed.

Insolvency The inability to pay one's debts when they fall due. In the case of individuals this may lead to bankruptcy and in the case of companies to liquidation. In both of these cases the normal procedure is for a specialist, a trustee in bankruptcy or a liquidator, to be appointed to gather and dispose of the assets of the insolvent and, to pay the creditors. Insolvency does not always lead to bankruptcy and liquidation. although it often does. An insolvent person may have valuable assets that are not immediately realizable.

Insolvency practitioner A person authorized to undertake insolvency administration as a liquidator, provisional liquidator, administrator, administrative receiver, or nominee or supervisor under a voluntary arrangement. Insolvency practitioners are members of the Insolvency Practitioners Association.

Insolvency Practitioners Association A professional body whose members act as liquidators, receivers, and trustees in insolvency and are identified by the letters MIPA (membership) or FIPA (fellowship).

Inspection and investigation of a company An inquiry into the running of a company made by inspectors appointed by the Department of Trade and Industry. Such an inquiry may be held to supply company members with information or to investigate fraud, unfair prejudice, nominee shareholdings, or insider dealing. The inspectors' report is usually published.

Inspector general In the USA, the federal office that performs audit and investigative activities on federal agencies, making periodic reports to Congress.

Inspector of Taxes A civil servant responsible to the Board of Inland Revenue for issuing tax returns and assessments, the conduct of appeals, and agreeing tax liabilities with taxpayers.

Instalment One of a series of payments, especially when buying goods hire purchase, settling a debt, or buying a new issue of shares.

Insurable interest The legal right to enter into an insurance contract. A person is said to have an insurable interest if the event insured against could cause that person a financial loss. For example, anyone may insure their own property as they would incur a loss if an item was lost, destroyed, or aged. If no financial loss would occur, no insurance be arranged. For example, a person cannot insure a next-door neighbour's property. The limit of an insurable interest is the value of the item concerned, although there is no limit on the amount of life assurance a person can take out on his or her own life, or that of a spouse, because the financial effects of death cannot be measured.

Insurable interest was made a condition of all insurance by the *Life Assurance Act (1774)*. Without an insurable interest, an insured person is unable to enforce an insurance contract (or lifeassurance contract) as it is the insurable interest that distinguishes insurance from a bet or wager.

Insurance A legal contract in which an insurer promises to pay a specified amount to another party, the insured, if a particular event (known as the peril), happens and the insured suffers a financial loss as a result. The insured's part of the contract is to promise to pay an amount of money, known as the premium, either once or at regular intervals. In order for an insurance contract to be valid, the insured must have an insurable interest. It is usual to use the word 'insurance' to cover events (such as a fire) that may or may not happen, whereas assurance refers to an event (such as death) that must occur at some time (*see also*lifeassurance). The main branches of insurance are: accident insurance, fire insurance, holiday and travel insurance, household insurance, liability insurance, livestock and bloodstock insurance, loss-of-profit insurance (*see*business-interruption policy), marine insurance, motor insurance, National Insurance, pluvial insurance, private health insurance, and property insurance. *See* also reinsurance.

Insurance broker A person who is registered with the Insurance Brokers Registration Council to offer advice on all insurance matters and arrange cover, on behalf of the client, with an insurer. An insurance broker's income comes from commission paid to him by insurers, usually in the form of an agreed percentage of the first premium or on subsequent premiums. *See also* Lloyd's.

Insurance Brokers Registration Council A statutory body established under the *Insurance Brokers Registration Act (1977)*. It is responsible for the registration and training of insurance brokers and for laying down rules relating to such matters as accounting practice, staff qualifications, advertising, and the orderly conduct and discipline of broking businesses.

Insurance Ombudsman *see***Financial Ombudsmen.**

Insurance policy A document that sets out the terms and conditions of insurance contract, stating the benefits payable and the premium required. *See also*life assurance.

Insurance premium *See***insurance; premium.**

Insurance tied agent An agent who represents a particular insurance company or companies. In life and pensions insurance, a tied agent represents only one insurer and is only able to advise the public on the policies offered by that one company. In general insurance (motor, household, holiday, etc.), a tied agent represents no more than six insurers, who are jointly responsible for the financial consequences of any failure or mistake the agent makes. In both cases the agent receives a commission for each policy that is sold and a further commission on each subsequent renewal of the policy. The commission is calculated as an agreed percentage of the total premium paid by the policyholder. The distinction between the two forms of insurance tied agent was a consequence of the *Financial Services Act (1987)* and the *General Insurance Selling Code (1989)* of the Association of British Insurers.

Insured A person or company covered by an insurance policy. In some policies that cover death, the alternative word assured may be used for the person who receives the payment in the event of the assured's death.

Insurer A person, company, syndicate, or other organization that underwrites an insurance risk.

Intangible asset (invisible asset) An asset that can neither be seen nor touched. The most common of these are goodwill, patents, trademarks, and copyrights. Goodwill is probably the most intangible and invisible of all assets as no document provides evidence of its existence and its commercial value is difficult to determine. However, it frequently does have very substantial value as the capitalized value of future profits, not attributable purely to the return on tangible assets. While goodwill is called either an intangible asset or an invisible asset, such items as insurance policies and less tangible overseas investments are usually called invisible assets. *Compare*fictitious asset.

Intangible property A property that cannot be possessed physically but that confers on its owner a legally enforceable right to receive a benefit, for example money. *See also* intangible asset.

Integrated accounts Accounting records kept in one set of books that contains both the financial accounts and the cost accounts of an organization in an integrated form. This avoids the necessity of reconciling separate financial and cost books and at the same time ensures that both records are based on the same data.

Integrated marketing communications Marketing communications in which all the elements of promotional mix are coordinated and systematically plumed to be in harmony with each other.

Integration The combination of two or more companies under the same control for their mutual benefit, by reducing competition, saving costs by reducing overheads, capturing a larger market share, pooling technical or financial resources, cooperating on research and development, etc. In horizontal (*or* lateral) integration the businesses carry out the same stage in the production process or produce similar products or services; they are therefore competitors. In a monopoly, horizontal integration is complete, while, in an oligopoly there is considerable horizontal integration. In vertical integration a company obtains control of its suppliers (sometimes called bakcward integration) or of the concerns that buy its products or services (forward integration).

Intelligent terminal A computer terminal that is able to process data without the help of the computer to which it is connected. An example is a point-ofsale terminal that adds up the cost of the items purchased and also informs the main stock computer that they are no longer in stock.

Intensive distribution A distribution strategy aimed at obtaining the maximum number of outlets to stock a product at the wholesale or retail level. Such products as sweets, chewing gum, disposable razors, soft drinks, and camera film are sold in numerous outlets (including supermarkets, newsagents, and petrol stations) to provide maximum brand exposure and consumer convenience.

Interactive Denoting a computer system in which the operator can communicate with the computer while it is running a program. The computer program prompts the operator when it needs information and stops running until the information is supplied. Interactive computer systems are also called conversational systems. *Compare*batch processing.

Interbank market 1. The part of the Londonmoney market in which banks lend to each other and to other large financial institutions. The London Inter Bank Offered Rate (LIBOR) is the rate of interest charged on interbank loans. Trading is over-the-counter and usually through brokers and dealers. The sums are large but the periods of the loans are very short, often overnight.

2. The market between banks in foreign currencies, including spot currencies and forward options.

Intercompany transactions (intragroup transaction) Transactions between the companies in a group. These may be in the form of charges or the transfer of goods or services. It is important in the preparation of consolidated financial statements that such transactions are eliminated or suitable adjustments made as they do not reflect transactions between the group and external parties. *See also*consolidation.

Inter-dealer broker A member of the London Stock Exchange who is only permitted to deal with market makers, rather than the public.

Interest The charge made for borrowing a sum of money. The rate of interest is the charge made, expressed as a percentage of the total sum loaned, for a stated period of time (usually one year). Thus, a rate of interest of 15% per annum means that for every £100 borrowed for one year, the borrower has to pay a charge of £15, or a charge in proportion for longer or shorter periods. In simple interest the charge is calculated on the sum loaned only, thus $I = Prt$, where I is the interest, P is the principal sum, r is the rate of interest, and t is the period. In compound interest, the charge is calculated on the sum loaned plus any interest accrued in previousperiods. In this case $I = P\ [(1 + r)n - 1]$, where n is the number of periods for which interest is separately calculated. Thus, if £500 is loaned for 2 years at a rate of 12% per annum, compounded quarterly, the value of n will be 4 x 2 = 8 and the value of r will be 12/4 = 3%. Thus, $I = 500\ [(1.03)^8 - 1]$ = £133.38, whereas on a simple-interest basis it would be only EM.

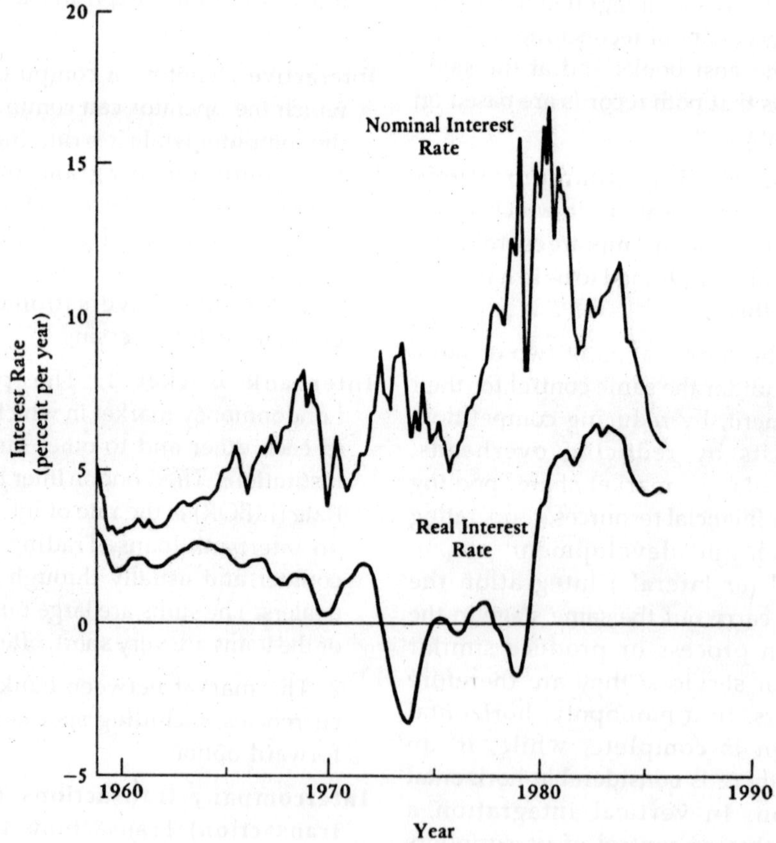

Fig. Interest rates

In general rates of interest depend on the money supply, the demand for loans, government policy, the risk of nonrepayment as assessed by the lender, and the period of the loan.

Interest arbitrage Transactions between financial centres in foreign currencies that take advantage of differentials in interest rates between the two centres and the difference between the forward and spot exchange rates. In some circumstances it is possible to make a profit by buying a foreign currency, changing it into the currency of the home market, lending it for a fixed period, and buying back the foreign currency on a forward basis.

Interest-bearing eligible liabilities (IBEL) Liabilities recognized by the Bank of England as being held by UK banks, e.g. net deposits. In times of strict monetary control the Bank of England requires other UK banks to deposit a percentage of these liabilities with it.

Interest cover (fixed-charge-coverage ratio) A ratio showing the number of times interest charges are covered by earnings before interest and tax. For example, a company with interest charges of £12 million and earnings before interest and tax of £36 million would have its interest covered three times. The ratio is one way of analysing gearing and reflects the vulnerability of a company to changes in interest rates or profit fluctuations. A highly geared company, which has a low interest cover, may find that an increase in the interest rate will mean that it has no earnings after interest charges with which to provide a dividend to shareholders.

Interest-in-possession trust A type of fixed-interest trust in which there is an entitlement to the income generated by the trust assets. The beneficiaries of an interest-in-possession trust, the life tenants, are entitled to the income arising for a fixed period or until their death. The capital in the trust then passes absolutely to the recipient known as the remainderman.

Interest rate The amount charged for a loan, usually expressed as a percentage of the sum borrowed. Conversely, the amount paid by a bank, building society. etc., to a depositor on funds deposited, again expressed as a percentage of the sum deposited.

Interest-rate futures A form of financial futures that enables investors, portfolio managers, borrowers, etc, to obtain protection against future movements in interest rates. Interest-rate futures also enable dealers to speculate on these movements. In the UK, interestrate futures are dealt in on the London International Financial Futures and Options Exchange, using contracts for three-month sterling, eurodollars, ECUs, etc., in the short term and long gilts, US Treasury bonds, ECU bonds, etc, in the long term. For long gilts, for example, the standard contract is for £50,000 of nominal value and the tick size is 0.01% of the nominal value.

Interest-rate guarantee An indemnity sold by a bank, or similar financial institution, that protects the purchaser against the effect of future movements in interest rates. It is similar to a forward-rate agreement, but the terms are specified by the customer.

Interest-rate margin 1. The amount charged to borrowers over and above the base rate. This margin is the bank's profit on the transaction but has to take account of risk of loss or default by the borrower.

2. The difference between the rate at which banks lend and the rate they pay for deposits. It may be a major indicator of banks' profitability.

Interest-rate option A form of option enabling traders and speculators to hedge themselves against future changes in interest rates.

Interest-rate policy The policy by which

governments influence interest rates through the supply of bonds. Higher rates of interest will discourage investment and reduce the demand for money, giving a downward impetus to both employment and prices. Lower interest rates will have the opposite effect.

Interes-rate swap A form of dealing between banks, security houses, and companies in which borrowers exchange fixed-interest rates for floating-interest rates, or vice versa. Swaps can be in the same currency (*see*currency interestrate swap) or cross-currency (*see*crossinterest-rate swap). *See also*rate anticipation swap.

Interest-sensitive Denoting an activity that is sensitive to changes in the general level of interest rates. For example, consumer goods usually bought on some form of hire-purchase basis are interest-sensitive because they cost more as interest rates rise.

Interim accounts *See***interim financial statements.**

Interim dividend A dividend paid midway through a financial year.

Interim financial statements (interim accounts; interim report) Financial statements issued for a period of less than a financial year. Although there are provisions under the *Companies Act (1985)* that refer to interim accounts in certain circumstances relating to the distribution of dividends, there are no legal requirements obliging companies to produce interim accounts on a regular basis. However, listed companies on the London Stock Exchange are required to prepare a half-yearly report on their activities and profit and loss during the first six months of each financial year. The interim financial statement must be either sent to the holders of the company's listed securities or advertised in at least one national newspaper not later than four months after the end of the period to which it relates. A copy of the interim financial statements must also be sent to the Company Announcements Office and to the competent authority of each other state in which the company's shares are listed. The vast majority of companies choose to send the interim statement to shareholders with a brief announcement of the headline figures reported in the press. There is no requirement for the interim statements to be audited. Although the stockexchange regulations require mainly profit information, there is a trend for the larger companies to also provide balance sheet and cash-flow statement. In the UK, the requirements only call for six-monthly financial statements, but some of the larger companies with interests in the USA follow the US practice of issuing reports quarterly.

Intermarket spread swap Swapping securities the hope of improving the spread between the yields of the securities.

Intermediation The activity of a similar financial institution, broker, etc., in acting as an intermediary between the two parties to a transaction; the intermediary can accept all or part of the credit risk or the other commercial risks. *Compare*disintermediation.

Intermodalism A method of using more than one form of transport in the delivery of goods. In the USA, for example, an increasing number of manufacturers and forwarding agents are delivering goods by combining low-cost long-haul rail transport with the flexibility of road transport to the final destination.

Internal audit An audit that an organization carries out on its own behalf, normally to ensure that its own internal controls are operating satisfactorily. Whereas an external audit is almost always concerned with financial matters, this may not necessarily be the case with an internal audit; internal auditors may also concern themselves with such matters as the observation of the safety and health at work regulations or of the equal opportunities legislation. It may also be used to detect any theft or fraud (*see also*internal control).

Internal control The measures an organization employs to ensure that opportunities for fraud or are minimized. Examples range from requiring more than one signature on certain documents, security arrangements for stock-handling, division of tasks, keeping of control accounts, use of special passwords, handling of computer files, etc. It is one of the principal concerns of an internal audit to ensure that internal controls are working properly so that the external auditors have faith in the accounts produced by the organization. Internal control should also reassure management of the integrity of its operations.

Internal control system A system of controls, both financial and non-financial set up by the management of a company to carry out the business of the company in an orderly and efficient manner. The system should ensure that management policies are adhered to, assets are safeguarded, and the records of the company's activities are both complete and accurate. The individual components of an internal control system are the individual internal controls.

Internal growth (organic growth) The means by which a business can grow using its own resources (*see*business strategy). A business will grow by increased market penetration at the expense of its competitors, by new product development, and by market development through seeking new applications and markets for existing products or services. All these means utilize a firm's core competencies and while some may have a relatively short lead tune (e.g. market penetration) others, notably product development, can involve significant lead times and development costs. Firms that have a successful record of innovation are usually more successful at internal growth. whereas others favour greater reliance on external growth.

Internal rate of return (IRRA) An interest rate that gives a net present value of zero when applied to a projected cash flow. This interest rate, where the present values of the cash inflows and outflows are equal, is the internal rate of return for a project under consideration, and the decision to adopt the project would depend on its size compared with the cost of capital. The approximate IRR can be computed by interpolation but most computer spreadsheet programs now include a routine enabling the ERR to be computed quickly and accurately. The ERR technique suffers from the possibility of multiple solution rates in some circumstances.

International Accounting Standards (IAS) Accounting standards issued by the International Accounting Standards Committee. Some of the advantages claimed for international standards are that financial statements prepared in different countries will be more comparable, multinational companies will find preparation of their accounts easier, listing on different stock exchanges can be achieved more simply, and financial statements will be of greater use to users. However, international standards are not mandatory; moreover, some permit such a degree of flexibility in accounting treatments that comparability is impaired. Countries in which the setting of accounting standards is well established have national accounting standards that usually deal with the same topics as international standards. Some countries, with a less well-developed procedure for setting standards, adopt international accounting standards or use them as a model for preparing their own.

International Bank for Reconstruction and Development (IBRD) A specialized agency working in coordination with the United Nations, established in 1945 to help finance postwar reconstruction and to help raise standards of living in developing countries, by making loans to governments or guaranteeing outside loans. It lends on broadly commercial terms, either for specific projects or for more general social purposes; funds are raised on the international capital markets. The Bank and its affiliates, the International Development Association and the International Finance Corporation, are often known as the World Bank; it is owned by the governments of 151 countries. Members must also be members of the International Monetary Fund. The headquarters of the Bank are in Washington, with a European office in Paris and a Tokyo office.

International Chamber of Commom (ICC) An international business organization that represents business interests in international affairs. Its office is in Paris. *See also* **Incoterms.**

international commodity agreements Agreements between governments that aim to stabilize the price of commodities. This is often important for producing nations, for whom the revenue from commodity sales may make a major contribution to the national income. Methods tried include government-financed buffer stocks and the imposition of price limits between which the price of the commodity is allowed to fluctuate.

International Development Association (IDA) An affiliate of the International Bank for Reconstruction and Development (IBRD) established in 1960 to provide assistance for poorer developing countries (those with a GNP of less than $835 per head). With the IBRD it is known as the World Bank. It is funded by subscription and transfers from the net earnings of IBRD. The headquarters of the IDA are in Washington, with offices in Paris and Tokyo, administered by IBRD staff.

International Finance Corporation (IFC) An affiliate of the International Bank for Reconstruction and Development (IBRD) established in 1956 to provide assistance for private investment projects. Although the IFC and IBRD are separate entities, both legally and financially, the IFC is able to borrow from the IBRD and reloan to private investors. The headquarters of the IFC are in Washington. Its importance increased in the 1980s after the emergence of the debt crisis and the subsequent reliance on the private sector.

International Fund for Agricultural Development (IFAD) A fund, proposed by the 1974 World Food Conference, that began operations in 1977 with the purpose of providing additional funds for agricultural and rural development in developing countries. It has some 158 member states. The headquarters of the IFAD are in Rome.

International Labor Organization (ILO) A special agency of the United Nations, established under the Treaty of Versailles in 1919, with the aim of promoting lasting peace through social justice. The ILO establishes international labour standards, runs a programme of technical assistance to developing countries, and attacks unemployment through its World Employment Programme. The ILO is financed by contributions from its member states. Its headquarters, The International Labour Office, is in Geneva.

International marketing *See***multinational marketing.**

International Monetary Fund (IMF) A specialized agency of the United Nations established in 1945 to promote international monetary cooperation and expand international trade, stabilize exchange rates, and help countries experiencing short-term balance of payments difficulties to maintain their exchange rates. The Fund assists members by supplying the amount of foreign currency it wishes to purchase in exchange for the equivalent amount of its own currency. The member repays this amount by buying back its own currency in a currency acceptable to the Fund, usually within three to five years (*see also*Special Drawing Rights). The Fund is financed by subscriptions from its members, the amount determined by an estimate of their means. Voting power is related to the amount of the subscription - the higher the contribution the higher the voting rights. The head office of the IMF is in Washington. *See also*conditionality.

International Organization for Securities Commission (IOSCO) A body formed in 1987 with the objective of establishing internationally agreed accounting standards to aid in multinational share offering by companies. Formerly critical of Accounting Standards issued by the International Accounting Standards Committee, it is now actively cooperating in the improvement of these standards.

International Petroleum Exchange (IPE) An exchange, founded in London in 1980, that deals in futures contracts and options (including traded options) in oil (gas oil, Brent crude oil, and heavy fuel oil) with facilities for making an exchange of futures for physicals (*see* actuals). There are 35 floor members of the exchange, which is located in St Katharine Dock and is shared with the London Commodity Exchange.

International Securities Market Association The eurobond market's trade association.

International Standards Organization (ISO) An organization founded in 1946 to standardize measurements and other standards for industrial commercial, and scientific purposes. The British Standards Institution is a member.

International Telecommunication Union (ITU) A special agency of the United Nations founded in Paris in 1865 as the International Telegraph Union. It sets up international regulations for telegraph, telephone, and radio services; promotes international cooperation for the improvement of telecommunications; and is concerned with the development of technical facilities. It is responsible for the allocation and registration of radio frequencies for communications. Its headquarters are in Geneva.

Interpolation Estimation of the value of an unknown quantity that lies between two of a series of known values. For example, one can estimate by interpolation the population of a country at any time between known 10-year census figures. This would normally be done on a graph, the shape of the curve between known points enabling a good estimate to be made of the interpolated value. **Extrapolation** involves estimating an unknown quantity that lies outside a series of known values. Thus one could obtain a figure for the population of a country five years after its last census, by extending the population curve after the last known value.

Interpreter A computer program that executes a program written in a programmming language without first translating it into machine code. The program runs as if it had been so translated, but much more slowly. An interpreter usually allows the operator to interrupt a program and inspect or alter the values of variables.

Intervention mechanism The action by central banks and governments to stabilize exchange rates of currencies in the European Monetary System by buying or selling currencies on the open market or by initiating currency swaps.

Interviewer error An error in marketing research that results from conscious or unconscious bias on the part of the interviewer. For example, the interviewer may guess incorrectly the age of the respondent or may assume that because someone is driving a car they own the car.

Intestate A person who dies without having made a will. The estate, in these circumstances, is divided according to the rules of intestacy. The division depends on the personal circumstances of the deceased. If there is a spouse then there is a fixed statutory legacy, with the remainder being split between an interest-in-possession trust for the remaining spouse and the children's absolute entitlement to the other half, if they are over 18. Where there is no surviving spouse the estate is divided between the children or their issue. Where there are no children then the split is rather more complicated and can include parents, brothers and sisters, grandparents, uncles, and aunts.

Intrapreneur An executive or senior manager of a large firm who uses the entrepreneurial approach to management. In Order to encourage entrepreneurship, it has become common for firms to encourage risk-taking and profit-orientated behaviour among its managers by organizing a business into a number of separate profit centres. Each centre is responsible for making a profit by intra-group trading and, where appropriate, by building up its external sales and clients. This approach has been used in the service departments of large companies, giving subsidiary companies the choice of using the internal services or of going outside to independent suppliers. This process of developing internal market for services has also been applied by the public sector; for example, the UK National Health Service has been encouraged to use this intrapreneurial approah. Critics, however, often suggest that decisions may then be taken too narrowly, rather than for the benefit of the whole organization.

Intrinsic value 1. The value something has because of its nature, before it has been processed in any way.

2. The difference between the market value of the underlying security in a traded option and the exercise price. An option with an intrinsic value is said to be in the money, an option with zero intrinsic is at the money, one with less intrinsic value than zero is out of themoney *See also*timevalue.

Introduction A method of issuing shares on the London Stock Exchange in which a broker or issuing house takes small quantities of the company's shares and issues them to clients at opportune It is also used by existing public companies that wish to issue additional shares. *Compare*issue by tender, offer for sale ; placing ; public issue.

Inventoriable costs Costs that can be included in the valuation of stocks, work in progress, or inventories according to Statement of Standard Accounting Practice 9. Stocks should be valued at the lower of cost or net realizable value and the costs incurred up to the stage of production reached. This effectively means that inventoriable costs for finished goods and work in progress include both fixed and variable production costs but exclude the selling and distribution costs.

Inventory 1. In the USA, the equivalent of the UKstock, i.e. the products or supplies of an organization on hand or in transit at any time. For a manufacturing company the types of inventory are raw materials, work in progress, and finished goods. An inventory count usually takes place at the end of the financial year to confirm that the actual quantities support the figures given in the books of account. The differences between the inventories at the beginning and the end of a period are used in the calculation of cost of sales for the profit and loss account and the end inventory is shown on the balance sheet as circulating capital

2. More broadly, the total accumulation of material resources within an operating system (not only what is held deliberately in stock). It includes all the transformed resources that are locked up in the system, such as work in progress, scrap, rejects, etc. All operations have some inventory, which represents locked-up capital. The task of operations is to ensure that levels are kept to the minimum without affecting the efficiency and effectiveness of the transformation process. Inventory, in this sense, falls into a number of different categories. Anticipatory inventory is produced in periods of relatively low demand in anticipation of a later increase in demand. This is only possible if the goods are non-perishable and is only economically viable if the costs of storage are less than the costs of changing production levels to match demand. Uncoupling inventory is the amount of material held to decouple elements of the transformation system, such as the raw-material supplier from the manufacturer, one machine from another, or the manufacturer from the customer. This provides an element of operational flexibility. Similar to uncoupling inventory is the buffer inventory, which protects against variations in supply, demand, and lead times. It represents a safety margin and is therefore directly related to the reliability of demand forecasting, etc. Pipeline inventory is that part of the inventory that is locked up in the system. Because all processes take time, the pipeline is a function of a transformation system rather than performing a function within it. If organizations have to subcontract part of their products or if elements of the system are widely separated geographically, the amount of resource locked up in the pipeline. can increase dramatically. Modern process-monitoring systems based on information technology are designed to ensure that the pipeline inventory in particular is kept to a minimum. Cycle inventory is a function of the economic order quantity (EOQ) calculation and represents a trade-off between minimizing the inventory and the costs of setting up and processes. Just-in-time techniques are designed to rely on a minimum EOQ. Clearly there is some overlap between these different types of inventory.

Inventory status record A datafile containing up-to-date information on the status of every item, component, etc., controlled by the material requirements planning system. This file contains the identification number, quantity in hand, safety stock level, quantity allocated, and procurement lead time of every item.

Inventory valuation (stock valuation) The valuation of stocks of raw material work in progress, and finished goods. According to Statement of Standard Accounting Practice 9, stocks should be valued at the lower of cost or net realizable value and the costs incurred up to the stage of production reached. This effectively means that finished goods and work in progress should include both fixed and variable production costs but exclude the selling and distribution costs. In the UK valuing stocks at cost, the first-in-first-out cost, or the average cost may be used, but not the last-in-first-out cost or the next-in-first-out cost. Marginal cost may be used as a basis of stock valuation for management accounting purposes but is unacceptable by Statement of Standard Accounting Practice 9 for financial accounting.

Investment 1. The purchase of capital goods, such as plant and machinery in a factory in order to produce goods for future

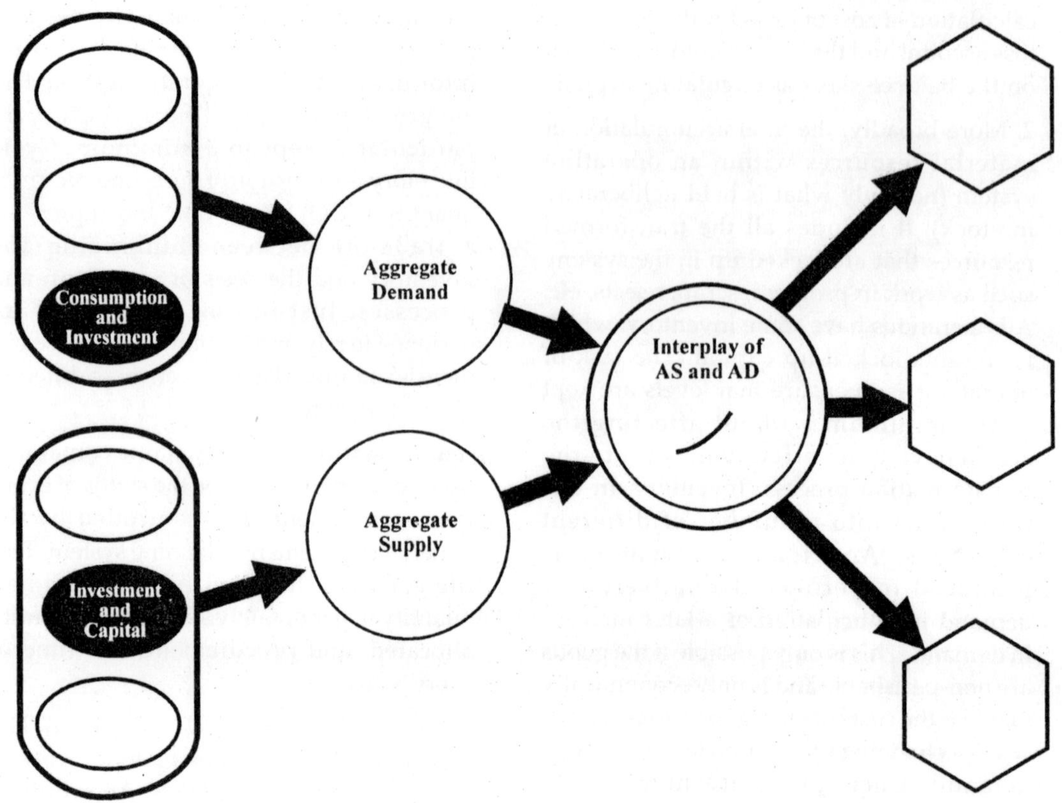

Fig. Consumption and Investment

consumption. This is known as capital investment, the higher the level of capital investment in an economy, the faster it will grow.

2. The purchase of assets, such as securities, works of art, bank and building-society deposits, etc., with a primary view to their financial return, either as income or capital gain. This form of financial investment represents a means of saving. The level of financial investment in an economy, will be related to such factors as the rate of interest, the extent to which investments are likely to prove profitable, and the general climate of business, confidence.

Investment analyst A person employed by stockbrokers, banks, insurance companies, unit trusts, pension funds, etc., to give advice on the making of investments, especially investments in securities, commodities, etc. Many pay special attention to the study of equities in the hope of being able to advise their employers to make profitable purchases of ordinary shares. To do this they use a variety of techniques, including a comparison of a company's present profits with its future trading prospects; this enables the analyst to single out the companies likely to outperform the general level of the market. This form of technical analysis is often contrasted with fundamental analysis, in which predicted future market movements are related to the underlying state of an economy and its expected trends. Analysts who rely on past movements to predict the future are called chartists.

Investment bank A US bank that fulfils many of the functions of a UK merchant bank. It is usually one that advises on mergers and acquisitions and provides finance for industrial corporations by buying shares in a company and selling them in relatively small lots to investors. Capital provided to companies is usually long-term and based on fixed assets. In the USA, commercial banks were excluded from selling securities for many years but the law was relaxed in the late 1980s, when certain safeguards were introduced including a ceiling on the value of transactions.

Investment bond A single-premium life-assurance policy in which an investment of a fixed amount is made (usually over £1000) in an asset-backed fund. Interest is paid at an agreed rate and at the end of the period the investment is returned with any growth. Investment bonds confer attractive tax benefits in some circumstances.

Investment club A group of investors who, by pooling their resources, are able to make more frequent and larger investments on a stock exchange, often being able to reduce brokerage and to spread the risk of serious loss. The popularity of investment clubs has waned with the rise of unit trusts and investment trusts, both of which have the advantages of professional management.

Incentive) A grant (investment incentive) A grant made to a company by a government to encourage investment in plant, machinery, buildings, etc. The incentives may take various forms, some being treated as a deduction for tax purposes, others being paid whether or not the company makes a profit.

Investment income 1. A person's income derived from investments.

2. The income of a business derived from its outside investments rather than from its trading activities.

Investment properties Properties owned by a company that holds investments as part of its business, such as an investment trust or a property-investment company. Investment properties may also include properties owned by a company whose main business is not the holding of investments. Such properties are strictly defined by Statement of Standard Accounting Practice 19, 'Accounting for Investment Properties', as being an interest in land and/or buildings: (1) in respect of which construction work and development have been completed; and (2) that is held for its investment potential any rental income being negotiated at arm's length. However, a property owned and occupied by a company for its own purposes is not an investment property, and a property let to and occupied by another company in the same group is not an investment property for the purposes of its own accounts or the group accounts. Investment properties should not be depreciated annually unless they are held on a lease. If they are leased they should be depreciated on the basis set out in Statement of Standard Accounting Practice 12, 'Accounting for Depreciation', at least over the period, when the unexpired term is 20 years or less. Investment properties should be included in the balance sheet at their open-market value, movements being taken to the investment revaluation reserve unless it is insufficient to cover a deficit, in which case it should be taken to the profit and loss account.

Investment revaluation reserve A reserve created by a company with investment properties, if these properties are included in the balance sheet at open-market value. Changes in the value of investment properties should be disdosed as movements on the investment revaluation reserve, unless the total of the investment revaluation reserve is insufficient to cover a deficit, in which case the amount by which the deficit exceeds the amount in the investment revaluation reserve should be charged to the profit and loss account. In the case of investment trust companies and property unit trusts it may not be appropriate to deal with these deficits in the profit and loss account; in these circumstances they should be shown prominently in the financial statements.

Investment tax credit In the USA, an incentive to investment in which part of the cost of an asset subject to depreciation is used to offset income tax falling due in the year of purchase.

Investment trust (investment company) A company that invests the funds provided by shareholders in a wide variety of securities. It makes its profits from the income and capital gains provided by these securities. The investments made are usually restricted to securities quoted on a stock exchange, but some will invest in unquoted companies. The advantages for shareholders are much the same as those with unit trusts, i.e. spreading the risk of investment and making use of professional managers. Investment trusts, which are not usually trusts in the usual sense, but private or public limited companies, differ from unit trusts in that in the latter the investors buy units in the fund but are not shareholders. Some investment trusts aim for high capital growth (capital shares), others for high income (income shares).

Invisible earnings (invisibles) The earnings from abroad that contribute to the balance of payments as a result of transactions involving services, such as insurance, banking, shipping, and tourism (often known as invisibles), rather than the sale and purchase of goods. Invisibles can play an important part in a nation's current account, although they are often difficult to quantify. The UK relies on a substantial invisible balance in its balance of payments.

Invoice A document stating the amount of money due to the organization issuing it for goods or services supplied. A commercial invoice will normally give a description of the goods and state how and when the goods were dispatched by the seder, who is responsible for insuring them in transit, and the payment terms.

Invoice discounting A form of debt discounting in which a business sells its invoices to a factoring house at a discount for immediate cash. The service does not usually include sales accounting and debt collecting

Irrecoverable advance corporation tax Advance corporation tax paid that cannot be set against the current year's gross corporation tax (GCT) or carried back against GCT arising on profits for the accounting periods beginning in the preceding six years, as it exceeds the maximum set-off available. The advance corporation tax is carried forward to set against future GCT liabilities arising on profits in future accounting periods. If profits are not expected in the foreseeable future or if future dividends are expected to be paid up to the maximum available for future years, the current advance corporation tax surplus is considered to be irrecoverable advance corporation tax.

Irredeemable securities (irredeemables) Securities, such as some government loan stock (*see*Consols) and some debentures, on which there is no date given for the redemption of the capital sum. The price of fixed-interest irredeemables on the open market varies inversely with the level of interest rates.

Irrevocable documentary acceptance credit A form of irrevocable confirmed letter of credit in which a foreign importer of UK goods opens a credit with a UK bank or the UK office of a local bank. The bank then issues an irrevocable letter of credit to the exporter, guaranteeing to accept bills of exchange drawn on it on presentation of the shipping documents. Once the letter of credit has been drawn up, the importer has to 'accept' that he or she will pay, by signing the acceptance.

ISLM Diagram The diagram which details the simultaneous determination of equilibrium values of the interest rate and the level of national icome because of conditions in the goods and money marketsl the integration of the real and monetary sectors of the economy by Hicks and Hansesn to reveal the determination of macroeconomic equilibirum.

The IS curve has been the schedule detailing the combination of interest rates (r) and levels of natinal income (Y) which ensure equilibrium inthe goods market. At al points along this schedule total with drawls, savings and taxation in the closed economy case, equal total injection, investment and government expenditure. The title IS curve reflects the simple case dealt with in Hick's initial exposition when only saving and investment had been considered. The LM curve has been the schedule which details the combination of interest rates and level of national income and ensure equilibrium in the money market. At all points along this schedule the demand for money market. At all point along this chedule the demand for money transaction (and precautionary motives if we subsume this into transactions demand) and speculative purposes has been equal to the exogenously determined supply of money. Initially this schedule called the L curve by Hicks, but Hansen christend it the LM curve, emphasizing that it represented a curve along which L, the demand for money=M, the supply of money.

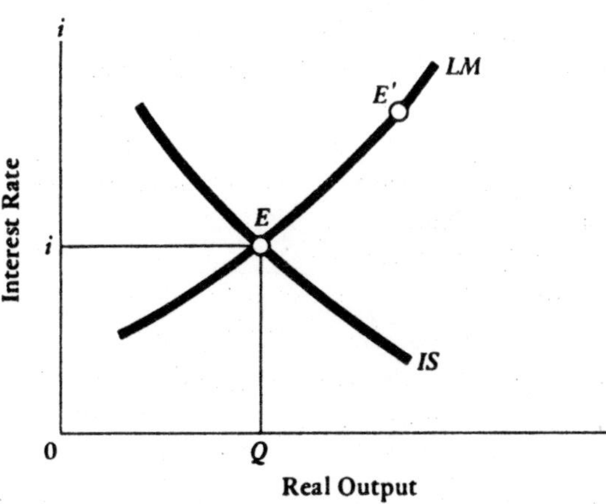

Fig. ISLM Curve

Islamic finance A form of business that is bound by strict religious rules; these rules normally prevent the making of a profit in association with nonIslamic organizations or individuals. It is now acceptable, however, for internal profits to be generated which may be achieved using an Islamic investment company and any of the following. Murabaha is a good vehicle for temporary idle funds, which are used to purchase goods from a supplier for immediate sale and delivery to the buyer, who pays a predetermined margin over cost on a deferred payment date. The term can be as short as seven days. Musharaka transactions involve participation with other parties in trade financing, leasing, real estate, and industrial projects. Net profits are shared in proportions agreed at the outset. Ijarah involves profit from rental income on real estate. Ijarawa-Iktina is leasing of large capital items, such as property or plant and machinery. Leasing is achieved by the equivalent of monthly rental payments, and at the expiry the lessee purchases the equipment.

Issue by tender (sale by tender) A method of issuing shares on the London Stock Exchange in which issuing house asks investors to tender for them. The stocks or shares are then allocated to the highest bidders. It is usual for the tender documents to state the lowest price acceptable. This method may be used for a new issue or for loan stock (*see*debenture), but is not frequently employed. *Compare*introduction ; offer for sale ; placing ; public issue.

Issued share A share that has been allotted by the directors of a company to an applicant and paid for in full by that applicant.

Issued share capital (subscribed share capital) The amount of the authorized share capital for which shareholders have subscribed. *See also* share capital.

issue price The price at which a new issue of shares is sold to the public. Once the issue has been made the securities will have a market price, which may be above (at a premium on) or below (at a discount on) the issue price (*see also***stag**). In an introduction, offer for sale, or public issue, the issue price is fixed by the company on the advice of its stockbrokers and bankers; in an issue by tender the issue price is fixed by the highest bidder; in a placing the issue price is negotiated by the issuing house or broker involved.

Issuing house A financial institution, usually a merchant bank that specializes in the flotation of private companies on a stock exchange. In some cases the issuing house will itself purchase the whole issue (*see*underwriter), thus ensuring that there is no uncertainty in the amount of money the company will raise by flotation. It will then sell the shares to the public, usually by an offer for sale, introduction, issue by tender, or placing.

Ierque note A certificate issued by a UK customs officer to the master of a ship, stating that he is satisfied that the cargo on his ship has been correctly entered on the manifest and that no unentered cargo has been found after searching the ship.

Ietsam Items of a ship's cargo that have been thrown overboard and sink below the surface of the water. *Compare*flotsam.

Iettisons Items of a ship's cargo that have been thrown overboard to lighten the ship in dangerous circumstances. If the items are insured they constitute a general average loss (*see*average).

J

Job An identifiable discrete piece of work carried out by an organization. For costing purposes a job is usually given a job number.

Job analysis A detailed study of a particular job, the tools and equipment needed to do it, and its relation to other jobs in an organization. The analysis should also provide the information needed to say how the job should best be done and the qualifications, experience, or aptitudes of the person best suited to doing it.

Jobber 1. A former dealer in stocks and shares, who had no contact with the general public, except through a stockbroker. Jobbers were replaced by market makers on the London Stock Exchange in the Big Bang of October, 1986.

2. A dealer who buys and sells commodities, etc, for his own account.

Jobbing A type of process in which small numbers of products are made to order. It involves high levels of skill and the movement of resources to the product. Examples of jobbing include prototype engineering, cabinet making, etc.

Jobbing backwards Looking back on a transaction or event and thinking about how one might have acted differently, had one known then what one knows now.

Job centre A UK government employment agency that seeks work for those out of a job, assists employers to find suitable employees, and provides training facilities for trades in which there are shortages of skilled workers. Most centres also offer occupational advice on retraining.

Job characteristics theory A theory that describes how job design affects motivation, performance, and satisfaction. Specific core job characteristics induce high levels of motivation and performance.

Job costing (job order costing; specific order costing) A costing process to assess the individual costs of performing each job. It is important in organizations in which the production of different products is carried out, and also in service organizations, in which the cost is required of each service provided.

Job design The process of putting together the various elements of work that constitute an operating system, to form jobs that individuals will perform (*see*work study). The process involves a balance between the needs of the individual to be healthy, safe, and motivated and the needs of the organization to be economic, efficient, and effective. There is no single solution: each compromise will reflect the strategy and management philosophy favoured by a particular organization.

Job enlargement An approach to job design that seeks to enhance staff motivation by increasing the number of related tasks given to an individual. In this way, repetition is reduced and the task can seem to play a more significant role within the overall process.

Job enrichment An approach to job design that increases motivation by allocating more related but different, tasks to an individual. These tasks should involve a greater level of autonomous decision making and personal control. For example, they can include giving production-line workers responsibility for preventive maintenance of their work station and for running their own statistical process control.

Job evaluation An assessment of the work involved in a particular job, the responsibilities borne by the person who does it, and the skills, experience, or qualifications required, all with a view to evaluating the appropriate remuneration for the job and the differentials between it and other work in the same organization. Job evaluation enables different jobs to be compared with agreed common standards. These standards may be specific to an organization or an industry or they may be universal.

Job number A number assigned to each job where job costing is in operation; it enables the costs to be charged to this number so that all the individual costs for a job can be collected.

Job rotation An approach to job design that reduces boredom, and thus increases motivation, by rotating staff through a range of jobs. This process also increase the flexibility of a system by introducing an element of multiskilling. The key is to make use of rotation with a minimum of disruption and learning-curve effects, and to retain the element of freshness.

Job satisfaction The sense of fulfilment and pride felt by people who enjoy their work and do it well. This feeling is enhanced if the significance of the work done and its value are recognized by those in authority (*see*empowerment ; self-actualization).

The factors that determine job satisfaction are investigated by organizational psychology because it is widely accepted that a satisfied workforce is more productive and compliant than a dissatisfied force. The general conclusion is that to reassure employees, managers should encourage the sense of community felt by everyone in a successful organization, in addition to broadening their jobs (*see*job enlargement ; job enrichment) and praising their work.

An absence of job satisfaction has been blamed for absenteeism, accident proneness, high labour turnover, poor industrial relations, and a demotivated workforce that produces shoddy work (*see*alienation).

Organizational psychologists have shown that apart from the human needs fulfilled by working, satisfaction is also related to the expectations aroused by the job: both the needs and the expectations require fulfilment if the job is to provide satisfaction. *See also*total qualitymanagement

Job sharing Dividing the work of one full-time employee between two or more part-time employees.

Joint account A bank account or a building-society account held in the names of two or more people, often husband and wife. On the death of one party the balance in the account goes to the survivor(s), except in the case of partnerships, executors' accounts, or trustees' accounts. It is usual for any of the holders of a joint account to operate it alone.

Joint and several liability A liability that is entered into as a group, on the understanding that if any of the group fail in their undertaking the liability must be shared by the remainder. Thus if two people enter into a joint and several guarantee for a bank loan, if one becomes bankrupt the other is liable for repayment of the whole loan.

Joint Industry Committee for National Readership (JICNAR) A committee formed in 1968 to run readership surveys, although continuous research and surveys have been undertaken since 1956. The committee comprises members of the Institute of Practitioners in Advertising, the Incorporated Society of British Advertisers Ltd, and the Newspaper and Periodical Contributors Committee. Its research is based on a stratified random sample involving 28 000 adult interviews in a continuous survey lasting 12 months. The research covers all major UK publications and is published twice yearly. It is heavily dependent on demographic data, but it also includes regional variations, television viewing, cinema attendance, commercial radio audiences, and special interests.

Joint Industry Committee for Radio Audience Research (JICRAR) A committee composed of representatives of the Institute of Practitioners in Advertising, the Incorporated Society of British Advertisers Ltd, and independent radio companies. The committee provides information on the listening habits of radio audiences. Its research is based on quota sampling of listeners aged 15 or over living in private households within a radio station's designated area. Samples of 300-1200 people, determined according to the population size of the area, record details of their radio listening (BBC or commercial) in 15minute periods in specially designed seven-day diaries. Other data about the respondents' media habits is also collected.

joint investment A security purchased by more than one person. The certificate will bear the names of all the parties, but only the first named will receive notices. To dispose of the holding all the parties must sign the transfer deed.

joint-life and last-survivor annuities Annuities that involve two people (usually husband and wife). A joint-life annuity begins payment on a specified date and continues until both persons have died. A last-survivor annuity only begins payment on the death of one of the two people and pays until the death of the other. *Compare*single-life pension.

join-stock bank A UK bank that is a public limited company rather than a private bank (which is a partnership). During the 19th century many private banks failed; the joint-stock banks became stronger, however, largely as a result of amalgamations and careful investment. In the present century, they became known as the commercial banks or high-street banks.

joint-stock company A company in which the members pool their stock and trade on the basis of their joint stock This differs from the earliest type of company, the merchant corporations or regulated companies of the 14th century, in which each member traded with his own stock, subject to the rules of the company. Joint-stock companies originated in the 17th century and some still exist, although they are now rare.

Joint supply (complementary supply) The supply of two or more separate commodities that are produced by the same process; examples include milk and butter, wool and mutton, and petrol and heavy oil. If the demand for one increases, the supply of the other will also increase, but its price will fall unless its demand also increases.

Joint tenants Two or more persons who jointly own an interest in land. When one dies, the other(s) takes his share by right of survivorship. No one tenant has rights to any particular part of the property and no tenant may exclude another.

Joint venture A commercial undertakentered into by two or more parties, usually in the short term. Joint ventures are generally governed by the *Partnership Act (1890)* but they differ from partnerships in that they are limited by time or by activity. Separate books are not usually kept and the joint venturers will have a profit- or loss-sharing ratio for the purpose of the joint venture only. Joint venturers often carry on their principal businesses independently of, and at the same time as, the joint venture. Joint ventures are becoming increasingly common as companies cooperate with each other in international markets, in order to share costs, exploit new technologies, or gain access to new markets.

Joint venturing In international marketing, an arrangement between a domestic company and a foreign host company to set up production and maketing facilities in the foreign market.

Journal A book of prime entry in which transfers to be made from one account to another are recorded. It is used for transfers not recorded in any other of the books of prime entry, such as the sales day book or the cash book.

Judgement creditor The person in whose favour a court decides, ordering the judgement debtor to pay the sum owed. If the judgement debtor fails to pay, the judgement creditor must return to the court asking for the judgement to be enforced.

Judgement sample A sample in marketing research selected because an interviewer thinks that particular respondents will answer the research questions. For example, an interviewer might interview a woman with a child because the questions relate to children's clothing.

Junk bond A bond that offers a high rate of interest because it carries a higher than usual probability of default. The issuing of junk bonds to finance the takeover of large companies in the USA is a practice that has developed rapidly over recent years and has spread elsewhere. *See*leveraged buy-out.

Just in time (JIT) An approach to manufacturing that aims to reduce waste (in terms of overproduction), queues and bottlenecks, the effects of poor product design, excess transport, buffer inventories, motion, and scrap. While traditional systems isolate operations from the environment, for example by uncoupling inventory, JIT recognizes that without protection processes can go wrong but, if they do, the reasons will have to be found and the process put right. It seeks to improve all performance variables, such as cost, quality, speed, dependability, and flexibility. The key elements of JIT are: respect for the contribution that employees can make; encouragement of personal responsibility and process ownership from a multiskilled workforce that is proactive in problem solving, a demand-pull system in which material only flows when needed; kanban; standardized product design; accelerated learning-curve effects; preventive rather than reactive maintenance; level scheduling; reduction of set-up times; efficient plant layout, eliminating places in which inventory can get lost; JIT purchasing involving close liaison with suppliers; and total quality management committed to continuous improvement JIT has been facilitated by the advent of cheap computing power and computer-aided manufacturing systems. It is easier to install JIT techniques from the outset, rather than attempting to introduce it piecemeal when a process is running.

K

Kaffirs An informal name for shares in South African gold-mining companies on the London Stock Exchange.

Kaizen The Japanese concept of continuous overall improvement. It is a philosophy that underlies total quality management and just-in-time techniques. Its prime consideration is to ensure that problems are continuously sought out and rectified and that opportunities for improvement, in any aspect of the system are exploited. Rather than concealing problems, staff are encouraged to bring them to the surface. The effect of this approach is to prevent the fear generated by a large-scale change that may be required if a system is allowed to stagnate and then has to be brought back into line.

Kamikaze pricing The practice of offering loans at exceptionally low interest rates to capture a larger share of the corporate banking market. It is named after the Japanese suicide pilots of World War II.

Kanban A Japanese method of controlling the movement of materials through a just-in-time system. Literally meaning 'visual record', kanban cards provide authority for a work station to produce or to transfer materials.

Kangaroos An informal name for Australian shares, especially in mining, land, and tobacco companies, on the London Stock Exchange.

Kappa The ratio of the dollar price change in the price of an option to a 1% change in the expected price volatility.

Karachi Stock Exchange The major securities exchange of Pakistan.

Keelage A fee paid by the owners of a ship for being permitted to dock in certain ports and harbours.

Keidanren The Japanese name for the Federation of Economic Organizations, which is the most powerful of the Japanese business organizations. It functions as the headquarters of the business community; its members are the major trade associations and the prominent companies.

Keiretsu A network of Japanese companies organized around a major bank. The term is also used outside of Japan to describe how a large corporation with many subsidiaries and associated firms can manipulate revenues. For example, firm A and B are controlled by firm C. Firm A is forced to buy its input from firm B at a high price. As a result, A is unprofitable and B is very profitable.

Keogh plan A US savings scheme to create a pension plan for self-employed people, in which tax is deferred until withdrawals are made. It can be held at the same time as a corporate pension or individual retirement account. Keogh plans originated with the selfemployment Individuals Retirement Act (1982).

Kerb market 1. The former practice of trading on the street after the formal close of business of the London Stock Exchange. **2.** Dealing on the London Metal Exchange in any metal after the formal ring trading in specified metals. **3.** Any informal market, such as one for dealing in securities not listed on a stock exchange.

Key man (or woman) insurance A life insurance policy purchased by a company to insure the life of a key executive. The company is the beneficiary in case of the executive's death.

Keyed advertisement An advertisement designed to enable the advertiser to know where a respondent saw it, for instance by including a code number of a particular department in the return address.

Keynesian economics An economic theory of British economist, John Maynard Keynes that active government intervention is necessary to ensure economic growth and stability.

Key-person assurance An insurance policy on the life of a key employee (male or female) of a company, especially the life of a senior executive in a small company, whose death would be a serious loss to the company. In the event of the key person dying, the benefit is paid to the company. In order that there should be an insurable interest, a loss of profit must be the direct result of the death of the key person.

Khoum A monetary unit of Mauritania, worth one fifth of an ouguiya.

Kickback In the context of finance, refers to compensation of dealers by sales finance companies for discounting installment purchase paper.
In the context of contracts, refers to secret payments made to insure that the contract goes to a specific firm.

Kicker An additional feature of a debt obligation that increases its marketability and attractiveness to investors.

Killer bees Those who aid a company in fending off a takeover bid, usually investment bankers who devise strategies to make the target less attractive or more difficult to acquire.

Kina (k) The standard monetary unit of Papua New Guinea, divided into 100 toea.

Kip (KN) The standard monetary unit of Laos, divided into 100 at.

Kiting Used in banking to refer to the practice of depositing and drawing checks at two or more banks and taking advantage of the time it takes for the second bank to collect funds from the first bank.
Also refers to illegally increasing the face value of a check by changing the numbers on the check.
In the context of securities, refers to the manipulation and inflation of stock prices.

Knock-for-knock agrrement An agreement between motor insurers that they will pay for their own policyholders' accident damage without seeking any contribution from the other insurers, irrespective of blame but provided that the relevant policies cover the risk involved. The agreement works on the principle that, having paid for their own policyholder's damage, insurer A refrains from claiming the payment back from insurer B, even though it may be clear that A's client was to blame for the accident. On another occasion the roles might be reversed; insurer B then refrains from claiming from A even though Ns client was responsible.

Knocking copy Advertising copy that attacks a rival product.

Knock-out option An option that- is worthless at expiration if the underlying commodity or currency price reaches a specific price level.

Know your customer An ethical foundation of securities brokers that an adviser who recommends the purchase or sale of any security to a customer, must believe that the recommendation is suitable for the customer, given the customer's financial situation.

Kobo A monetary unit of Nigeria, worth one hundredth of a naira.

Kondratieff Wave An economic theory of the Soviet economist Kondratieff stating that the economies of the western world are susceptible to major up-and-down "supercycles" lasting 50 to 60 years.

Korea Stock Exchange The major securities market of Korea.

Koruna The standard monetary unit of the Czech Republic and Slovakia, divided into 100 haleru.

Króna (ISK) The standard monetary unit of Iceland, divided into 100 aurar.

Krona (Skr) The standard monetary unit of Sweden, divided into 100 öre.

Krone The standard monetary unit of Denmark (Dkr), the Faeroe Islands (Fkr), Greenland (Dkr), and Norway (Nkr), divided into 100 øre.

Kroon The standard monetary unit of Estonia, divided into 100 centes.

Krugerrand A South African coin containing 1 troy ounce of gold. Minted since 1967 for investment purposes, it enables investors to evade restrictions on holding gold. Since 1975 an import licence has been required to bring them into the UK. In the 1980s their popularity waned for political reasons and because other countries have produced similar coins. *See*britannia coins.

Kruggerand A gold coin minted by the republic of South Africa that typically sells for slightly higher prices than the market value of the gold it contains.

Kuala Lumpur Commodities Exchange (KLCE) The Malaysian commodity exchange for trading futures in crude palm oil, crude palm kernel oil, tin, rubber, and cocoa.

Kuala Lumpur Options and Financial Futures Exchange (KLOFFE) Established in 1995, the Kuala Lumpur Options and Financial Futures Exchange offers equity derivative products based on underlying instruments traded on the Kuala Lumpur Stock Exchange (KLSE).

Kuala Lumpur Stock Exchange (KLSE) Established in 1973, the Kuala Lumpur Stock Exchange (KLSE) is the only stock exchange in Malaysia.

Kuna The standard monetary unit of Croatia, divided into 100 lipas.

Kurtosis. Measures the fatness of the tails of a probability distribution. A fat-tailed distribution has higher-than-normal chances of a big positive or negative realization. Kurtosis should not be confused with skewness, which measures the fatness of one tail. Kurtosis is sometimes referred to as the volatility of volatility.

Kuru A monetary unit of Turkey, worth one hundredth of a lira.

Kwacha 1. (Mk) The standard monetary unit of Malawi divided into 100 tambala.

2. (k) The standard monetary unit of Zambia, divided into 100 ngwee.

Kwanza The standard monetary unit of Angola, divided into 100 lwei.

Kyat The standard monetary unit of Myanmar (Burma) divided into 100 pyas.

L

Laari A monetary unit of the Maldives, worth one hundredth of a rufiyaa.

Labor union US name for a trade union.

Labour cost (wages costs) Expenditure on wages paid to those operators who are both directly and indirectly concerned with the production of the product, service, or cost unit. *See also*direct labour cost indirect costs.

Labour turnover rate The ratio, usually expressed as a percentage, of the number of employees leaving an organization or industry in a stated period to the average number of employees working in that organization or industry during the period.

Labour-intensive Denoting an industry or firm in which the remuneration paid to employees represents a higher proportion of the costs of production than the cost of raw materials or capital equipment (in the reverse case the industry or firm is said to be capitalintensive). While publishing is a labourintensive industry, printing is capital-intensive.

Laches (Norman French lasches: negligence) Neglect and unreasonable delay in enforcing an equitable right. If a plaintiff with full knowledge of the facts takes an unnecessarily long time to bring an action (e.g. To set aside a contract obtained by fraud) the court will be of no assistance; hence the maxim "the law will not help those who sleep on their rights". No set period is given but if the action is covered by limitation-of-actions legislation, the period given will not be shortened. Otherwise the time allowed depends on the circumstances.

Ladder strategy A bond portfolio construction strategy that invests approximately equal amounts in every maturity within a given range.

Lady Macbeth strategy A strategy used in takeover battles in which a third party makes a bid that the target company would favour, i.e. It appears to act as a white knight, but subsequently changes allegiance and joins the original bidder.

Laesio enormis (Latin: extraordinary injury) A doctrine of Roman law that is included in some European legal systems but not in English law. It states that the price given in a contract by way of consideration must be fair and reasonable or the contract may be rescinded.

Laffer curve A curve conjecturing that economic output will increase if marginal tax rates are cut. Named after economist Arthur Laffer.

Laft-survivor policy 1. An assurance policy on

the lives of two people, the sum assured being paid on the death of the last to die. *See also*joint-life and lastsurvivor annuities.

2. A contract (formerly called a tontine) in which assurance is arranged by a group of people, who all pay premiums into a fund while they are alive. No payment is made until only one person from the group is left alive. At that point the survivor receives all the policy proceeds. Contracts of this kind are not available in the UK because of the temptation they provide to members to murder their fellows, in order to be the last survivor.

Lag response of prepayments A delay of typically about three months between the time the weighted-average coupon of an MBS pool crosses the threshold for refinancing and observation of an acceleration in prepayment speed is observed.

Lagging indicators Economic indicators that follow rather than precede the country's overall pace of economic activity. See also: Leading indicators and coincident indicators.

Laisse-faire Doctrine that a government should not interfere with business and economic affairs.

Laissez-faire economy An economy in which government intervention is kept to a minimum and market forces are allowed to rule. The term, attributed to the French merchant J Gourlay (1712-59), is best translated as "let people do as they think best".

Lakh In India, Pakistan, and Bangladesh 100,000. It often refers to this number of rupees.

Lambda The ratio of a change in the option price to a small change in the option volatility. It is the partial derivative of the option price with respect to the option volatility.

Lame duck A company that has national prestige and is a large employer of labour (especially in an area of high unemployment) or is a leader in a particular technology but, being unable to meet foreign competition, is unable to survive without government support.

Land certificate A document that takes the place of a title deed for registered land. It is granted to the registered proprietor of the land and indicates ownership. It will usually be kept in the Land Registry if the land is subject to a mortgage.

Land In law, the part of the earth's surface that can be owned, including all buildings, trees, minerals, etc, attached to or forming part of it. It also includes the airspace above the land necessary for its reasonable enjoyment. In law, a firstfloor flat can constitute a separate piece of land, although it has no contact with the soil just as a cellar may be a separate estate in land.

Landing account An account prepared by a public warehousing company to the owner of goods that have recently been landed from a ship and taken into the warehouse. The account, which is usually accompanied by a weight note, states the date on which the warehouse rent begins and also the quantitity of goods and their condition (from the external appearance of the packages).

Landing charges The charges incurred in disembarking a cargo, or part of it, from a ship at the port of destination.

Landing order A document given to an importer by the UK Customs and Excise, enabling him to remove newly imported goods from a ship or quay to a bonded warehouse, pending payment of duty or their re-export.

Landlord A property owner who rents property to a tenant.

Landlord and tenant The parties to a lease. The landlord grants the lease to the tenant. The landlord may be the freehold owner of the land or may be a tenant of a superior landlord (*see*head lease). The law of landlord and tenant is that governing the creation, termination, and regulation of leases. Many leases, both business and residential, are subject to statutory control. This means that the right of the landlord to end a lease and evict tenants may be restricted and the rent may be controlled. It is therefore essential to consider statutory rights when considering the relation between landlord and tenant, as well as those given by the lease.

Lapping In the USA, the fraudulent practice of concealing a shortage of cash by delaying the recording of cash receipts. In the UK it is referred to as teeming and lading. There are a number of variations, but essentially the cashier conceals the theft of cash received from the first customer by recording the cash received from the second customer as attributable to the first, and so on with subsequent customers. The cashier hopes to be in a position to replace the cash before the dishonesty is discovered. As such hopes are frequently based on attempts at gambling, the deception is often discovered.

Lapsed option An option that no longer has any value because it has reached its expiration date without being exercised.

Large-cap A stock with a high level of capitalization, usually at least Rs.5 billion market value.

Lari The standard monetary unit of Georgia.

Last split After a stock split, the number of shares distributed for each share held and the date of the distribution.

Last trading day The final day under an exchange's rules during which trading may take place in a particular futures or options contract. Contracts outstanding at the end of the last trading day must be settled by delivery of underlying physical commodities or financial instruments, or by agreement for monetary settlement depending upon futures contract specifications.

Last-in-first-out cost (LIFO cost) A method of valuing units of raw material or finished goods issued from stock by using the latest unit value for pricing the issues until all the quantity of stock received at that price is used up. The next earliest price is then used for pricing the issues, and so on. Because the issues are based on a LIFO cost, the valuation of closing stocks is described as being on the same LIFO basis. The method may also be used in process costing to value the work in process at the end of an accounting period. This method of costing is not normally acceptable for stock valuation in the UK.

Latin American Free Trade Area (LAFTA) An economic grouping that became the Latin American Integration Association in 1981.

Launder To move illegally acquired cash through financial systems so that it appears to be legally acquired.

Laundering money Processing money acquired illegally (as by theft, drug dealing, etc) so that it appears to have come from a legitimate source. This is often achieved by paying the illegal cash into a foreign bank and transferring its equivalent to a bank with a good name in a hard-currency area.

Law of large numbers The mean of a random sample approaches the mean (expected value) of the population as the sample grows.

Law of one price An economic rule stating that a given security must have the same price no matter how the security is created. If the payoff of a security can be synthetically created by a package of other securities, the implication is that the price of the package and the price of the security whose payoff it replicates must be equal. If it is unequal, an arbitrage opportunity would present itself.

Lay days The number of days allowed to a ship to load or unload without incurring demurrage. Reversible lay days permit the shipper to add to the days allowed for unloading any days he has saved while loading. Lay days may be calculated as running days (all consecutive days), working days (excluding Sundays and public holidays), or weather working days (working days on which the weather allows work to be carried out).

Lead manager The commercial or investment bank with the primary responsibility for organizing syndicated bank credit or bond issue. The lead manager recruits additional lending or underwriting banks, negotiates terms of the issue with the issuer, and assesses market conditions.

Lead regulator A leading self-regulatory organization that over sees compliance with a particular section of the law, such as the NYSE, ASE, or NASDAQ.

Lead The first named underwriting syndicate on a Lloyd's insurance policy. When a broker seeks to cover a risk he will first try to get a large syndicate to act as lead, which encourages smaller syndicates to cover a share of the risk. The premium rate is calculated by the lead; if others wish to join in the risk they have to insure at that rate. On a collective policy the lead insurer is the first insurer on the schedule of insurers; he issues the policy, collects the premiums, and distributes the proportions to the coinsurers.

Lead underwriter The head of a syndicate of financial firms that are sponsoring an initial public offering of securities or a secondary offering of securities. Could also apply to bond issues.

Leader A stock or group of stocks that is the first to move in a market upsurge or downturn.

Leading and lagging Techniques often used at the end of a financial year to enhance a cash position and reduce borrowing. This is achieved by arranging for the settlement of outstanding obligations to be accelerated (leading) or delayed (lagging).

Leading indicator A change in a measurable economic factor that is evident before the economy starts to follow a specific trend.

Leading Strategy used by a firm to accelerate payments, normally in response to exchange rate expectations.

Leading the market In the context of general equities, this is a stock or group of stocks moving with the market as a whole, but moving in advance of the general market.

League tables A ranking of lenders and advisors according to the underwriting, final take, or number of project finance loans or advisory mandates.

Leakage Release of information selectively or not before official public announcement.

Lean back A period of cautious inaction by a government agency before intervening in a market. For example, a central bank might allow a lean-back period to elapse to allow exchange rates to stabilize, before intervening in the foreign exchange market.

Learning curve A technique that takes into account the reduction in time taken to carry out production as the cumulative output rises. The concept is based on a doubling of output, so that a 70% learning curve means that the cumulative average time taken per unit falls to 70% of the previous cumulative average time as the output doubles. The cumulative average time per unit is measured from the very first unit produced. The formula for the learning curve is: Y = ax-b, Where y = cumulative average time per unit of production, a = the time taken to produce the first unit, x = cumulative number of units manufactured to date, and b = the learning coefficient.

Lease A contract between the owner of a specific asset, the lessor, and another party, the lessee, allowing the latter to hire the asset. The lessor retains the right of ownership but the lessee acquires the right to use the asset for a specific period of time in return for the payment of specific rentals or payments. Statement of Standard Accounting Practice 21, Accounting for Leases and Hire Purchase Contracts', classifies leases into operating leases and finance leases with differing accounting treatments.

Leaseback (renting back) An arrangement in which the owner of an asset (such as land or buildings) sells it to another party but immediately enters into a lease agreement with the purchaser to obtain the right to use the asset. Such a transaction is a method for raising funds and can affect the financial statements of a company, depending on whether a finance lease or an operating lease is entered into.

Leased department retailer An independent retailer who owns the merchandise stocked and displayed but who leases floor space from another (usually larger) retailer, often using their own retailing name. This arrangement is sometimes called a shop within a shop.

Leasehold An asset providing the right to use property under a lease agreement.

Leasehold land Land held under a lease. The land will eventually revert to the freehold owner, although there has been some statutory modification of this right to repossession (e.g. In the Rent Acts). This is the most common way for blocks of offices to be owned. The landlord maintains possession of the common parts and creates separate leases for each office. The ownership of each office may subsequently change as leases are assigned.

Leases and executory contracts Executory contracts are contracts where obligations are yet to be performed by both parties. The bankruptcy code gives special treatment to unexpired leases and executory contracts by allowing the debtor, in many cases, to choose whether it wants to continue with the agreement (assume) or terminate the agreement (reject). A contract to sell goods where the goods have not yet been delivered is an executory contract. A contract for the sale of goods where the goods have been delivered and the receiving party has yet to pay, is not an executory contract. Most leases of personal property or real estate are executory contracts. A collective bargaining agreement is an executory contract and is given special treatment.

Leasing Hiring equipment, such as a car or a piece of machinery, to avoid the capital cost involved in owning it. In some companies it is advantageous to use capital for other purposes and to lease some equipment, paying for the hire out of income. The equipment is then an asset of the leasing company rather than the lessor. Sometimes a case can be made for leasing rather than purchasing, on the grounds that some equipment quickly becomes obsolete.

Ledger A collection of accounts of a similar type. Traditionally, a ledger was a large book with separate pages for each account. In modern systems, ledgers may consist of separate cards or computer records. The most common ledgers are the nominal ledger containing the impersonal accounts, the sales (*or* debtors) ledger containing the accounts of an organization's customers, and the purchase (*or*creditors) ledger containing the accounts of an organization's suppliers.

Legal A computerized database maintained by the NYSE to keep track of enforcement actions, audits, and complaints against member firms. This term is not an acronym but is referred to in capitals.

Legal bankruptcy A legal proceeding for liquidating or reorganizing a business.

Legal capital In the USA, the amount of stockholders' equity, which cannot be reduced by the payment of dividends.

Legal defeasance The deposit of cash and permitted securities, as specified in the bond indenture, into an irrevocable trust sufficient to enable the issuer to discharge fully its obligations under the bond indenture.

Legal entity A person or organization that can legally enter into a contract, and may therefore be sued for failure to comply with the terms of the contract.

Legal investments Investments that a regulated entity is permitted to make under the rules and regulations that govern its investing.

Legal monopoly A government-regulated firm that is legally entitled to be the only company offering a particular service in a particular area.

Legal opinion A statement, usually written by a specialized law firm, required for a new municipal bond issue stating that the issue is legally acceptable.

Legal reserve The minimum amount of money that building societies, insurance companies, etc., are bound by law to hold as security for the benefit of their customers.

Legal risk The risk associated with the impact of a defect in the documentation on cash flow or debt service.

Legal tender, i.e. It must be accepted but only up to specified limits of payment; or **unlimited legal tender,** i.e. Acceptable in settlement of debts of any amount. Bank of England notes and the £2 and £1 coins are unlimited legal tender in the UK. Other Royal Mint coins are limited legal tender; i.e. Debts up to £10 can be paid in 50p and 20p coins; up to £5 by 10p and 5p coins; and up to 20p by bronze coins.

Legislative risk The risk that new or changed legislation will have a large positive or negative effect on an investment.

Lehman Brothers Adjustable-Rate Mortgage Index A benchmark index that includes all agency-guaranteed securities with coupons that periodically adjust based on a spread over a published index.

Lempira The standard monetary unit of Honduras, divided into 100 centavos.

Lend To provide money temporarily on the condition that it or its equivalent will be returned, often with an interest fee.

Lendable funds The pool of funds available to borrows; typically categorized by currency and maturity.

Lender liability A developing body of law which includes liability for breach of contract and other wrongs where banks and other lenders may be made to pay damages to customers. A bank which terminates a long standing credit relationship without warning, where no default exists, may be held liable for damages.

Lender of last resort A country's central bank with responsibility for controlling its banking system. In the UK, the Bank of England fulfils this role, lending to discount houses, either by repurchasing Treasury bills, lending on other paper assets, or granting direct loans, charging the base rate of interest. Commercial banks do not go directly to the Bank of England; they borrow from the discount houses.

Lending at a premium A loan from one broker to another of securities to cover a customer's short position, with a borrowing fee included. A fee is unusual since securities are normally lent freely between brokers.

Less-developed countries (LDCs) Also known as emerging markets. Countries who's per capita GDP is below a World Bank-determined level.

Lessee An entity that leases an asset from another entity.

Lessor A person granting a lease; landlord. *See*landlord and tenant.

Letter of Administration A certificate issued by the Court evidencing the appointment of the Administrator of an Estate.

Letter of comfort A letter to a bank from the parent company of a subsidiary that is trying to borrow money from the bank. The letter gives no guarantee for the repayment of the projected loan but offers the bank the comfort of knowing that the subsidiary has made the parent company aware of its intention to borrow, the parent also usually supports the application, giving, at least, an assurance that it intends that the subsidiary should remain in business and that it will give notice of any relevant change of ownership.

Letter of credit (documentary credit) A letter from one banker to another authorizing the payment of a specified sum to the person named in the letter on certain specified conditions (*see*letter of indication). Commercially, letters of credit are widely used in the international import and export trade as a means of payment. In an export contract, the exporter may require the foreign importer to open a letter of credit at the importer's local bank (the issuing bank) for the amount of the goods. This will state that it is to be negotiable at a bank (the negotiating bank) in the exporter's country in favour of the exporter; often, the exporter (who is called the beneficiary of the credit) will give the name of the negotiating bank. On presentation of the shipping documents (which are listed in the letter of credit) the beneficiary will receive payment from the negotiating bank.

An irrevocable letter of credit cannot be cancelled by the person who opens it or by the issuing bank without the beneficiary's consent, whereas a revocable letter of credit can. In a confirmed letter of credit the negotiating bank guarantees to pay the beneficiary, even if the issuing bank fails to honour its commitments (in an unconfirmed letter of credit this guarantee is not given). A confirmed irrevocable letter of credit therefore provides the most reliable means of being paid for exported goods. However, all letters of credit have an expiry date, after which they can only be negotiated by the consent of all the parties.

A circular letter of credit is instruction from a bank to its correspondent banks to pay the beneficiary a stated sum on presentation of a means of identification. It has now been replaced by traveller's cheques. Although the term 'letter of credit' is still widely used, in 1983 the International Chamber of Commerce recommended documentary credit as the preferred term for these instruments.

Letter of credit (LOC) A form of guarantee of payment issued by a bank on behalf of a borrower that assures the payment of interest and repayment of principal on bond issues.

Letter of Guarantee A letter from a bank to a brokerage firm which states that a customer (who has written a call option) does indeed own the underlying stock and the bank will guarantee delivery if the call is assigned. Thus the call can be considered covered. Not all brokerage firms accept letters of guarantee. Also: letter issued to Option Clearing Corporation by member firms covering a guarantee of any trades made by one of its customers, (a trader or broker on the exchange floor).

Letter of indemnity 1. A letter stating that the organization issuing it will compensate the person to whom it is addressed for a specified loss. *See also* indemnity.

2. A letter written to a company registrar asking for a replacement for a lost share certificate and indemnifying the company against any loss that it might incur in so doing. It may be required to be countersigned by a bank.

3. A letter written by an exporter accepting responsibility for any losses arising from faulty packing, short weight, etc., at the time of shipment. If this letter accompanies the shipping documents, the shipping company will issue a clean bill of lading, even if the packages are damaging, enabling the exporter to negotiate the documents and receive payment without trouble.

Letter of indication (letter of identification) A letter issued by a bank to a customer to whom a letter of credit has been supplied. The letter has to be produced with the letter of credit at the negotiating bank; it provides evidence of the bearer's identity and a specimen of his or her signature. It is used particularly with a circular letter of credit carried by travellers, although traveller's cheques are now more widely used.

Letter of intent A letter in which a person formally sets out his or her intentions to do something, such as signing a contract in certain circumstances, which are often specified in detail in the letter. The letter does not constitute either a contract or a promise to do anything, but it does indicate the writer's serious wish to pursue the course set out.

Letter of licence A letter from a creditor to a debtor, who is having trouble raising the money to settle the debt. The letter states that the creditor will allow the debtor a stated time to pay and will not initiate proceedings against the debtor before that time. *See also*arrangement.

Letter of regret A letter from a company, or its bankers, stating that an application for an allotment from a new issue of shares has been unsuccessful.

Letter of Testamentary A certificate issued by the court evidencing the appointment of an executor of estate.

Letter stock Privately placed common stock, so-called because the SEC requires a letter from the purchaser stating that the stock is not intended for resale.

Letters of administration An order authorizing die person named (the administrator) to distribute the property of a deceased person, who has not appointed anyone else to do so. The distribution must be in accordance with the deceased's will or the rules of intestacy if he or she died intestate.

Level output strategy An approach to aggregate planning that aims to regulate production at a constant level regardless of fluctuation in demand. It involves storing the product in times of low demand and is not therefore option available to the providers of services. The benefits include maintaining a stable workforce and the optimum use of capacity. The disadvantages include the costs of storage and the difficulty of responding to major changes in demand. *Compare*chase demand strategy.

Level pay The characteristic of the scheduled principal and interest payments due under a mortgage such that total monthly payment of P&I is the same while characteristically the principal payment component of the monthly payment becomes gradually greater while the monthly interest payment becomes less.

Leveraged company A company that has debt in its capital structure.

Leveraged equity Stock in a firm that relies on financial leverage. Holders of leveraged equity face the benefits and costs of using debt.

Leveraged investment company An investment company or mutual fund entitled to borrow capital for its operations. Also, an investment company that issues both income shares and capital shares.

Leveraged lease A lease arrangement under which the lessor borrows a large proportion of the funds needed to purchase the asset and grants the lender a lien on the assets and a pledge of the lease payments to secure the borrowing.

Leveraged portfolio A portfolio that includes risky assets purchased with funds borrowed.

Leveraged recapitalization Often used in risk arbitrage. A public company takes on significant additional debt with the purpose of either paying an extraordinary dividend or repurchasing shares, leaving the public shareholders with a continuing interest in a more financially leveraged company. Popular form of shark repellent See: Stub.

Leveraged required return The required return on an investment when the investment is financed partially by debt.

Levered portfolio Investment at least partially financed by borrowing.

Liabilities General term for what the business owes. Liabilities are long-term loans of the type used to finance the business and short-term debts or money owing as a result of trading activities to date . Long term liabilities, along with Share Capital and Reserves make up one side of the balance sheet equation showing where the money came from. The other side of the balance sheet will show Current Liabilities along with various Assets, showing where the money is now.

Liability A financial obligation, or the cash outlay that must be made at a specific time to satisfy the contractual terms of such an obligation.

Liability funding strategies Investment strategy that select assets so that cash flows will equal or exceed the client's obligations.

liability insurance A form of insurance policy that promises to pay any compensation and court costs the policyholder becomes legally liable to pay because of claim for injury to other people or damage to their property as a result of the policyholder's negligence. Policies often define the areas in which they will deal with liability, e.g. personal liability or employers' liability.

Liability swap An interest rate swap used to alter the cash flow characteristics of an institution's liabilities so as to provide a better match with its assets.

Licencing In international marketing, an agreement by which a company (the licensor) permits a foreign company (the licensee) to set up a business in the foreign market using the licensor's manufacturing processes, patents, trademarks, trade secrets, etc., in exchange for payment of a fee or royalty.

License agreement A contract by which a domestic company (the licensor) allows a foreign company (the licensee) to market its products in a foreign country in return for royalties, fees, or other forms of compensation.

Licensed dealer A dealer licensed by the Department of Trade and Industry to provide investment advice and deal in securities, either as an agent or principal. Licensed dealers are not members of the London Stock Exchange and are not covered by its compensation fund.

Licensing Arrangement in which a local firm in the host country produces goods in accordance with another firm's (the licensing firm's>) specifications; as the goods are sold, the local firm can retain part of the earnings.

Lien The right of one person to retain possession of goods owned by another until the possessor's claim against the owner have been satisfied. In a general lien, the goods are held as security for all the outstanding debts of the owner, whereas in a particular lion only the claim of the possessor in respect of the goods held must be satisfied. Thus an unpaid seller may in some contracts be entitled to retain the goods until receiving the price, a carrier may have a lien over goods being transported, and a repairer over goods being repaired. Whether a lien arises or not depends on the terms of the contract and usual trade practice. Tins type of lien is a possessory lien, but sometimes actual posession of the goods is not necessary. In an equitable lien, for example, the claim exists independently of possession. If a purchaser of the property involved is given notice of the Hen it binds him; otherwise he will not be bound. Similarly a maritime lien, which binds a ship or cargo in connection with some maritime liability, does not depend on possession and can be enforced by arrest and sale (unless security is given). Examples of maritime liens are the lien of a salvor, those of seamen for their wages and of masters for their wages and outgoings, that of a bottomry or respondentia bondholder (*see*hypothecation), and that over a ship at fault in a collision in which property has been damaged.

Life annuity An annuity that ceases to be paid on the death of a specified person, which may or may not be the annuitant.

Life annuity An annuity that pays a fixed amount for the lifetime of the annuitant.

Life assurance An insurance policy that pays a specified amount of money on the death of the life assured or, in the case of an endowment assurance policy, on the death of the life assured or at the end of an agreed period, whichever is the earlier. Life assurance grew from a humble means of providing funeral expenses to a means of saving for oneself or one's dependants, with certain tax advantages. With-profits policies provide sums of money in excess of the sum assured by the addition of bonuses. Unit-linked policies invest the premiums in funds of assets, by means of buying units in the funds.

Life expectancy The length of time that an average person is expected to live, which is used by insurance companies use to make projections of benefit payouts.

Life insurance An insurance policy that pays a monetary benefit to the insured person's survivors after death.

Lifeboat 1. A fund set up to rescue dealers on an exchange in the event of a market collapse and the ensuing insolvencies.

2. The rescue of a company that is in financial difficulty by new or restructured loans from its group of bankers.

Life-cycle costing The approach to determining the total costs of a fixed asset that takes into account all the costs likely to be incurred both in acquiring it and in operating it over its effective life. For example, the initial cost to an airline of an aircraft is only part of the costs relevant to the decision to purchase it. The operating costs over its effective life are also relevant and would therefore be part of the decision - making data. This is an aspect of terotechnology.

Lifestyle business A small business run by owners with a personal interest in the product. The business reflects their lifestyles and provides them with a comfortable income, but is not a growth venture. Compilers of reference books could fall into this category.

Lifestyle segmentation Segmentation of a market based on lifestyles. More companies are segmenting their markets by Hfestyles, as these are increasingly seen as good predictors of consumer behaviour. Most companies use off-the-shelf research-agency classifications, because of the high cost and complexity of developing their own.

Lifestyle The activities, interests, opinions, and values of individuals as they affect their mode of living. Lifestyle reflects more than social class or personality; it provides a profile of a person's whole pattern of behaviour and interaction with the world. Lifestyles are important in marketing because they influence the way in which people elect to make purchases. The technique for measuring lifestyle is part of psychographic segmentation It involves measuring primary characteristics, known as. AIO dimensions (activities, interests, and opinions). Lifestyle classifications have been developed out of this, particularly the Values and Lifestyles Typology (VALS) which classifies consumers into nine lifestyle groups. VALS 2 classifies people into eight groups according to their consumption tendencies, i.e. How they spend their time and money.

Lifted Refers to over-the-counter trading. Having an offer taken in a stock, followed by the market maker raising the offer price.

Likuta (plural makuta) A monetary unit of Zaire, worth one hundredth of a zaïre.

Limit 1. An order given by an invester to a stockbroker or commodity broker restricting a particular purchase to a stated maximum price or a particular sale to a stated minimum price. Such a limit order will also be restricted as to time; it may be given firm for a stated period or firm until cancelled.

2. The maximum fluctuations (up or down) allowed in certain markets over a stated period (usually one day's trading). In some

volatile circumstances the market moves the limit up (or down). The movement of prices on the Tokyo Stock Exchange is limited in this way as it is on certain US commodity markets.

Limit Cycles An attractor for non-linear dynamic systems which has periodic cycles or orbits in phase space. An example is an undamped pendulum which will have a closed circle orbit equal to the amplitude of the pendulum's swing. See: Attractor, Phase Space.

Limit on close order An order to buy or sell stock at the closing price only if the price is at a predetermined level or better.

Limit order An order to buy a stock at or below a specified price or to sell a stock at or above a specified price. For instance, you could tell a broker "Buy me 100 shares of XYZ Corp at Rs.8 or less" or to "sell 100 shares of XYZ at Rs.10 or better." The customer specifies a price and the order can be executed only if the market reaches or betters that price. A conditional trading order designed to avoid the danger of adverse unexpected price changes.

Limit order book A record of unexecuted limit orders maintained by the specialist. These orders are treated equally with other orders in terms of priority of execution.

Limit order information system The electronic system supplying information about securities traded on participating exchanges so that the best securities prices can be found.

Limit price The highest price that established sellers in a market can charge for a product, without inducing a new seller to enter the market. The limit price is usually lower than the monopoly price but higher than the competitive price. Limit pricing enables established sellers to make higher profits than they would if they sold at the competitive price and it also ensures that new firms will not enter the market and drive the prices down to competitive levels.

Limit up, limit down The maximum price change allowed for a commodity futures contract per trading day.

Limitation of actions Statutory rules limiting the time within which civil actions can be brought. Actions in simple contract and tort must be brought within six years of the cause of action. There are several special rules, including that for strict liability actions for defective products, in which case the period is three years from the accrual of the cause of action or (if later) the date on which the plaintiff knew, or should have known, the material facts, but not later than ten years from the date on which the product was first put into circulation.

Limitation on asset dispositions A bond covenant that restricts in some way a firm's ability to sell major assets.

Limitation on conversion Applies mainly to convertible securities. Possible delay in convertibility. More frequently, the right to convert may be terminable prior to a redemption date, preventing the holder from receiving a final coupon or dividend. See: Accrued interest.

Limitation on sale-and-leaseback A bond covenant that restricts in some way a firm's ability to enter into sale-and-leaseback transactions, financing techniques that could affect creditor thinness.

Limited company A company in which the liability of the members in respect of the company's debts is limited. It may be limited by shares, in which case the liability of the members on a winding-up is limited to the amount (if any) unpaid on their shares. This is by far the most common type of registered company. The liability of the members may alternatively be limited by guarantee ; in this case the liability of members is limited by the memorandum to a certain amount, which the members undertake to contribute on winding-up. These are usually societies, clubs, or trade associations. Since 1980 it has not been possible for such a company to be formed with a share capital or converted to a company limited by guarantee with a share capital.

Limited liability Limitation of possible loss to what has already been invested.

Limited partnership A partnership that includes one or more partners who have limited liability.

Limited payment policy Life insurance providing full life protection but requiring premiums for only part of the customer's lifetime.

Limited price order Used in the context of general equities. See: Limit order.

Limited recourse A term describing a type of loan in which the lender has limited or no claim against the parent company if the collateral is insufficient to repay the debt. See:Nonrecourse.

Limited recourse financing A loan made to a company specifically set up by a developer to manage a particular property. In case of default, the lender has no recourse to the other assets of the developer.

Limited risk The risk inherent in options contracts, which is much lower than that of a futures contract, which has unlimited risk. The maximum loss in buying a call option, for example, is the premium paid for the option.

Limited warranty A warranty with certain conditions and limitations on the parts covered, type of damage covered, and/or time period for which the agreement is good.

Limited-liability instrument A security, such as a call option, in which the owner can lose only the initial investment.

Limited-service merchant wholesaler (limited-function wholesaler) A wholesaler that offers less than full service for customers but charges lower prices than a full-service merchant wholesaler.

Limited-tax general obligation bond A general obligation bond that is limited as to revenue sources.

Line and staff management A system of management used in large organization in which there are two separate hierarchies; the line management side consists of line managers with responsibility for deciding the policy of and running the organization's main activities (such as manufacturing, sales, etc), while the staff management, and its separate staff managers, are responsible for providing such supporting services as warehousing, accounting, transport, personnel management, and plant maintenance.

Line filling Adding products to an existing line of products in order to leave no opportunities for competitors. Line filling can be horizontal or vertical. In horizontal line filling, a video manufacturer may produce machines with a variety of features, such as half-speed copying, multi-recording memory, etc, at the top end of the price range, and relatively cheap cost effective machines at the opposite end; a competitor can therefore only compete on price. In vertical line-filling a manufacturer may produce a wide variety of brand names within a single product line; some detergent manufacturers do this.

Line of credit An informal loan arrangement between a bank and a customer allowing the customer to borrow up to a prespecified amount.

Line production (mass production) A type of process in which high volumes of identical or very similar, products are made in a set sequence of operations. It involves high capital expenditure and depends on efficient plant layouts and procedures, as well as good product design to minimize unit costs. The levels of skill required are not usually high. In recent years a number of organizations have experimented with systems that trade off the benefits of full line production in exchange for the increased levels of motivation following from giving staff more ownership of the product, e.g. Through cellular layout. The introduction of computerized control systems has enabled some manufacturers to produce apparently unique products without losing the benefits of the line process. For example, some

Japanese car manufacturers have put together a relatively small range of standard components, trims, finishes, etc, to the customer's specification, which enables them to claim that no two of their cars are identical.

Linear cost function Cost behaviour that, when plotted on a graph against activity levels, results in a straight line. For example, total fixed cost levels and variable costs per unit of activity will both result in a straight line horizontal to the x-axis when activity, production, or sales is plotted on the x-axis. Total variable costs will also result in a straight line and is thus a linear cost function.

Linear depreciation Depreciation charges that, when plotted on a graph against time on the x-axis, result in a straight line, as a constant amount per annum is written off the assets concerned Both the straight-line method of depreciation and the rate per unit of production method when the depreciation charge is plotted against production levels, result in linear depreciation.

Linear programming A modelling technique that determines an optimal solution for attaining an objective by taking into consideration a number of constraints. The objective function, often to optimize profits or minimize costs, is expressed as an equation and the constraints are also expressed in mathematical terms. Where only two products and few constraints are involved a solution may be obtained graphically More than two products requires a computer program.

Linear regression A statistical technique for fitting a straight line to a set of data points.

Linking method Method for calculating rates of return that multiplies one plus the interim rate of return.

Lipper Mutual Fund Industry Average The average level of performance for all mutual funds, as reported by Lipper Analytical Services.

Liquid asset Asset that is easily and cheaply turned into cash-notably, cash itself and short-term securities.

Liquid instrument A negotiable instrument that the purchaser is able to sell before maturity.

Liquid ratio (acid-test ratio; quick ratio) A ratio used for assessing the liquidity of a company, it is the ratio of the liquid (quick) assets, i.e. The circulating capital less the stock to the current liabilities. The answer is expressed either as a percentage or as *x*:1. For example, a company with current assets of £25,000 including stock of £15,000 and liabilities of £12,000 will have a liquid ratio of:(£25,000 - £15,000)/£12,000 = 0.83, i.e. 83% or 0.83:1. This may be interpreted as the company having 83 pence of liquid or current assets for every £1 of current liabilities. If, for some reason, the company is obliged to repay the current liabilities immediately there would be insufficient liquid assets to allow it to do so. The company may therefore be forced into a hurried sale of stock at a discount to raise finance. Although there is no rule of thumb, and there are industry differences, a liquid ratio significantly below 1:1 will give rise to concern. The liquid ratio is regarded as an acid test of its solvency and is therefore sometimes called the add-test ratio.

Lira 1. (l; lit) The standard monetary unit of Italy and San Marino, divided into 100 centesimi

2. (lt) The standard monetary unit of Turkey, divided into 100 kurus. **3.** (Lm) The standard monetary unit of Malta, divided into 100 cents.

Liquidation (winding-up) The distribution of a company's assets among its creditors and members prior to its dissolution. This brings the life of the company to an end. The liquidation may be voluntary (*see*creditor's voluntary liquidation; members' voluntary liquidation) or by the court (*see*compulsory liquidation).

Liquidation committee A committee set up by

creditors of a company being wound up in order to consent to the liquidator exercising certain of his powers. When the company is unable to pay its debts, the committee is usually composed of creditors only; otherwise it consists of creditors and contributories.

Liquidation When a firm's business is terminated, assets are sold, proceeds pay creditors and any leftovers are distributed to shareholders. Any transaction that offsets or closes out a Long or short position.

Liquidator A person appointed by a court, or by the members of a company or its creditors, to regularize the company's affairs on a liquidation (windingup). In the case of a members' voluntary liquidation, it is the members of the company who appoint the liquidator. In a creditors' voluntary liquidation, the liquidator may be appointed by company members before the meeting of crediton or by the creditors themselves at the meeting; in the former case the liquidator can only exercise his or her powers with the consent of the court. If two liquidators are appointed, the court resolves which one is to act. In a compulsory liquidation, the court appoints a provisional liquidator after the winding-up petition has been presented; after the order has been granted, the court appoints the official receiver as liquidator, until or unless another officer is appointed.

The liquidator is in a relationship of trust with the company and the creditors as a body; a liquidator appointed in a compulsory liquidation is an officer of the court, is under statutory obligations, and may not profit from the position. A liquidator must be a qualified insolvency practitioner.

Liquidity A market is liquid when it has a high level of trading activity, allowing buying and selling with minimum price disturbance. Also a market characterized by the ability to buy and sell with relative ease.

Liquidity preference hypothesis The argument that greater liquidity is valuable, all else equal. Also, the theory that the forward rate exceeds expected future interest rates.

Liquidity ratio Indicates the company's ability to pay its short term debts, by measuring the relationship between current assets (ie those which can be turned into cash) against the short-term debt value. (current assets/ current liabilities) Also referred to as the Current Ratio.

Liquidity risk The risk that arises from the difficulty of selling an asset. It can be thought of as the difference between the "true value" of the asset and the likely price, less commissions.

Liquidity theory of the term structure A biased expectations theory that asserts that the implied forward rates will not be a pure estimate of the market's expectations of future interest rates because they embody a liquidity premium.

List renting The practice of renting a list of potential customers to an organization involved in direct-mail selling or to a charity raising funds. When the terms of hiring restrict the use of the list to one mail shot, it is usual for the owner of the list to carry out the mailing, so that the hirer cannot copy it for subsequent use.

Listing broker In the context of equity, when a stock is traded in exchange it is said to be listed. A licensed real estate broker who completes a listing of a property for sale.

Listing In the context of real estate, written agreement between a property owner and a real estate broker that gives the broker permission to find a buyer or tenant for some property. See: Listing broker.

Listing requirements Requirements, including minimum shares outstanding, market value, and income, that are laid down by an exchange for any stock to be listed for trading.

Lists closed The closing of the application lists for a new issue on the London Stock Exchange, after a specified time or after the issue has been fully subscribed.

Litas The standard monetary unit of Lithuania.

Litigation 1. The taking of legal action. The person who takes it is called a

Litigant 2. The activity of a solicitor when dealing with proceedings in a court of law.

Little Board The colloquial name for the American Stock Exchange, the New York market for smaller company stocks and bonds. *Compare*big Board

Livery company One of some eighty chartered companies in the City of London that are descended from medieval craft guilds. Now largely social and charitable institutions, livery companies owe their name to the elaborate ceremonial dress (livery) worn by their officers. Several support independent schools (e.g. Merchant Taylors, Haberdashers, Mercers) and although none are now trading companies, some still have some involvement in their trades (e.g. Fishmongers). In 1878 they joined together to form the City and Guilds of London Institute, which founded the City and Guilds College of the Imperial College of Science and Technology and has been involved in other forms of technical education.

Livestock and bloodstock insurance An insurance policy covering the owners against financial losses caused by the death of an animal. Policies may be widened to include cover for treatment fees for certain specified diseases or lost profits for stud animals. Insurances can be arranged to cover single animals or for whole herds.

Living benefits Life insurance benefits from which the insured can draw cash while still living, usually in the case of some high-cost illness.

Living trust A trust that an individual establishes during the individual's lifetime, enabling the person to control the assets contributed to the trust. Also known as an inter vivos trust.

Living will A document specifying the kind of medical care a person wants-or does not want-in the event of terminal illness or incapacity.

Lloyd's A corporation of underwriters (Lloyd's underwriters) and insurance brokers (Lloyd's brokers) that developed from a coffee shop in Tavern Street in the City of London in 1689. It takes its name from the proprietor of the coffee shop, Edward Lloyd. By 1774 it was established in the Royal Exchange and in 1871 was incorporated by act of parliament. It now occupies a building in Lime Street (built in 1986 by Richard Rogers). As a corporation, Lloyd's itself does not underwrite insurance business; all its business comes to it from some 260 Lloyd's brokers, who are in touch with the public, and is underwritten by some 279 syndicates of Lloyd's underwriters, who are approached by the brokers and who do not, themselves, contact the public.

Lloyd's agent A person situated in a port to manage the business of Lloyd's members, keeping the corporation informed of shipping movements and accidents, arranging surveys, helping in the settlement of marine insurance claims, and assisting masters of ships. A Lloyd's agent will be found in every significant port in the world.

Lloyd's List and Shipping Gazette A daily newspaper published by Lloyd's, founded in 1734 and formerly known as Lloyd's List It gives details of the movements of ships and aircraft, accidents, etc. Lloyd's Loading List published weekly by Lloyd's, lists ships loading in British and continental ports, with their closing dates for accepting cargo. It also gives general news on the insurance market.

Lloyds of London A marketplace in London for underwriting syndicates.

Load fund A mutual fund with shares sold at a price including a large sales charge – typically 4% to 8% of the net amount indicated. Some "no-load" funds have distribution fees permitted by article 12b-1 of the Investment Company Act; these are typically 0. 25%. A "true no-load" fund has neither a sales charge nor Freddie Mac program, the aggregation that the fund purchaser receives some investment advice or other service worthy of the charge.

Load spread option A method of allocating the annual sales charge on load funds, often through percentage deductions from a customer's periodic fixed payments.

Load The sales fee charged to an investor when shares are purchased in a load fund or annuity. See: Back-end load; front-end load; level load.

Loading The process of allocating jobs or orders to work centres within a production facility. If there is a range of jobs there could be a range of possible routes through the various work centres. Building up a profile of the load on each location enables approximate completion dates for each job to be estimated and bottlenecks in the system to be identified. Loading is based on average tunes and does not take into account job interference, etc Forward loading begins with the present date and loads jobs forwards Backward loading begins with the completion date and works backwards.

Load-to-load Arrangement whereby the customer pays for the last delivery when the next one is received.

Loan account An account opened by a bank in the name of a customer to whom it has granted a loan, rather than an overdraft facility. The amount of the loan is debited to this account and any repayments are credited; interest is charged on the full amount of the loan less any repayments. The customer's current account is credited with the amount of the loan. With an overdraft facility, interest is only charged on the amount of the overdraft, which may be less than the full amount of the loan.

Loan creditor A person or institution that has lent money to a business. For example, when a bank loan is obtained the bank becomes a loan creditor.

Loan note A form of loan stock (*see* debenture in which an investor takes a note rather than cash as the result of a share offer to defer tax liability. The yield is often variable and may be linked to the London Inter Bank Offered Rate. Loan notes are not usually marketable but are usually repayable on demand.

Loan syndication Group of banks sharing a loan. See: syndicate.

Loan-to-value ratio (LTV) The ratio of money borrowed on a property to the property's fair market value.

Local authority bill A bill of exchange drawn on a UK local government authority.

Local expectations hypothesis (LEH) Theory that bonds similar in all aspects except maturity will have the same holding-period rate of return.

Location strategy The process of choosing where to locate a unit producing goods or services. This must take account of a range of variables to achieve the minimum cost that optimizes the overall strategy of the particular organization. There are three basic strategies. In a product-based strategy, independent facilities are used to produce each product in a company's range. This permits dedicated design and organization of facilities, which may vary from product to product. A marketbased strategy gives priority to locating as close to customers as possible. This is clearly important in the case of services and for organizations offering a quick response to customer requirements, e.g. In fast-moving consumer goods. A vertically differentiated strategy locates different elements of the process at different sites, according to the economic characteristics of each element. These considerations are particularly important for transnational and global operations.

Location-specific advantages Advantages (natural and created) that are available only or primarily in a particular place.

Lock Used in the context of general equities. Make a market both ways (bid andoffer) either on the bid, offering, or an in-between price only. Locking on the offering occurs to attract a seller, since the trader is willing to pay (and ask) the offering side when others only ask it. Locking on the bid side attracts buyers for similar reasons. Typically, the sell side requires a plus tick to comply with short sale rules.

Locked in When an investor is unable to take advantage of preferential tax treatment because of time remaining on a required holding period. Also, a commodities position in which the market has a limit up or limit down day and investors are unable to move in to or out of the market.

Locked market A market is locked if the bid price equals the ask price. This can occur, for example, if the market is brokered and one side pays brokerage only, in over-the-counter trading the initiator of the transactions. Highly competitive market environment with inside bid and offering at the same price. Often occurs when an OTC dealer has not updated the market.

Lockup Agreement Where the purchaser of securities agrees not to sell the securities for a certain period of time (the lockup time).

Loco Denoting a price for goods that does not include any loading or transport charges; i.e. The goods are located in a specified place, usually the seller's warehouse or factory, and the buyer has to pay all the charges involved in loading, transporting, or shipping them to their destination. A loco price might be quoted in the form: *£100 per tonne, loco Islington factory*.

Locus poenitentiae (Latin: an opportunity to repent) An opportunity for the parties to an illegal contract to reconsider their positions, decide not to carry out the illegal act, and so save the contract from being void. Once the illegal purpose has been carried out, no action in law is possible.

Lognormal distribution Pattern of frequency of occurrence in which the logarithm of the variable follows a normal distribution. Lognormal distributions are used to describe returns calculated over periods of a year or more.

Lombard rate Applies mainly to international equities. Interest rate the German Bundesbank uses as an upper limit to the day-to-day money rate, since no bank will pay higher rates in the money market than it has to pay for very short-term recourse to Lombard credit.

Loss adjuster A person appointed by an insurer to negotiate an insurance claim. The loss adjuster, who is independent of the insurer, discusses the claim with both the insurer and the policyholder, producing a report recommending the basis on which the claim should be settled. The insurer pays a fee for this service based on the amount of work involved for the loss adjuster, not on the size of the settlement. *Compare*loss assessor.

Loss assessor A person who acts on behalf of the policyholder in handling a claim. A fee is charged for this service, which is usually a percentage of the amount received by the policyholder. *Compare*loss adjuster.

Loss ratio The total of the claims paid out by an insurance company, underwriting syndicate, etc., expressed as a percentage of the amount of premiums coming in in the same period. For example, if claims total £2M and premiums total E4K the result is a 50% loss ratio. Insurers use this figure as a guide to the profitability of their business when they are reconsidering premium rates for a particular risk.

Low ball Slang for making an offer well below the fair value of an asset in hopes that the seller may be desperate to sell.

Low In the context of general equities, this is a specific minimum limit required by a seller in execution an order ("I'll sell 50 with an eighth low."); implies a not-held limit order. Antithesis of top.

Low reliefs Relief available to sole traders, partnerships, and companies making losses, as adjusted for tax purposes. Capital allowances can create a trading loss or can enhance it. Trading losses can be carried forward to set against future trading profits. For sole traders and partnerships, trading losses can be set against other income for the year of the loss and for the previous year. Partners can decide individually how to use their share of the losses. Special rules apply to trading losses in the early years of a trade. These losses can be carried back three years to a period before the trade commenced. From 1991-92 it has been possible to set trading losses against capital gains if the loss cannot be used first by setting against other income during the year. Terminal-loss relief is available when a trade is permanently discontinued and a loss is made during the last 12 months. For companies, a trading loss can be set off against profits of the three previous accounting periods, provided the company was carrying on the same trade during that period. Terminal-loss relief is available for companies. Capital losses can be set against capital gains in the same period. Any surplus capital loss that cannot be utilized during the current year must be carried forward to set against future capital gains. Capital losses cannot be set against other income, unlike trading losses.

Low-coupon bond refunding Refunding of a low-coupon bond with a new, higher-coupon bond.

Low-involvement product A product that involves the consumer in little or no trouble or deliberation when making a purchase. Such products are invariably cheap; manufacturers try to make them more interesting and appealing by means of advertising with the intention of developing a brand loyalty. The chimpanzees used in TV advertisements for PG Tips are an example of this form of advertising. *Compare*high-involvement product.

Low-load fund A mutual fund that charges a sales commission of 3.5% or less for the purchase of shares.

Ltd The usual abbreviation for limited. This must appear in the name of a private limited company.

Lump sum 1. A sum of money paid all at once, rather than in instalments.

2. A sum of money paid for freight, irrespective of the size of the cargo.

3. An insurance benefit, such as a sum of money paid on retirement or redundancy or to the beneficiaries on the death of an insured person. Retirement pensions can consist of a lump sum plus a reduced pension.

4. A form of damages; a lump-sum award is given in tort cases.

Lump-sum distribution A single payment that represents an employee's interest in a qualified retirement plan. The payment must be prompted by retirement (or other separation from service), death, disability, or attainment of age 59-1/2, and must be made within a single tax year to avoid the federal government's 10% penalty tax.

Macaroni defense A tactic used by a corporation that is the target of a hostile takeover bid involving the issue of a large number of bonds that must be redeemed at a higher value if the company is taken over.

Macaulay duration The weighted-average term to maturity of the cash flows from the bond, where the weights are the present value of the cash flow divided by the price.

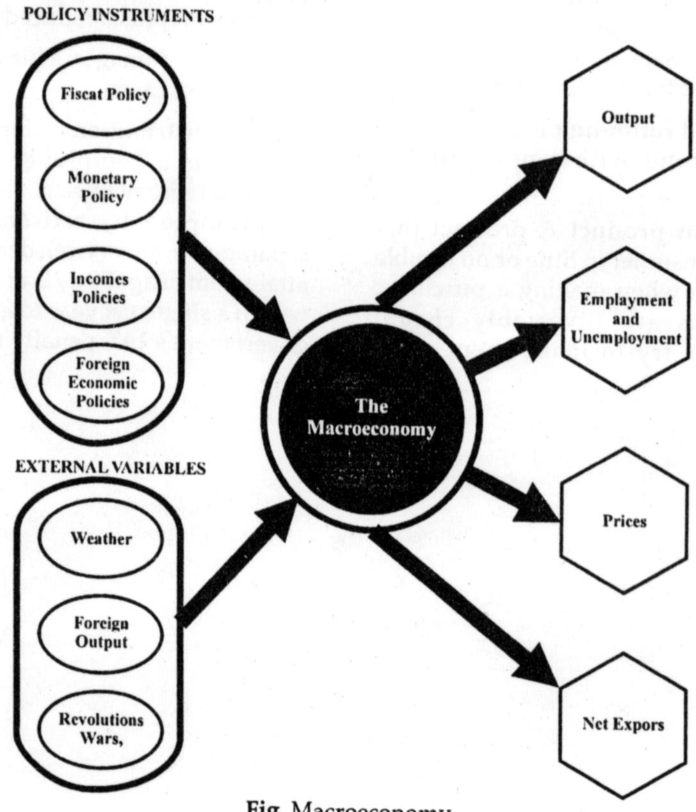

Fig. Macroeconomy

Machine-down time The period during which a machine cannot be used, usually because of breakdown. If a machine is 'down', clearly production is not taking place, but it is customary in costing to attribute costs to the down time.

Machine-readable Denoting any form of data that can be input directly into a computer.

Macroeconomics The study of economic factors and their relationships to, for example, money, employment, interest rates, government spending, investment, and consumption. Modern macroeconomics is largely concerned with the role, if any, that a government should play in a national economy.

Macroenvironment The broad sweep of the environment that creates the forces that shape every business and non-profit marketer. The physical environment, social and cultural forces, demographic make-up, economic climate, prevailing political beliefs, and current legal infrastructure are components of the macroenvironment. Generally, most companies have no control over these factors.

Magic of diversification The effective reduction of risk (variance) of a portfolio, achieved without reduction to expected returns through the combination of assets with low or negative correlations (covariances).

Magnetic ink character recognition (MICR) A type of magnetic ink used on cheques and other documents to enable them to be automatically sorted and the characters to be read and fed into a computer.

Magnuson-Moss Warranty Act A US federal law requiring that guarantees provided by sellers should be made available to buyers before a purchase is made and that they specify who the warrantor is, what products (or parts of products) are covered, what the warrantor must do if the product is defective, how long the warranty applies, and the obligations of the buyer.

Mail Delay Time a payment spends in the postal system before delivery.

Mail float Refers to the part of the collection and disbursement process where checks are trapped in the postal system.

Mail shot Selling, advertising, or fundraising material sent to all the names on a mailing list.

Mail survey Marketing research conducted by mail. As respondents have time to consider their answers, this form of research has advantages over both telephone surveys and face-to-face surveys. However, it does suffer from low returns.

Mail-order house A firm that specializes in selling goods direct to customers by post. Orders are obtained from an illustrated catalogue supplied by the firm or by agents, who introduce the catalogue. The low costs of selling, especiary the absence of retail premises, enable the mail-order houses to offer goods at competitive prices.

Main market The premier market for the trading of equities on the London Stock Exchange. For this market the listing requirements are the most stringent and the liquidity of the market is greater than in the unlisted securities market and the other junior markets. A company wishing to enter this market must have audited trading figures covering at least five years and must place 25% of its shares in public hands. The main market currently deals in some 7000 secutities.

Main product The product of a process that has the greatest economic significance. Other products of secondary economic importance are regarded as by-products; however, if all products have equal economic significance they are regarded as joint products. *See also* process costing.

Maintenance margin requirement A sum,

usually smaller than -but part of the original margin, which must be maintained on deposit at all times. If a customer's equity in any futures position drops to, or under, the maintenance margin level, the broker must issue a margin call for the amount at money required to restore the customer's equity in the account to the original margin level.

Maintenance margin The dollar amount required to be kept at the OCC throughout the life of an option contract; percentage of the dollar amount of securities that must always be kept as margin.

Make whole provision Related to the lump-sum payments made when a loan or bond is called, equal to the NPV of future loan or coupon payments not paid because of the call. The payment can be significant and negate the attractiveness of a call.

Making a price On the London Stock Exchange, the quoting by a market maker of a selling price and a buying price for a particular security, usually without knowing whether the broker or other person asking for a price wishes to buy or sell. Having made a price, the market maker is bound to buy or sell at the prices quoted, though the quantity may be limited if the broker stipulates the quantity by saying so at the time of making the price.

Malpractice insurance In the USA, liability insurance taken out by an accountant against legal action in connection with professional services. There have been a number of very high awards made to plaintiffs and this has greatly increased the cost to the accountant of obtaining insurance cover. One solution to this may be for accountants to form corporate bodies rather than partnerships, thus reducing their exposure to personal liability.

Managed account An investment portfolio one or more clients entrusted to a manager who decides how to invest it.

Managed currency (managed floating) A currency in which the government controls, or at least influences, the exchange rate. This control is usually exerted by the central bank buying and selling in the foreign-exchange market. *See also*clean floating

Managed float Also known as "dirty" float, this is a system of floating exchange rates with central bank intervention to reduce currency fluctuations.

Managed fund A fund made up of investments in a wide range of securities, that is managed by a life-assurance company to provide low risk investments for the smaller investor, usually in the form of investment bonds, unit trusts, or unit-linked saving plans (*see* unit-linked policy). The fund managers will have a stated investment policy favouring a specific category of investments.

Management 1. The running of an organization or part of it. Management has two main components: an organizational skill including the ability to delegate, and an entrepreneurial sense. The organizational skill involving the principles and techniques of management, is taught at colleges and business schools; the entrepreneurial sense, recognizing and making use of opportunities, predicting market needs and trends, and achieving one's goals by sustained drive, skilful negotiation, and articulate advocacy, are not so easily taught, although contact with the market place in association with a successful entrepreneur will encourage an inherent ability to develop.

Management audit An independent review of the management of an organization, carried out by a firm of management consultants specializing in this type of review. The review will cover all aspects of running the organization, including the control of production, marketing, sales, finance, personnel warehousing, etc.

Management buy-in The acquisition of a company by a team of managers, usually specially formed for the purpose, often backed by a venture-capital organization.

Their normal target is the small family-owned company, which the owners wish to sell, or occasionally unwanted subsidiary of a public company.

Management buy-out (MBO) The acquisition of a company or a subsidiary by the existing management. It is frequently used as a means of divestment by companies seeking to focus on their core activities. The new owner-managers of the buy-out frequently improve its performance as they usually are well aware of any remedial action required and have a serious incentive in the form of their equity stake. Additional capital is provided by financial institutions and venture capitalists and also in many cases by allowing other employees to buy shares. In a few instances, all employees have participated in a buyout; these are sometimes called employee buy-outs. Some very large companies have been bought out by management, but many of these have suffered from unrealistically high levels of debt (in the form of loan capital), which has restricted their successful recovery and longer-term viability.

Management by wandering around (MBWA) The practice by managers of spending a significant part of their time listening to staff problems and ideas while wandering around an office or plant. The opportunity to communicate with managers can be seen as providing an important contribution to staff motivation. are paid by the unit holders.

Management consultant A professional adviser who specializes in giving advice to organizations on ways for improving their efficiency and hence their profitability. They come into an organization as total outsiders, uninfluenced by either internal politics or personal relationships, and analyse the way the business is run. At the end of a period, during which two or more members of the consultant firm have spent a considerable tune in the organization, they provide a detailed report, giving their suggestions for improving efficiency. Their advice usually spans board-level policy-making and planning, the use of available resources (department by department), the best use of manpower, and a critical assessment of industrial relations, production, marketing, and sales.

Management Consultants' Association (MCA) An association founded in 1956 to establish and maintain the professional standards of the management consultancy profession in the UK.

Management contract An agreement by which a company will provide its organizational and management expertise in the form of services.

Management fee An investment advisory fee charged by the financial adviser to a fund typically on the basis of the fund's average assets, but sometimes determined on a sliding scale that declines as the dollar amount of the fund increases.

Management information system (MIS) An information system designed to provide financial and quantitative information to all the levels of management in an organization. Most modern management information systems provide the data from an integrated computer database, which is constantly updated from all areas of the organization in a structured way. Access to the data is usually restricted to the areas regarded as useful to particular managers, access to confidential information is limited to top management. An MIS should run parallel to the configuration of the physical and organizational structures. The application of information technology has enabled MIS to drive these structures.

Management is usually broken down into the categories formalized in line and staff management: the line managers organize the production of the goods or oversee the services provided by the organization, while the staff management provides such support as personnel management, transport management, service management, etc. *See also*institute of Management.

2. The people involved in the running of an organization.

Management letter A letter written by an auditor to the management of a client company at the end of the annual audit to suggest any possible improvements that could be made to the company's accounting and internal control system or to communicate any other such information that the auditor believes would be of benefit to the client.

Manager A medium-level participant established according to final take.

Managing director (MD) The company director responsible for the day-today running of a company. Second in the hierarchy only to the chairman, if there is one, the managing director is the company's chief executive, a title that is becoming increasingly popular in both the USA and the UK. In the USA the president is often the equivalent of the MD. If a company has more than one MD, they are known as joint managing directors.

Manat The standard monetary unit of Azerbaijan and of Turkmenistan, divided into 100 gopik.

Mandate 1. A written authority given by one person (the mandator) to another (the mandatory) giving the mandatory the power to act on behalf of the mandator. It comes to an end on the death, mental illness, or bankruptcy of the mandator.

2. A document instructing a bank to open an account in the name of the mandator (customer), giving details of the way it is to be run, and providing specimen signatures of those authorized to sign cheques, etc.

Man-hour The amount of work done by one man in one hour. It is sometimes convenient in costing a job to estimate the number of man-hours it will take.

Manifest A list of all the cargo carried by a ship or aircraft. It has to be signed by the captain (or first officer) before being handed to the customs on leaving and arriving at a port or airport.

Manipulation Dealing in a security to create a false appearance of active trading, in order to bring in more traders. Illegal.

Manufactured housing securities (MHS) Loans on manufactured homes-that is, factory-built or prefabricated housing, including mobile homes.

Manufacturer brand A brand name created by a manufacturer. Kellogs and Polaroid are examples.

Manufacturing cost of finished goods (total production cost of finished goods) An item computed in a manufacturing account by the addition of prime cost and manufacturing overhead, adjusted by the opening and closing work in progress for the period.

Manufacturing costs (manufacturing expenses) Items of expenditure incurred to carry out the manufacturing process in an organization. They include direct materials, direct labour, direct costs (such as subcontract costs), and manufacturing overhead

Manufacturing overhead (production overhead) The costs of production that cannot be traced directly to the product or cost unit. Apart from the direct costs, all other costs incurred in the manufacturing process are the manufacturing overhead; examples include depreciation of machinery, factory rent and business rates, cleaning materials, and maintenance expenses.

Manufacturing profit (or loss) The difference between the value of the goods transferred from a manufacturing account to a trading account at a price other than the production cost of finished goods, and the production cost of finished goods.

Mareva injunction An order of the court preventing the defendant from dealing with specified assets. Such an order will be granted in cases in which the plaintiff can show that there will be a substantial risk that any judgement given against the defendant will be worthless, because the defendant will sell assets to avoid paying it. It is usually granted to prevent assets leaving the jurisdiction of the English courts, but may in exceptional circumstances extend to assets abroad.

Margin account (Stocks) A leverageable account in which stocks can be purchased for a combination of cash and a loan. The loan in the margin account is collateralized by the stock and, if the value of the stock drops sufficiently, the owner will be asked to either put in more cash, or sell a portion of the stock. Margin rules are federally regulated, but margin requirements and interest may vary among broker/dealers.

Margin Allows investors to buy securities by borrowing money from a broker. The margin is the difference between the market value of a stock and the loan a broker makes. Related: Security deposit (initial). In the context of hedging and futures contracts, the cash collateral deposited with a trader or exchanged as insurance against default.

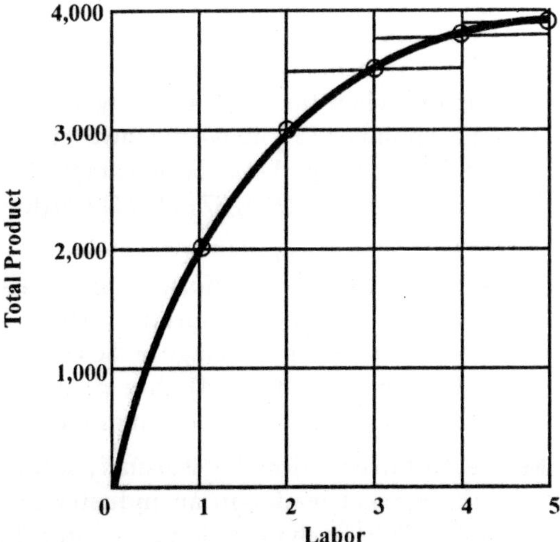

Fig. Marginal Product

Margin call A call to a client from his commodity broker or stockbroker to increase his margin, i.e. The amount of money or securities deposited with the broker as a safeguard. This usually happens if the client has an open position in a market that is moving adversely for this position.

Margin department The department in a brokerage firm that monitors customers' margin accounts, ensuring that all short sales, stock purchases, and other positions are covered by the margin account balance.

Margin of safety The difference between the level of activity at which an organization breaks even and a given level of activity greater than the breakeven point. The margin of safety may be expressed in the same terms as the breakeven point, i.e. Sales value, number of units, or percentage of capacity.

Margin requirement (options) The amount of cash an uncovered (naked) option writer is required to deposit and maintain to cover his daily position valuation and reasonably foreseeable intraday price changes.

Margin stock Any stock listed on a national securities exchange, any over-the-counter security approved by the SEC for trading in the national market system, or appearing on the Board's list of over-the-counter margin stock and most mutual funds.

Marginal cost pricing The setting of product selling prices based on the charging of marginal costs only to the product. The approach is only likely to be used in exceptional circumstances, such as when competition is intensive, as its application to the complete range of products is likely to cause the business to make losses by its failure to cover its fixed costs.

Marginal cost The variable costs per unit of production. The variable costs are usually regarded as the direct costs plus the variable overheads. Marginal cost represents the additional cost incurred as a result of the production of one additional unit of production.

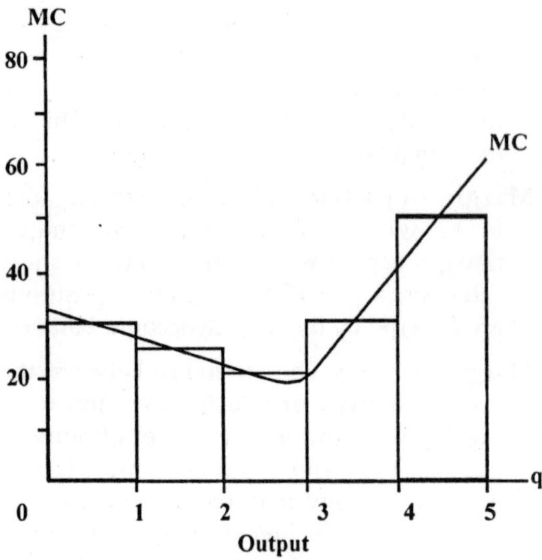

Fig. Marginal Cost

Marginal costing (direct costing; variable costing) A costing and decision-making technique that charges only the marginal costs to the cost units and treats the fixed costs as a lump sum to be deducted from the total contribution, in obtaining the profit or loss for the period. In some cases, inventory valuation is also at marginal cost, although this approach does not conform to Statement of Standard Accounting Practice 9 and is used for internal reporting purposes only.

Marginal rate of tax The rate of corporation tax that applies to the profits of a company between the lower limit for corporation tax (£300,000 for Emancial year 1996-97) and the upper limit (£1,500,000 for that year).

Marital trust A trust created to allow one spouse to transfer, during life or upon death, an unlimited amount of property to his/her spouse without incurring gift or estate tax.

Marked cheque A cheque that the bank on which it is drawn has marked "good for payment". This practice has been replaced in the UK by bank drafts, although it is still used in the USA, where such cheques are called certified checks.

Marker barrel A standard oil price based on one barrel of Saudi Arabian oil or some other internationally recognized oil (e.g. North Sea Brent crude). It sets standard prices for oil in the rest of the world.

Market analysis An analysis of technical corporate and market data used to predict movements in the market.

Market clearing Total demand for loans by borrowers equals total supply of loans from lenders. The market, any market, clears at the equilibrium rate of interest or price.

Market conversion price Also called conversion parity price, the price that an investor effectively pays for common stock by purchasing a convertible security and then exercising the conversion option. This price is equal to the market price of the convertible security divided by the conversion ratio.

Market correction A relatively short-term drop in stock market prices, generally viewed as bringing overpriced stocks back to a level closer to companies' actual values.

Market cycle The period between the 2 latest highs or lows of the S&P 500, showing net performance of a fund through both an up and a down market. A market cycle is complete when the S&P is 15% below the highest point or 15% above the lowest point (ending a down market). The dates of the last market cycle are: 12/04/87 to 10/11/90 (low to low).

Market Eye A financial information service based in the U.K. sponsored by the ISE (International Stock Exchange of the UK and the Republic of Ireland) that provides current market and statistical information.

Market follower A firm that is usually second to the market leader in an industry and wants to hold its share without upsetting the status quo. Market followers do not want to challenge the leader, but can maintain their market share with much lower investment costs than the leader.

Market forces The forces of supply and demand that in a free market determine the quantity available of a particular product or service and the price at which it is offered. In general, a rise in demand will cause both supply and price to increase, while a rise in supply will cause both a fall in price and a drop in demand, although many markets have individual features that modify this simple analysis. In practice, most markets are not free, being influenced either by restrictions on supply or by government intervention that can affect demand, supply, or price.

Market index Market measure that consists of weighted values of the components that make up certain list of companies. A stock market tracks the performance of certain stocks by weighting them according to their prices and the number of outstanding shares by a particular formula.

Market leader The firm in a particular industry with the largest market share. The market leader usually dominates other firms in price changes, the introduction of new products, distribution coverage, and promotion spending. This firm may or may not be admired by other companies, but generally these companies accept the dominance.

Market maker A dealer in securities on the London Stock Exchange who undertakes to buy and sell securities as a principal and is therefore obliged to announce prices at which he will buy or sell a particular security at a particular time.

Market manager An individual responsible for administering all the marketing activities of a company that relate to a particular market, including forecasting, product planning, and pricing.

Market model The market model says that the return on a security depends on the return on the market portfolio and the extent of the security's responsiveness as measured by beta. The return also depends on conditions that are unique to the firm. The market model can be graphed as a line fitted to a plot of asset returns against returns on the market portfolio. This relationship is sometimes called the single-index model.

Market model This relationship is sometimes called the single-index model. The market model says that the return on a security depends on the return on the market portfolio and the extent of the security's responsiveness as measured, by beta. In addition, the return will also depend on conditions that are unique to the firm. Graphically, the market model can be depicted as a line fitted to a plot of asset returns against returns on the market portfolio.

Market Neutral In the context of hedge funds, a style of management that has long and short equity exposure with nearly exposure on average to fluctuations in the market. However, the on average qualification is important. The risk of the longs and the shorts could fluctuate through time leading to negative returns when the market falls sharply.

Market Not Held Order Also a market order, but the investor is allowing the floor broker to use his own discretion as to the exact timing of the execution. If the floor broker expects a decline in price and he is holding a "market not held buy order", he (she) may wait to buy, figuring that a better price will soon be available. There is no guarantee that a "market not held order" will be filled.

Market objectives The objectives to be fulfilled by a marketing plan. Marketing objectives are determined by interpreting the relevant parts of a corporate plan

Market order Used in the context of general equities. Order to buy or sell a stated amount of a security at the most advantageous price obtainable after the order is represented in the trading crowd. You cannot specify special restrictions such as all or none (AON)

or good 'til canceled order (GTC) on market orders.

Market out clause A clause that may appear in an underwriting firm commitment that releases it from its purchase requirement if there are negative securities market developments.

Market overhang The theory that, in certain situations, institutions wish to sell their shares but postpone the sale because large orders under current market conditions would drive down the share price and that the consequent threat of securities sales will tend to retard the rate of share price appreciation. Support for this theory is largely anecdotal.

Market penetration 1. The process of entering a market to establish a new brand or product. Market penetration may be achieved by offering the brand or product at a low initial price to familiarize the public with its name. This is known as market-penetration pricing.

2. A measure of the extent to which a market has been penetrated equal to the ratio of all the owners of that brand or product to the total number of potential owners, usually expressed as a percentage.

Market Performance Committee (MPC) A group of NYSE market oversight specialists who monitor specialists' efficiency in maintaining fair prices and orderly markets.

Market portfolio A portfolio consisting of all assets available to investors, with each asset held in proportion to its market value relative to the total market value of all assets.

Market positioning Arranging for a product to occupy a clear, distinctive, and desirable place relative to competing products in the minds of target consumers.

Market price of risk A measure of the extra return, or risk premium, that investors demand to bear risk. The reward-to-risk ratio of the market portfolio.

Market price of risk A measure of the extra return, or risk premium, that investors demand to bear risk. The reward-to-risk ratio of the market portfolio.

Market price The price of a raw material product, service, security, etc, in an open market. In formal markets, such as a stock exchange, commodity market, foreign-exchange market, etc., there is often a margin between the buying and selling price; there are, therefore, two market prices. In these circumstances the market prices often quoted are the average of the buying and selling price (*see*middle price).

Market prices The amount of money that a willing buyer pays to acquire something from a willing seller, when a buyer and seller are independent and when such an exchange is motivated by only commercial consideration.

Market research A technical analysis of factors such as volume, price trends, and market breadth that are used to predict price movement.

Market return The return on the market portfolio.

Market segmentation The division of a market into homogeneous groups of consumers, each of which can be expected to respond to a different marketing mix. There are numerous ways of segmenting markets, the more traditional being by age, sex, family size, income, occupation, and social class; more recently geodemographic segmentation, which identifies housing areas in which people share a common lifestyle and will be more likely to buy certain types of products, has become more popular. Another frequently used method is **benefit segmentation** ; for example, the toothpaste market can be segmented into those primarily concerned with dental protection and those concerned with a fresh taste. Once a segment has been identified, the marketer. Can then develop a unique marketing mix to reach it, for example by advertising only in the newspapers read by that market segment. Therefore, to be of practical value a market segment must be large enough to warrant the development costs.

Market segmentation theory or preferred habitat theory A biased expectations theory that asserts that the shape of the yield curve is determined by the supply of and demand for securities within each maturity sector.

Market segmentation theory or preferred habitat theory A biased expectations theory that asserts that the shape of the yield curve is determined by the supply of and demand for securities within each maturity sector.

Market share The share of the total sales of all brands or products competing in the same market that is captured by one particular brand or product, usually expressed as a percentage. For example, if brands A, B, and C are the competing brands of a product and in a particular month they achieved sales of £48,000,£62,000, and £90,000, respectively, brand Ns market share would be (48,000/ (48,000 + 62,000 + 90,000)) × 100 = 24%.

Market sweep A second offering following a tender offer, allowing institutional investors to obtain a controlling interest at a price higher than the original offer.

Market targeting The process of evaluating each market segment and selecting the most attractive segments to enter with a particular product or product line.

Market timer A money manager who assumes he or she can forecast when the stock market will go up and down.

Market timing Asset allocation in which the investment in the market is increased if one forecasts that the market will outperform T-bills.

Market timing costs Costs that arise from price movement of the stock during the time of the transaction which is attributed to other activity in the stock.

Market Usually refers to the equity market. "The market went down today" means that the value of the stock market dropped that day.

Market value (1) The price at which a security is trading and could presumably be purchased or sold. (2) What investors believe a firm is worth; calculated by multiplying the number of shares outstanding by the current market price of a firm's shares.

Market value ratios Ratios that relate the market price of the firm's common stock to selected financial statement items.

Market value The value of an asset if it were to be sold on the open market at its current market price. When land is involved it may be necessary to distinguish between the market value in its present use and that in some alternative use; for example, a factory site may have a market value as a factory site, and be so valued in the company's accounts, which may be less than its market value as building land.

Market value-weighted index An index of a group of securities computed by calculating a weighted average of the returns on each security in the index, where the weights are proportional to outstanding market value.

Market/product matrix A matrix containing the four possible combinations of old and new products with old and new markets. The purpose of the matrix is to classify alternative opportunities in terms of basic strategies for growth.

Market penetration: a weak marketing strategy – generally based on lower prices.

Market development: the company searches for new markets and new market segments – generally a good medium-term strategy. Product development: the company offers new or modified products to existing markets – generally a good medium-term strategy. Product diversification: the company moves into new markets with new products – generally a good long-term strategy, although there is the risk of error.

Market-based corporate governance system Organization of a corporation whereby the supervisory board represents a dispersed set of largely equity shareholders.

Marketed claims Claims that can be bought and sold in financial markets, such as those of stockholders and bondholders.

Marketing audit A review of an organization's marketing capabilities based on a structured appraisal of a marketing department's internal strengths and weaknesses, which helps the company decide how best to respond to external opportunities and threats. Sales are analysed, the effectiveness of the marketing mix is assessed, and, externally, the marketing environment is monitored. When the marketing audit has been completed, a marketing plan is drawn up based on its results.

Marketing costs The costs incurred by an organization in carrying out its marketing activities. These would include sales promotion costs, salesmen's salaries, advertising, and point-of-sale promotional material, such as display stands.

Marketing environment The combined influence of all the factors external to a company that could affect its sales. These factors include: cultural traditions, technological developments, competitors' activity, government policies, and changes in distribution channels. The marketing environment cannot be controlled by the company and, because it may change frequently, requires constant monitoring.

Marketing implementation The process that converts marketing strategies and plans into action in order to accomplish marketing objectives.

Marketing intelligence system A network of diverse sources that provides data about the marketing environment; it forms part of an organization's datacollection system. The data should include customers, sales personnel, suppliers, retailers, agencies, competitors' activities, and government bodies.

Marketing research The systematic collection and analysis of data to resolve problems concerning marketing, undertaken to reduce the risk of inappropriate marketing activity. Data is almost always collected from a sample of the target market, by such methods as observation, interviews, and audit of shop sales. Interviews (*see*open-ended question; structured interview) are the most common technique, and can be carried out face-to-face, by telephone, or by post. When the results have been analysed (usually by computer), recommendations regarding the original problem can be made. *See also*qualitative marketing research; quantitative marketing research. Market research, which is often used synonymously with the preferred term 'marketing research', is also sometimes used to refer to techniques with the restricted objective of discovering the size of the market for a particular brand or product.

Marketing strategy A plan identifying what marketing goals and objectives will be pursued to sell a particular product or product line and how these objectives will be achieved in the time available.

Market-skimming pricing Setting a high price for a new product in order to skim the maximum revenues, layer by layer, from the market segments willing to pay the high price. Using this highprice strategy, the company makes fewer but more profitable sales.

Marking up or down The amount by which a securities dealer raises or lowers the price of a stock or bond due to changes in demand and supply.

Markings The official number of bargains that have taken place during a working day on the London Stock Exchange.

Markka (Fmk) The standard monetary unit of Finland, divided into 100 pennïa.

Markovian Dependence The condition where observations in a time series are dependent on previous observations in the near term. Markovian dependence dies quickly, while long-memory effects like Hurst dependence, decay over very long time periods.

Markowitz diversification A strategy that seeks to combine in a portfolio assets with returns that are less than perfectly positively correlated, in an effort to lower portfolio risk (variance) without sacrificing return. Related: Naive diversification.

Markowitz efficient frontier The graphical depiction of the Markowitz efficient set of portfolios representing the boundary of the set of feasible portfolios that have the maximum return for a given level of risk. Any portfolios above the frontier cannot be achieved. Any below the frontier are dominated by Markowitz efficient portfolios.

Markowitz model A method of selecting the optimum investment portfolio, devised by H. M. Markowitz (1927-). It makes use of the concepts of a market portfolio (containing all available investments in amounts proportional to the total market value of each individual security) and a graph (the capital market line) of alternative combinations of risk and return, resulting from investing fixed sums in the market portfolio.

Married Put Strategy A put and stock are considered to be married if they are bought on the same day, and the position is designated at that time as a hedge.

Marrried Put and Stock The simultaneous purchase of stock and the corresponding number of put options. This is a limited risk strategy during the life of the puts because the stock can be sold at the strike price of the puts.

Mass customization A business strategy that combines line (or mass) production with computerization to produce customized products. It can be regarded as the ultimate in market segmentation – in that a product can be customized for every individual that constitutes the market.

Mass service technology The type of technology used in labour-intensive service industries, with low customer contact, e.g. University teaching.

Master production schedule (MPS) A schedule developed from firm orders or from forecasts of demand. The MPS is a key input into a material requirements planning (MRP) system. The MPS identifies how many of each end product need to be produced and when they need to be produced by within each planning period. This data forms one of three key inputs enabling an MRP system to produce detailed timings for component manufacture and material purchases. It summarizes what the business plans to make and drives the operation in terms of what is assembled, what is made, and what is bought in. In addition it forms the basis for planning the utilization of labour and equipment and determines the requirements for materials and cash.

Match fund A bank is said to match fund a loan or other asset when it does so by buying (taking) a deposit of the same maturity. The term is commonly used in the Euromarket.

Matched and lost The outcome of the flip of a coin used to determine which of two brokers who are locked in competition for equal trades may actually execute the trades.

Matched bargain A transaction in which a sale of a particular quantity of stock is matched with a purchase of the same quantity of the same stock. Transactions of this kind are carried out on the London Stock Exdmnge by matching brokers.

Matched book A bank runs a matched book when the distribution of maturities of its assets and liabilities are equal.

Matched maturities The coordination by a financial_institution of the maturities of its assets (loans) and liabilities (deposits) in order to enable it to meet itsobligations at the required times.

Matched orders Used for listed equity securities. Participate in equal amounts of a trade at a certain price, particularly when two parties have the same level of priority on the exchange floor (this requires standing in the trading crowd).

Matched Sale Purchase Transactions Transcations in which the Federal Reserve sells a government security to a dealer or a foregin central bank and agrees to buy back the security to a dealer or a foreign central bank and agrees to buy back the security on a specified date (usually within seven days) at eh same price (the reverse of a repurchase agreement). Such transaction allow the Federal Reserve to temporarily absorb excess reserves from the banking system, limiting the ability of banks to make new loans and investments.

Matched sale transaction Applies mainly to convertible securities. Procedure whereby the Federal Reserve Bank of New York sells government securities to a nonbank dealer against payment in federal funds. The agreement requires the dealer to sell the securities back by a specified date, which ranges from 1 to 15 days. The Fed pays the dealer a rate of interest equal to the discount rate. These transactions, also called reverse repurchase agreements, decrease the money supply for temporary periods by reducing dealers' bank balances and thus excess reserves.

Matched sale-purchase agreement (MSP) A device used by the US Federal Reserve in which it sells money-market instruments for immediate effect and couples the sale with the forward purchase of the same instrument, to facilitate the distribution of reserves of the banking system.

Match-fund A bank is said to match-fund a loan or other asset when it does so by buying (taking) a deposit of the same maturity. The term is commonly used in the Euromarket.

Matching concept The accounting principle that requires the recognition of all costs that are associated with the generation of the revenue reported in the income statement.

Material adverse change A clause in a loan agreement or bank facility stating that the loan will become repayable if there should a material change in the borrower's credit standing. The clause can be contentious because it is not always clear what constitutes a material change.

Material fact 1. Any important piece of information that a person seeking insurance must disclose to the insurer to enable the insurer to decide whether or not to accept the insurance and to calculate the premium. What is and what is not material may have to be decided by a court of law, but very often it is obvious. While a person cannot be penalized for not revealing facts that are not known or cannot reasonably be expected to be known, an insurance contract can become void if the facts have been deliberately concealed from the insurer if they could have influenced the acceptance of the risk, the premium charged, or the exceptions made. **2.** Relevant information about a company that must be made public in its prospectus, if it is seeking a flotation on a stock exchange.

3. Any information that could be provided by a witness in court proceedings and could influence the decision of the court.

Materiality The extent to which an item of accounting information is material. Information is considered material if its omission from a financial statement could influence the decision making of its users. Materiality is therefore not an absolute concept but is dependent on the size and nature of an item and the particular circumstances in which it arises.

Materials cost The expenditure incurred by an organization on direct materials or indirect materials. The expenditure on direct materials is part of prime cost and that on indirect materials is a manufacturing overhead.

Materials management The management of the flows of materials through an organization's immediate supply chain; this includes purchasing, inventory management, operations planning and control, and distribution to first-tier customers. It also includes the management of information flows. Effectively, for every unit of material flow in one direction through the supply chain, there is an equivalent information flow in the opposite direction. Many of the recent improvements in materials management have resulted from the use of computers and information technology in the management of information flows.

Matrix organization A management structure in an organization in which some employees report to two (or more) managers in different departments. It is often used if two separate areas of the external environment demand management attention. For example, if a wholesaler of an organization's product requires extra credit, the distribution management and the finance management would both need to be consulted. In effect, this would involve the employees concerned in a multiple-authority structure rather than a simple chain of command (see diagram).

Maximum allowable cycle time The maximum time that can elapse between consecutive items leaving a production process, if the designated capacity of the process is to be achieved.

Maximum investment plan (MIP) A unit-linked endowment policy marketed by a life-assurance company that is designed to produce maximum profit rather than life-assurance protection. It calls for regular premiums, usually over ten years, with options to continue. These policies normally enable a tax-free fund to be built up over ten years and, because of the regular premiums, pound cost averaging can be used, linked to a number of markets.

Maximum price fluctuation The maximum amount the contract price can change, up or down, during one trading session, as fixed by exchange rules in the contract specification.

MBS depository A book-entry depository for GNMA securities. The depository was initially operated by MBSCC and is now a separately incorporated, participant-owned, limited-purpose trust company organized under the State of New York Banking Law.

MBWA Abbreviation for management by wandering around.

MCA Abbreviation for Management Consultants' Association.

Mean absolute deviation (MAD) A measure of forecast error, for example when carrying out exponential smoothing of time-series data. MAD is the average forecast error, either positive or negative, calculated as the sum of the absolute value of forecast error for all periods, divided by the total number of periods evaluated.

Mean deviation In statistics, the arithmetic mean of the deviations (all taken as positive numbers) of all the numbers in a set of numbers from their arithmetic mean. For example, the arithmetic mean of 5, 8, 9, and 10 is 8, and therefore the deviations from this mean are 3, 0, 1, and 2, giving a mean deviation of 15.

Mean reversion The idea that stock prices revert to a long term level. Hence, if there is a shock in prices (unexpected jump, either up or down), prices will return or revert eventually to the level before the shock. The time it takes to revert is often referred to as the time to reversion. If the process is very persistent, it might take a long time to revert to the mean. The key difference between a mean-reverting process and a random walk is that after the shock, the random walk price process does not return to the old level.

Means test Any assessment of the income and capital of a person or family to determine their eligibility for benefits provided by the state or a charity.

Mean-variance analysis Evaluation of risky prospects based on the expected value and variance of possible outcomes.

Mean-variance criterion The selection of portfolios based on the means and variances of their returns. The choice of the higher expected return portfolio for a given level of variance or the lower variance portfolio for a given expected return.

Measured daywork A method of assessing wages in which a daily production target is set and a daily wage agreed on this basis. If the target is reached the worker receives the agreed daily wage; if it is not reached (or is exceeded) the worker is paid pro rata.

Measurement error Errors in measuring an explanatory variable in a regression, which leads to biases in estimated parameters.

Mechanical observation Observation techniques involving mechanical observers either in conjunction with, or in place of, human observers. Examples include the motion-picture or video camera, the audimeter, the psychogalvanometer, the eye-camera, and the pupilometer. None of these would be used without the knowledge and permission of the respondent. Mechanistic organization An organization characterized by clearly specified roles and clear definitions of the rights and obligations of individuals within the hierarchical structure. Interaction and communication are largely vertical This type of structure is appropriate if the environment is certain. Compare organic organization.

Media (mass media) The means, such as television, radio, newspapers, and magazines, by which advertisers, politicians, etc, communicate with large numbers of members of the general public.

Median A form of mean in which a set of numbers is arranged in an ascending or descending scale and the middle number (if there are an odd number in the set) or the arithmetic mean of the middle two numbers (if there are an even number) is taken as the median.

Median market cap The midpoint of market capitalization (market price multiplied by the number of shares outstanding) of the stocks in a portfolio. Half the stocks in the portfolio will have higher market capitalizations; half will have lower.

Medium-term financial assistance (MTFA) A loan available to member states of the European Union experiencing monetary difficulties due to adverse balance of payments. The donor countries may attach conditions on granting these loans, which usually have a term of two to five years.

Medium-term note (MTN) A corporate debt instrument that is continuously offered to investors over a period of time by an agent of the issuer. Investors can select from maturity bands of: 9 months to 1 year, more than 1 year to 18 months, more than 18 months to 2 years, etc., up to 30 years.

Medium-term note A corporate debt instrument that is continuously offered to investors over a period of time by an agent of the issuer. Investors can select from the following maturity bands: 9 months to 1 year, more than 1 year to 18 months, more than 18 months to 2 years, etc., up to 30 years.

Meeting of creditors Also known as the section 341 meeting or first meeting of creditors. This is an opportunity for the trustee and the creditors to question the debtor about its pre- and post-petition activities, assets, debts, etc. An individual debtor appears in person; a corporation is represented by an executive officer; both must answer questions under oath.

Meff Renta Variable Stock index and equity derivatives market in Spain trading futures and options on the Iberian Exchange (IBEX)-35 index and on individual stocks.

Member bank A bank that belongs to a central banking or clearing system. In the UK a member bank is a commercial bank that is a member of the Association for Payment Clearing Services. In the USA it is a commercial bank that is a member of the Federal Reserve System.

Member firm A firm of brokers or market makers that is a member of the London Stock Exchange (International Stock Exchange). Banks, insurance companies, and overseas securities houses can now become corporate members.

Member of a company A shareholder of a company whose name is entered in the register of members. Founder members (see founders' shares) are those who sign the memorandum of association; anyone subsequently coming into possession of the company's shares becomes a member.

Member short sale ratio The total shares sold short by NYSE members divided by total short sales, which is used to analyze market expectations and bullish or bearish trends.

Members' voluntary liquidation (members' voluntary winding-up) The winding-up of a company by a special resolution of the members in circumstances in which the company is solvent. Before making the winding-up resolution, the directors must make a declaration of solvency. It is a criminal offence to make such a declaration without reasonable grounds for believing that it is true. When the resolution has been passed, a liquidator is appointed; if, during the course of the winding-up, the liquidator believes that the company will not be able to pay its debts, a meeting of creditors must be called and the winding-up is treated as a members' compulsory liquidation.

Membership or a seat on the exchange A limited number of exchange positions that enable the holder to trade for the holder's own accounts and charge clients for the execution of trades for their accounts. Related: member firm.

Memorandum of association An official document setting out the details of a company's existence. It must be signed by the first subscribers and must contain the following information (as it applies to the company in question): the company name; a statement that the company is a public company; the address of the registered office; the objects of the company; a statement of limited liability; the amount of the guarantee; and the amount of authorized share capital and its division.

Menu A list of choices displayed by a computer. When many options are available the user may be presented first with a main menu, from which more detailed menus can be selected. A welldesigned menu system can make a complicated program simple to use.

Mercantile agency An organization that supplies credit ratings and reports on firms that are prospective customers.

Mercantile law The commercial law, which includes those aspects of a country's legal code that apply to banking, companies, contracts, copyrights, insolvency, insurance, patents, the sale of goods, shipping, trademarks, transport, and warehousing.

Merchandising The promotion by a retailer of selected products. Commonly used techniques include displays to encourage impulse buying, free samples and gifts, and temporary price reductions. Merchandising policy is usually designed to influence the retailer's sales pattern and is influenced by such factors as the firm's market, the speed at which different products sell, margins, and service considerations. Sometimes merchandising is used to attract customers into the shop rather than to promote the product itself.

Merchant A trader who buys goods for resale,

acting as a principal and usually holding stocks. Typically a merchant sells goods in smaller lots than he or she buys and often exports goods or is involved in entrepôt trade.

Merchantable quality An implied condition respecting the state of goods sold in the course of business. Such goods should be as fit for their ordinary purpose as it is reasonable to expect, taking into account any description applied to them, the price (if relevant), and all the other relevant circumstances. The condition does not apply with regard to defects specifically drawn to the buyer's attention or defects that should have been noticed in an examination of the goods before the contract was made.

Merchant bank A bank that formerly specialized in financing foreign trade, an activity that often grew out of its own merchanting business. This led them into accepting bills of exchange and functioning as accepting houses. More recently they have tended to diversify into the field of hire-purchase finance, the granting of long-term loans (especially to companies), providing venture capital, advising companies on flotations and takeover bids, underwriting new issues, and managing investment portfolios and unit trusts. Many of them are old-established and some offer a limited banking service. Their knowledge of international trade makes them specialists in dealing with the large multinational companies. They are most common in Europe, but some merchant banks have begun to operate in the USA. Several UK merchant banks were taken over in the 1990s either by the commercial banks or by large overseas banks.

Merchant wholesaler An independently owned business that functions as a wholesaler and takes title to the merchandise it handles, i.e. It takes-goods into stock and supplies its customers from stock, rather than acting as an agent or distributor for a manufacturer. Merger A combination of two or more businesses on an equal footing that results in the creation of a new reporting entity formed from the combining businesses. The shareholders of the combining entities mutually share the risks and rewards of the new entity and no one party to the merger obtains control over another. Under Financial Reporting Standard 6, 'Acquisitions and Mergers', to qualify as a merger a combination must satisfy four criteria:

No party is the acquirer or acquired;

All parties to the combination participate in the management structure of the new entity;

The combining entities are relatively equal in terms of size;

The consideration received by the equity shareholders of each party consists primarily of equity shares in the combined entity, any other consideration received being relatively immaterial.

In the UK, approval of the Monopolies and Mergers Commission may be required and the merger must be conducted on lines sanctioned by the City Code on Takeovers and Mergers.

Merchant bank A British term for a bank that specializes not in lending out its own funds, but in providing various financial services such as accepting bills arising out of trade, underwriting new issues, and providing advice on acquisitions, mergers, foreign exchange, portfolio management, etc.

Merger reserve A reserve credited in place of a share premium account when merger relief is made use of. Goodwill on consolidation may be written off against a merger reserve (unlike the share premium account).

Micro country risks Country or political risks that are specific to an industry, company, or project within a host country.

Microeconomics Analysis of the behavior of individual economic units such as companies, industries, or households.

Microenvironment The environment close to a company, which affects its ability to serve its customers. This includes the company itself, firms in its market channel customer markets, and competitors.

Microprocessor In computer technology, an integrated circuit that contains on one chip the arithmetic, logical and control functions of the central processing unit. Microprocessors are used as control systems in many devices, including car electronics and cameras, as well as in personal computers.

Middle price (mean price) The average of the offer price of a security, commodity, currency, etc., and the bid price. It is the middle price that is often quoted in the financial press.

Middleman A person or organization that makes a profit by trading in goods as an intermediary between the producer and the consumer. Middlemen include agents, brokers, dealers, merchants, factors, wholesalers, distributors, and retailers. They earn their profit by providing a variety of different services, including finance, bulk buying, holding stocks, breaking bulk, risk sharing, making a market and stabilizing prices, providing information about products (to consumers) and about markets (to producers), providing a distribution network, and introducing buyers to sellers.

Migrant worker A worker who has come from another country or region to work Attracted by better wages in the country or region to which he has migrated, he may or may not have brought his family with him. In many cases he will not have done so and will therefore be sending a substantial proportion of his earnings back to his native country This can have an adverse effect on the balance of payments of countries that attract substantial numbers of migrant workers and conversely can be advantageous to countries from which many workers migrate. Migrant workers can also cause problems with the trade unions in the countries to which they migrate.

Minicomputer A computer that is less powerful than a mainframe but more powerful than a personal computer. It is unable to handle as many people or programs at the same time as a mainframe, but it is adequate for a wide range of business and scientific applications

Minority interest The interest of individual shareholders in a company more than 50% of which is owned by a holding company. For example, if 60% of the ordinary shares in a company are owned by a holding company, the remaining 40% will represent a minority interest. These shareholders will receive their full share of profits in the form of dividends although they will be unable to influence company policy as they will always be outvoted by the majority interest held by the holding company.

Misfeasance summons An application to the court by a creditor, contributory, liquidator, or the official receiver during the course of winding up a company. The court is asked to examine the conduct of a company officer and order restitution to be made if the officer is found guilty of breach of trust or duty.

Mismatch 1. A floating-rate note in which the coupon is paid monthly but the interest rate paid is that applicable to a note of longer maturity.

2. The state of the typical books of a bank that borrows short-term and lends long-term, in which assets and liabilities are not matched. This is acceptable as long as deposits keep coming in.

Misrepresentation An untrue statement of fact, made by one party to the other in the course of negotiating a contract, that induces the other party to enter into the contract. The person making the misrepresentation is called the representor, and the person to whom it is made is the representee. A false statement of law, opinion, or intention does not constitute a misrepresentation; nor does a statement of fact known by the representee to be untrue. Moreover, unless the representee relies on the statement so that it becomes an inducement (though not necessarily the only inducement) to enter into the contract, it is not a misrepresentation. The remedies for misrepresentation vary according to the degree of culpability of the representor. If guilty of fraudulent misrepresentation (i.e. If the truth of a statement is not honestly believed, which is not the same as saying that it is known to be false) the representee may, subject to certain limitations, set the contract aside and may also sue for damages. If guilty of negligent misrepresentation (i.e. If the statement is believed without reasonable grounds for doing so) the representee may also rescind (see rescission) the contract and sue for damages. If the representor has committed merely an innococent misrepresentation (one reasonably believed to be true) the representee is restricted to rescinding the contract.

Missionary sales personnel Salespersons who visit prospective customers, distribute information to them, and handle their questions and complaints but do not usually take orders from them. Sales personnel who perform these functions tend to be employed by an agency and are hired out to the selling company.

Mix variance 1. A variance that arises in standard costing as a direct materials variance, when a number of raw materials are used in production. It measures the loss or gain arising when the actual mix of materials used is different from the standard mix of materials specified.

2. A sales margin mix variance for an organization that sells a number of alternative products. It measures the loss or gain that arises from the actual mix of sales volumes being different from the standard mix of sales volumes specified.

Mixed economy An economy in which some goods and services are produced by the government and some by private enterprise. A mixed economy lies between a command economy and a complete laissez-faire economy. In practice, however, most economies are mixed; the significant feature is whether an economy is moving towards or away from a more laissez-faire situation. Although most western economies retreated from laissez-faire policies in the post-war era this trend has tended to be reversed since the 1980s.

Modified duration The ratio of Macaulay duration to (1 + y), where y = the bond yield. Modified duration is inversely related to the approximate percentage change in price for a given change in yield.

Modified duration The ratio of Macaulay duration to (1 + y), where y = the bond yield. Modified duration is inversely related to the approximate percentage change in price for a given change in yield.

Modified rebuy A situation in which a buyer is unhappy with current suppliers or products and is looking at competitors' products and prices rather than automatically rebuying the same product as before.

Monetary / non-monetary method Under this translation method, monetary items (e.g. cash, accounts payable and receivable, and long-term debt) are translated at the current rate while non-monetary items (e.g. inventory, fixed assets, and long-term investments) are translated at historical rates.

Monetary assets Assets, such as cash and debtors, that have a fixed monetary exchange value and are not affected by a change in the price level. If there are no regulations requiring companies to account for changing price levels, monetary assets remain in the financial statements at their original amounts. If the principle of accounting for changes in price levels is applied, the monetary assets will be indexed.

Monetary indicators Economic indicators of the effects of monetary policy, such as the condition of the credit market.

Monetary reform The revision of a country's currency by the introduction of a new currency unit or a substantial change to an existing system. Examples included decimalization of the UK currency (1971) and the change from the austral to the peso in Argentina (1992).

Money A medium of exchange that functions as a unit of account and a store of value. Originally it enhanced economic development by enabling goods to be bought and sold without the need for barter. However, throughout history money has been beset by the problem of its debasement as a store of value as a result of inflation. Now that the supply of money is a monopoly of the state, most governments are committed in principle to stable prices. The word 'money' is derived from the Latin *moneta*, which was one of the names of Juno, the Roman goddess whose temple was used as a mint.

Money at call and short notice One of the assets that appears in the balance sheet of a bank. It includes funds lent to discount houses, money brokers, the stock exchange, bullion brokers, corporate customers, and increasingly to other banks. 'At call' money is repayable on demand whereas 'short notice' money implies that notice of repayment of up to 14 days will be given. After cash, money at call and short notice are the banks' most liquid assets. They are usually interest-earning secured loans but their importance lies in providing the banks with an opportunity to use their surplus funds and to adjust their cash and liquidity requirements.

Money broker In the UK, a broker who arranges short-term loans in the money market, i.e. Between banks, discount houses, and dealers in government securities. Money brokers do not themselves lend or borrow money, they work for a commission arranging loans on a day-to-day and overnight basis. Money brokers also operate in the eurobond markets.

Moneylender A person whose business it is to lend money, other than pawnbrokers, friendly or building societies, corporate bodies with special powers to lend money, banks, or insurance companies. The Consumer Credit Act (1974) replaces the earlier Moneylenders Acts and requires all moneylenders to be registered, to obtain an annual licence to lend money, and to state the true annual percentage rate (APR) of interest at which a loan is made.

Money market 1. The UK market for short-term loans in which money brokers arrange for loans between the banks, the government, the discount houses and the accepting houses, with the Bank of England acting as lender of last resort. The main items of exchange are bills of exchange, Treasury bills, and trade bills. The market takes place in and around Lombard Street in the City of London. Private investors, through their banks, can place deposits in the money market at a higher rate of interest than bank deposit accounts, for sums usually in excess of £10,000. **2.** The foreign-exchange market and the bullion market in addition to the shortterm loan market.

Money-market deposit account (MMDA) A high-yielding savings account introduced in the USA in 1982 to allow deposit-taking institutions to compete for savers' funds with the money markets. As long as the account has a balance of more than Rs.1000 there is no regulatory limit on the account. Balances below Rs.1000 attract a lower rate of interest. Restrictions on the account apply to withdrawals (three a month) and transfers for bill payment (three a month).

Money market line An agreement between a

bank and a company that entitles the company to borrow up to a certain limit each day in the money markets, on a short-term basis (often overnight or in some cases up to one month). *See*uncommitted facility.money-market unit trust (cash unit tmst) A unit trust that invests in money-market instruments in order to provide investors with a risk-free income.

Money-purchase pension scheme A pension scheme in which payment is based directly on contributions made during the retired person's working life, rather than a percentage of his or her final salary. *See*personal pension schemes.money supply The quantity of money issued by a country's monetary authorities (usually the central bank). If the demand for money is stable, the widely accepted quantity theory of money implies that increases in the money supply will lead directly to an increase in the price level, i.e. To inflation. Since the 1970s most western governments have attempted to reduce inflation by controlling the money supply. This raises two issues: (i) how to measure the money supply; (ii) how to control the money supply (*see* interest-rate policy). In the UK various measures of the money supply have been used, from the very narrow M0 to the very broad M5. They are usually defined as: M0 – notes and coins in circulation plus the banks' till money and the banks' balances with the Bank of England; M1 – notes and coins in circulation plus private-sector current accounts and deposit accounts that can be transferred by cheque; M2 – notes and coins in circulation plus non-interest-bearing bank deposits plus building society deposits plus National Savings accounts; M3 – M1 plus all other private-sector bank deposits plus certificates of deposit; M3c – M3 plus foreign currency bank deposits; M4 – M1 plus most private-sector bank deposits plus holdings of money-market instruments (e.g. Treasury bills); M5 – M4 plus building society deposits.

Money Market Demand Account (M.M.D.A.) An account that pays interest based on short-term interest rates. Same as a Money Market Deposit Account

Money market demand account An account that pays interest based on short-term interest rates.

Money market fund A mutual fund that invests only in short term securities, such as bankers' acceptances, commercial paper, repurchase agreements and government bills. The net asset value per share is maintained at Rs.1.00. Such funds are not federally insured, although the portfolio may consist of guaranteed securities and/or the fund may have private insurance protection.

Money market Money markets are for borrowing and lending money for three years or less. The securities in a money market can be U.S.government bonds, Treasury bills and commercial paper from banks and companies.

Money market yield A bond quotation convention based on a 360-day year and semiannual coupons. See: Bond equivalent yield.

Money purchase plan A defined benefit contribution plan in which the participant contributes some part and the firm contributes at the same or a different rate. Also called and individual account plan.

Monopolies and Mergers Commission (MMC) A commission established in 1948 as the Monopolies and Restrictive Practices Commission and reconstructed under its present title by the Fair Trading Act (1973). It investigates questions referred to it on unregistered monopolies relating to the supply of goods in the UK, the transfer of newspapers, mergers qualifying for investigation under the Fair Trading Act, and uncompetitive practices and restrictive labour practices, including public-sector monopolies as laid down in the provisions of the Competition Act (1980).

Monopoly A market in which there is only a single seller (producer). If there is a single seller and a single buyer the situation is called bilateral monopoly. While governments usually try to eliminate monopolies, regulate them, or nationalize them, there is some reason to believe that natural monopolies exist (e.g. Electricity distribution).

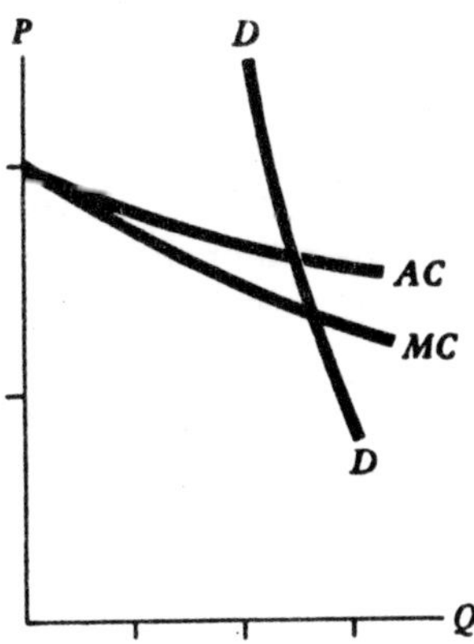

Fig. Natural Monopoly

Monopsony The existence of only one buyer in a market, forcing sellers to accept a lower price than the socially optimal price

Mortgage duration A modification of standard duration to account for the impact on duration of MBSs of changes in prepayment speed resulting from changes in interest rates. Two factors are employed: one that reflects the impact of changes in prepayment speed or price.

Mortgage life insurance A life insurance policy that pays off the remaining balance of the insured person's mortgage at death.

Mortgage pass-through security Also called a passthrough, a security created when one or more mortgage holders form a collection (pool) of mortgages and sells shares or participation certificates in the pool. The cash flow from the collateral pool is "passed through" to the security holder as monthly payments of principal, interest, and prepayments. This is the predominant type of MBS traded in the secondary market.

Mortgage servicing The collection of monthly payments and penalties, record keeping, payment of insurance and taxes, and possible settlement of default, involved with a mortgage loan.

Mortgage-Backed Securities Clearing Corporation (MBSCC) "Founded" in 1979, MBSCC is the sole provider of automated post-trade comparison, netting, risk management and pool notification services to the mortgage-backed securities market. The organization is a registered clearing agency with the Securities and Exchange Commission and majority-owned by its members — MBS dealers, inter-dealer brokers and other non-broker/dealers. MBSCC provides its specialized services to major market participants active in various Government National Mortgage Association (GNMA), Fannie Mae(FNMA) and Federal Home Loan Mortgage Corporation (FHLMC) MBS programs.

Mortgagee in possession A mortgagee who has exercised the right to take possession of the mortgaged property; this may happen at any time, even if there has been no default by the mortgagor. However, the mortgage deed may contain an agreement not to do this unless there is default and a court order will be needed to obtain possession in the case of a dwelling house. The court may adjourn the hearing to allow the mortgagor time to pay. The mortgagee will either receive the rents and profits if the property has been let or will manage the property. The mortgagee will be liable to account strictly for any actions and is not entitled to reap any personal benefit beyond repayment of the interest and the principal debt. The mortgagee must carry out reasonable repairs and must not damage the property.

Most-favoured-nation clause A clause in a trade agreement between two countries stating that each will accord to the other the same treatment as regards tariffs and quotas as they extend to the most favoured nation with which each trades. Both GATT and the EU have used this concept.

Motivation The mental processes that arouse, sustain, and direct human behaviour. It reflects the changes to their physiological and psychological conditions as a result of their previous and current experiences. Motivation may stem from processes taking place within an individual (intrinsic motivation) or.

From the impact of factors acting on the individual from outside (extrinsic motivation); in most cases these two influences are continually interacting.

The vocabulary associated with motivation is large; such terms as purpose, desire, need, goal, preference, perception, attitude, recognition, achievement, and incentive are commonly used. Many of these drives can act on an individual simultaneously, causing varying degrees of conflict. A child deciding between buying chocolate and ice cream is in conflict. An employee who wants to disagree with the boss but also wants to keep his job is in conflict.

In a business context, some understanding of motivations is important in consumer buying behaviour and in the setting of organizations. Considerable importance has been given to the hierarchy of needs investigated by Abraham Maslow (1908-70), although many now regard human needs to be too variable to be confined to a fixed hierarchy. The more flexible ERG theory (existence, relatedness, and growth theory) of Clayton Alderfer (1940-) focuses on three groups of needs that form a hierarchy: existence needs (physical and material wants); relatedness needs (the desire for interpersonal relationships and deeper relationships with the important people in the person's life); and growth needs (desires to be creative and productive). The theory suggests that these needs change their position in the hierarchy as circumstances change (see diagram). A need does not have to be fully satisfied for upward movement to occur; a downward movement can occur when a need is not satisfied. In the deficiency cycle at the bottom of the hierarchy the individual can become obsessed with satisfying existence needs. The enrichment cycle leads a person to learn and grow, generally in multiple environments (i.e. Work home, etc.). For consumers, when a need is activated it becomes a motive, although the needs and motives involved in the decision to buy a new washing machine (rather than have the old one repaired) may depend on quite simple financial calculations rather than conflict between deep psychological needs.

In an organizational setting it is also argued that, besides goals, ambitions, and achievement, there is a need for success to be recognized by others and a need to develop and progress.

Motivational research A form of marketing research in which the motivations for consumers preferring one product rather than another are studied. It may form the basis of a plan to launch a competitive product or of a campaign to boost sales of an existing product.

Motor insurance A form of covering loss or damage to motor vehicles and any legal liabilities for bodily injury or damage to other people's property. Drivers have a legal obligation to be covered against third-party claims (*see third-party insurance*), but most drivers or owners of vehicles have a comprehensive insurance, providing wide coverage of all the risks involved in owning a motor vehicle. An intermediate cover is known as third-party, fire, and theft.

Moving average A series of arithmetic means calculated from data in a time series, which reduces the effects of temporary seasonal variations. For example, a moving average of the monthly sales figures for an organization might be calculated by averaging the 12 months from January to

December for the December figure, the 12 months from February to January for the January figure, and so on.

Moving average Used in charts and technical analysis, the average of or commodity prices constructed in a period as short as a few days or as Long as several years and showing trends for the latest interval. As each new variable is included in calculating the average, the last variable of the series is deleted.

Multichannel marketing Marketing one's products or services using two or more marketing channels, often to reach more than one market segment (*see*market segmentation). Retail chains, for example, besides using the shops to distribute their products, quite often also use catalogue selling.

Multifactor CAPM A version of the capital asset pricing model derived by Robert Merton that includes extra-market sources of risk referred to as factors. Related: arbitrage pricing theory

Multifamily loans Loans usually represented by conventional mortgages on multi-family rental apartments.

Multilateral netting A method of reducing bank charges in which the subsidiaries of a group offset their receipts and payments with each other, usually monthly, resulting in a single net intercompany payment or receipt made by each subsidiary to cover the period concerned. This saves both on transaction costs and paperwork *See also*bilateral netting; netting.

Multinational corporation (MNC) A firm that operates in more than one country.

Multiperiod immunization A portfolio strategy in which a portfolio is created that will be capable of satisfying more than one predetermined future liability regardless of interest rate changes.

Multiple Arbitrage In the context of hedge funds, a style of management where by the fund employs more than one arbitrage strategy. Portfolio manager opportunistically allocates capital among the various strategies in order to create the best risk/reward profile for the overall fund. Common strategies include merger arbitrage, convertible arbitrage, fixed income arbitrage, long/short equities pairs trading, and volatility arbitrage. In the context of equity and private equity investment, this refers to an investment in a firm where by standard multiples (earnings/ price, book/price) indicate the price is far cheaper than industry averages.

Multiple listing An agreement used by a broker who is a member of a multiple-listing organization, providing the exclusive right to sell with an additional authority and obligation on the part of the listing broker to distribute the listing to the other brokers.

Multiple regression The estimated relationship between a dependent variable and more than one explanatory variable

Multiple-choice question A question requiring a respondent to select an answer from a list of answers provided with the question. For example, the respondent may be asked to name the brand of cigarettes purchased in the last week from a list of major brands. *Compare*open-ended question.

Multiplier The investment multiplier which quantifies the overall effects of investment spending on total income. The deposit multiplier which shows the effects of a change in bank deposits on the total amount of outstanding credit and the money supply.

Multirule system A technical trading strategy that combines mechanical rules, such as the CRISMA (cumulative volume, relative strength, moving average) Trading System of Pruitt and White.

Municipal bond State or local governments offer muni bonds or municipals, as they are called, to pay for special projects such as highways or sewers. The interest that investors receive is exempt from some income taxes.

Mutual fund Mutual funds are pools of money that are managed by an investment company. They offer investors a variety of goals, depending on the fund and its investment charter. Some funds, for example, seek to generate income on a regular basis. Others seek to preserve an investor's money. Still others seek to invest in companies that are growing at a rapid pace. Funds can impose a sales charge, or load, on investors when they buy or sell shares. Many funds these days are no load and impose no sales charge. Mutual funds are investment companies regulated by the Investment Company Act of 1940.

Mutual fund Mutual funds are pools of money that are managed by an investment company. They offer investors a variety of goals, depending on the fund and its investment charter. Some funds, for example, seek to generate income on a regular basis. Others seek to preserve an investor's money. Still others seek to invest in companies that are growing at a rapid pace. Funds can impose a sales charge, or load, on investors when they buy or sell shares. Many funds these days are no load and impose no sales charge. Mutual funds are investment companies regulated by the Investment Company Act of 1940. Related: open-end fund, closed-end fund.

Mutual offset A system, such as the arrangement between the Chicago Mercantile Exchange (CME) and Singapore International Monetary Exchange (SIMEX), which allows trading positions established on one exchange to be or transferred on another exchange.

Naive diversification A strategy whereby an investor simply invests in a number of different assets and hopes that the variance of the expected return on the portfolio is lowered. Related: Markowitz diversification.

Naked call writing Selling (writing) a call option on equities that one does not own. A person may do this if he or she expects the price of a particular share to fall or remain unchanged. It is, however, a dangerous strategy because if the price rises the shares will have to be purchased at the, market price in order to deliver them, thus involving an unlimited risk.

Naked option strategies An unhedged strategy making exclusive use of one of the following: Long call strategy (buying call options), short call strategy (selling or writing call options), Long put strategy (buying put options), and short put strategy (selling or writing put options). By themselves, these positions are called naked strategies because they do not involve an offsetting or risk-reducing position in another option or the underlying security. Related covered option strategies.

NASDAQ National Association of Securities Dealers Automatic Quotation System. An electronic quotation system that provides price quotations to market participants about the more actively traded common stock issues in the OTC market. About 4,000 common stock issues are included in the NASDAQ system.

National Association of Securities Dealers Automatic Quotation System (Nasdaq) An electronic quotation system that provides price quotations to market participants about the more actively traded common stock issues in the OTC market. About 4000 common stock issues are included in the Nasdaq system.

National Market System (NMS) Refers to over-the-counter trading. System of trading OTC stocks under the sponsorship of the NASD. Must meet certain criteria for size, profitability and trading activity. More comprehensive information is available for NMS stocks than for non-NMS stocks traded OTC (high, low, and last-sale prices, cumulative volume figures, and bid and ask quotations throughout the day). This is due to the fact that market makers must report the actual price and number of shares in each transaction within 90 seconds verses nonreal-time reporting for non-NMS stocks (thus, last sales prices and minute-to-minute volume updates are not possible).

National Quotation Bureau A service that publishes bid and offer quotes from market makers in OTC transactions.

Nationalization The process of bringing the assets of a company into the ownership of the state. Examples of industries nationalized in the past in the UK include the National Coal Board and British Rafl. Historically, nationalization has been achieved through compulsory purchase, although this need not necessarily be the case. Nationalization has often been pursued as much for political as economic ends and the economic justifications themselves are varied. One argument for nationalization is that if a company possesses a natural monopoly, the profits it earns should be shared by the whole population through state ownership. Another argument might be that particular industries are strategically important for the nation and therefore cannot be entrusted to private enterprise. In the 1980s and 1990s Conservative governments have tended to reverse Labour's nationalization of the 1950s, 1960s, and 1970s with a series of privatization measures, on the grounds flat competition increases efficiency and reduces prices.

Natural justice The minimum standard of fairness to be applied by a court when resolving a dispute. The main rules of natural justice include: (1) the right to be heard – each party to the dispute should be given an opportunity to answer any allegations made by the other party; (2) the rule against bias -the person involved in settling the dispute should act impartially, in particular by disclosing any interest in the outcome of the dispute. The rules of natural justice apply equally in judicial as well as in administrative proceedings. Alleging a breach of natural justice is the method commonly used to challenge an administrative decision before the courts.

Natural wastage The method by which an organization can contract without making people redundant, relying on resignations, retirements, or deaths. If the time is available, this method of reducing a workforce causes the least tension.

NAV Abbreviation for net asset value.

Near money An asset that is immediately transferable and may be used to settle some but not all debts, although it is not as liquid as banknotes and coins. Bills of exchange are examples of near Money. Near money is not included in the money supply indicators.

Nearby futures contract When several futures contracts are considered, the contract with the closest settlement date is called the nearby futures contract. The next futures contract is the one that settles just after the nearby futures contract. The contract farthest away in time from settlement is called the most distant futures contract.

Negative amortization A loan repayment schedule in which the outstanding principal balance of the loan increases, rather than amortizing, because the scheduled monthly payments do not cover the full amount required to amortize the loan. The unpaid interest is added to the outstanding principal, to be repaid later.

Negative covenant A bond covenant that limits or prohibits certain actions unless the bondholders agree.

Negative income tax (NIT) A means of targeting social security benefits to those most in need. The payments would be made through the income-tax system by granting personal allowances to tax payers so that the standard rate of income tax on these allowances would constitute a minimum amount required for living. Those with high incomes would obtain that amount as an incometax relief, while those with incomes lower than the allowance would have a negative income-tax liability and be paid the appropriate sums. The principal objection to the system is that to cover the needs of the disadvantaged the wealthier would obtain excessively high allowances.

Negative obligation A New York Stock Exchange rule that governs the behavior of specialists. Negative obligation is the mandate of the specialists not trade for the specialist's firm's own account when enough public investor orders exist to match up naturally – without intervention. An example of violating negative obligation is Trading Ahead. Also see positive obligation.

Negative Pledge An agreement in which the borrower agrees not to pledge any of its assets as security and/or not to incur further indebtedness.

Negligence A tort in which a breach of a **duty of care** results in damage to the person to whom the duty is owed. Such a duty is owed by manufacturers to the consumers who buy their products, by professional persons to their clients, by a director of a company to its shareholders, etc. A person who has suffered loss or injury as a result of a breach of the duty of care can claim damages in tort.

Negotiable Order of Withdrawal Account (NOW) An interest-earning account on which chechs may be drawn. Withdrawals from NOW accounts may be offered by commerical banks, mutual savings banks, and savings and loan associations and may be owned only by individuals and certain nonprofit organizations and govermental units.

Negotiated markets Markets in which each transaction is separately negotiated between buyer and seller (i.e., an investor and a dealer).

Negotiated sale Situation in which the terms of an offering are determined by negotiation between the issuer and the underwriter rather than through competitive bidding by underwriting groups.

Negotiated underwriting A securities offering process in which the purchase price paid to the issuer and the public offering price are determined by negotiation rather than through competitive bidding.

Net asset value (NAV) The value of a share in a company calculated by dividing the amount for the net assets of the company by the number of shares in issue. The net asset value is frequently below the market price of a share because financial statements do not reflect the present values of all assets because of the monetary measurement convention and the historical cost accounting convention, whereas the market price may reflect them.

Net asset value arbitrage For a number of assets, the most recent transaction price at 4PM ET does not fully reflect all available market information. One example is international equities that trade on exchanges that are located in different time zones and close 2-15 hours before U.S. markets. In addition, domestic small-capitization equities and high-yield and convertible bonds often trade infrequently and have wide bid-ask spreads. This can cause the most recent transaction price to be much different from the price that one would see in a liquid market at 4 PM, even for assets that trade on exchanges that are open at that time. Investors can take advantage of mutual funds that calculate their NAVs using stale closing prices by trading based on recent market movements. For example, if the U.S. market has risen since the close of overseas equity markets, investors can expect that overseas markets will open higher the following morning. Investors can buy a fund with a stale-price NAV for less than its current value, and they can likewise sell a fund for more than its current value on a day that the U.S. market has fallen. Similar opportunities exist when the values of infrequently or illiquidly-traded domestic assets have recently changed. Also known as Stale Price Arbitrage.

Net assets The difference between total assets on the one hand and current liabilities and noncapitalized long-term liabilities on the other hand.

Net current assets Current assets less current liabilities. The resultant figure is also known as working or circulating capital as it represents the amount of the organization's capital that is constantly being turned over in the course of its trade. *See also*net assets.

Net Denoting an amount remaining after specific deductions have been made. For example, net profit before taxation is the profit made by an organization after the deduction of all business expenditure but before the deduction of the taxation charge.

Net income The company's total earnings, reflecting revenues adjusted for costs of doing business, depreciation, taxes and other expenses.

Net interest cost (NIC) The total amount of interest that will be paid on a debt obligation by a corporate or municipal bond issuer.

Net investment income per share Income received by an investment company from dividends and interest on investments less administrative expenses, divided by the number of outstanding shares.

Net investment The addition to the stock of capital goods in an economy during a particular period (the grow investment) less capital consumption (i.e. Depreciation).

Net lease A lease arrangement under which the lessee is responsible for all property taxes, maintenance expenses, insurance, and other costs associated with keeping the asset in good working condition.

Net margin The gross margin less all the other costs of an organization in addition to those included in the cost of goods sold.

Net operating loss carrybacks The application of losses to offset earnings in previous years.

Net operating margin The ratio of net operating income to net sales.

Net position The value of the position subtracting the initial cost of setting up the position. For example, if 100 options where purchased for Rs.1 each and the option is currently trading for Rs.9, the value of the net position is Rs.900 - Rs.100 = Rs.800.

Net present value (NPV) In discounted cash flows, the difference between the present values of the cash outflows and the present values of the cash inflows. The NPV is the application of discount factors, based on a required rate of return to each year's projected cash flow, both in and out, so that the cash flows are discounted to present values. If the NPV is positive, the required rate of return is likely to be earned and the project should be considered; if it is negative, the project should be rejected.

Net profit 1. (not profit before taxation) The profit of an organization when all receipts and expenses have been taken into account. In trading organizations, net profit is arrived at by deducting from the gross profit all the expenses not already taken into account in arriving at the gross profit.

2. (net profit after taxation) The final profit of an organization, after all appropriate taxes have been deducted from the net profit before taxation.

Net profit margin Net income divided by sales; the amount of each sales dollar left over after all expenses have been paid.

Net profit Net profit can mean different things so it always needs clarifying. Net strictly means 'after all deductions' (as opposed to just certain deductions used to arrive at a gross profit or margin). Net profit normally refers to profit after deduction of all operating expenses, notably after deduction of fixed costs or fixed overheads. This contrasts with the term 'gross profit' which normally refers to the difference between sales and direct cost of product or service sold (also referred to as gross margin or gross profit margin) and certainly before the deduction of operating costs or overheads. Net profit normally refers to the profit figure before deduction of corporation tax, in which case the term is often extended to 'net profit before tax' or PBT.

Net realizable value (NRV) The sales value of

the stock of an organization less the additional costs likely to be incurred in getting the stocks into the hands of the customer. It is the value placed on the closing stock according to the requirements of Statement of Standard Accounting Practice 9, when the NRV is lower than cost.

Net residual value (disposal value) The expected proceeds from the sale of an asset, net of the costs of sale, at the end of its estimated useful life. It is used for computing the straight-line method and diminishing-balance method of depreciation, and also for inclusion in the final year's cash inflow in a discounted cash flow appraisal.

Net sales transaction Refers to over-the-counter trading. Securities deal in which the quoted prices include commissions (i.e., OTC); looked at another way, the buyer and seller do not pay fees or commissions in addition to the print or quotation prices.

Net salvage value The after-tax net cash flow for terminating the project.

Net tangible assets The tangible assets of an organization less its current liabilities. In analysing the affairs of an organization the net tangible assets indicate its financial strength in terms of being solvent, without having to resort to such nebulous (and less easy to value) assets as goodwill.

Net working capital Current assets minus current liabilities. Often simply referred to as working capital.

Net worth The value of an organization when its liabilities have been deducted from the value of its assets. Often taken to be synonymous with net assets (i.e. The total assets as shown by the balance sheet less the current liabilities), net worth so defined can be misleading in that balance sheets rarely show the real value of assets; in order to arrive at the true net worth it would normally be necessary to assess the true market values of the assets rather than their book values. It would also be necessary to value goodwill, which may not even appear in the balance sheet.

Net-profit ratio The proportion that net profit bears to the total sales of an organization. This ratio is used in analysing the profitability of organizations and is an indicator of the extent to which sales have been profitable.

Netting The process of setting off matching sales and purchases against each other, especially sales and purchases of futures, options, and forward foreign exchange. This service is usually provided for an exchange or market by a clearing house.

Network modelling Techniques that enable complex projects to be scheduled, taking into account the precedence of each activity.

Neutral Describing an opinion that is neither bearish not bullish. Neutral option strategies are generally designed to perform best if there is little or no net change in the price of the underlying stock or index. See also Bearish and Bullish.

Neutral hedge Hedge that is expected to yield a dollar-neutral result of the combined position, regardless of price change in any part of the hedge securities. For any convertible trading at a premium, this ratio is less than 100%. The higher the convertible premium, the lower a ratio must be to be neutral. See: Delta.

Neutral period In the Euromarket, a period over which Eurodollars are sold is said to be neutral if it does not start or end on either a Friday or the day before a holiday.

Neutral period In the Euromarket, a period over which Eurodollars are sold is said to be neutral if it does not start or end on either a Friday or the day before a holiday.

New account report A broker's document including information about a new client. See: Know your customer.

New Community instrument (NCI) Loans organized on the international money market by the European Investment Bank.

New York Cotton Exchange (NYCE) A historic commodities exchange in New York which traded futures and options on cotton, frozen concentrated orange juice, and potatoes, as well as interest rate, currency, and index futures and options. In June 2004, the NYCE merged with the Coffee, Sugar and Cocoa Exchange to form the New York Board of Trade. As a result of this merger, all previous exchanges and subsidiaries ceased to exist, including the Coffee, Sugar, & Cocoa Exchange, the New York Cotton Exchange, the Citrus Associates of the New York Cotton Exchange, the New York Futures Exchange (NYFE), and the FINEX Exchange. All markets are now referred to as the New York Board of Trade or NYBOT.

News out Refers to over-the-counter trading. A news story concerning the stock being considered has recently been posted on one of the news services, such as the Dow Jones News Service or Reuters. A courtesy standard in trading is to mention that "news is out," in case the other party is unaware of the new development.

Next day settlement Transaction in which the contract is settled the day after the trade is executed. See: Settlement date.

Ngultrum The standard monetary unit of Bhutan, divided into 100 chetrum.

Ngwee A monetary unit of Zambia, worth one hundredth of a kwacha.

Night safe A service provided by commercial banks, enabling customers to deposit cash with the bank after banking hours. By using this service shopkeepers, etc., can avoid keeping large sums overnight. A special wallet provided by the bank is inserted into a safe in the outside wall of the bank to which customers are given the key. The following day the wallet is opened, either by the customer or by a bank clerk, and the customer's account is credited accordingly.

Nikkei The common term for the Nihon Keizai newspaper, Japan's leading financial newspaper. The Nikkei usually refers to the price-weighted average of 225 stocks of the first section of the Tokyo Stock Exchange.

Nil basis The basis upon which the earnings per share of a company is calculated taking into account only the constant elements in the company's tax charge. *Compare*net basis.

Nine-bond rule An NYSE rule requiring that orders for nine bonds or fewer stay on the floor for one hour to seek a market.

No Adjournment Within the text on the proxy, card are the words: "Shares will be voted at this annual meeting or at any adjournment thereof." If a securityholder strikes out this phrase, the proxy cannot be counted at any adjournment (reconvening) of the meeting.

No book Used for listed equity securities. Not much, if any, stock is being bid for or offered at the present time by customers or the specialist.

No load mutual fund An open-end investment company, shares of which are sold without a sales charge. There can be other distribution charges, however, such as Article 12B-1 fees. A true "no load" fund will have neither a sales charge nor a distribution fee.

No par value capital stock In the USA, stock (shares) that have no par value or assigned value printed on the stock certificate. An advantage of this stock is that it avoids a contingent liability to stockholders in the event of a stock discount. For accounting purposes, on the issue of no par value capital stock, cash is debited and a capital stock account credited with the total proceeds received. No premium account is required.

No Substitution Within the text on a proxy card are the words: "The shareholder appoints certain people (collectively, the proxy committee) with full power of substitution to vote the shares." If the security holder strikes out this phrase, the proxy cannot be voted if there is a change in the designated proxy committee.

No-action letter A letter from the Securities and Exchange Commission agreeing that the commission will take no civil or criminal action against a party, regarding a specific activity.

Noah Effect The tendency of persistent time series (0.50<H<1.00) to have abrupt, and discontinuous changes. The normal distribution assumes continuous changes in a system. However, a time series which exhibits Hurst statistics may abruptly change levels, skipping values either up or down. Mandelbrot coined the term "Noah effect" after the biblical story of the deluge. See: Joseph Effect, Hurst Exponent, Persistence, Anti-persistence.

Nominal account A ledger account that is not a personal account in that it bears the name of a concept, e.g. Light and heat, bad debts, investments, etc; rather than the name of a person. These accounts are normally grouped in the nominal ledger.

Nominal cash flow A cash flow expressed in nominal terms if the actual dollars to be received or paid out are given.

Nominal dollars Dollars that are not adjusted for inflation.

Nominal exchange rate The actual foreign exchange quotation in contrast to the real exchange rate, which has been adjusted for changes in purchasing power.

Nominal group technique A qualitative forecasting technique involving a panel of experts, who are given copies of a question requiring a forecast and are asked to write down a list of their ideas without talking to their colleagues. Each expert, in turn, shares one of their ideas, which is written up by a facilitator. When all the ideas have been listed the panel members discuss them. Similar ideas are grouped or merged; when the discussion has finished the panel ranks the remaining ideas, again without discussion. The individual rankings are then combined mathematically.

Nominal interest rate The interest rate on a fixed-interest security calculated as a percentage of its par value rather than its market price.

Nominal ledger The ledger containing the nominal accounts and real accounts necessary to prepare the accounts of an organization. This ledger is distinguished from the personal ledgers, such as the sales and purchase ledgers, which contain the accounts of customers and suppliers respectively.

Nominal quotation Used in the context of general equities. Bid and offer prices given by a market maker for the purpose of valuation, not as an invitation to trade; must be specifically identified as such by prefixing the quotes FYI (for your information) or FVO (for valuation only).

Nonaccredited investor Wealthy, sophisticated investors who do not meet SEC net worth requirements. These investors require less protection because of large financial resources, but only 35 nonaccredited investor can be included per investment.

Nonclearing member An exchange member firm that is not able to clear transactions, and must pay another member firm to carry out its clearing operations.

Noncompetitive bid In a Treasury auction, bidding for a specific amount of securities at the price, whatever it may turn out to be, equal to the average price of the accepted competitive bids.

Noncontributory pension plan A pension plan that is fully paid for by the employer, requiring no employee contributions.

Non-cumulative preferred stock Preferred stock whose holders must forgo dividend payments when the company misses a dividend payment. Related: Cumulative preferred stock

Noncurrent asset Any asset that is expected to be held for the whole year, not sold or exchanged, such as real estate, machinery, or a patent.

Non-dischargeable debt Certain debts are not included in the debtor's discharge. Some are automatically excluded. Others to be excluded require action by the creditor. • Obtaining credit by fraud or a debt for alimony, maintenance or child support are types of debts that may be non-dischargeable.

Non-Discretionary Proposal A proposition on a proxy card requiring a response from the beneficial owner which does not fall under the Ten Day Rule. Therefore, the broker cannot vote on behalf of the beneficial owner, it can only vote after specific instructions have been received from the beneficial owner.

Non-financial services Include such things as freight, insurance, passenger services, and travel.

Non-marketable securities UK government securities that cannot be bought and sold on a stock exchange (*compare*marketable security). Nonmarketable securities include savings bonds and National Savings Certificates, tax-reserve certificates, etc, all of which form part of the national debt.

Non-price competition A form of competition in which two or more producers sell goods or services at the same price but compete to increase their share of the market by such measures as advertising, sales promotion campaigns, improving the quality of the product or service, delivery time, and (for goods) improving the packaging, offering free servicing and installation, or giving free gifts of unrelated products or services.

Non-probability sampling A marketing research sample chosen on the basis of convenience or personal judgement.

Non-production overhead costs The indirect costs of an organization that are not classified as manufacturing overhead. They include administration overheads, selling overhead, distribution overhead and (in some cases) research and development costs.

Nonproductive loan A loan that increases spending power, but is used in business that does not directly increase the economy's output, such as a leveraged buyout loan.

Non-profit marketing Applying the concept of marketing to organizations that are not profit-orientated. Examples include symphony orchestras, museums, and charities.

Non-profitability sampling A technique in marketing research in which the researcher uses his own discretion, instead of random selection, in choosing a representative sample of people to interview. This reduces the statistical validity of the results, which are then not sufficiently reliable to use as a basis for estimating profits.

Non-profit-making marketing (social marketing) Organized marketing used by non-profit-making organizations, including those involved in education, health care, charity, social causes, politics, and religious doctrines.

Nonpublic information Information about a company that is not known by the general public, which will have a definite impact on the stock price when released. See: Insider trading.

Nonpurpose loan A loan with securities pledged as collateral, but which is not to be used in securities trading or transactions.

Nonqualified plan A retirement plan that does not meet the IRS requirements for favorable tax treatment.

Nonqualifying annuity An annuity that does not fall under an IRS-approved pension plan. Contributions are made with after-tax dollars, but earnings can accumulate tax-deferred until withdrawal.

Non-qualifying policy A UK lifeassurance policy that does not satisfy the qualification rules contained in Schedule 15 of the Income and Corporation Taxes Act (1988) .

Nonqualifying stock option An employee stock option that does not satisfy IRS qualifying rules and therefore is liable for taxation upon exercise .

Nonrated A bond that has not been rated by a large rating agency, usually because the issue is too small.

Non-recourse finance A bank loan in which the lending bank is only entitled to repayment from the profits of the project the loan is funding and not from other resources of the borrower.

Nonrecourse In the case of default, the lender has no ability to claim assets over and above what the limited partners contributed.

Nonrecourse loan A loan for which no partner or related person bears the economic risk of loss. For example, if a partnership fails to repay a nonrecourse loan, the lender has no recourse against any partner except to foreclose of the assets used to secure the loan.

Nonrecurring charge A one-time expense or credit shown in a company's financial statement.

Nonrefundable Not permitted, under the terms of indenture, to be refundable.

Non-reproducible assets A tangible asset with unique physical properties, like a parcel of land, a mine, or a work of art.

Non-resident Being resident in a country other than the country in question. Some taxpayers take active steps to ensure that they are considered nonresident for UK tax purposes as this will affect their exposure to UK tax.

Non-revolving bank facility A loan from a bank to a company in which the company has a period (often several years) in which to make its drawdowns, as well as flexibility with regard to the amount and timing of the drawdowms, but once drawn an amount takes on the characteristics of a term loan.

Non-sampling errors All the errors that can occur in marketing research, except the sampling error (*see*population). These can include a badly phrased questionnaire, an error made by the interviewer, an error made by the respondent, etc.

Non-statutory accounts Any balance sheet or profit and loss account dealing with a financial period of a company that does not form part of the statutory accounts. Prior to the *Companies Act (1989)* they were known as abridged accounts.

Nonsterilized intervention Taking an action in the foreign exchange market without adjusting for changes in money supply.

Nonsystematic risk Nonmarket or firm-specific risk factors that can be eliminated by diversification. Also called unique risk or diversifiable risk. Systematic risk refers to risk factors common to the entire economy.

Non-tariff trade barrier A nonmonetary barrier to a foreign product or service, such as a bias against a foreign company's product standards or a rejection of a foreign country's political record.

Non-taxable income Income that is specifically exempt from tax. This includes the first £70 of interest from a National Savings Bank ordinary account, increase in value of National Savings Certificates, premium bond prizes, the capital part of the yearly amount received from a purchased life annuity, some social security benefits, prizes and betting winnings (including the National Lottery), savings related share option scheme (SAYE) account bonuses, shares allotted by an employer under an approved profit-sharing scheme, profitrelated pay up to £4000, statutory redundancy pay, and maintenance payments made under arrangements after 14 March 1988.

Nontradables Goods and services produced and consumed domestically that are not close substitutes to import or export goods and services.

Non-tradables Refer to goods and services produced and consumed domestically that are not close substitutes to import or export goods and services.

Nonvoting stock A security that does not entitle the holder to vote on the corporation's resolutions or elections.

No-par-value stock A stock with no par value given in the charter or stock certificate.

Normal annuity form The manner in which retirement benefits are paid out.

Normal backwardation theory Holds that the futures price will be bid down to a level below the expected spot price.

Normal backwardation theory Holds that the futures price will be bid down to a level below the expected spot price.

Normal distribution The symmetrical bell-shaped frequency curve formed when the number of occurrences of a range of values is plotted on the vertical axis against a series of classes plotted on the horizontal axis. The bell shape indicates that extreme values (both large and small) occur infrequently, while the more frequent occurrences are clustered around the arithmetic mean value, which in a normal distribution is also equal to the median and the mode (the value that occurs most frequently in a set of data).

Normal growth firms Companies whose earnings grow at a constant rate.

Normal investment practice The investment history of a customer, which is used as a benchmark to test the bona fide public offerings requirement of the allocation of a hot issue.

Normal loss The loss arising from a manufacturing or chemical process through waste, seepage, shrinkage, or spoilage that can be expected, on the basis of historical studies, to be part of that process. It may be expressed as a weight or volume or in other units appropriate to the process. It is usually not valued but if it is, a notional scrap value is used. It is axiomatic in costing that normal losses are part of the normal manufacturing cost, whereas the cost of abnormal losses should not be borne by the good output.

Normal Market Size (NMS) A classification system for trading in securities, which replaced the alpha, beta, gamma, and delta classification used on the London Stock Exchange in January 1991. The old system had developed in a way that, contrary to the original intention, had made it a measure of corporate virility. NMS is the minimum size of the package of shares in a company traded in normal-sized market transactions; there are 12 categories. The main purpose of the system is to fix the size of transactions in which market makers are obliged to deal, and to set a basis on which the bargains should be published.

Normal portfolio A customized benchmark that includes all the securities from which a manager normally chooses, weighted as the manager would weight them in a portfolio.

Normal probability distribution A probability distribution for a continuous random variable that is forms a symmetrical bell-shaped curve around the mean.

Normal probability distribution A probability distribution for a continuous random variable that forms a symmetrical bell-shaped curve around the mean. This distribution has no skewness or excess kurtosis.

Normal random variable A random variable that has a normal probability distribution.

Normal retirement The age or number of working years after which a pension plan beneficiary can retire and receive unreduced benefits immediately.

Normalized earnings Earnings that have been adjusted in order to take into account the effect of cycles in the economy.

Normalizing method Making a change in the income account equivalent to the tax savings realized through the use of different depreciation methods for shareholder and income tax purposes, thus washing out the benefits of the tax savings reported as final net income to shareholders.

North American Free Trade Agreement (NAFTA) A regional trade pact among the United States, Canada, and Mexico.

Not negotiable Words marked on a bill of exchange indicating that it ceases to be a negotiable instrument, i.e. Although it can still be negotiated, the holder cannot obtain a better title to it than the person from whom it was obtained, thus providing a safeguard if it is stolen. A cheque is the only form of bill that can be crossed 'not negotiable'; other forms must have it inscribed on their faces.

Notary public A legal practitioner, usually a solicitor, who is empowered to attest deeds and other documents and notes (*see*noting) dishonoured bills of exchange.

Note agreement A contract for privately placed debt.

Note agreement A contract for privately placed debt.

Note Debt instruments with initial maturities greater than one year and less than 10 years.

Note issuance facility (NIF) An agreement by which a syndicate of banks indicates a willingness to accept short-term notes from borrowers and resell these notes in the Eurocurrency markets.

Notes to the accounts (notes to financial statements) Information supporting that given on the face of a company's financial statements. Many notes are required to be given by law, including those detailing capital assets, investments, share capital debentures, and reserves. Other information may be required by accounting standards or be given to facilitate the users' understanding of the company and its current and future performance; social and environmental information, for example, falls into this category.

Notes to the financial statements A detailed set of notes immediately following the financial statements in an annual report that explain and expand on the information in the financial statements.

Not-for-profit An organization established for charitable, humanitarian, or educational purposes that is exempt from some taxes and in which no one in profits or losses.

Notice day A day on which notices of intent to deliver pertaining to a specified delivery month may be issued.

Notice of Meeting The legal one-page notice to security holders stating the date, time and place of the shareholder meeting. This page is normally attached to the front of the proxy statement.

Notice Period The time during which the buyer of a futures contract can be called upon to accept delivery. Typically, the 3 to 6 weeks preceding the expiration of the contract.

Notification date The day the option is either exercised or expires.

Noting 1. The procedure adopted if a bill of exchange has been dishonoured by non-acceptance or by non-payment. Not later than the next business day after the day on which it was dishonoured, the holder has to hand it to a notary public to be noted. The notary re-presents the bill; if it is still unaccepted or unpaid, the circumstances are noted in a register and also on a notarial ticket, which is attached to the bill. The noting can then, if necessary, be extended to a protest. 2. In advertising research, noticing a particular advertisement when first looking through a newspaper or magazine in which it appears. *See*noting score

Noting score The average number of readers noting a particular advertisement or editorial item in a newspaper or magazine expressed as a percentage of the total readership.

Notional income Income that is not received although it might be deemed to be properly chargeable to income tax. An example is the rental income one could receive from one's own home, if one were not living in it. This was formerly taxed under Schedule A. A further example might be interest foregone on the grant of an interest-free loan.

Notional principal amount In an interest rate swap, the predetermined dollar principal on which the exchanged interest payments are based.

Not-sufficient-funds check A bank check having insufficient funds to back it.

Nouveau Marche An equity market unit of the Paris Bourse that deals solely in innovative, high-growth companies.

Novation The replacement of one legal agreement by a new obligation, with the agreement of all the parties. For example, on exchanges using the London Clearing House (LCH), transactions between members are novated by the LCH, so that one contract is created between the buyer and LCH while a matching contract is created between the seller and LCH

Nudum pactum (Latin: nude contract) An agreement that is unenforceable in British law because no consideration is mentioned.

Numbered account A bank account identified only by a number. This service, offered by some Swiss banks at one time, encouraged funds that had been obtained illegally to find their way to Switzerland. A numbered account is now more frequently used to ensure legitimate privacy.

NYSE Abbreviation for New York Stock Exchange.

NYSE composite index Composite index covering price movements of all new world common stocks listed on the New York Stock Exchange. It is based on the close of the market on December 31, 1965, at a level of 50.00, and is weighted according to the number of shares listed for each issue. Print changes in the index are converted to dollars and cents so as to provide a meaningful measure of changes in the average price of listed stocks. The composite index is supplemented by separate indexes for four industry groups: industrial, transportation, utility, and finance.

Objective (mutual funds) The fund's investment strategy category as stated in the prospectus. There are more than 20 standardized categories.

Objective and task method A method of setting a promotional budget in which the marketer decides the objective to be accomplished and the tasks necessary to achieve the objective. The budget is decided by estimating the costs of carrying out the tasks.

Objectivity An accounting concept attempting to ensure that any subjective actions taken by the preparer of accounts are The aim of the rules and regulations required to achieve objectivity is that users should be able to compare financial statements for different companies over a period with some confidence that the statements have been prepared on the same basis. One of the major advantages claimed for historical cost accounting is that it is objective, but necessarily some subjective decisions will have been made.

Obligation bond A municipal bond with a face value greater than the value of the underlying property. The difference is designed to compensate the lender for costs exceeding the mortgage value.

Obligatory expenditure The spending of the European Union, governed by such treaties as the Common Agricultural Policy and the European Regional Development Fund. The European Parliament can add to the amount spent but if it wishes to reduce it, it must obtain the agreement of the Council of Ministers, or reject the whole of the Union Budget in that year.

Obligor A person who has an obligation to pay off a debt.

Observation research Marketing research in which a respondent's actions are monitored without direct interaction, without using an interviewer.

Observational Noise The error between the true value in a system and its observed value due to imprecision in measurement. Also called Measurement Noise.

Ocean bill of lading Receipt for a shipment by boat, that includes freight charges and title to the merchandise.

Obsolescence A fall in the value of an asset as a result of its age. For example, plant and equipment may not have actually worn out but may have become out of date because technology has advanced and more efficient plant has become available. It also applies to consumer durables (*see*consumer goods), in which a change of style may render a serviceable piece of equipment, such as a car or washing machine, out of date. This can result in the fall in the value of such items held in stock by a company. Builtin obsolescence or planned obsolescence is a deliberate policy adopted by a manufacturer to limit the durability of his product in order to encourage the consumer to buy a replacement more quickly than he otherwise might have to. The morality of this technique has been frequently questioned but is usually defended on the grounds that many western economies depend on strong consumer demand; if such consumer durables as cars and washing machines were built to last for their purchaser's lifetime, consumer demand would be reduced to a level that would create enormous unemployment.

Occupational hazard A risk of accident or illness at one's place of work. Dangerous jobs usually command higher salaries than those involving no risks, the increase being known as danger money.

Odd lot A trading order for less than 100 shares of stock.

Odd lot dealer A broker who combines odd lots of securities from multiple buy or sell orders into round lots and executes transactions in those round lots.

Odd-even pricing The pricing of a product so that the price ends in an odd number of pence, which is not far below the next number of pounds. For example, £4.99 might be used in preference to £5.00 in order to make the product appear cheaper (at least to marketing personnel if not to consumers).

Odd-Lot Buy Back An offer made by the corporation or its agent to purchase shares from odd-lot shareholders.

Odd-Lot Resale An offer made by the corporation or its agent to purchase shares from odd-lot shareholders and immediately resell them in the market, usually in round-lots to institutions, thus saving the corporation the expense of merely buying shares back.

Odd-lot theory The theory that profits can be made by making trades contrary to odd-lot trading patterns, since odd-lot investors have poor timing. This theory is no longer popular.

Off balance sheet finance (OBSF) A method of financing a company's activities so that some or all of the finance and the corresponding assets do not appear on the balance sheet of the company. By making use of OBSF a company can enhance its accounting ratios, such as the gearing ratio and return on capital employed, and also avoid breaking any agreements it has made with the banks in respect of the total amount it may borrow. It has been possible for companies, by drawing up complex legal agreements, to conduct off balance sheet finance and thus mislead the user of the accounts. The accounting profession has attempted to counter these practices by emphasizing that accounting should reflect the commercial reality of transactions and not simply their legal form. The recent Financial Reporting Standard 5, 'Reporting the Substance of Transactions', provides specific guidance for certain transactions, such as factoring and consignment stock, for which companies have previously used off balance sheet finance.

Offer by prospectus An offer to the public of a new issue of shares or debentures made directly by means of a prospectus, a document giving a detailed account of the aims, objects, and capital structure of the company, as well as its past history. The prospectus must conform to the provisions of the *Companies Act* .

Offer document A document sent to the shareholders of a company that is the subject of a takeover bid. It gives details of the offer being made and usually provides shareholders with reasons for accepting the terms of the offer.

Offer price The price at which a security is offered for sale by a market maker and also the price at which institution will sell units in a unit trust.

Offer The price at which a seller makes it known that he is willing to sell something. If there is an acceptance of the offer a legally binding contract has been entered into. In law, an offer is distinguished from an invitation to treat, which is an invitation by one person or firm to others to make an offer. An example of an invitation to treat is to display goods in a shop window.

Offer wanted Used in the context of general equities. Notice by a potential buyer of a security that he or she is looking for supply from a potential seller of the security, often requiring a capital commitment. Antithesis of bid wanted.

Offering date Date on which a new set of stocks or bonds will first be sold to the public.

Offering memorandum A document that outlines the terms of securities to be offered in a private placement.

Offering scale The range of prices offered by the underwriter of a serial bond issue with different maturities.

Offerings Often refers to initial public offerings. When a firm goes public and makes an offering of stock to the market.

Off-floor order Used for listed equity securities. (1) Order to buy or sell a security that originates off the floor of an exchange; customer orders originating with brokers, as distinguished from orders placed by floor members trading for their own accounts. Exchange rules require that an off-floor order be executed before orders initiated on the floor. Upstairs order. Antithesis of on-floor order; (2) order not handled on the floor but instead upstairs.

Offshore company 1. A company not registered in the same country as that in which the persons investing in the company are resident.

2. A company set up in a foreign country or tax haven by a financial institution with the object of benefiting from tax laws or exchange control regulations in that country.

Offshore finance subsidiary A wholly owned affiliate incorporated overseas, usually in a tax haven country, whose function is to issue securities abroad for use in either the parent's domestic or its foreign business.

Offshore financial centres Centres that provide advantageous deposit and lending rates to non-residents because of low taxation, liberal exchange controls, and low reserve requirements for banks. Some countries have made a lucrative business out of offshore banking; the Cayman Islands is currently one of the world's largest offshore centres. In Europe, the Channel Islands and the Isle of Man are very popular. The USA and Japan have both established domestic offshore facilities enabling non-residents to conduct their business under more liberal regulations than domestic transactions. Their objective is to stop funds moving outside the country.

Offshore fund 1. A fund that is based in a tax haven (offshore tax haven) outside the UK to avoid UK taxation. Offshore funds operate in the same way as unit trusts but are not supervised by the Department of Trade and Industry.

2. A fund held outside the country of residence of the holder.

Off-the-peg research Marketing research that uses existing data, rather than a fresh investigation of a market.

Off-the-shelf company A company that is registered with the Registrar of Companies although it does not trade and has no

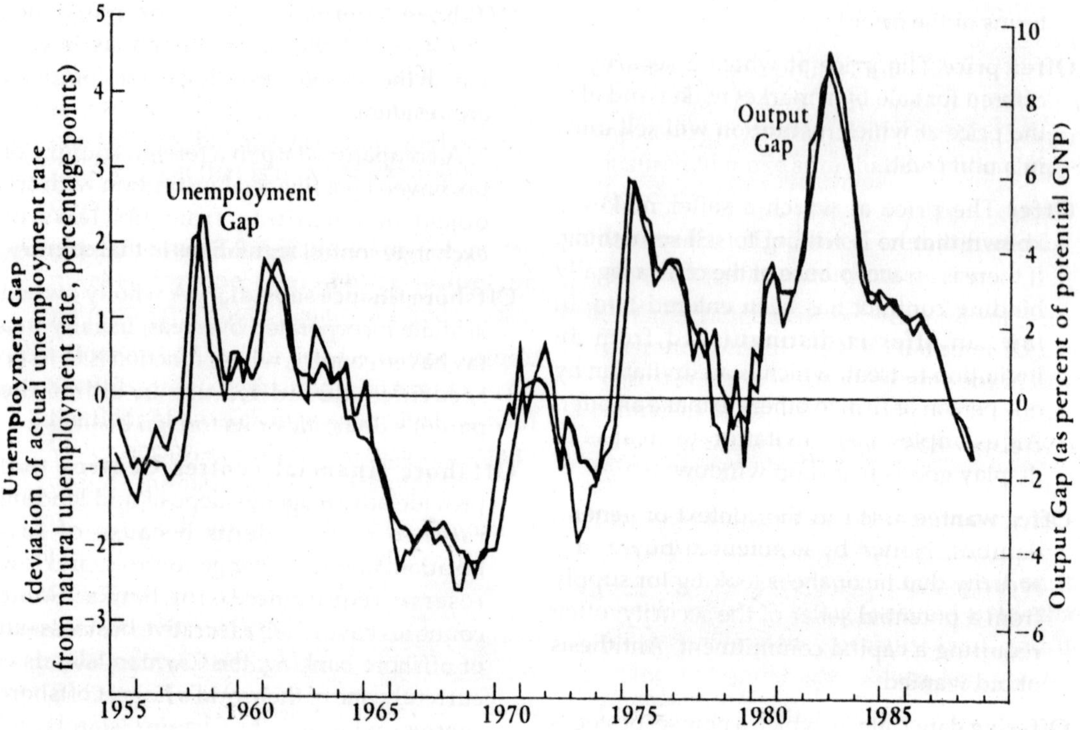

Fig. Okun's Law

directors. It can, however, be sold and reformed into a new company with the minimum of formality and expense. Such companies are easily purchased from specialist brokers.

Okun's Law Refers to a relationship between rel economic growth and changes in the rate of unemployment. Accounting to this law for every 2.5 percentage point growth in real gross national product, above the trend rate that is sustained for a year, the unemployment rate declines by 1 percentage point. The law has been simply a rule of thumb for assessing the implications of rel growth for unemployment.

Old-line factoring Factoring arrangement that provides collection, insurance, and finance for accounts receivable.

Oligopoly A market in which relatively few sellers supply many buyers. Each seller recognizes that prices can be controlled to a certain extent and that competitors' actions will influence profits.

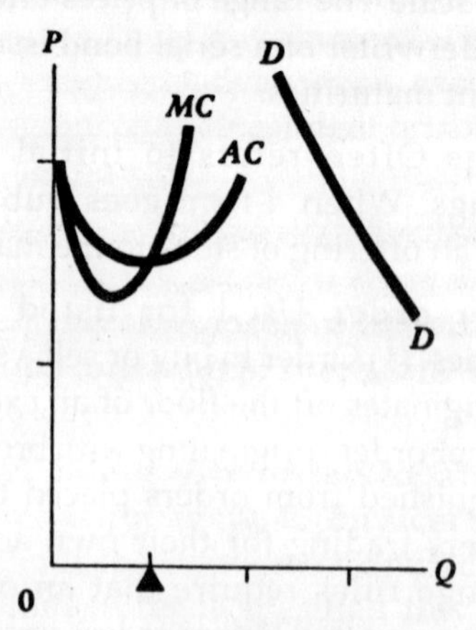

Fig. Oligopoly

Oligopsony A Market characterized by a small number of large buyers who control all purchases and therefore the market price of a good or service.

Omitted dividend A dividend that was scheduled to be declared, but that is not voted by the Board of Directors probably because the company is experiencing financial difficulties.

Omnibus account An account carried by one futures commission merchant with another futures commission merchant in which the transactions of two or more persons are combined and carried in the name of the originating broker, rather than designated separately. Related: Commission house.

Omnibus Proxy A list issued by depositories detailing their participants, and their holdings, and authorizing the participants to vote their proxies directly. This type of proxies are issued by Cede & Co. and by certain bank custodians.

Omnibus research Marketing research surveys based on multipart questionnaires sent out regularly to a panel of respondents. Space on the questionnaire is available to companies that have specific marketing research needs, especially those having a limited number of questions to ask, which would not alone justify setting up a separate research study.

On approval (on appro) The practice of allowing potential buyers to take possession of goods in order to decide whether or not they wish to buy them. The potential buyer is a bailee (*see*bailment) and is obliged to return the goods in perfect condition after deciding not to purchase them. Many department stores and mail-order firms allow their customers to take goods on appro.

On balance Used for listed equity securities. Left over after pairing off other market buy and sell orders, usually before the opening of a stock or market but at times at the close (especially during index expirations). See: Imbalance of orders.

On Board Ocean Bill of Lading An ocean bill of lading bearing an on board notation, or words indicating that the merchandise is located aboard the vessel for transportation. These notations must be initialed or signed by an authorized employee or agent of the ship line.

On board Used in the context of general equities. Long.

On Carriage Freight costs arising after the cost of principal international freight costs. These are usually inland freight charges for delivery within the buyer's country.

On demand Denoting a bill of exchange that is payable on presentation. An uncrossed cheque is an example of such a bill.

On stream Denoting that a specified investment or asset is bringing in the income expected of it.

On the close order A market order that is to be executed as close as possible to the closing price of the day.

On the money Used in the context of general equities. In-line, or at the same price, as the last sale.

On the opening order A market order that is to be executed at the price of the first trade of the day.

On the print Used in the context of general equities. To participate in a block trade that has already transpired, as if that customer had been part of the trade originally; often used by a new party looking to participate in a trade that has just happened. See: Open on the print.

On the run The most recently issued (and typically the most liquid) government bond in a particular maturity range.

On the sidelines An investor who decides not to invest due to market uncertainty.

On the take Used in the context of general equities. Price moving upward, because more buyers are taking offerings, causing offerings to vanish and be replaced by higher ones. Antithesis of come in, get hit.

One-way market (1) A market in which only one side, the bid or asked, is quoted or firm. (2) A market that is moving strongly in one direction.

One-year money Money placed on the money market for one year, i.e. It cannot be withdrawn without penalty for one year.

On-floor order Used for listed equity securities. Security order originating with a member on the floor of an exchange when dealing with his or her own account, versus an upstairs order. Antithesis of off-floor order.

OPEC (Organization of Petroleum Exporting Countries) A cartel of oil-producing countries.

Open (good-til-cancelled) order An individual investor can place an order to buy or sell a security. That open order stays active until it is completed or the investor cancels it.

Open account Arrangement whereby sales are made with no formal debt contract. The buyer signs a receipt, and the seller records the sale in the sales ledger.

Open contracts Contracts that have been bought or sold without completion of the transaction by subsequent sale or purchase, or by making or taking actual delivery of the financial instrument or physical commodity.

Open cover (open policy) A marine cargo insurance policy in which the insurer agrees to cover any voyage undertaken by the policyholder's vessel(s) or any cargo shipped by a particular shipper. A policy condition requires a declaration to be completed on a weekly, monthly, or quarterly basis indicating the vessels involved, the commodities carried and the voyages undertaken. Insurers use the declaration to calculate the premium. The main advantage of a policy of this kind is that the insurer can be confident that all cargos have insurance cover without the need to notify insurers of the details before the voyage. In some cases there may be a policy limit above which the insurance ceases. On such policies it is necessary to keep a running total of the sums insured to make sure that the policy limit is not exceeded.

Open credit 1. Unlimited credit offered by a supplier to a trusted client.

2. An arrangement between a bank and a customer enabling the customer to cash cheques up to a specified amount at other branches or banks. This practice is less used since credit cards and cash cards were introduced.

Open depending on the floor Used for listed equity securities. Having room for a customer buyer or seller contingent on the results of a trade being executed on the floor (i.e., satisfying the specialist book and the orders the trader opened up). See: Open on the print, subject.

Open economy An economy in which a significant percentage of its goods and services are traded internationally. The degree of openness of an economy usually depends on the amount of overseas trade in which the country is involved or the political policies of its government. Thus the UK economy is relatively open, as the economy is significantly dependent on foreign trade, while the US economy is relatively closed as overseas trade is not as important to its economy.

Open general licence An import licence for goods on which there are no import restrictions.

Open indent An order to an overseas purchasing agent to buy certain goods, without specifying the manufacturer. If the manufacturer is specified this is a closed indent.

Open interest The total number of derivative contracts traded that not yet been liquidated either by an offsetting derivative transaction or by delivery. Related: liquidation

Open on the print Used in the context of general equities. Block trader's term for a block trade that has been completed with an institutional client and printed on the consolidated tape, but leaves the block trader with stock available (because the trader has taken a long or short position to complete the trade) for new customers who are on the opposite side of the market to the initiating customer.

Open order (good-till-cancelled, GTC order) Order to buy or sell a security that stays active until it is completed or the investor cancels it.

Open outcry Quoting prices, making offers, bids, and acceptances, and concluding transactions by word of mouth in a commodity market or financial futures exchange, usually in a trading pit.

Open Policy A marine cargo insurance policy issued to cover various unspecified exports over the life of the policy.

Open position (naked position) A trading position in which a dealer has commodities, securities, or currencies bought but unsold or unhedged (*see* hedging), or sales that are neither covered nor hedged. The former position is a bull position ; the latter, a bear position. In either case the dealer is vulnerable to market fluctuations until the position is closed or hedged. See also option.

Open repo A repurchase agreement with no definite term. The agreement is made on a day-to-day basis, and either the borrower or the lender may choose to terminate. The rate paid is higher than on overnight repo and is subject to adjustment if rates move.

Open system A system that is affected by influences from outside its system boundary and that is able to affect other systems. *Compare***closed system.**

Open up Used in the context of general equities. Disclose more information (e.g., the exact price and quantity of one's potential interest). See: Put pants on it.

Open-end credit Revolving line of credit that is extended with every purchase or cash advance.

Open-end fund Also called a mutual fund, an investment company that stands ready to sell new shares to the public and to redeem its outstanding shares on demand at a price equal to an appropriate share of the value of its portfolio, which is computed daily at the close of the market.

Open-end fund Used in the context of general equities. Mutual fund that continually creates new shares on demand. Mutual fund shareholders buy the funds at net asset value and may redeem them at any time at the prevailing market prices. Antithesis of closed-end fund.

Open-end lease A lease agreement that provides for an additional payment at the expiration of the lease to adjust for any change in the value of the property.

Open-end mortgage Mortgage against which additional debts may be issued. Related: closed-end mortgage.

Open-end trust A form of unit trust in which the managers of the trust may vary the investments held without notifying the unit holders. Open-end trusts are used in the USA.

Opening Bank A bank which establishes a letter of credit.

Opening prices The bid prices and offer prices made at the opening of a day's trading on any security or commodity market. The opening prices may not always be identical to the previous evening's closing prices, especially if any significant events have taken place during the intervening period.

Opening purchase A transaction in which the purchaser's intention is to create or increase a long position in a given series of options.

Opening sale A transaction in which the seller's intention is to create or increase a short position in a given series of options.

Opening stock The stock held by an organization at the beginning of an accounting period as raw materials, work in progress, or finished goods. The closing stocks of one period become the opening stocks of the succeeding period and it is necessary to establish the level of closing stocks so that the cost of their creation is not charged against the profits of that period but brought forward as opening stocks to be charged against the profits of the succeeding period.

Opening transaction Applies to derivative products. (1)Buy or sell transaction that creates a position out of a flat one (writing an option short or buying an option long). Antithesis of closing transaction. (2) First transaction of the day in a stock.

Opening, the The period at the beginning of the trading session officially designated by the exchange during which all transactions are considered made "at the opening". **Related:** Close, the

Open-market operation Purchase or sale of government securities by the monetary authorities to increase or decrease the domestic money supply.

Open-market operations The purchase or sale by a government of bonds (gilt-edged securities) in exchange for money. This is the main mechanism by which monetary policy in developed economies operates. To buy (or sell) more bonds the government must raise (or lower) their price and hence reduce (or increase) interest rates.

Open-market purchase operation A systematic program of repurchasing shares of stock in market transactions at current market prices, in competition with other prospective investors.

Open-market value (OMV) The value of an asset equal to the amount that a willing purchaser would be prepared to pay a willing vendor. For capital gains tax purposes, the open-market value refers to such a valuation made on 31 March 1982, the date from which indexation is calculated for assets acquired before 31 March 1982.

Open-outcry The method of trading used at futures exchanges, typically involving calling out the specific details of a buy or sell order, so that the information is available to all traders.

Open-pricing agreement An agreement between firms operating in an oligopolistic market in which prices and intended price changes are circulated to those taking part in the agreement in order to avoid a price war.

Operating budget A forecast of the financial requirements for the future trading of an organization, including its planned sales, production, cash flow, etc. An operating budget is normally designed for a fixed period, usually one year, and forms the plan for that period's trading activities. Any divergences from it are usually monitored and, if appropriate, changes can then be made to it as the period progresses.

Operating cash flow Earnings before depreciation minus taxes. Measures the cash generated from operations, not counting capital spending or working capital requirements.

Operating cycle The average time intervening between the acquisition of materials or services and the final cash realization from those acquisitions.

Operating doctrine Stock-control policies that specify when and how much stock should be reordered.

Operating exposure Degree to which exchange rate changes, in combination with price changes, will alter a company's future operating cash flows.

Operating lease A lease under which an asset

is hired out to a lessee or lessees for a period that is substantially shorter than its useful economic life. Under an operating lease, the ownership of the leased asset remains with the lessor. Statement of Standard Accounting Practice 21, 'Accounting for Leases and Hire Purchase Contracts', defines an operating lease as a lease other than a finance lease.

Operating leverage Fixed operating costs, so-called because they accentuate variations in profits.

Operating profit (or loss) Revenue from a firm's regular activities less costs and expenses and before income deductions.

Operating profit margin The ratio of operating margin to net sales.

Operating rate The percentage of total production capacity of a company, industry, or country that is being used.

Operating risk The inherent or fundamental risk of a firm, without regard to financial risk. The risk that is created by operating leverage. Also called business risk.

Operating statement A financial and quantitative statement provided for the management of an organization to record the performance achieved by that area of the operation for which the management is responsible, for a selected budget period. An operating statement may include production levels, costs incurred, and (where appropriate) revenue generated, all compared with budgeted amounts and the performance in previous periods.

Operation job card A card or form on which an employee writes up the details of a particular task and the length of time it took to complete. It is used in work studies.

Operational research (operations research) A scientific method of approaching industrial and commercial problems in order to arrive at the most efficient and economic method of achieving the desired objective. It basicary consists of making a clear statement of the problem, designing a model to represent the possible solutions using different strategies, and applying the solutions obtained from the analysis of the model to the real problem. It makes use of game theory, critical-path analysis, simulation techniques, etc.

Operationally efficient market Also called an internally efficient market, one in which investors can obtain transactions services that reflect the true costs associated with furnishing those services.

Operations A set of activities concerned with transforming resource inputs into desired outputs, i.e. Goods or services. Every purposeful organization has desired outputs and, as every output requires inputs of some sort, all purposeful organizations are involved in operations. Traditionally, the word has been used to refer exclusively to manufacturing processes, but it applies equally to the service sector and to nonprofit-making organizations. In some organizations it is possible to identify a single coherent sub-system as performing the operations function. In others, responsibility for various elements may be split between different linked departments.

Operations management The management of an operating system to achieve the desired organizational outputs within an optimum balance of economy, efficiency, and effectiveness.

Operations manager A manager with responsibility for running the resource inputs that constitute an operating system. Most managers have an operational function within their overall job description, e.g. A human resource manager (*see*human-resource management) has to make use of inputs consisting of cash, skilled personnel, documentation, etc., to produce an output consisting of an appropriately skilled workforce.

Operations strategy A plan to transform an organization's overall strategic objectives into operational deliverables. It involves the design of the product or service and the processes by which the product or service is produced; the way in which production is managed and controlled; and the design of processes for the constant improvement of the operation. If, at the strategic level the organization decides to compete in the market on the basis of superior quality, the operations strategy has to plan to produce a package incorporating the specified quality within stated financial constraints. If the strategy is to compete on cost, the plan must produce the product cheaply within the Emits of acceptable quality.

Opinion poll A poll using marketing research techniques that is specifically concerned with political opinion. Public opinion polls are usually commissioned by the media (especially television and newspapers) in order to keep their readers or viewers informed. Private opinion polls are usually commissioned by the political parties in order to help develop campaign strategies.

Opinion shopping A practice prohibited by the SEC which involves attempts by a corporation to obtain reporting objectives by following questionable accounting principles with the help of a pliable auditor willing to go along with the desired treatment.

Opportunity cost of capital Expected return that is foregone by investing in a project rather than in comparable financial securities.

Opportunity cost The income or benefit foregone as the result of carrying out a particular decision, when resources are limited or when mutually exclusive projects are involved. For example, the opportunity cost of building a factory on a piece of land is the income foregone by not constructing an office block on this particular site. Similarly, the income foregone by not constructing a factory if an office block is constructed represents the opportunity cost of an office block. Opportunity cost is an important factor in decision making (*see*cost-benefit analysis), although it represents costs that are not recorded in the accounts of the relevant organization.

Opportunity costs The difference in the performance of an actual investment and a desired investment adjusted for fixed costs and execution costs. The performance differential is a consequence of not being able to implement all desired trades. Most valuable alternative that is given up.

Opportunity line Slope of a graph representing portfolios achieved by combining different levels of borrowing and lending with a single risky portfolio. Sometimes called investment opportunity set.

Opportunity set The possible expected return and standard deviation pairs of all portfolios that can be constructed from a given set of assets.

Optical character recognition (OCR) The recognition of printed characters by fight-sensitive optical scanners. The scanner recognizes the shape of a letter by scanning it with a very fine point of light. It then uses a computer to compare the pattern of reflected light with the patterns of the letters of the alphabet stored in its memory.

Optimal contract The contract that balances the three types of agency costs (contracting, monitoring, and misbehavior) against one another to minimize the total cost.

Optimal portfolio An efficient portfolio most preferred by an investor because its risk/ reward characteristics approximate the investor's utility function. A portfolio that maximizes an investor's preferences with respect to return and risk.

Optimal redemption provision Provision of a bond indenture that governs the issuer's ability to call the bonds for redemption prior to their scheduled maturity date.

Optimization approach to indexing An

approach to indexing that seeks to optimize some objective, such as to maximize the portfolio yield, to maximize convexity, or to maximize expected total returns.

Optimum Leverage Ratio The borrowing level that maximizes the value of the firm. The cost of capital to the firm is minimized at that same level.

Option account A brokerage account that is approved to hold option positions or trades.

Option agreement A form that an options investor opening an option account fills out guarantees the investor will follow trading regulations and has the financial resources to settle possible losses.

Option cycle The cycle of option expiration months. The most common cycles are: January, April, July, and October (JAJO); February, May, August, and November (FMAN); and March, June, September, and December (MJSD).

Option dealer A dealer who buys and sells either traded options or traditional options on a stock exchange, commodity exchange, or currency exchange.

Option elasticity The percentage increase in an option's value given a 1% change in the value of the underlying security.

Option Gives the buyer the right, but not the obligation, to buy or sell an asset at a set price on or before a given date. Investors, not companies, issue options. Investors who purchase call options bet the stock will be worth more than the price set by the option (the strike price), plus the price they paid for the option itself. Buyers of put options bet the stock's price will go down below the price set by the option. An option is part of a class of securities called derivatives, so named because these securities derive their value from the worth of an underlying investment.

Option holder A person who has an option that has not been exercised.

Option margin The margin requirement for options described in Regulation T and in brokers' individual policies.

Option money The price paid for an option. The cost of a call option is often known as the **call money** and that for a put option as the **put money**.

Option mutual fund A mutual fund that buys and sells options for aggressive or conservative investment.

Option premium The option price.

Option price Also called the option premium; the price the buyer of the options contract pays for the right to buy or sell a security at a specified price in the future.

Option Pricing Curve A graphical representation of the projected price of an option at a fixed point in time. It reflects the amount of time value premium in the option for various stock prices, as well. The curve is generated by using a mathematical model. The delta (or hedge ratio) is the slope of a tangent line to the curve at a fixed stock price. See also Delta and Hedge Ratio

Option seller Also called the option writer , the party who grants a right to trade a security at a given price in the future.

Option series A group of options on the same underlying security with the same exercise price and maturity month.

Option spread The trading of options of the same class at the same time in order to profit from changes in the size of the spread between different options.

Option The right to buy or sell a fixed quantity of a commodity, currency, or security at a particular date at a particular price (the exercise price). Unlike futures, the purchaser of an option is not obliged to buy or sell at the exercise price and will only do so if it is profitable; if the option is allowed to lapse, the purchaser loses only the initial purchase price of the option (the option money). In London, options in securities (stock options) and financial futures, are bought and sold through the London International Financial Futures and Options Exchange (LIFFE), while options in commodity futures are bought and sold on the London Commodity Exchange.

Option to purchase 1. A right given to shareholders to buy shares in certain companies in certain circumstances at a reduced price.

2. A right purchased or given to a person to buy something at a specified price on or before a specified date. Until the specified date has passed, the seller undertakes not to sell the property to anyone else and not to withdraw it from sale.

Option-adjusted spread (OAS) (1) The spread over an issuer's spot rate curve, developed as a measure of the yield spread that can be used to convert dollar differences between theoretical value and market price. (2) The cost of the implied call embedded in a MBS, defined as additional basis-yield spread. When added to the base yield spread of an MBS without an operative call produces the option-adjusted spread.

Optional dividend A dividends that the shareholder can elect to receive either in cash or in stock.

Optional payment bond A bond whose principal and/or interest may be paid in foreign or domestic currency at the discretion of the bondholder.

Options Clearing Corporation (OCC) Applies to derivative products. Financial institution that is the actual issuer and guarantor of all listed option contracts.

Options contract A contract that, in exchange for the option price, gives the option buyer the right, but not the obligation, to buy (or sell) a financial asset at the exercise price from (or to) the option seller within a specified time period, or on a specified date (expiration date).

Options contract multiple A constant, set at Rs.100, that when multiplied by the cash index value gives the dollar value of the stock index underlying an option. That is the dollar value of the underlying stock index = Cash index value x Rs.100 (the options contract multiple).

Options on physicals Interest rate options written on fixed income securities, as opposed to those written on futures contracts.

Oral contract A contract not recorded on paper or on computer, but made vocally which is usually enforceable.

Order Book Official The exchange employee in charge of keeping a book of public limit orders on exchanges utilizing the "marker-maker" system, as opposed to the "specialist system", of executing orders. See also Market-Marker and Specialist.

Order imbalance Orders of one kind for a stock not offset by the opposite orders, which causes a wide spread between bid and offer prices.

Order Instruction to a broker/dealer to buy, sell, deliver, or receive securities or commodities that commits the issuer of the "order" to the terms specified. See: indication, inquiry, bid wanted, offer wanted.

Order of business The sequence of the items on the agenda of a business meeting. It is usual to adopt the following order: apologies for absence; reading and signing of the minutes of the last meeting; matters arising from these minutes; new correspondence received; reading and adoption of reports, accounts, etc.; election of officers and auditors; any special motions; any other business; date of next meeting.

Order Parameter In a nonlinear dynamic system, a variable-acting link a macrovariable, or combination of variables-that summarizes the individual variables that can affect a system. In a controlled experiment, involving thermal convection, for example, temperature can be a control parameter; in a large complex system, temperature can be an order parameter, because it summarizes the effect of the sun, air pressure, and other atmospheric variables. See: Control parameter.

Order room The brokerage firm department receives and processes all orders to buy and sell securities.

Order splitting Breaking up orders so that they can be processed as small orders for execution by SOES. Prohibited by NASD.

Order ticket A form detailing an order instruction that a customer gives an account executive.

Ordering Costs Costs that occur when an order is placed regardless of the size of the order.

Order-qualifying criteria (entrylevel criteria) The characteristics that a product or service has to have before customers will consider buying it, i.e. Before it can gain entry to the market.

Order-winning criteria The characteristics of a product or service that are most important in deciding whether or not a customer will make a purchase. These are the characteristics that positively differentiate a particular product from similar products in the market. Clearly order-winning criteria vary from customer to customer.

Ordinary activities Any activities undertaken by an organization as part of its business and any related activities in which it engages in furtherance of, incidental to, or, arising from, these activities. This definition is based on that given in Financial Reporting Standard 3, 'Reporting Financial Performance. Ordinary activities include the effects on the reporting entity of any event in the various environments in which it operates, including the political, regulatory, economic, and geographical environments, irrespective of the frequency or unusual nature of the events. *See also* extraordinary items.

Ordinary income The income derived from the regular operating activities of a firm or individual.

Ordinary interest Interest based on a 360-day year instead of a 365-day year, resulting in what can be a significant difference.

Ordinary resolution A resolution that is valid if passed by a majority of the votes cast at a general meeting of a British company. No notice that the resolution is to be proposed is required.

Ordinary shares Apples mainly to international equities. Shares of non-U.S. companies traded in their individual home markets. Usually cannot be delivered in the.

Organic growth Refers to growth achieved by internal investments of the firm. This could be the day to day business of the firm or a division of the firm starting a new business from scratch. This is distinguished from growth by acquisition or merger which involves an outside firm.

Organic organization An organization characterized by roles that are not well defined, tasks being redefined as individuals interact; there is little reliance on authority, with control and decision making decentralized, and communication is both lateral and vertical This type of organization, which is a feature of the computer software industry, is appropriate if the external environment is uncertain.

Organization and methods (O & M) A form of work study involving the or of procedures and controls in a business and the methods of implementing them in management terms. It is usually applied to office procedures rather than factory production.

Organization chart A chart illustrating the structure of an organization; in particular it will show for which function of the business each manager is responsible and the chain of responsibility throughout the organization. Some organization charts include managers by name; others show the management positions in the structure.

Organization for Economic Cooperation and Development (OECD) An organization formed in 1961, replacing the Organization for European Economic Cooperation (OEEC), to promote cooperation among industralized member countries on economic and social policies. Its objectives are to assist member countries in formulating policies designed to achieve high economic growth while maintaining financial stability, contributing to world trade on a multilateral basis, and stimulating members' aid to developing countries. Members are Australia, Austria, Belgium. Canada, Denmark, Finland, France, Germany, Greece, Iceland, Ireland, Italy, Japan, Luxembourg, Mexico, the Netherlands, New Zealand, Norway, Portugal, Spain, Sweden, Switzerland, Turkey, UK, and USA. Its headquarters are in Paris.

Organization for Economic Cooperation and Development (OECD) An organization of industrialized countries formed to promote the economic health of its members and to contribute to worldwide development.

Organization of Petroleum Exporting Countries (OPEC) A cartel of oil-producing countries.

Organization of the Petroleum Exporting Countries (OPEC) An organization created in 1960 to unify and coordinate the petroleum policies of member countries and to protect their interests, individually and collectively. Present members are Algeria, Gabon, Indonesia, Iran, Iraq, Kuwait, Libya, Nigeria, Qatar, Saudi Arabia, UAE, and Venezuela.

Organizational theory The study of the design, structure, and process of decision making in organizations (*see* organizational design). The sociology of social systems, and the relationships within them, is central to organizational theory (*see*organizational behaviour).

Considerable work has been carried out on the way in which decisions are made, both by individuals and groups (committees) within organizations. The extent to which managers are influenced by organizational politics, rather than being guided by reason and economics, in making their decisions is an aspect of considerable importance.

Organized exchange A securities marketplace where purchasers and sellers regularly gather to trade securities according to the formal rules adopted by the exchange.

Organized market A formal market in a specific place in which buyers and sellers meet to trade according to agreed rules and procedures. Stock exchanges, financial futures exchanges, and commodity markets are examples of organized markets.

Original cost The cost of an item at the time of purchase or creation. This applies particularly to capital assets in which depreciation by the straight-line method uses the original cost as a basis for the calculation. *See also*historical cost.

Original goods Natural products that have no economic value until factors of production are applied to them. They include virgin land, wild fruit, natural waterways, etc.

Original Issue Discount securities (OIDS) Bonds on which the coupon rate is set considerably below the yield to maturity at the time of issuance so that the bonds are issued at a discount from a par value.

Original margin The margin needed to cover a specific new position. Related: Margin, security deposit (initial).

Original maturity Maturity at issue. For example, a five-year note has an original maturity of five years; one year later it has a maturity of four years.

Originally, the Internet was mainly used by the academic community, but the wider availability of personal computers has led to an enormous growth in the number of individual users. This has resulted in increasing commercial exploitation in such areas as electronic shopping, advertising, and marketing research.

Originator A bank, savings and loan, or mortgage banker that initially made a mortgage loan that is part of a pool. Also, an investment bank that has worked with the issuer of a new securities offering from the beginning and is usually appointed manager of the underwriting syndicate.

Osaka Securities Exchange (OSE) Established after World War II, one of the three major securities markets in Japan.

Oslo Stock Exchange An exchange founded in 1819 and trading stocks, bonds, and stock options that is considered the options market of Norway.

Other capital In the balance of payments, other capital is a residual category that groups all the capital transactions that have not been included in direct investment, portfolio investment, and reserves categories. It is divided into long-term capital and short-term capital and, because of its residual status, can differ from country to country. Generally speaking, other long-term capital includes most non-negotiable instruments of a year or more like bank loans and mortgages. Other short-term capital includes financial assets of less than a year such as currency, deposits, and bills.

Other current assets Value of non-cash assets, including prepaid expenses and accounts receivable, due within 1 year.

Other long-term liabilities Value of leases, future employee benefits, deferred taxes, and other obligations not requiring interest payments that must be paid over a period of more than one year.

Other sources Amount of funds generated during the period from operations by sources other than depreciation or deferred taxes. Part of free cash flow calculation.

Ouguiya (UM) The standard monetary unit of Mauritania, divided into 5 khoums.

Out of the name Used in the context of general equities. To no longer have an active trading profile/ position in the stock.

Outlay cost The expenditure incurred as the initial cost of a project or activity. The outlay cost may include both capital expenditure and expenditure on working capital, such as stocks of raw material. Outlay tax *See*expenditure tax.

Out-of-favor industry or stock An unpopular industry or stock that usually has a low price-earnings ratio.

Out-of-pocket costs The incremental costs incurred as the result of a particular decision being pursued. For example, an organization with limited cash resources may make a decision to pursue an investment alternative that would not have been the first choice had it not offered the lowest level of out-ofpocket costs.

Out-of-the-money option A call option is "out of the money" if the strike price is greater than the market price of the underlying security. That is, you have the right to purchase a security at a price higher than the market price, which is not valuable. A put option is out of the money if the strike price is lower than the market price of the underlying security.

Outperform In general, this means to do better than some particular benchmark. Mutual Fund XYZ is said to outperform the S&P500 if its return exceeds the S&P500 return. However, this language does not take risk into account. That is, one might have a higher return than the benchmark in a particular year because of higher risk exposure. Outperform is also a term used by analysts to describe the prospects of a particular company. Usually, this means that the company will do better than its industry average.

Output tax The VAT that a trader adds to the price of the goods or services he supplies. He must account for this output tax to HM

Customs and Excise, having first deducted the input tax.

Outright rate Actual forward rate expressed in dollars per currency unit, or vice versa.

Outside broker A stockbroker who is not a member of a stock exchange but acts as an intermediary between the public and a stockbroker who is a member.

Outside director In the USA, a person who is a member of the board of directors of a company but is not an employee of the company and has no executive responsibilities. Such directors are appointed because they offer either wide business experience or specialist knowledge not available from the executive directors.

Outside of you Used for listed equity securities. Another order bidding for or offering stock at the same price that the trader has put on the floor himself, represented by another broker in the trading crowd. These orders may have different price limits (possible top or low on floor mentioned to floor broker but not announced in the crowd). See: Matching orders.

Outsourcing The buying in of components, sub-assemblies, finished products, and services from outside suppliers rather than by supplying them internally. A firm may decide to buy in rather than supply internally because it lacks the expertise, investment capital, or physical space required to do so. It may also be able to buy in more cheaply or more quickly than manufacturing inhouse. More recently this decision may be taken in the process of business process re-engineering or downsizing, when specific activities have been redefined as noncore activities. In certain instances, computer, legal, and personnel services have been outsourced as well as the more traditional manufactured components.

Outstanding Dividends Dividend checks which have been mailed to shareholders of record but not yet cashed. Funds are held until the check is paid, reissued or escheated to the state as abandoned property.

Outstanding share capital Issued share capital less the par value of shares that are held as the company's treasury stock.

Outstanding shares Shares that are currently owned by investors.

Outstanding shares Shares that are currently owned by investors.

Outwork Work carried out in a person's own home or workshop rather than on a company's own premises. Outworkers are still used in the upmarket end of the garment trade.

Overage Apples mainly to convertible securities. Difference between how much common stock one party must sell and the other wishes to buy for the same amount of convertible in a swap.

Overall market price coverage Total assets less intangibles divided by the total of the market value of the security issue and the book value of liabilities and issues having a prior claim. This is used to determine how much of the market value of a certain class of securities would be covered in liquidation.

Overbought Used in the context of general equities. Technically too high in price, and hence a technical correction is expected. See: Heavy. Antithesis of oversold.

Overbought-oversold indicator An indicator that attempts to define when prices have moved too far and too fast in either direction and thus are vulnerable to reaction.

Overcapitalization A condition in which an organization has too much capital for the needs of its business. If a business has more capital than it needs it is likely to be overburdened by interest charges or by the need to spread profits too thinly by way of dividends to shareholders. Businesses can now reduce overcapitalization by repaying long-term debts or by buying their own shares.

Overdraft A loan made to a customer with a cheque account at a bank or building society, in which the account is allowed to go into debit, usually up to a specified limit (the overdraft limit). Interest is charged on the daily debit balance. This is a less costly way of borrowing than taking a bank loan (providing the interest rates are the same) as, with an overdraft, credits are taken into account.

Overdraft checking account A checking account associated with a line of credit that allows a person to write checks for more that the actual balance in the account, with a finance charge on the overdraft.

Overdraft Provision of instant credit by a lending institution.

Over-entry certificate A document issued by HM Customs and Excise stating that imported goods have paid excessive duty on entry into the country, which may be reclaimed. If too little duty has been paid, past-entry duty is claimed.

Overfunded pension plan A pension plan that has a positive surplus (i.e., assets exceed liabilities).

Overhang Used in the context of general equities. Sizable block of securities or commodities contracts that, if released on the market, would put downward pressure on prices; prohibits buying activity that would otherwise translate into upward price movement. Examples include shares held in a dealer's inventory, a large institutional holding, a secondary distribution still in registration, and a large commodity position about to be liquidated.

Overhanging Bond A concertible bond issue that investors do not convert into common stock because the stock has not appreciated in value.

Overhead (overhead cost) The indirect costs of an organization, usually classified as manufacturing overhead, administration overheads, selling overhead, distribution overhead, and research and development costs.

Overhead An expense that cannot be attributed to any one single part of the company's activities.

Overhead The expenses of a business that are not attributable directly to the production or sale of goods.

Overhead variance A variance arising because overhead costs differ from those budgeted for.

Overheating The state of an economy during a boom, with increasing aggregate demand leading to rising prices rather than higher output. Overheating reflects the inability of some firms to increase output as fast as demand; they therefore choose to profit from the excess demand by raising prices.

Overinsurance The practice of insuring an item for a greater amount than its value. This is pointless as insurers are only obliged to pay the full value (usually the replacement value) of an insured item and no more, even if the sum insured exceeds this value. If insurers find a policyholder has overinsured an item, the premium for the cover above the true value is returned.

Overinvestment Excessive investment of capital, especially in the manufacturing industry towards the end of a boom as a result of over-optimistic expectations of future demand. When the boom begins to fade, the manufacturer is left with surplus capacity and can therefore make no further capital investments, which itself creates unemployment and fuels the imminent recession.

Overissue An excess of issued shares over authorized shares.

Overlap the market Used in the context of general equities. Create a crossed market by expressing a willingness to sell on the bid side of the market and buy on the offer side.

Overlapping debt The portion of debt of political subdivisions or neighboring special districts that a municipality is responsible for.

Overlay strategy A strategy of using futures for asset allocation by pension sponsors to avoid disrupting the activities of money managers.

Overnight delivery risk A risk brought about because differences in time zones between settlement centers require that payment or delivery on one side of a transaction be made without knowing until the next day whether the funds have been received in an account on the other side. Particularly apparent where delivery takes place in Europe for payment in dollars in New York.

Overnight loan A loan made by a bank to a bill broker to enable the broker to take up bills of exchange. Initially the loan will be repayable the following day but it is usually renewable. If it is not, the broker must turn to the lender of last resort, i.e. The Bank of England in the UK.

Overnight position A broker-dealer's position in a security at the end of a trading day.

Overproduction Theory of Business Cycles The theory ascribes the causes of cycles to excessive expansion of productive capacity whenever demand increases.

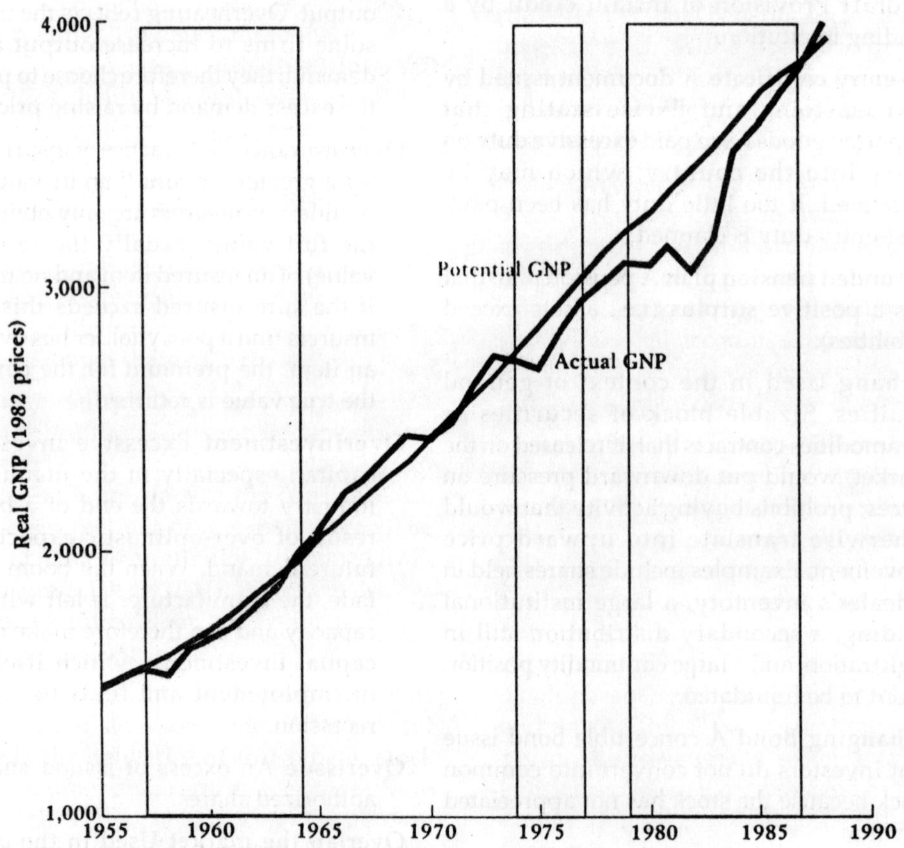

Fig. Periods of overproduction an underproduction

Overreaching Used in the context of general equities. Creating artificial volume in a stock through activity not generated by normal/ natural buyers and sellers in the .

Overreaction hypothesis The supposition that investors overreact to unanticipated news, resulting in exaggerated movements in stock prices followed by corrections.

Overrun In the context of project financing, the amount of capital expenditures or funding above the original estimate to complete the project.

Overseas company A company incorporated outside the UK that has a branch or a subsidiary company in the UK. Overseas companies with a place of business in the UK have to make a return to the Registrar of Companies, giving particulars of their memorandum of association, directors, and secretary as well as providing an annual balance sheet and profit-and-loss account.

Overseas investment Investment by the government, industry, or members of the public of a country in the industry of another country. For members of the public this is often most easily achieved by investing through foreign stock exchanges.

Overseas-income taxation Income that has been subject to taxation outside the jurisdiction of the UK tax authorities. When the same income is subject to taxation in more than one country, relief for the double tax is given either under the provisions of the double taxation agreement with the country concerned or unilaterally.

Overshooting A jump in the value of an asset followed by a slow adjustment to equilibrium, caused by a change in expectations. It is usually applied to analysis of the real exchange rates and has been used to explain the extreme volatility of exchange rates in the 1970s and 1980s (for example, between 1980 and 1984 the US dollar rose by around 70% against foreign currencies and between 1985 and 1987 fell by 40%). Overshooting provides an argument for more government intervention in the determination of exchange rates.

Overside delivery The unloading of a cargo over the side of a ship into lighters.

Oversold 1. Having sold more of a product or service than one can produce or purchase. **2.** Denoting a market that has fallen too fast as a result of excessive selling. It may therefore be expected to have an upward reaction.

Oversold Used in the context of general equities. Technically too low in price, and hence a technical correction is expected. Antithesis of overbought.

Oversubcription A situation that arises when there are more applications for a new issue of shares than there are shares available. In these circumstances, applications have to be scaled down according to a set of rules devised by the company issuing the shares or their advisors. Alternatively some companies prefer to allocate the shares by ballot (*see*allotment). Oversubscription usually occurs because of the difficulty in arriving at an issue price that will be low enough to attract sufficient investors to take up the whole issue and yet will give the company the capital. Speculative purchases by stags also make it difficult to price a new issue so that it is neither oversubscribed or undersubscribed. In the case of undersubscription, which is rare, the underwriter has to take up that part of the issue that has not been bought by the public. Undersubscription can occur if some unexpected event occurs after the announcement of the issue price but before the issue date.

Oversubscribed issue Investors are not able to buy all of the shares or bonds they want, so underwriters must allocate the shares or bonds among investors. This occurs when a new issue is underpriced or in great demand because of growth prospects.

Oversubscription The excess number of shares or bonds that investors want to buy but are not available due to high demand.

Over-the-counter market (OTC) A decentralized market (as opposed to an exchange market) where geographically dispersed dealers are linked together by telephones and computer screens. The market is for securities not listed on a stock or bond exchange.

Over-the-Counter Option An option traded off-exchange, as opposed to a listed stock option. The OTC option has a direct link between buyer and seller, has no secondary market, and has no standardization of striking prices and expiration dates. See also Secondary Market.

Overtime ban A form of industrial action in which employees refuse to work overtime, thus causing considerable dislocation to normal working but clearly less than would occur in a total strike. Overtime bans are usually supported and often organized by unions.

Overtime Hours worked in excess of an agreed number per week or per day. The payment made for overtime work is usually higher than the basic rate of pay. Unions therefore have an interest in making the basic working week as short as possible, so that those working longer hours earn as much as possible. Employers, on the other hand, want to make the working week as long as possible.

Overtrading Excessive broker trading in a discretionary account. Underwriters persuade brokerage clients to purchase some part of a new issue in return for the purchase by the underwriter of other securities from the clients at a premium. This premium is offset by the underwriting spread.

Overvalued A stock price that is seen as too high according to the company's price-earnings ratio, expected earnings, or financial condition.

Overweight Usually refers to recommendation that leads an investor to increase their investment in a particular security or asset class. The increase is usually with respect to a benchmark. Suppose that U.S. equities compose 40% of the benchmark portfolio. If one thinks the U.S. will outperform, the investor may increase the exposure to U.S. equity to more than 40%.

Overwithholding Deducting and paying too much tax that may be refunded to the taxpayer or applied against the next period's obligation.

Own foreign offices U.S. reporting institutions' parent organizations, branches, and/or majority owned subsidiaries located outside the United States.

Own shares purchase The purchase or redemption of its own shares by a company; this is permitted subject to certain conditions set out in the *Companies Act*. For example, redeemable shares may only be redeemed if they are fully paid. If the redemption or purchase of a company's own shares would lead to a reduction of its capital, a capital redemption reserve will need to be created. In certain cases private companies may reduce their capital in this way .

Owners' equity The funds of an organization that have been provided by its owners, i.e. Its total assets less its total liabilities. The balance-sheet value of the owners' equity is unlikely to be equal to its market value.

Ownership Rights over property, including rights of possession, exclusive enjoyment, destruction, etc. In UK common law, land cannot be owned outright, as all land belongs to the Crown and is held in tenure by the 'owner'. However, an owner of an estate in land in fee simple is to all intents and purposes an outright owner. In general, ownership can be split between different persons. For example, a trustee has the legal ownership of trust property but the beneficiary has the equitable or beneficial ownership. If goods are stolen, the owner still has ownership but not possession. Similarly, if goods are hired or pledged to someone, the owner has ownership but no immediate right to possession.

Ownership-specific advantages Property rights or intangible assets, including patents, trademarks, organizational and marketing expertise, production technology, and management and general organizational abilities, that form the basis for a company's advantage over other firms.

P

P/E ratio Assume XYZ Co. sells for Rs.25.50 per share and has earned Rs.2.55 per share this year; Rs.25. 50 = 10 times Rs.2. 55 XYZ stock sells for 10 times earnings. P/E = Current stock price divided by trailing annual earnings per share *or* expected annual earnings per share.

PA 1. Abbreviation for personal account, used to denote a transaction made by a professional investment adviser for his own account rather than for the firm for which he works.

2. Abbreviation for personal assistant.

3. Abbreviation for power of attorney.

4. Abbreviation for particular average.

Pa'anga The standard monetary unit of Tonga, of divided into 100 seniti.

Package A set of computer programs designed to be sold to a number of users. They are used in computerized accounting systems for purchase and sales ledgers, stock control, payroll records, etc. Buying a software package from firms specializing in them saves users a considerable amount of time and money in developing their own.

Package deal An agreement that encompasses several different parts, all of which must be accepted. A package deal may have involved either or both parties in making concessions on specific aspects of the package in order to arrive at a compromise arrangement.

Package mortgage A mortgage on a house and property in the house.

Packaging 1. The design of wrappers or containers for a product.

2. The wrappers or containers themselves. 3. An activity that combines several operations and enables the packager to deliver a finished or near-finished product to a selling organization. For example, many non-fiction books are produced by book packagers, who write, illustrate, typeset, and print books, which they deliver to publishers, who sell them to the public under their own imprint.

Pac-Man strategy Takeover defense strategy in which the prospective acquiree retaliates against the acquirer's tender offer by launching its own tender offer for the other firm.

Paid-in capital Capital received from investors in exchange for stock, but not stock from capital generated from earnings or donated. This account includes capital stock and contributions of stockholders credited to accounts other than capital stock. It would also include surplus resulting from recapitalization.

Paid-up policy An endowment assurance policy in which the assured has decided to stop paying premiums before the end of the policy term. This results in a surrender value, which instead of being returned in cash to the assured is used to purchase a single-premium whole (of) life policy. In this way the life assurance protection continues (for a reduced amount), while the policyholder is relieved of the need to pay further premiums. If the original policy was a with-profits policy, the bonuses paid up to the time the premiums ceased would be included in the surrender value. If it is a unit-linked policy, capital units actually allocated would be allowed to appreciate to the end of the term.

Paid-up share capital (fully paid capital) The total amount of money that the shareholders of a company have paid to the company for their fully paid shares. *See*share capital.

Paired comparisons A technique used in marketing research in which consumers are presented with pairs of competing products and asked to choose the one they prefer. This technique can be used to compare, say, a number of brands of soap, giving respondents two at a time to compare. The number of times each brand is selected as a preference in a large number of such tests will reveal an order of brand preference. *Compare***monadic testing**.

Paired off Used for listed equity securities. Matched buy and sell market orders, usually pertaining to the pre-opening market picture in a stock, or MOC orders (especially relating to futures/options expirations).

Paired shares Stock of two companies under the same management that are sold as one unit with one certificate.

Pairoff A buyback to offset and effectively liquidate a prior sale of securities.

Paisa 1. A monetary unit of India, Nepal, and Pakistan, worth one hundredth of a rupee. 2. A monetary unit of Bangladesh, worth one hundredth of a taka.

Paper profit A profit shown by the books or accounts of an organization, which may not be a realized profit because the value of an asset has fallen below its book value, because the asset, although nominally showing a profit, has not actually been sold, or because some technicality of book-keeping might show an activity to be profitable when it is not. For example, a share that has risen in value since its purchase might show a paper profit but this would not be a real profit since the value of the share might fall again before it is sold.

Par banking The US practice of one bank paying the full face value of a cheque drawn on another bank, without imposing a charge for encashment. Members of the Federal Reserve System are obliged to comply with the rules of par banking.

Par bond A security or financial instrument that is bought and sold at its face value, rather than at a discount or premium.

Par Equal to the nominal or face value of a security. A bond selling at par is worth an amount equivalent to its original issue value or its value upon redemption at maturity.

Par of exchange The theoretical rate of exchange between two currencies in which there is equilibrium between the supply and demand for each currency. The par value lies between the market buying and selling rates.

Para A Yugoslavian monetary unit, worth one hundredth of a dinar.

Parallel bonds Fixed income instruments denominated in the respective currencies of the countries where they are placed.

Parallel loan A process whereby two companies in different countries borrow each other's currency for a specific period of time, and repay the other's currency at an agreed maturity for the purpose of reducing foreign exchange risk. Also referred to as back-to-back loans.

Parallel processing A method of computing in

which two or more parts of a program are executed simultaneously rather than sequentially. Strictly, parallel processing is only possible on computers with more than one arithmetic and logical unit in the central processing unit, but it is often simulated on other machines by such techniques as time sharing.

Parallel shift in the yield curve A shift in economic conditions in which the change in the interest rate on all maturities is the same number of basis points. In other words, if the three month T-bill increases 100 basis points (one %), then the 6-month, 1-year, 5-year, 10-year, 20-year, and 30-year rates all increase by 100 basis points as well. Related: Non-parallel shift in the yield curve.

Parameter A representation that characterizes a part of a model (e.g. a growth rate), the value of which is determined outside of the model. **See:** exogenous variable.

Parent company A company that controls subsidiaries through its ownership of voting stock, as well as runs its own business.

Parenting The way in which the directors of the main holding company of a large group manage their relationships with the subsidiary companies or business units. The method by which control is exercised varies, with some firms concentrating on financial control and others on the broader strategic targets. *See* strategic styles.

Pareto's Rule (80/20 rule) A rule originally applied by the economist Vilfredo Pareto (1848-1923) to income distribution, i.e. 80% of a nation's income will only benefit 20% of the population. It can be widely extended; for example: 80% of value is locked up in 20% of inventory; 80% of car breakdowns can be fixed by 20% of the spare parts range; 80% of a company's profits will come from 20% of its product range; and 80% of system failures will be caused by 20% of possible causes. The rule can be used as a guide to managers: in deciding how to spend. Limited funds, priority should be given to the most significant 20% of problems. Once they are eliminated the 80/20 rule will again apply to the remainder.

Pari passu (Latin: with equal step) Ranking equally. When a new issue of shares is said to rank pari passu with existing shares, the new shares carry the same dividend rights and winding-up rights as the existing shares. A pari passu bank loan is a new loan that ranks on level par with older loans.

Parity For convertibles, level at which a convertible security's market price equals the aggregate value of the underlying common stock; value/worth of the convertible bond considered only as an equity instrument (Conversion ratio times common price). See: Conversion value. For international parity, U.S.Rs. price of a foreign stock's last sale in an overseas market (Local currency stock price times forex rate times ADR ratio). For listed parity, condition whereby no party has floor priority, and matching thus occurs. For options parity, dollar amount by which an option is in the money. See: Intrinsic value.

Parking Putting money into safe investments such as money market investments while deciding where to invest the money.

Parkinson's laws The propositions made by the British writer Cyril Northcote Parkinson (1909-93) in his book *Parkinson's Law*. Facetious but true, they include such aphorisms as: work expands to fill the time available in which to do it; expenditure rises to meet income; and subordinates multiply at a fixed rate that is independent of the amount of work produced. They were devised with large organizations in mind and it is to such organizations that they apply, unless managers take steps to ensure that they do not.

Partial compensation Incomplete payment for the delivery of goods to one party by buying back a certain amount of product from the same party.

Partial Used in the context of general equities. Trade whose size is only part of the total customer indication/order, usually made to avoid a compromise in price and also to get some business instead of losing the customers inquiry/order to a competitor.

Partial Vote When only a portion of the total shares in an account is voted. For example, a broker has 1,000 shares and sends out a card to each of four shareholder clients. If only three of the four client cards are returned to the broker, the broker will submit only 3/4ths(750 shares) of the total 1,000 shares to vote. If the fourth card arrives later, an additional vote can be counted.

Participant A party of a funding. It usually refers to the lowest rank or smallest level of funding.

Participant risk The risk associated with the credit of the participants and possibility of non-performance.

Participate but do not initiate Used for listed equity securities. "Participate in the side of the market indicated by the order, but do not initiate the interest that causes the trade to take place." This kind of order can cause one to "miss stock" because the broker or investor is at the mercy of the player who does initiate the trade. See: Market order go along, percentage order.

Participating buyer/seller Used for listed equity securities. (1) Customer willing to buy/sell in line with market. (2) Buyer/seller who goes along with another buyer/seller in a percentage order.

Participating convertible preferred stock Preferred stock that can be converted into common stock at the option of the holder. In contrast, to the usual preferred stock, the value of the preferred stock is refunded to the holder. That is, one gets conversion plus the value of the stock.

Participating dividend Dividend received from ownership of participating preferred stock.

Participating fees The portion of total fees in a syndicated credit that go to the participating banks.

Participating GIC A guaranteed investment contract whose policyholder is not guaranteed a crediting rate, but instead receives a return based on the actual experience of the portfolio managed by the life insurance company.

Participating life insurance policies Life insurance that pays to policyholders depending on the company's success as provided by few claims and profitable underwritings and investments.

Participating preference shares carry additional rights to dividends, such as a further share in the profits of the company, after the ordinary shareholders have received a stated percentage. *See also***preferred ordinary share.**

Participation certificates (PC) Used in the context of general equities. Investments representing an interest in a pool of funds or in other instruments, such as foreign securities, that allow participation in the rise or fall of a security or group of securities.

Participator Any person having an interest in the capital or income of a company, e.g. A shareholder, loan creditor, or any person entitled to participate in the distributions of the company.

Partner Business associate who shares equity in a firm.

Partnership agreement A written agreement among partners detailing the terms and conditions of participation in a business ownership arrangement.

Partnership Shared ownership among two or more individuals, some of whom may, but do not necessarily, have limited liability. See: general partnership, limited partnership, and master limited partnership.

Pass the book The process of transferring responsibility for a brokerage firm's trading account from one office to another around the world in order to benefit from trading 24 hours a day.

Passing a name The disclosure by a firm of brokers of the name of the principal for whom they are acting. In some commodity trades, if brokers disclose the names of their buyers to the sellers, they do not guarantee the buyers' solvency, although they may do so in some circumstances. However, if brokers do not pass their principals' names, it is usual for them to guarantee their solvency. Thus, to remain anonymous, a buyer may have to pay an additional brokerage.

Passing off Conducting a business in a manner that misleads the public into thinking that one's goods or services are those of another business. The commonest form of passing off is marketing goods with a design, packaging, or trade name that is very similar to that of someone else's goods. It is not necessary to prove an intention to deceive; innocent passing off is actionable.

Passive income Income (such as investment income) that does not come from active participation in a business. Specified by the U.S. tax code.

Passive investing Putting money into a profitable business opportunity that is deemed passive by the IRS and thus benefits from tax deductions.

Passive investment management Buying a well diversified portfolio to represent a broad-based market index without attempting to search out mispriced securities.

Passive portfolio strategy A strategy that involves minimal expectational input, and instead relies on diversification to match the performance of some market index. A passive strategy assumes that the marketplace will reflect all available information in the price paid for securities, and therefore, does not attempt to find mispriced securities. Related: active portfolio strategy

Pass-through coupon rate The interest rate paid on a securitized pool of assets, which is less than the rate paid on the underlying loans by an amount equal to the servicing and guaranteeing fees.

Pass-through coupon rate The interest rate paid on a securitized pool of assets, which is less than the rate paid on the underlying loans by an amount equal to the servicing and guaranteeing fees.

Pass-through rate The net interest rate passed through to investors after deducting servicing, management, and guarantee fees from the gross mortgage coupon.

Pass-through securities A pool of fixed income securities backed by a package of assets (i.e., mortgages) where the holder receives the principal and interest payments. Related: Mortgage pass-through security

Pataca The standard monetary unit of Macao, divided into 100 avos.

Patent agent A member of the Chartered Institute of Patent Agents, who gives advice on obtaining patents and prepares patent applications.

Patent The grant of an exclusive right to exploit an invention. An applicant for a patent (usually from the inventor or the inventor's employer) must show that the invention is new, is not obvious, and is capable of industrial application. An expert known as a patent agent often prepares the application, which must describe the invention in considerable detail. The Patent Office publishes these details if it grants a patent. A patent remains valid for 20 years from the date of application (the priority date) provided that the person to whom it has been granted (the patentee) continues to pay the appropriate fees. During this time, the patentee may assign the patent or grant licences to use it. Such transactions are registered in a public register at the Patent

Office. If anyone infringes the monopoly, the patentee may sue for an injunction and damages or an account of profits

Path dependent option An option whose value depends on the sequence of prices of the underlying asset rather than just the final price of the asset.

Pathfinder prospectus An outline prospectus concerning the flotation of a new company in the UK; it includes enough details to test the market reaction to the new company but not its main financial details or the price of its shares. Pathfinder prospectuses are known in the USA as red herrings.

Pattern A technical chart formation used to make market predictions by following the price movements of securities.

Pawnbroker A person who lends money against the security of valuable goods used as collateral. Borrowers can reclaim their goods by repaying the loan and interest within a stated period. However, if the borrower defaults the pawnbroker is free to sell the goods. The operation of pawnbrokers is governed by the *Consumer Credit Act (1974)*.

Pay and file A procedure for companies introduced in the UK for accounting periods ended after 30 September 1993. Corporation tax is due nine months after the end of the accounting period. The company must file a detailed return within twelve months of the end of the accounting period. Under this system there is more responsibility placed on the company to finalize the accounts and accompanying tax computations in order to pay the correct amount of tax by the due date. Previously, estimated assessments were issued by the Inland Revenue if the finalized computations were not available. These enabled companies to defer payment, for a limited period, without incurring interest charges. The new pay and file system means that it is expensive for companies to be late in finalizing their corporate tax position.

Payable date The date when dividends or capital gains are paid to shareholders or reinvested in additional shares.

Payable through drafts A method of making payment that is used to maintain control over payments made on behalf of the firm by personnel in noncentral locations. The payer's bank delivers the payable through draft to the payer, which must approve it and return it to the bank before payment can be received.

Payable to bearer Describing a bill of exchange in which neither the payee or endorsee are named. A holder, by adding his name, can make the bill payable to order.

Pay-as-you-go basis A method of paying income tax in which the employer deducts a portion of an employee's monthly salary to remit to the IRS.

Payback period method A method of capital budgeting in which the time required before the projected cash inflows for a project equal the investment expenditure is calculated; this time is compared to a required payback period to determine whether or not the project should be considered for approval. If the projected cash inflows are constant annual sums, after an initial capital investment the following formula may be used: Payback (years) = initial capital Investment/annual cash inflow. Otherwise, the annual cash inflows are accumulated and the year determined when the cumulative inflows equal the investment expenditure.

Payback period The period required for a project to repay its initial outlay. This is usually projected on the assumption that any further recoveries are then pure profit. This is a relatively unsophisticated method of appraising a capital project (*compare*discounted cash flow; net present value), although it is frequently used in industry.

Payback The length of time it takes to recover the initial cost of a project, without regard to the time value of money.

Pay-down In a Treasury refunding, the amount by which the par value of the securities maturing exceeds that of those sold. In the context of general equities, paying a lower price in an accumulation of stock. Antithesis of pay-up.

Paydown In a Treasury refunding, the amount by which the par value of the securities maturing exceeds that of those sold.

Paye Pay-as-you-earn. This means of collecting tax arose under Schedule E of the UK income-tax legislation on wages and salaries. Because it is often difficult to collect tax at the end of the year from wage and salary earners, the onus is placed on employers to collect the tax from their employees as payments are made to them. There is an elaborate system of administration to ensure that broadly the correct amount of tax is deducted week by week or month by month and that the employer remits the tax collected to the Inland Revenue very quickly. Although technically called payas-you-earn, the system would be better called pay-as-you-get-paid.

Payee A person or organization to be paid. In the case of a cheque payment, the payee is the person or organization to whom the cheque is made payable.

Payer The person making a payment to a payee.

Paying agent An agent who makes principal and interest payments to bondholders on behalf of the issuer.

Paying banker The bank on which a bill of exchange (including a cheque) has been drawn and which is responsible for paying it if it is correctly drawn and correctly endorsed (if necessary).

Paying-in book A book of slips used to pay cash, cheques, etc., into a bank account. The counterfoil of the slip is stamped by the bank if the money is paid in over the counter.

Payment date The date on which each shareholder of record will be sent a check for the declared dividend.

Payment float Company-written checks that have not yet cleared.

Payment in advance (prepayment) Payment for goods or services before they have been received. In company accounts, this often refers to rates or rents paid for periods that carry over into the next accounting period.

Payment in due course The payment of a bill of exchange when it matures (becomes due).

Payment terms The agreed way in which a buyer pays the seller for goods. The commonest are cash with order or cash on delivery; prompt cash (i.e. Within 14 days of delivery); cash in 30, 60, or 90 days from date of invoice; letter of credit; cash against documents; documents against acceptance; or acceptance credit.

Payment-In-Kind (PIK) bond A bond that gives the issuer an option (during an initial period) either to make coupon payments in cash or in the form of additional bonds.

Payments pattern Describes the collection pattern of receivables. The pattern might describe the probability that a 72-day-old account will still be unpaid when it is 73 days-old.

Payments System Collective term for mechanisms (both paper-backed and electronic) for moving funds, payments and money among financial institutions throughout the nation. The Federal Reserve plays a major role in the nation's payments system through distribution of currency and coin, processing of checks, electronic transfer of funds and the operation of automated clearinghouses that transfer funds electronically among depository intitutions; various private organizations also perform payments system functions.

Payoff profile The slope of a line graphed according to the value of an underlying asset on the x-axis and the value of a position taken to hedge against risk exposure on the y-axis. Also used with changes in value. See: Risk profile.

Payout period The time period during which withdrawals from a retirement account or annuity are paid.

Payout ratio Generally, the proportion of earnings paid out to the common stockholders as cash dividends. More specifically, the firm's cash dividend divided by the firm's earnings in the same reporting period.

Payout ratio Generally, the proportion of earnings paid out to the common stockholders as dividends. Morespecifically, the firm's cash dividend divided by the firm's earnings in the same reporting period.

Payroll tax A tax based on the total of an organization's payroll. The main function of such a tax would be to discourage high wages or over-employment. The current employers' Class 1 National Insurance contributions constitute such a tax.

Pay-to-play Attempts by municipal bond underwriting businesses to gain influence with political officials who decide which underwriters are awarded the municipality's business.

Pay-up The loss of cash resulting from a swap into higher price bonds or the need/willingness of a bank or other borrower to pay a higher rate of interest to get funds.

Pecking-order view (of capital structure) The argument that external financing transaction costs, especially those associated with the problem of adverse selection, create a dynamic environment in which firms have a preference, or pecking-order of preferred sources of financing, when all else is equal. Internally generated funds are the most preferred, new debt is next, debt-equity hybrids are next, and new equity is the least preferred source.

Pegged exchange rate Exchange rate whose value is pegged to another currency's value or to a unit of account.

Pegging Making transactions in a security, currency, or commodity in order to stabilize or target its value through market intervention.

Penalty An arbitrary pre-arranged sum that becomes payable if one party breaches a contract or undertaking. It is usually expressly stated in a penalty clause of the contract. Unlike liquidated damages, a penalty will be disregarded by the courts and treated as being void. Liquidated damages will generally be treated as a penalty if the amount payable is extravagant and unconscionable compared with the maximum loss that could result from the breach. However, use of the words 'penalty' or 'liquidated damages' is inconclusive as the legal position depends on the interpretation by the courts of the clause in which they appear.

Penalty tax A federal tax that can be applied if a plan holder does not meet certain requirements when making withdrawals from a tax-advantaged retirement plan (for instance, if the plan holder has not reached age 59-1/2). This penalty tax is owed in addition to any income taxes due.

Penetration strategy A marketing strategy based on low prices and extensive advertising to increase a product's market share. For penetration strategy to be effective the market will have to be large enough for the seller to be able to sustain low profit margins.

Pension liabilities Future liabilities resulting from pension commitments made by a corporation. Accounting for pension liabilities varies widely by country.

Pension parachute A form of poison pill providing that in the event of a hostile takeover attempt, any excess pension plan assets can be used to benefit pension plan participants. This prevents the raiding firm from using the pension assets to finance the takeover. In the context of corporate governance, these provisions prevent an acquirer from using surplus cash in the pension fund of the target in order to finance an acquisition. Surplus funds are required to remain the property of the pension fund and to be used for plan participants' benefits. replacement of it with a life insurance company-sponsored fixed annuity plan.

Pensionable earnings The part of an employee's salary that is used to calculate the final pension entitlement. Unless otherwise stated, overtime, commission, and bonuses are normally excluded.

Pensioneer trustee A person authorized by the Superannuation Funds Office of the Inland Revenue to oversee the management of a pension fund in accordance with the provisions of the Pension Trust Deed.

PEP mortgage A mortgage in which the borrower repays only the interest on the loan to the lender, but at the same time puts regular sums into a personal equity plan (PEP). When the peps mature they are used to repay the capital. The PEP mortgage is similar to an endowment mortgage, except that the PEP mortgage does not provide any life assurance cover. The advantages are that PEP funds are untaxed, unlike endowments, and that the borrower has the flexibility of being able to change the PEP provider each year.

Per capita debt The total bonded debt of a municipality divided by the population of the municipality.

Per mille (per mill; per mil) (Latin: per thousand) Denoting that the premium on an insurance policy is the stated figure per £1000 of insured value. Per mille is also used to mean 0.1% in relation to interest rates.

Percent to double Percentage that the stock price has to rise (fall) to double the price of the call (put).

Percentage order Used for listed equity securities. Market limited price order to buy/sell a specified percentage (usually 50%) of shares traded (sometimes after a fixed number of shares of the stock have already traded). See: participating buyer/seller, "Participate but do not initiate."

Percentage premium Applies mainly to convertible securities. Premium over parity of a convertible bond divided by parity.

Perfect competition An idealized market environment in which every market participant is too small to affect the market price by acting on its own.

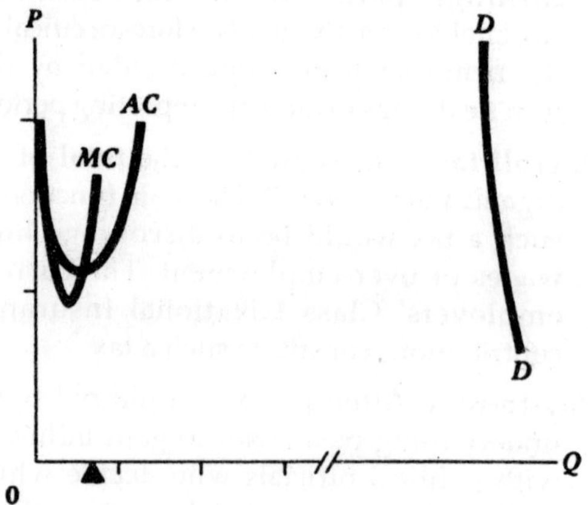

Fig. Perfect Competition

Perfect hedge A financial result in which the profit and loss from the underlying asset and the hedge position are equal.

Perfect market assumptions Conditions under which the law of one price holds. The assumptions include frictionless markets, rational investors, and equal access to market prices and information.

Perfect market view (of dividend policy) Analysis of a decision on dividend policy, in a perfect capital market environment, that shows the irrelevance of dividend policy.

Perfected first lien A first attachment on an asset that is duly recorded with the relevant government body so that the lender will be able to act on it should the borrower default.

Perfectly competitive financial markets Markets in which no trader has the power to change the price of goods or services.

Perfect capital markets are characterized by the following conditions: 1) trading is costless, and access to the financial markets is free, 2) information about borrowing and lending opportunities is freely available, 3) there are many traders, and no single trader can have a significant impact on market prices.

Performance attribution analysis The decomposition of a money manager's performance results to explain the reasons why those results were achieved. This analysis seeks to answer the following questions:

(1) What were the major sources of added value?

(2) Was short-term factor timing statistically significant?

(3) Was market timing statistically significant? And

(4), Was security selection statistically significant?

Performance evaluation The assessment of a manager's results, which involves, first, determining whether the money manager added value by outperforming the established benchmark (performance measurement) and, second, determining how the money manager achieved the calculated return (performance attribution analysis).

Performance evaluation The evaluation of a manager's performance which involves, first, determining whether the money manager added value by outperforming the established benchmark (performance measurement) and, second, determining how the money manager achieved the calculated return (performance attribution analysis).

Performance index A risk-adjusted measure of how well a portfolio has performed.

Performance measures Ways in which the performance of whole organizations or parts of organizations, such as profit centres, cost centres, divisions, departments, and sections, and the managers responsible for these parts of the business, can be measured. Performance measures, which may be in the form of quantitative, qualitative, or financial measures, include measures based on profitability or comparison with budgets and standard-cost-based measures as well as the figures for previous periods.

Performance shares Shares of stock given to managers on the basis of performance as measured by earnings per share and similar criteria. A control device used by shareholders to tie management to the self-interest of shareholders.

Performance standard A standard, used in standard costing, of the level of performance to be achieved during a period. For example, a standard performance for direct labour of two standard hours to complete a task would be combined with the rate per standard hour for labour to create the standard direct labour cost for the task.

Performance stock High-growth stock in a company that retains earnings for further growth and therefore pays no dividends, but that an investor feels has significant future potential.

Period of digestion The time period of often high volatility after a new issue is released when the trading price of the security is established by the market.

Period-certain annuity An annuity that provides guaranteed payments to an annuitant for a specified period of time.

Periodic payment plan Accumulation of capital in a mutual fund by making regular payments on a monthly or quarterly basis.

Periodic rate The monthly effective interest rate. For example, the periodic rate on a credit card with an 18% annual percentage rate is 1.5% per month.

Permanent Current Assets The minimum level

of current assets that a firm needs to continue operation. Because some level is always maintained, they are called permanent current assets.

Permanent establishment A fixed place of business that a trader of one country has in another country, thus rendering him liable to that second country's taxation. It will include a place of management, branch, office, factory or workshop, and long-term construction projects but it may exclude some sales offices and storage depots. The concept is important in double-taxation treaties.

Permanent interest bearing share (PIBS) A fixed-interest non-redeemable security that pays interest at a rate fixed at issue. This is often 10-13.5%, giving investors a high yield for perpetuity. However, these shares carry the risks associated with fixed-interest securities, being the last to be paid out should an issuing building society go into liquidation

Permissible capital payment (PCP) A payment made out of capital when a company is redeeming or purchasing its own shares and has used all available distributable profits as well as the proceeds of any new issue of shares. *See*own shares purchase.

Perpetual annuity The receipt or payment of a constant annual amount in perpetuity. Although the word annuity refers to an annual sum, in practice the constant sum may be for periods of less than a year. The present value of an annuity is obtained from the formula: $P = a \times 100/i$, Where P = present value, a = annual sum, and i = interest rate.

Perpetual inventory A method of continuous stock control in which an account is kept for each item of stock; one side of the account records the deliveries of that type of stock and the other side records the issues from the stock. Thus, the balance of the account at any time provides a record of either the number of items in stock or their values, or both. This method is used in large organizations in which it is important to control the amount of capital tied up in the running of the business. It also provides a means of checking pilferage. Less sophisticated organizations rely on annual stocktaking to discover how much stock they have.

Perpetual succession The continued existence of a corporation until it is legally dissolved. A corporation, being a separate legal person, is unaffected by the death or other departure of any member but continues in existence no matter how many changes in membership occur.

Perpetuity A constant stream of identical cash flows without end, such as a British consol.

Perquisites Personal benefits, including direct benefits, such as the use of a firm car or expense account for personal business, and indirect benefits, such as up-to-date office decoration.

Personal account An account in a ledger that bears the name of an individual or of an organization; it records the state of indebtedness of the named person to the organization keeping the account or vice versa. Personal accounts are normally kept in the sales (or total debtors) ledger and the purchases (or total creditors) ledger.

Personal allowances Sums deductible from taxable income under an income-tax system to allow for personal circumstances. The principal allowances in the UK system are a personal allowance to which all taxpayers are entitled and a married couple's allowance, both of which are age-related. Additional personal allowances include a single parent's allowance, an age allowance, and an additional personal allowance. The personal allowance, to which everyone is entitled, is £3765. For those aged between 65 and 74 this is increased to £4910, and for those aged 75 and over it is £5090.

Personal article floater Insurance policy

attachment designed to cover specified personal valuables.

Personal assistant (PA) A person who is appointed to help a manager or director and who has wider responsibilities than a secretary.

Personal exemption Amount of money a taxpayer can exclude from personal income for each member of the household in calculation of a tax obligation.

Personal financial planning Financial planning for individuals, which involves analysing their current financial position, predicting their short-term and long-term needs, and recommending a financial strategy. This may involve advice on pensions, the provision of independent school fees, mortgages, life assurance, and investments.

Personal identification number (PIN) A number memorized by the holder of a cash card, credit card, or multifunctional card and used in automated teller machines and electronic funds transfer at point of sale to identify the card owner. The number is given to the cardholder in secret and is memorized so that if the card is stolen it cannot be used. The number is unique to the cardholder. Banks will not honour phantom withdrawals if it can be shown that the PIN number was known to a party other than the customer to whom it was allocated.

Personal income Total income received from all sources, including wages, salaries, or rents, and the like.

Personal inflation rate The inflation rate as it affects a specific individual.

Personal ledger A ledger containing personal accounts, for example the debtors' ledger and the creditors' ledger.

Personal loan A loan to a private person by a bank or building society for domestic purposes, buying a car, etc. There is usually no security required and consequently a high rate of interest is charged. Repayment is usually by monthly instalments over a fixed period. This is a more expensive way of borrowing from a bank than by means of an overdraft.

Personal property (personalty) Any property other than real property (realty). This distinction is especially used in distinguishing property for inheritance tax. Personal property includes money, shares, chattels, etc.

Personal representative A person whose duty is to gather in the assets of the estate of a deceased person, to pay his liabilities, and to distribute the residue. Personal representatives of a person dying testate are known as executors; those of a person dying intestate are known as his administrators.

Personal selling Person-to-person interaction between a buyer and a seller in which the seller's purpose is to persuade the buyer of the merits of the product, to convince the buyer of his or her need for it, and to develop with the buyer an ongoing customer relationship.

Personal tax view (of capital structure) The argument that the difference in personal tax rates between income from debt and income from equity eliminates the disadvantage of the double taxation (corporate and personal) of income from equity.

Personal trust An interest in an asset held by a trustee for the benefit of another person.

Personnel selection The method of choosing the most suitable applicant for a vacancy within an organization. Ideally, a job description is written, detailing the essential qualifications and experience required for the job together with any other relevant information. The job description enables a job specification to be created, showing how the vacancy would fit in the structure of the organization, the salary level, and any fringe benefits. A search for the right person then begins, using advertisements in newspapers, trade papers, or employment agencies; if the vacancy is sufficiently senior a head hunter may be employed (*see*employment agency). Initial selection is then made by inviting letters of application (or using application forms) supported by a curriculum vitae and listing the applicants that most closely match the job specification. Screening interviews may then be held, which may include practical tests so that a short list of final candidates can be drawn up. Second interviews are sometimes required and, when appropriate, psychological tests are used. The number of interviews and the extent of testing depend on the level of the position within the organization. Interviews may be conducted by a single person or by a panel; applicants may be seen singly or in groups. Moreover the extent to which the personnel department is involved in the selection will depend on the particular organization.

Peseta The standard monetary unit of Spain and Andorra, divided into 100 céntimos.

Pesewa A monetary unit of Ghana, worth one hundredth of a cedi.

Peso 1. The standard monetary unit of Argentina (Rs.), Chile (ChRs.), Colombia (ColRs.), Cuba (CUP), the Dominican Republic (RDRs.), Guinea-Bissau (PG), Mexico (MexRs.), and the Philippines; it is divided into 100 centavos.

2. (nurRs.) The standard monetary unit of Uruguay, divided into 100 centésimos.

Petrodollars Deposits by countries that receive dollar revenues from the sale of petroleum to other countries; the term commonly refers to OPEC deposits of dollars in the Eurocurrency market.

Petty cash The amount of cash that an organization keeps in notes or coins on its premises to pay small items of expense. This is to be distinguished from cash, which normally refers to amounts held at banks. Petty-cash transactions are normally recorded in a petty-cash book, the balance of which should agree with the amounts of petty cash held at any given time.

Phantom stock plan An incentive scheme that awards management bonuses based on increases in the market price of the company's stock.

Phantom withdrawals The removal of funds from bank accounts through automated teller machines (atms) by unauthorized means and without the knowledge or consent of the account holder. Banks maintain that such withdrawals are not possible without the collusion, intended or unintended, of account holders by divulging their personal identification numbers (pins) to a third party or lending their ATM card to someone else. Many disputes have arisen over so-called phantom withdrawals but no case has so far been proved.

Phase space A graph which shows all possible states of a system. In phase space we plot the value of a variable against possible values of the other variables at the same time. If a system had three descriptive variables, we plot the phase space in three dimensions, with each variable taking one dimension.

Physical asset Actual property such as precious metals or real estate. Also called real or tangible assets.

Physical capital Items such as plant and machinery, buildings, and land that can be used to produce goods and services. It is compared to financial capital, i.e. Money, and human capital.

Physical completion The state in which a project is physically functioning, but not yet fully generating cash flow.

Physical distribution management The management of the link between the producer of a product and the immediate customer. *See*logistics.

Physical distribution The tasks, involved in

planning, implementing, and controlling the flow of materials and final goods from their points of origin to their final destinations to meet the requirements of customers at a profit.

Physical option An option whose underlying security is a physical commodity that is not stock or futures. The physical commodity itself (a currency, treasury debt issue, commodity) - underlies that option contract. See also index option.

Physical price The price of a commodity that is available for delivery. In most forward contracts (*see*forward dealing) and futures contracts the goods are never actually delivered because sales and purchases are set off against each other.

Physical verification A procedure auditors use to ensure that inventory recorded in the book is correct by actually checking out the physical inventory.

Piastre A monetary unit of Egypt, Lebanon, the Sudan, and Syria, worth one hundredth of a pound.

Pickup bond A bond with a relatively high coupon that is close to the date at which it is callable, meaning that a fall in interest rates will most likely cause early redemption of the bond at a premium.

Pictogram A diagram providing quantitative information, using stylized drawings of people and things, in which the number of such individual drawings represents the actual number of people or things involved.

Pie chart A diagram providing a visual representation of the proportions into which something is divided. It consists of a circle divided in sectors, the area of each sector representing the proportion of a specified component.

Pie model of capital structure A model of the debt-equity ratio of the firms, graphically depicted in slices of a pie that represent the value of the firm in the capital markets.

Piece rate A payment scheme in which an employee is paid a specific price for each unit made. The rate is therefore directly related to output and not to time (*compare*time rate). This method is often combined with a basic salary and takes the form of a productivity bonus, in which individual effort is rewarded. In modern factories, many tasks have been, making this method of payment less common, although it is simple to operate and popular with employees. Piece rate, to be attractive to workers, depends on speed of production, therefore the standard of quality and safety may suffer if adequate precautions are not taken.

Piggyback registration When a securities underwriter allows existing holdings of shares in a corporation to be sold in combination with an offering of new public shares.

Piggyback service A distribution service in which loaded truck trailers, or other sealed containers, are carried by rail to destinations from which they can be taken by truck to their final destination. Similar services are fishyback service, involving transport of containers by water, and birdyback service, involving transport of containers by air.

PIK (Payment-in-kind) securities Highly bonds or preferred stock that pay interest or dividends through additional bonds or preferred stock.

Pilot production The small-scale production of a new product in a pilot plant. The object is to check and, if necessary, improve the production method. The product itself is often used in a marketing exercise to monitor its reception by the market and, in some cases, to improve it.

Pilot study A small-scale marketing research study conducted as a trial so that any problems can be eliminated before a full study is undertaken. For example, a pilot study might show that changes are needed in a questionnaire because questions are ambiguous, miss the point, etc.

Pink form (preferential form) An application form in a flotation that is printed on pink paper and usually distributed to employees of the company to give them preference in the allocation of shares. Up to 10% of a share issue can be set aside under London Stock Exchange rules for applications from employees or from shareholders in a parent company that is floating a subsidiary.

Pipeline The underwriting process that must be completed with the SEC before a security can be offering for sale to the public.

Piracy 1. An illegal act of violence, detention, robbery, or revenge committed on a ship or aircraft. It excludes acts committed for political purposes and during a war. In marine insurance it extends to include any form of plundering at sea.

2. Infringement of copyright. The usual remedy is for the copyright holder to obtain an injunction to end the infringement. This may not be possible in a foreign country and many British and American books are pirated in foreign countries by copying them photographically and printing them cheaply.

Pit committee A committee of the exchange that determines the daily settlement price of futures contracts.

Pivot Price level established as being significant by market's failure to penetrate or as being significant when a sudden increase in volume accompanies the move through the price level.

Place The marketing of new securities, usually through sales to institutional investors. See: Float.

Placement A bank depositing Eurodollars with (selling Eurodollars to) another bank is often said to be making a placement.

Placement ratio The percentages of last week's new municipal bond offerings that have been bought from the underwriters, according to the Bond Buyer newspaper.

Placing The sale of shares by a company to a selected group of individuals or institutions. Placings can be used either as a means of flotation or to raise additional capital for a quoted company (*see also*pre-emption rights; rights issue). Placings are usually the cheapest way of raising capital on a stock exchange and they also allow the directors of a company to influence the selection of shareholders. The success of a placing usually depends on the placing power of the company's stockbroker. Placings of public companies are sometimes called public placings (*compare* private placing). In the USA a placing is called a placement.

Plain vanilla A term that refers to a relatively simple derivative financial instrument, usually a swap or other derivative that is issued with standard features.

Plan The Chapter 11 plan must contain, among other things, a division of the claims into various classes, a description of the treatment of each claim, a comparison of the treatment under the Chapter 11 plan to projected treatment under Chapter 7 (liquidation) and the means for carrying out the plan. Once a plan is proposed and a disclosure statement approved, the plan will be distributed to creditors for voting. The court, in each case, will set rules for the balloting process.

The plan must provide for a distribution to the debtor's creditors of all the debtor's projected disposable income over the term of the plan. The plan's term is a minimum of three years, but no longer than five years. The plan must describe the treatment of all classes of creditors, the amount of the payments to be made by the trustee, and the length of time that the plan will be in effect. In return, the debtor pays into the plan the projected disposable income.

Plan for reorganization A plan for

reorganizing a firm during the Chapter 11 bankruptcy process.

Plan participants Employees or other beneficiaries who are eligible to receive benefits from a company's employee benefit plan.

Plan sponsors The entities that establish pension plans, including private business entities acting for their employees; state and local entities operating on behalf of their employees; unions acting on behalf of their members; and individuals representing themselves.

Planned capital expenditure program Budgeted or projected outlays for major expenditures on permanent or fixed assets as outlined in the corporate financial plan.

Planned financing program Budgeted or projected ways need for reasons or to obtain short-term and long-term financing as outlined in the corporate financial plan.

Planned location of industry Government intervention in the location of new industries in an area or country. As old industries in a region, such as shipbuilding, gradually decrease, new industries are required to take their place to avoid high unemployment and skilled workers moving away. Most governments provide assistance to those wishing to set up new industries in these areas.

Planning blight Difficulty in selling or developing a site, building, etc., because it is affected by a government or local-authority development plan. Planning blight may be ended by a compulsory purchase by the government or local authority, but it may continue indefinitely if the government plans fail to mature or if the site itself is not required by the development but is rendered unsaleable or less valuable by its proximity to a development.

Planning One of the functions of management accounting in which plans for the future activities and operations of an organization are incorporated into its budgets, etc. *See also* business plan.

Plastic money A colloquial name for a credit card, cash card, or multifunctional card. It is often shortened to plastic

Player Used in the context of general equities. Customer or trader who is actively involved in a particular stock or the market in general.

Playing the market Trading in high, uncalculated risk usually refers to actions of amateur investors.

Plaza Accord Agreement among country representatives in 1985 to implement a coordinated program to weaken the dollar.

Pledge An article given by a borrower (pledgor) to a lender (pledgee) as a security for a debt. It remains in the ownership of the pledgor although it is in the possession of the pledgee until the debt is repaid. *See also*pawnbroker.

Plow back To reinvest earnings in a business rather than pay out them out as dividends. Common practice in high-growth companies.

Plug A variable that handles financial slack in the financial plan.

Plus a match Used for listed equity securities. Floor indication that someone is on the floor with equal priority standing who wants to buy/sell at least the same number of shares at the same price as one's own order. Outside. See: Matched orders. Compare to ahead.

Point and figure chart A price-only chart that takes into account only whole integer changes in price, i.e., a 2-point change. Point and figure charting disregards the element of time and is solely used to record changes in price.

Point and figure chart A price-only chart that takes into account only whole integer Point Attractor In non-linear dynamics, an attractor where all orbits in phase space are drawn to one point, or value. Essentially, any system which tends to a stable, single valued equilibrium will have a point attractor. A pendulum which is damped by friction will always stop, so its phase space will always be drawn to the point where velocity and position are equal to zero. See: Attractor, Phase Space.

Point The smallest unit of price change quoted, or one one-hundredth of a percent. Related: Minimum price fluctuation and tick.

Poison pill A tactic used by a company that, fears an unwanted takeover by ensuring that a successful takeover bid will trigger some event that substantially reduces the value of the company. Examples of such tactics include the sale of some prized asset to a friendly company or bank or the issue of securities with a conversion option enabling the bidder's shares to be bought at a reduced price if the bid is successful. Poison pills are used all over the world but were developed in the USA. *See also*porcupine provisions; staggered directorships.

Poison put A covenant allowing the bondholder to demand repayment in the event of a hostile takeover.

Policy asset allocation A long-term asset allocation method, in which the investor seeks to assess an appropriate long-term "normal" asset mix that represents an ideal blend of controlled risk and enhanced return.

Policy limit The maximum dollar amount of coverage provided by an insurance company for a certain policy.

Policy loan A loan often made at a below-market interest rate from an insurance company to a policyholder that is secured by the cash surrender value of a life insurance policy.

Policy proof of interest (PPI) An insurance policy (usually marine insurance) in which the insurers agree that they will not insist on the usual requirement that the insured must prove an insurable interest existed in the subject matter before a claim is paid. The possession of the policy is all that is required. These policies are a matter of trust between insurer and insured as they are not legally enforceable. They are issued for convenience in cases in which an insurable interest exists but is extremely difficult to prove or in which an insurable interest might come into existence later in the voyage.

Policyholder An individual who owns an insurance policy.

Policyholder loan bonds Packaged loans acquired by policyholders that are secured by the cash surrender value of the policies, and are offered by a broker/dealer as bonds.

Political and charitable contributions Donations for political or charitable purposes made by an organization. Under the *Companies Act* a disclosure of such a donation has to be made by companies that are not wholly owned subsidiary undertakings of another British company and which have on their own or with their subsidiaries given in the financial year in aggregate more than £200. Charitable purposes is taken to mean purposes that are exclusively charitable. A donation for political purposes is taken to mean the giving of money either directly or indirectly to a political party of the UK or any part of it, or to a person who is carrying on activities likely to affect support for a political party. The total amounts given for both political and charitable purposes must be separately disclosed. If the payments are for political purposes, where applicable, the following information must be provided: (a) the name of each person to whom money exceeding £200 in amount has been given for those purposes and the amount given; and (b) if more than £200 has been given as a donation or subscription to a political party, the identity of the party and the amount given must be disclosed.

Political risk insurance The risk associated with possible negative events such as expropriation of assets, changes in tax policy, restrictions on the exchange of foreign currency, or other changes in the business climate of a country.

Pollution permits Permits issued by a government that allow the owner to pollute the environment legally. These permits may be bought and sold; firms that improve their environmental efficiency would reduce their need for permits and be in a position to sell them to less environmentally efficient rivals. In the USA, as part of the *Clean Air Act,* trading in these permits has cut sulphur dioxide emissions faster and more cheaply than expected. The use of pollution permits appears to have two main advantages over traditional environmental regulations. First, it gives firms a financial incentive to reduce emissions for less than it would cost to buy permits, and secondly, by leaving it to companies to decide how and when to reduce emissions, it reduces not only the cost of compliance but also the bureaucracy required to enforce environmental regulations.

Pollution The disposal into the environment of substances that cannot be made harmless by normal biological processes. As part of a wider and increasing concern for environmental issues, the costs of industrialization are being assessed in terms of discharges into the air and water, the disposal and longterm effects of wastes, recycling, energy efficiency, land use and degradation, and the depletion of raw materials. Legislation has been enacted, increasingly, to limit levels of pollution and in 1996 an Environmental Agency was established in the UK to coordinate the previously separate activities of various agencies. Most other advanced industrial countries have introduced legislation and similar agencies. International targets have been agreed to eliminate the use of cfcs in aerosols and as a refrigerant (because they deplete the ozone layer). The adoption of the 'Polluter Pays Principle' (the fining of major companies for pollution incidents) and increasing awareness of green consumerism has meant that many leading companies have now issued environmental policy statements and some publish annual environment audits.

Pool In capital budgeting, the concept that investment projects are financed out of a pool of bonds, preferred stock, and common stock, and a weighted-average cost of capital must be used to calculate investment returns. In insurance, a group of insurers who share premiums and losses in order to spread risk. In investments, the combination of funds for the benefit of a common project, or a group of investors who use their combined influence to manipulate prices.

Pooling of interests An accounting method for reporting acquisitions accomplished through the use of equity. The combined assets of the merged entity are consolidated using book value, as opposed to the purchase method, which uses market value. The merging entities' financial results are combined as though the two entities have always been a single entity.

Population (universe) In marketing research, any complete group of entities sharing a common set of characteristics. For example, a population could be all the companies in the car-supplying industries or all the consumers in the Southwest of England. A population is the group from which a sample is taken.

Population hypotheses In marketing research assumptions made by a researcher or manager about some characteristic of a population being reviewed. Typical hypotheses could be that for a particular population one method of advertising (e.g. TV) is more effective than another (e.g. Radio) or that one brand of cereal was more popular in a selected area of the country than in other areas. On the basis of such a hypothesis, data is collected, evaluated, and accepted or rejected according to the probability that it is true.

Population pyramid A diagram that illustrates the distribution of a population by age. The youngest and most numerous age group forms a rectangle at the base of the pyramid; the oldest and least numerous is a small rectangle at its apex.

Porcupine provisions (shark repellents.) Provisions made by a company to deter takeover bids. They include poison pills, staggered directorships, etc.

Port mark The markings on the packages of goods destined for export, giving the name of the overseas port to which they are to be shipped or sent by air freight.

Position A market commitment; the number of contracts bought or sold for which no offsetting transaction has been entered into. The buyer of a commodity is said to have a long position and the seller of a commodity is said to have a short position . Related: open contracts

Position audit A systematic assessment of the current situation of an organization. This normally involves drawing up a report (either internally or through external consultants) of the strengths and weaknesses of the organization and the opportunities or threats that it faces (*see*swot). The position audit is a vital tool in designing the future strategies of a large organization.positioning A marketing strategy that will position a company's products and services against those of its competitors in the minds of consumers. To achieve positioning success it is suggested that there are four basic competitive strategies that a company can follow. The three winning strategies are:

Cost leadership - the company tries to achieve lowest costs of production and distribution;

Differentiation - making use of specific marketing mixes (*see*differentiated marketing);

Focus - paying attention to a few market segments.

The fourth strategy is a losing strategy in which a company pursues a middle-ofthe-road path. Companies that try to be good at everything are rarely particularly good at anything.

Position limits Applies to derivative products. Maximum position available in any one future or option contract for a given institution. For "bona fide" futures hedgers, there are no position limits.

Position sheet Used in the context of general equities. List of long and short positions for an individual trader or desk, at times accompanied by the trades from the previous trading session that brought these closing positions.

Positive convexity A property of option-free bonds that the price appreciation for a large downward change in interest rates will be greater (in absolute terms) than the price depreciation for the same downward change in interest rates.

Positive covenant (of a bond) A bond covenant that specifies certain actions the firm must take. Also called and affirmative covenant.

Positive obligation A New York Stock Exchange rule that governs the behavior of specialists. Positive obligation is the mandate of the specialists to step in and act as either the buyer or the seller public investor orders exist do not match up naturally. Also known as affirmative_obligation. Related: negative_obligation.

Possession Actual physical control of goods or land. Possession has a wide variety of meanings in English law, depending on the nature of the property and the circumstances. For example, a person may still have possession of goods that have been lost or mislaid, provided that they are not abandoned and that no-one else has physical possession of them. It is sometimes used in the sense of a right to possession, e.g. The right of a person who has pledged goods to get them back. Possession may be divorced from ownership; for example, a thief has legal possession of stolen goods but not ownership. Possession usually requires intention to possess – a person cannot possess something without knowing that he or she does so.

Possessions corporation A type of corporation permitted under the US tax code whose branch operation in a US possession can obtain tax benefits as though it were operating as a foreign subsidiary.

Post-audit A set of procedures for evaluating a capital budgeting decision after the fact.

Post-date To insert a date on a document that is later than the date on which it is signed, thus making it effective only from the later date. A postdated (or forward-dated) cheque cannot be negotiated before the date written on it, irrespective of when it was signed. *Compare*ante-date

Postponement option The option of postponing a project without eliminating the possibility of undertaking it.

Postponing income Purposely delaying receipt of income to a later year in order to reduce current tax liability.

Post-testing Testing that takes place after an advertisement has appeared to determine whether or not it has met the objectives set for it by the company that paid for it.

Posttrade benchmarks Prices after the decision to trade.

Pound 1. (£) The standard UK monetary unit, divided into 100 pence. It dates back to the 8th century AD when Offa, King of Mercia, coined 240 pennyweights of silver from 1 pound of silver. When sterling was decimalized in 1971 it was divided into 100 newly defined pence. The pound is also the currency unit of the Falkland Islands (Fk£) and Gibraltar (Gib£). *See also*punt

2. The standard monetary unit of Egypt (LE), Lebanon (LL), the Sudan (lsd), and Syria (LS), divided into 100 piastres. **3.** (£C) The standard monetary unit of Cyprus, divided into 100 cents.

Pound cost averaging A method of accumulating capital by investing a fixed sum of money in a particular share every month (or other period). When prices fall the fixed sum will buy correspondingly more shares and when prices rise fewer shares are bought. The result is that the average purchase price over a period is lower than the arithmetic average of the market prices at each purchase date (because more shares are bought at lower prices and fewer at higher prices).

Power of attorney (PA) A formal document giving one person the right to act for another. A power to execute a deed must itself be given by a deed. An attorney may not delegate these powers unless specifically authorized to do so.

Power styles Management styles in using power to achieve results, affect the behaviour of employees, and overcome resistance. Three styles are recognized. Impression management is a competitive power style, usually through deceit and manipulation, to overcome any opposition. Consensus and charismatic power styles are forms of the collective use of power. The consensus power style involves joint decision making and joint problem solving, whereas the charismatic style makes use of the manager's personality to inspire the members of an organization to work together for a common purpose. The transactional style can be either competitive or collective, depending on the decision maker. Cooperation is needed for any transaction to take place: if that transaction involves the use and allocation of scarce resources it could become very competitive.

Pre Carriage Usually freight charges for port or airport delivery arising before the principal international carriage.

Pre shipment Finance Short term funding for inventory and production costs associated with manufacturing goods being exported.

Preacquisition profit The retained profit of one company before it is taken over by another company. Preacquisition profits should not be distributed to the shareholders of the acquiring company by way of dividend, as such profits do not constitute income to the parent company but a partial repayment of its capital outlay on the acquisition of the shares.

Preapproach A step in the industrial selling process in which a salesperson learns as much as possible about a prospective customer before making a sales call.

Prearranged trading Possibly fraudulent practice whereby commodities dealers carry out risk-free trades at predetermined prices to acquire tax advantages.

Preauthorized checks (PAC) Checks that are authorized by a payer in advance, and written either by the payee or by the payee's bank and then deposited in the payee's bank account.

Preauthorized electronic debits (PAD) Debits to a bank account in advance by the payer. The payer's bank sends payment to the payee's bank through the Automated Clearing House (ACH) system.

Preautomation analysis An analysis that should precede the replacement of a manual system by automation. Research suggests that large-scale investments in automation frequently fail to provide the expected benefits, because they are not used as an opportunity to eliminate unnecessary elements of the process.

Preceding-year basis (PYB) A basis for assessing the profits made by a business in which the profits assessed in any given fiscal year are based on the accounts that ended during the previous tax year. For example, in an ongoing business with accounts drawn up to 30 April, the profits assessed in 1996-97 will be based on the profits as set out in the accounts for the year ended 30 April 1995 (the accounts ending in the previous tax year, 6 April 1995 to 5 April 1996).

Precious metals Gold, silver, platinum, and palladium, which are used for their intrinsic value or for their value in production. These may be traded either in their physical state or by way of futures and options contracts, mining company stocks, bonds, mutual funds, or other instrument.

Precompute Method of charging interest in which the annual interest is either deducted from the face amount of the loan when the funds are distributed or is added to the total amount and divided into the regular payments.

Predetermined motion-time standards (PMTS) A work measurement technique using libraries of pre-established times for basic human motions, to construct standard times for jobs or processes.

Pre-emption rights A principle, established in company law, according to which any new shares issued by a company must first be offered to the existing shareholders as the legitimate owners of the company. To satisfy this principle a company must write to every shareholder, involving an expensive and lengthy procedure. Newer methods of issuing shares, such as vendor placings or bought deals, are much cheaper and easier to effect, although they violate pre-emption rights. In the USA pre-emption rights have now been largely abandoned but controversy is still widespread in the UK.

Preference dividend A dividend payable to the holders of preference shares. Preference dividends not paid in previous periods will only be due to the holders of cumulative preference shares.

Preference Refers to over-the-counter trading. Selection of a dealer to handle a trade despite the dealer's market not being the best available. Often the "preferenced dealer" will then move his market in line.

Preference share A share in a company yielding a fixed rate of interest rather than a variable dividend. A preference share is an intermediate form of security between an ordinary share and a debenture. Preference shares, like ordinary shares but unlike debentures, usually confer some degree of ownership of the company. However, in the event of liquidation, they are paid off after debt capital but before ordinary share capital. Preference shares may be redeemable (*see*redeemable shares) at a fixed or variable date; alternatively they may be undated. Sometimes they are convertible. The rights of preference shareholders vary from company to company and are set out in the articles of association. Voting rights are normally restricted, often only being available if the interest payments are in arrears.

Preference stock A security that ranks junior to preferred stock but senior to common stock in the right to receive payments from the firm; essentially junior preferred stock.

Preferential duty A specially low import duty imposed on goods from a country that has a trade agreement of a certain kind with the importing country. For example, in the days of the British Empire, imports from member countries were granted Imperial Preference on imported materials, which later became Commonwealth Preference.

Preferential transfer Certain transfers or payments made to creditors within 90 days before the bankruptcy is filed may be reversed and recovered by the bankruptcy court. Where the creditor is an insider (relative, shareholder, etc.) the 90 days is extended to one year. The recovery of these preferential payments is a method to provide an equal distribution among creditors of the same class.

A debtor owes many creditors but pays one "friendly creditor" in full just before filing the bankruptcy petition. This payment could most likely be recovered by the trustee to be spread among all creditors.

Preferred dividend coverage Net income after interest and taxes (before common stock dividends) divided by preferred stock dividends.

Preferred equity redemption stock (PERC) Preferred stock that converts automatically into equity at a stated date. A limit is placed on the value of the shares the investor receives.

Preferred habitat theory A biased expectations theory that believes the term structure reflects the expectation of the future path of interest rates as well as risk premium. The theory rejects the assertion that the risk premium must rise uniformly with maturity, but instead profits that to the extent that the demand for and supply of funds do not match for a given maturity range, some participants will shift to maturities showing the opposite imbalances, as long as they are compensated by an appropriate risk premium whose magnitude will reflect the extent of aversion to either price or reinvestment risk.

Preferred shares Preferred shares give investors a fixed dividend from the company's earnings and entitle them to be paid before common shareholders. See: Preferred stock.

Preferred stock A security that shows ownership in a corporation and gives the holder a claim, prior to the claim of common stockholders, on earnings and also generally on assets in the event of liquidation. Most preferred stock pays a fixed dividend that is paid prior to the common stock dividend, stated in a dollar amount or as a percentage of par value. This stock does not usually carry voting rights. Preferred stock has characteristics of both common stock and debt.

Preferred stock agreement A contract for preferred stock.

Preferred stock ratio Preferred stock at par value divided by total capitalization, which gives the portion of capitalization that consists of preferred stock.

Preliminary estimate The second estimate of GDP released about two months after the measurement period.

Preliminary prospectus An initial or tentative version of a prospectus.

Premature distribution A distribution from an IRA before the owner reaches age 59-1/2. Generally, a 10% penalty tax is owed on such a distribution. Also known as an early distribution or an early withdrawal.

Premium A bond sold above its par value. (2) The price of an option contract; also, in futures trading, the amount by which the futures price exceeds the price of the spot commodity. (3) For convertibles, amount by which the price of a convertible exceeds parity, and is usually expressed as a percentage. Suppose a stock is trading at $45, and the bond is convertible at a $50 stock price and the convertible bond trading at 105. A similar bond without the conversion feature trades at $90. In this case, the premium is $15, or 16.66%=(105-90)/90. If the premium is high, the bond trades like any fixed income bond; if low, like a stock. See: Gross parity, net parity. (4) For futures, excess of fair value of future over the spot index, which in theory will equal the Treasury bill yield for the period to expiration minus the expected dividend yield until the future's expiration. (5) For options, price of an option in the open market (sometimes refers to the portion of the price that exceeds parity). (6) For straight equity, price higher than that of the last sale or inside market. Related: Inverted market premium payback period. Also called break-even time; the time it takes to recover the premium per share of a convertible security.

Premium bond A bond that is sellingfor more than its par value.

Premium income The income received by an investor who sells an option.

Premium raid An attempt to acquire a large portion of a company's stock to gain control by offering stockholders a premium over the market value for their shares.

Pre-Money Valuation The value of a company just before its most recent round of financing. Related: Post-Money Valuation

Prepackaged bankruptcy A bankruptcy in which a debtor and its creditors pre-negotiate a plan of reorganization and then file it along with the bankruptcy petition.

Prepaid interest An asset account showing interest that has been paid in advance, which is expensed and charged to the borrower's P & L statement.

Prepayment penalty A fee a borrower pays a lender when the borrower repays a loan before its scheduled time of maturity.

Prepayment speed Also called speed, the estimated rate at which mortgagors pay off their loans ahead of schedule, critical in assessing the value of mortgage pass-through securities.

Prepayments Payments made in excess of scheduled mortgage principal repayments.

Prerefunded bond Refunded bond.

Prerefunding Procedure of floating a second bond at a lower interest rate in order to pay off the first bond at the first call date and to reduce overall borrowing costs.

Presale order An order to purchase part of a new municipal bond issue that is accepted by an underwriting syndicate before an official public offering.

Present value The amount of cash today that is equivalent in value to a payment, or to a stream of payments, to be received in the future. To determine the present value, each future cash flow is multiplied by a present value factor. For example, if the opportunity cost of funds is 10%, the present value of $100 to be received in one year is $100 x [1/(1 + 0.10)] = $91.

Present Value Components Analysis An analytical tool that establishes a base NPV for a project that can then be adjusted for the incremental NPV effect of separate elements of the project's overall potential sales.

Present value factor Factor used to calculate an estimate of the present value of an amount to be received in a future period. If the opportunity cost of funds is 10% over next year, the factor is [1/(1 + 0.10)].

Present value of growth opportunities Net present value (NPV) of investments the firm is expected to make in the future.

Present Value Index (PVI) The ratio of the NPV of a project to the initial outlay required for it. The index is an efficiency measure for investment decisions under capital rationing.

President Highest-ranking officer in a corporation after the chief executive officer.

Pre shipment Finance Short term funding for inventory and production costs associated with manufacturing goods being exported.

Presidential election cycle theory A theory that stock market trends can be predicted and explained by the four-year presidential election cycle.

Pre-sold issue An issue that is sold out before the coupon announcement.

Pre-tax contribution Payment to an account made with funds from a worker's paycheck before federal income taxes are deducted.

Pretax earnings or profits Net income before federal income taxes are subtracted.

Pretax rate of return Gain on a security before taxes.

Pre-trade benchmarks Prices occurring before or at the decision to trade.

Previous balance method Method of calculating finance charges based on the account balance at the end of the previous month.

Price of admission Used in the context of general equities. Cost to become a player in a stock in an inordinately aggressive market (i.e.,locking on one side, size or price concessions); trader becomes aggressive in order to break the domination of customer activity by another dealer.

Price-book ratio Compares a stock's market value to the value of total assets less total liabilities (book value). Determined by dividing current stock price by common stockholder equity per share (book value), adjusted for stock splits. Also called Market-to-Book.

Price change Increase or decrease in the closing price of a security compared to the previous day's closing price.

Price compression The limitation of the price appreciation potential for a callable bond in a declining interest rate environment, based on the expectation that the bond will be redeemed at the call price.

Price continuity Minimal price changes due to transactions.

Price discovery process The process of determinin the prices of assets in the marketplace through the interactions of buyers and sellers.

Price-earnings ratio Shows the multiple of earnings at which a stock sells. Determined by dividing current stock price by current earnings per share (adjusted for stock splits). Earnings per share for the P/E ratio are determined by dividing earnings for past 12 months by the number of common shares outstanding. Higher multiple means investors have higher expectations for future growth, and have bid up the stock's price.

Price effect Impact of a change in interest rates on bond prices.

Price elasticities The percentage change in quantity divided by a percentage change in the price. Answers the question: How much will the demand for my product decrease if I raise prices by 10%?

Price gap A term used when the price of a stock rockets or dives in a direction away from its last price range, such as a stock with a trading range of $10-$12 that closes at $12 and climbs to $14 the next day.

Price give Used in the context of general equities. Willingness of a buyer or seller to negotiate on price, within reason, from the price at the last sale or the indicated level. See: Takes price.

Price immunization Portfolio protection strategy that focuses on the market value of assets and liabilities.

Price impact costs Related: Market impact costs

Price leadership A price charged by the dominant producer that becomes the price adopted by all the other producers.

Price range The interval between the high and low prices over which a stock has traded over a particular period of time.

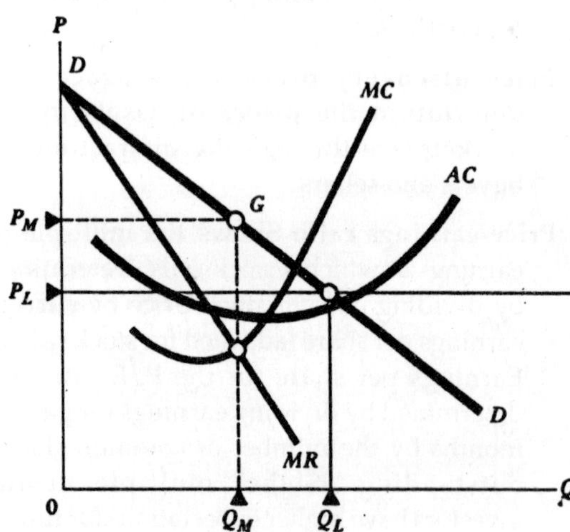

Fig. Price Model

Price risk The risk that the value of a security (or a portfolio) will decline in the future. Or, a type of mortgage pipeline risk created in the production segment when loan terms are set for the borrower in advance of setting terms for secondary market sale. If the general level of rates rises during the production cycle, the lender may have to sell the originated loans at a discount.

Price-sales ratio Determined by dividing current stock price by revenue per share (adjusted for stock splits). Revenue per share for the P/S ratio is determined by dividing revenue for past 12 months by number of shares outstanding.

Price-specie flow mechanism Adjustment mechanism under the classic gold standard allowing disturbances in the price level in one country to be wholly or partly offset by a countervailing flow of specie (gold coins) that would act to equalize prices across countries and automatically bring international payments into balance.

Price spread An options strategy that involves buying and selling two options on the same security with the same expiration month, but with different exercise prices.

Price Elasticity of Demand Means the responsiveness of the quantity demanded of a good to its own price. In order to avoid the measure of clasticity being sensitive to the units in which quantities and prices have been measured, the elasticity of demand has been expressed as the percentage change in demand that takes place in response to a percentage change in price.

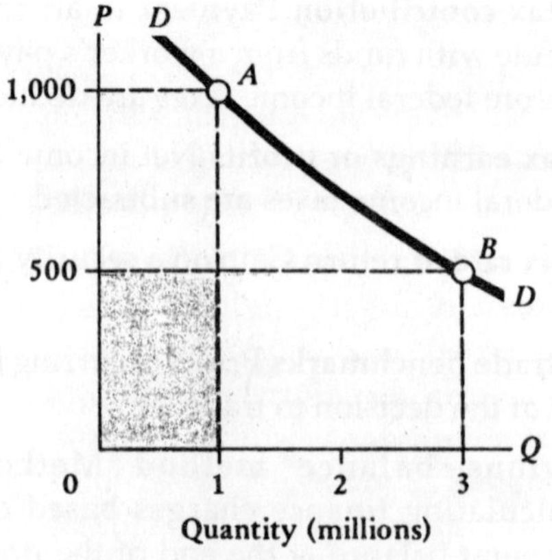

Fig. Price Elasticity of Demand

Price support Government intervention to set an artificially high price through the use of a price floor designed to aid producers.

Price takers Individuals who respond to rates and prices by acting as though prices have no influence on them.

Price uncertainty Chance that the future price of an asset will change.

Price value of a basis point (PVBP) Also called the dollar value of a basis point; a measure of the change in the price of a bond if the required yield changes by one basis point.

Price-volume relationship A relationship espoused by some technical analysts that signals continuing rises or falls in security prices that are related to changes in volume traded.

Price-weighted index An index giving a greater influence to higher-valued stocks by weighting all component stocks by their price.

Prices (of equity) Price of a share of common stock on the date shown. Highs and lows are based on the highest and lowest intraday trading price.

Priced out The market hasalready incorporated information, such as a low dividend, into the price of a stock.

Pricey Term used for an unrealistically low bid price or unrealistically high offer price.

Pricing efficiency Also called external efficiency; a market characteristic that prices at all times fully reflect all available information that is relevant to the valuation of securities.

Primary dealer Usually refers to the select list of securities firms that are authorized to deal in new issues of government bonds.

Primary distribution Sale of a new issue of stock or bonds, as distinguished from a secondary distribution.

Primary earnings per (common) share Earnings available for the payment of dividends to common stockholders divided by the number of common shares outstanding.

Primary market Where a newly issued security is first offered. All subsequent trading of this security occurs is done in the secondary market.

Primary offering Direct/Sale of a firm's newly issued shares by the firm to investors.

Primary trend General movement in price data that lasts 4 to 4 1/2 years.

Prime Stands for Prescribed Right to Income and Maximum Equity, a certificate that entitles the owner to the dividend/income from an underlying security, but not to the capital appreciation of that security.

Prime paper The highest-quality, investment-grade debt of corporations as decided by rating agencies such as Moody's.

Prime rate The interest rate at which banks lend to their best (prime) customers. More often than not, a bank's most creditworthy customers borrow at rates below the prime rate.

Prime rate fund A mutual fund that buys portions of corporate loans from banks and pays the interest to shareholders.

Primitive security An instrument such as a stock or bond for which payments depend only on the financial status of the issuer.

Principal The total amount of money being borrowed or lent. (2) The party affected by agent decisions in a principal-agent relationship.

Principal-agent relationship Occurs when one person, an agent, acts on the behalf of another person, the principal.

Principal amount The face amount of debt; the amount borrowed or lent. Often called principal.

Principal Exchange-Rated-Linked Securities (PERLS) A debt instrument with its principal and interest denominated in U.S. dollars, but with principal repayment depending on the exchange rate of the U.S. dollar against a foreign currency.

Principal Finance Usually refers to the area within an investment bank that deals with high grade fixed income. This group will not just trade bonds on the secondard market but will be actively involved in the debt financing of new projects. Many firms have Principal Finance Officers.

Principal Finance Officer The head of an investment bank's Principal Finance division or a person that overseas the principal finance dealings of a firm. These dealings usually involve high grade bonds that are used to finance new projects for firms.

Principal-only (PO) A mortgage-backed security (MBS) whose holder receives only principal cash flows on the underlying mortgage pool. All the principal distribution due from the underlying collateral pool is paid to the registered holder of the stripped MBS on the basis of the current face value of the underlying collateral pool.

Principal stockholder A stockholder who owns 10% or more of the voting stock of a company. Such stockholders must report all trading in the stock to the SEC pursuant insider trading rules.

Principle of diversification That portfolios of different sorts of assets differently correlated with one another will have negligible unsystematic risk. In other words, unsystematic risks disappear in diversified portfolios, and only systematic risks persist, those related to particular assets.

Print Used in the context of general equities. As a verb execute a trade, evidenced by its printing on the ticker tape. As a noun, a trade.

Prior-lien bond A bond usually arising from reorganization with precedence over another bond of the same issuing company that is equally secured.

Prior-preferred stock Preferred stock that has a higher claim on all dividends and assets in liquidation than claims of other preferred stock.

Priority Used for listed equity securities. System used in an auction market, in which the first bid or offer price is executed before other bid and offer prices, even if subsequent orders are larger. NYSE rules stipulate that the bid made first should be executed first, or if two bids came in at once, the bid for the large number of shares receives "priority." The bid not executed is then turned to the broker, who informs the customer that the trade was not completed because there was stock ahead. See: Standing.

Private Export Funding Corporation (PEFCO) Company that mobilizes private capital for financing the export of big-ticket items by US firms by purchasing at fixed interest rates the medium- to long-term debt obligations of importers of US products.

Private Investment in Public Equity (PIPE) Occurs when private investors take a sizable investment in publicly traded corporations. This usually occurs when equity valuations have fallen and the company is looking for new sources of capital. This is a means by which a public company gets additional access to the equity markets in express mode— they already have public shares trading and this is an additional offering to investors under a securities purchase agreement, the issuer promises to register the shares typically via a resale registration statement within so many days after the closing.

Private-label pass-throughs Related: Conventional pass-throughs.

Private letter ruling A ruling by the IRS in response to a request for interpretation of a tax law.

Private limited partnership A limited partnership with no more than 35 participants that is not registered with the SEC.

Private market value (PMV) The break-up market value of all divisions of a company if divisions were each independent and established their own market stock prices.

Private Mortgage Insurance (PMI) Policy protecting the holder against loss resulting from default on a mortgage loan.

Private placement The sale of a bond or other security directly to a limited number of investors. For example, sale of stocks, bonds, or other investments directly to an institutional investor like an insurance company, avoiding the need for SEC registration if the securities are purchased for investment as opposed to resale. Antithesis of public offering.

Private-purpose bond A municipal bond allowing more than 10% of the proceeds go to private activities.

Private unrequited transfers Resident immigrant workers' remittances to their country of origin as well as, e.g., gifts, dowries, inheritances, prizes, charitable contributions.

Privatization The transfer of government-owned or government-run companies to the private sector, usually by selling them.

Pro forma capital structure analysis A method of analyzing the impact of alternative possible capital structure choices on a firm's credit statistics and reported financial results, especially to determine whether the firm will be able to use projected tax shield benefits fully.

Pro forma financial statements A firm's financial statements as adjusted to reflect a projected or planned transaction. "What-if" analysis.

Pro forma statement A financial statement showing the forecast or projected operating results and balance sheet, as in pro forma income statements, balance sheets, and statements of cash flows.

Probability The relative likelihood of a particular outcome among all possible outcomes.

Probability density function The function that describes the change of certain realizations for a continuous random variable.

Probability distribution A function that describes all the values a random variable can take and the probability associated with each. Also called a probability function.

Probability function A measure that assigns a likelihood of occurrence to each and every possible outcome.

Proceeds Money received by the seller of an asset.

Proceeds sale OTC securities sale whose revenue is used to buy another security.

Processing Delay Time a selling firm takes to record receipt of a payment and deposit it.

Producer Price Index (PPI) Index measuring changes in wholesale prices, published by the US Bureau of Labor Statistics every month.

Product cycle The time it takes to bring new and/or improved products to market.

Product cycle theory Theory suggesting that a firm initially establish itself locally and expand into foreign markets in response to foreign demand for its product; over time, the MNC will grow in foreign markets; after some point, its foreign business may decline unless it can differentiate its product from competitors.

Product Differentiation A source of competitive advantage that depends on producing some item that is regarded to have unique and valuable characteristics.

Product risk A type of mortgage pipeline risk that occurs when a lender has an unusual loan in production or inventory but does not have a sale commitment at a prearranged price.

Production In the context of project financing, a defined portion of the proceeds of production up to a dollar amount.

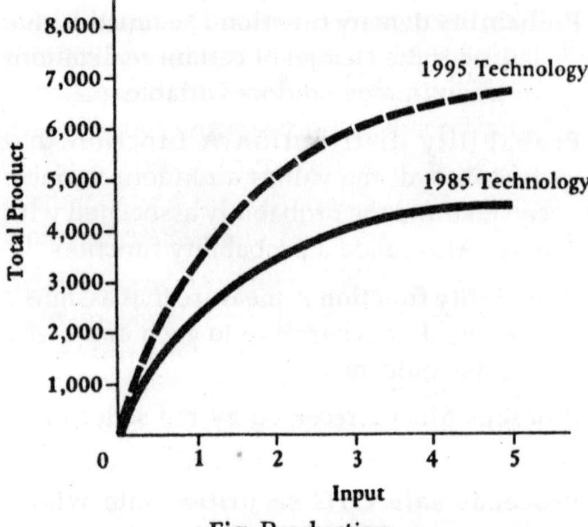

Fig. Production

Production Cost Advantage A source of competitive advantage that depends on producing some product or service at the lowest cost.

Production-flow commitment An agreement by the loan purchaser to allow a monthly loan quota to be delivered in batches.

Production Loan A project financing where the repayment is linked to the production, often on a dollar/unit basis.

Production payment financing A method of nonrecourse asset-based financing in which a specified percentage of revenue realized from the sale of the project's output is used to pay debt service.

Production possibilities schedule The maximum amount of goods (i.e., food and clothing) that a country is able to produce given its labor supply.

Production rate The coupon rate at which a pass-through security guaranteed by Ginnie Mae is issued.

Productivity The amount of output per unit of input, such as the quantity of a product produced per hour of capital employed.

Profile buyer/seller Trader trying to get involved in a stock who presents self as a buyer/seller to draw a call from a customer. That is the trader has nothing real, or natural.

Profit Revenue minus cost. The amount one makes on a transaction.

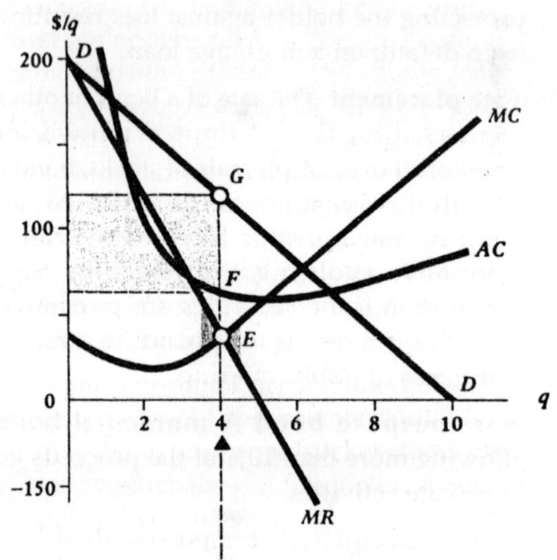

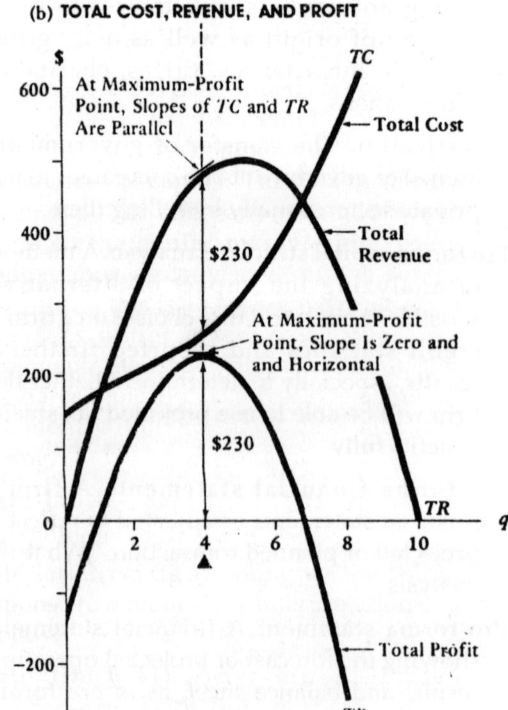

Fig. Profit Mazimization

Profit center A division of an organization held responsible for producing its own profits.

Profit forecast A prediction of future profits of a company, which may affect investment decisions.

Profit Graph A graphical representation of the potential outcomes of a strategy. Dollars of profit or loss are graphed on the vertical axis, and various stock prices are graphed on the horizontal axis. Results may be depicted at any point in time, although the graph usually depicts the results at expiration of the options involved in the strategy.

Profit margin Indicator of profitability. The ratio of earnings available to stockholders to net sales. Determined by dividing net income by revenue for the same 12-month period. Result is shown as a percentage. Also known as net profit margin.

Profit Range The range within which a particular position makes a profit. Generally used in refernce to strategies that have two break-even points - an upside break-even and a downside break-even. The price range between the two break-even points would be the profit range.

Profit-sharing plan An incentive system providing that employees share in companys profits through a cash fund or a deferred plan used to buy stock or bonds.

Profit Table A table of results of a particular strategy at some point in time. This is usually a tabular compilation of the data drawn on a profit graph. See also Profit Graph.

Profit taking Action by short-term securities traders to cash in on gains created by a sharp market rise, which pushes prices down temporarily but implies an upward market trend. See: Ring the [cash] register.

Profitability index The present value of the future cash flows divided by the initial investment. Also called the benefit-cost ratio.

Profitability ratios Ratios that focus on how well a firm is performing. Profit margins measure performance with relation to sales. Rate of return ratios measure performance relative to some measure of size of the investment.

Pro forma A financial projection based on assumptions. Also, refers to a statement of income and balance sheets that exclude non-recurring items.

Pro forma Earnings Often used in two ways. First, pro forma earnings refers to projections of earnings. This is often used internally or on a road show for an IPO. Second, it refers to a way of reporting earnings that excludes non-recurring items such as restructuring charges, extraordinary items.

Pro forma Invoice A quotation in the form of a ninvoice prepared by the seller that details items which would appear on a commercial invoice if an order results.

Program trades Orders requiring the execution of trades in a large number of different stocks at as near the same time as possible. Also called basket trades. Related: Block trade

Program trading Trades based on signals from computer programs, usually entered directly from the trader's computer in to the market's computer system and executed automatically. Applies to derivative products. A process of electronic execution of trading of a basket of stocks simultaneously, for index arbitrage, portfolio restructuring, or outright buy/sell interests. See: super dot. 0

Progress payments Periodic payments to a supplier, contractor, or subcontractor for work as it is completed as desired, in order to reduce working capital requirements.

Progress review A periodic review of a capital investment project to evaluate its continued economic viability.

Progressive tax system A tax system that taxes the wealthy at a higher percentage rate than the less wealthy.

Progressive taxation Characterizes a convex tax schedule that results in a higher effective tax rate on higher income levels. Increases for some increases in income, but never decreases with an increase in income.

Project The asset constructed with or owned via a project financing, which is expected to produce cash flow at a debt-service coverage ratio sufficient to repay the project financing.

Project contracts In the context of project financing, the suite of agreements underlying the project.

Project Finance Loan Program Program under which banks, the Ex-Im Bank, or a combination of both may extend long-term financing for capital equipment and related services for major projects.

Project financing A form of asset-based financing in which a firm finances a discrete set of assets on a stand-alone basis.

Project link An econometric model forecasting and describing the effects of changes in different economies on other economies.

Project loan certificate (PLC) A primary program of Ginnie Mae for securitizing FHA-insured and coinsured multifamily, hospital, and nursing home loans.

Project loans Usually FHA-insured and HUD-guaranteed mortgages on multiple-family housing complexes, nursing homes, hospitals, and other special development.

Project loan securities Securities backed by a variety of FHA-insured loans-primarily multifamily apartment buildings, hospitals, and nursing homes.

Project notes (PN) Notes issued by municipalities to finance federally sponsored programs in urban renewal and housing and guaranteed by the U.S. Department of Housing and Urban Development.

Projected benefit obligation (PBO) A measure of a pension plan's liability at the calculation date assuming that the plan is ongoing and will not terminate in the foreseeable future. Related: Accumulated benefit obligation.

Projected maturity date With CMOs, the date at the end of the estimated cash flow window where final payment is made.

Projection The use of econometric models to forecast the future performance of a company, country, or other financial entity using historical and current information.

Promissory note Written pledge to pay.

Property inventory A list of personal property with corresponding values and initial costs often used to substantiate insurance claim and tax losses.

Property rights Rights of individuals and companies to own and use property as they see fit and to receive the stream of income that their property generates.

Property tax A tax levied on real property based on its use and its assessed value.

Proportional representation A method of stockholder voting that allows minority shareholders and groups of small shareholders to have a better chance of getting representation on a Board of Directors than under statutory voting.

Proprietary trading Principal trading in which firm seeks direct gain rather than commission dollars.

Proprietorship An unincorporated business that is owned and operated by only one person who has complete liability for all assets, and complete rights to all profits.

Pro rata Shared or divided according to a ratio or in proportion to participation.

Proration Refers to the division between cash and stock in a takeover offer. Often a takeover is a combination of cash and the acquiring firm's equity. Shareholders can elect to take cash or equity. After the election is made, the stock is prorated. For example, if the takeover offer was 500 million in cash and 500 million in shares, if everybody elected cash, then the maximum cash for each shareholder is 50%. If 75% elected to receive cash, 25% of the shareholders would get 100% equity and the other 75% would get 75% cash and 25% equity. The proportions are complicated to compute if the new shares are worth more than the old shares. In this case, small shareholders (with say 100 shares) might receive 100% cash because it is disadvantageous to have a lot less than 100 shares.

Prospective Earnings Growth (PEG Ratio) Based on forecasts from proprietary sources such as Institutional Brokers' Estimate System (IBES), First Call, or Zach's. Growth is forecast of earnings minus current earnings divided by current earnings. Forward-looking measure rather than typical earnings growth measures, which look back in time (historical).

Prospectus Formal written document to sell securities that describes the plan for a proposed business enterprise, or the facts concerning an existing one, that an investor needs to make an informed decision. Prospectuses are used by mutual funds to describe fund objectives, risks, and other essential information.

Protect Assure the salesperson or trader that interest, buy or sell, will be attended to, given any change in the trading circumstances, as follows: At a price: If the stock trades at a certain price or price range, the trader will show this market to the salesperson and thus allow participation under these favorable circumstances. Floor protection: Representation of a client on the floor of the exchange-so that if size were to trade at his price or a better price, salesperson wouldparticipate. Volume (OTC): If a certain amount of volume trades (that parallels the protectee's interest), trader assures salesperson of reasonable participation in the trading activity. The extent of this protection depends on liquidity, number of market makers, and other aspects of the stock.

Protected Strategy A position that has limited risk. A protected short sale (short stock, long call) has limited risk, as does a protected straddle write (short straddle, long out-of-the-money combination). See also Combination and Straddle.

Protectionism Notion that governments should protect domestic industry from import competition by means of tariffs, quotas, and other trade barriers.

Protective covenant A part of an indenture or loan agreement that limits certain actions a company may take during the term of the loan to protect the lender's interests.

Protective put buying strategy A strategy that involves buying a put option on the underlying security that is held in a portfolio. Related: Hedge option strategies.

Protest Instructions given to a collecting bank that drafts falling due for payment are to be formally presented to the drawee by a notary, who is to formally record any default.

Prototype plan A qualified retirement plan sponsored by a financial institution. It may be adopted by executing a written agreement. A prototype is generally more flexible than the IRS Form 5305 or 5305-A and may have additional special features. Also called a master pension plan.

Provision for income taxes An amount on the P & I statement that estimates a company's total income tax liability for the year.

Provisional call feature A stipulation in a convertible issue that allows the issuer to call the issue during the noncall period if the price of the stock reaches a certain level. In the case of convertible securities, right of an issuer to accelerate the first redemption date if the underlying common should trade at or above a certain level for a sustained period. Most typical terms are 150% of conversion price for 20 consecutive days. Note that under these circumstances the security has appreciated, at a minimum, 50% since being issued.

Proxy Authorization, whether written or electronic, that shareholders' votes may be cast by others. Shareholders can and often do give management their proxies, delegating the right and responsibility to vote their shares as specified.

Proxy Committee A group of individuals appointed by the board of directors of the companies to formally represent the shareholders who send in proxy cards, to vote the represented shares in accordance with the shareholders' instructions.

Proxy Committee Ballot The ballot signed and submitted at the meeting by the Proxy Committee. It is the legal voting of shares represented by proxies assigned to the Proxy Committee and should always be completed.

Proxy contest A battle for the control of a firm in which a dissident group seeks, from the firm's other shareholders, the right to vote those shareholders' shares in favor of the dissident group's slate of directors. Also called proxy fights.

Proxy fight Often used in risk arbitrage. Technique used by an acquiring company to attempt to gain control of a takeover target. The acquirer tries to persuade the shareholders of the target company that the present management of the firm should be ousted n favor of a slate of directors favorable to the acquirer, thus enabling the acquiring company to gain control of the company without paying a premium price.

Proxy Fight Competition of outside group with management for stockholders' proxies in order to accumulate votes to elect a new board of directors.

Proxy Solicitor A specialist (firm) hired to gather proxy votes.

Proxy statement Document intended to provide shareholders with information necessary to vote in an informed manner on matters to be brought up at a stockholders' meeting. Includes information on closely held shares. Information required by the SEC that must be provided to shareholders who wish to vote for directors and on other company decisions by proxy.

Proxy vote Vote cast by one person or entity on behalf of another.

Prudent-man rule A common law standard against which those investing the money of others fiduciaries are judged.

P&S Purchase and sale statement. A statement provided by the broker showing change in the customer's net ledger balance after the offset of any previously established positions.

PSA Prepayment Rate The Bond Market Trade Association's Mortgaged Asset-Backed Securities Division's prepayment model based on an assumed rate of prepayment each month of the then unpaid principal balance of a pool of mortgages. PSA is used primarily to derive an implied prepayment speed of new production loans. 00% PSA assumes a prepayment rate of 2% per month in the first month following the date of issue, increasing at 2% percentage points per month thereafter until the 30th month. Thereafter, 100% PSA is the same as 6% CPR (Constant prepayment rate).

PSSG Financial ratio defined as stock price divided by sales over sales growth. Often used in the valuation of Internet stocks. Related: PREG.

Public Book (of order) The orders to buy or sell, entered by the public, that are generally away from the current market. The order book official or specialist keeps the public book. Market-Makers on the CBOE can see the highest bid and lowest offer at any time. The specialist's book is closed (only he knows at what price and in what quantity the nearest public orders are). See also Market-Maker and Specialist.

Public Company A company that has held an initial public offering and whose shares are traded on a stock exchange or in the over-the-counter market. Public companies are subject to periodic filing and other obligations under the federal securities laws.

Public debt Issues of debt by governments to compensate for a lack of tax revenues.

Public housing authority bond Bonds of local public housing agencies that are secured by the federal government and whose proceeds are used to provide low-rent housing.

Public limited partnership A limited partnership with an unlimited number of partners that is registered with the SEC and is available for public trading by broker/ dealers.

Public offering Used in the context of general equities. Offering to the investment public, after compliance with registration requirements of the SEC, usually by an investment banker or a syndicate made up of several investment bankers, at a price agreed upon between the issuer and the investment bankers. Antithesis of private placement. See: Primary distribution and secondary distribution.

Public offering price The price of a new issue of securities at the time that the issue is offered to the public.

Public ownership The portion of a company's stock that is held by the public.

Public-purpose bond A specific type of municipal bond used to finance public projects such as roads or government buildings. Interest on municipal bonds is federal income tax-free.

Public Securities Association(PSA) The trade association for primary dealers in US government securities, including MBSs. In 1997, they became known as the Bond Market Association.

Public securities offering A securities issue placed with the public through an investment or commercial bank.

The Public Individual investors who trade single securities independently or invest in intermediaries such as mutual funds, as opposed to professional investors.

Public Utility Holding Company Act of 1935 Legislation intended to eliminate many holding company abuses by reorganizing the financial structures of holding companies in the gas and electric utility industries and regulating their debt and dividend policies.

Public warehouse Storage facility operated by an independent warehouse company on its own premises.

Publicly held Describes a company whose stock is held by the public, whether individuals or business entities.

Publicly traded assets Assets that can be traded in a public market, such as the stock market.

Puke Slang for a trader selling a position, usually a losing position, as in, "When in doubt, puke it out."

Pull Used in the context of general equities. See: Cancel.

Pullback The downward reversal of a prolonged upward price trend.

Pulling in their horns Investors selling off positions after a stock or bond market has increased sharply or setting up hedging positions to guard against a negative turn of the market.

Purchase Buy; be long; have an ownership position.

Purchase accounting Method of accounting for a merger that treats the acquirer as having purchased the assets and assumed the liabilities of the acquiree, which are then written up or down to their respective fair market values. The difference between the purchase price and the net assets acquired is attributed to goodwill.

Purchase agreement Used in connection with project financing; an agreement to purchase a specific amount of project output per period.

Purchase fee A charge assessed by an intermediary, such as a broker-dealer or a bank, for assisting in the sale or purchase of a security.

Purchase fund Resembles a sinking fund, except that money is used to purchase bonds only if they are selling below their par value.

Purchase group See: Underwriting syndicate

Purchase loan A consumer loan taken to finance a purchase.

Purchase method Accounting for an acquisition using market value for the consolidation of the two entities' net assets on the balance sheet. Generally, depreciation/amortization will increase for this method (due to the creation of goodwill) compared to the pooling method resulting in lower net income.

Purchase-money mortgage A mortgage given by a buyer in lieu of cash when the buyer is unable to borrow commercially for the purchase of property.

Purchase order A written order to buy specified goods at a stipulated price.

Purchase and sale A method of securities distribution in which a firm purchases securities from the issuer for its own account at a stated price and then resells them, as contrasted with a best-efforts sale.

Purchasing power The amount of credit available for securities trading in a margin account, after taking margin requirements into consideration.

Purchasing power of the dollar The amount of goods and services that can be exchanged for a dollar as compared with amount of a previous time period.

Purchasing power parity The notion that the ratio between domestic and foreign price levels should equal the equilibrium exchange rate between domestic and foreign currencies.

Purchasing power risk Related: Inflation risk

Pure discount bond A bond that will make only one payment of principal and interest. Also called a zero-coupon bond or a single-payment bond.

Pure expectations theory A theory that asserts that forward rates exclusively represent the expected future rates. In other words, the entire term structure reflects the market's expectations of future short-term rates. For example, an increasing slope to the term structure implies increasing short-term interest rates. Related: Biased expectations heories.

Pure index fund A portfolio that is managed so as to perfectly replicate the performance of the market portfolio.

Pure monopoly A market in which only one firm has total control over the entire market for a product due to some sort of barrier to entry for other firms, often a patent held by the controlling firm.

Pure play A company involved in only one line of business.

Pure yield pickup swap Moving to higher yield-bonds.

Purpose credit Credit used for the purpose of buying, carrying or trading in securities.

Purpose loan A loan that is backed by securities and that is used to buy other securities under certain government regulations.

Purpose statement A form filed by a borrower that describes the use of a loan backed by securities, and guarantees that the funds lent will not be used illegally to buy securities against Federal Reserve regulations.

Put An option granting the right to sell the underlying futures contract. Opposite of a call.

Put bond A bond that the holder may choose

either to exchange for par value at some date or to extend for a given number of years. If the price is above par, the put is a "premium put."

Put-call parity Applies to derivative products. Option pricing principle that says, given a stock's price, a put and call of the same class must have a static price relationship because arbitrage opportunities or activities will always reestablish such a relationship.

Put-call parity relationship The relationship between the price of a put and the price of a call on the same underlying security with the same expiration date, which prevents arbitrage opportunities. Holding the underlying stock and buying a put will deliver the exact payoff as buying one call and investing the present value (PV) of the exercise price. The call value equals C = S + P - PV(k).

Put-call ratio The ratio of the volume of put options traded to the volume of call options traded, which is used as an indicator of investor sentiment (bullish or bearish).

Put guarantee letter A bank's letter certifying that the person writing a put option has sufficient funds in an account to cover the exercise price if required.

Put on Used for listed equity securities. Trade, or cross, a block of stock at the designated price and quantity. See: Print.

"Put it on " Used for listed equity securities. "Go to the floor to transact." See: Print.

Put option This security gives investors the right to sell (or put) a fixed number of shares at a fixed price within a given period. An investor, for example, might wish to have the right to sell shares of a stock at a certain price by a certain time in order to protect, or hedge, an existing investment.

Put an option To exercise a put option.

"Put pants on it " Used in the context of general equities. "Elaborate on your intentions or your inquiry," especially with respect to size, side, and price.

Put price The price at which an asset will be sold if a put option is exercised. Also called the strike or exercise price of a put option.

Put provision Gives the holder of a floating-rate bond the right to redeem the note at par on the coupon payment date.

Put ratio backspread A complex options strategy adopted when one believes a stock price will decline but wants to protect against it rising.

Put to seller Exercise a put option; require that the option writer to purchase the stock at the strike price.

Put swaption A financial instrument giving the buyer the right, or option, to enter into a swap as a floating-rate payer. The writer of the swaption therefore becomes the floating-rate receiver/fixed-rate payer.

Pyramid scheme An illegal, fraudulent scheme in which a con artist convinces victims to invest by promising an extraordinary return but instead simply uses newly invested funds to pay off any investors who insist on terminating their investment.

Pyramiding A type of stock swap option exercise in which a small number of previously-owned shares is surrendered to the company to pay a portion of the exercise price, for which a slightly larger number of option shares may be purchased, which are then immediately surrendered back to the company to pay additional amounts of the exercise price, and so on until the full option price has been paid and the optionee is left with just the number of shares equal to the option spread. With the advent of broker-assisted "Cashless Exercise/Same Day Sale" programs (see above), pyramiding has fallen out of favor.

Quadratic programming Variant of linear programming in which the objective function is quadratic rather than linear. In portfolio selection, we often minimize the variance of the portfolio (which is a quadratic function) subject to constraints on the mean return of the portfolio.

Qualification period A period of time during the first few months or weeks of a new policy when an insurance company will not reimburse a policyholder for a claim in order to allow the insurance company time to find any fraudulent information in the application.

Qualified Domestic Relations Order (QDRO) A judgment, decree, or order that gives a pension plan participant access to retirement assets that must be used to pay an ex-spouse or dependent children.

Qualified endorsement A signature on the back of a negotiable instrument transferring the amount to some other party but that includes wording that limits the endorser's liability.

Qualified opinion An auditor's opinion expressing certain limitations of an audit.

Qualified plan or trust A tax-deferred plan allowing employer and employee contributions that build up savings, which are paid out at retirement or on termination of employment. Tax is paid only when amounts are drawn from the trust.

Qualified retirement plan A retirement plan established by employers for their employees that meets the requirements of Internal Revenue Code Section 401(a) or 403(a) and is eligible for special tax considerations. The plan may provide for employer contributions, as in a pension or profit-sharing plan, as well as employee contributions. Employers can deduct plan contributions made on behalf of eligible employees on the business's tax return as business expenses. Plan earnings are not taxed to the employee until withdrawn.

Qualified Terminable Interest Property Trust (Q-TIP) A trust that allows a surviving spouse to receive income generated from the trust, while the actual distribution of the trust's assets is made to other beneficiaries such as the grantor's children.

Qualified total distribution A payment representing an employee's interest in a qualified retirement plan. The payment must be prompted by retirement (or other separation from service), death, disability, or attainment of age 59-1/2. Payment can be in installments as long as the complete distribution is made within a single tax year.

Qualifying annuity An annuity allowable as

investment for a qualified plan or trust.

Qualifying share Shares of common stock that a person must hold in order to qualify as a director of the issuing corporation.

Qualifying stock option A benefit granted by a corporation that allows employees to purchase shares at a discount price.

Qualitative analysis An analysis of the qualities of a company that cannot be measured concretely, such as management quality or employee morale.

Qualitative research Traditional analysis of firm-specific prospects for future earnings. It may be based on data collected by the analysts, there is no formal quantitative framework used to generate projections.

Quality of earnings Increased earnings due to increased sales and cost controls, as compared to artificial profits created by inflation of inventory or other asset prices.

Quality option Gives the seller choice of deliverables in Treasury bond and Treasury note futures contracts. Also called the swap option. Related: Cheapest to deliver issue.

Quantitative analysis An assessment of specific measurable securities or investment factors, such as cost of capital, value of assets; and projections of sales, costs, earnings, and profits. Combined with more subjective or qualitative considerations (such as management effectiveness), quantitative analysis can enhance investment decisions and portfolios.

Quantity risk Occurs when the quantity of an asset to be hedged is uncertain.

Quality spread Difference between Treasury securities and non-Treasury securities that are identical in all respects except for quality rating. For instance, the difference between yields on Treasuries and those on single A-rated industrial bonds. Also called credit spread.

Quant A person with numerical and computer skills who carries out quantitative analyses of companies.

Quantize To convert an asset or liability into a currency other than the regular trading currency.

Quantitative analysis An analysis of the mathematically measurable figures of a company, such as the value of assets or projected sales.

Quantitative research Use of advanced econometric and mathematical valuation models to identify the firms with the best possible prospectives. Antithesis of qualitative research.

Quantos Currency options with a guaranteed exchange rate that enable buyers who like an asset, German bonds for example, but not the asset's pricing currency, to arrange payment in a different currency for a fee.

Quarter stock Stock with a par value of $25 per share.

Quarterly Occurring every three months.

Quarterly financing February 15, May 15, August 15 and November 15, or next working day offerings of several "coupon" security issues. Quarterly issues currently consist of a 3-year note, a 10-year note, and a 30-year bond. The Treasury sometimes offers additional amounts of outstanding long-term notes or bonds, rather than selling new security issues. See: Reopening.

Quasi-public corporation A corporation that is operated privately, but is supported by the government in its operations and that often traded publicly.

Quay A landing place or pier, usually of solid construction, where vessels berth to load or unload cargo.

Quick assets Current assets minus inventories.

Quick ratio Indicator of a company's financial strength (or weakness). Calculated by taking current assets less inventories, divided by current liabilities. This ratio provides information regarding the firm's liquidity and ability to meet its obligations. Also called the Acid test ratio.

Quid pro quo An arrangement allowing a firm to use research from another firm at no cost in exchange for executing all of its trades with the firm that provides the research.

Quiet period Time period an issuer is "in registration" with the SEC and may not promote its forthcoming issue.

Quorum The minimum number of people who must be present or must provide a proxy to vote at a meeting in order to make a valid decision.

Quotation Highest bid and lowest offer (asked) price currently available on a security or a commodity.

Quotation board The electronic board at a brokerage firm displaying prices other financial data.

Quote rule Rule requiring market makers to publish quotations for any listed security when a quotation represents more than 1% of the aggregate trading volume for that security.

Quoted price The price at which the last trade of a particular security or commodity took place.

R

Radar alert Close monitoring of trading patterns in a company's stock by senior managers to uncover unusual buying activity that might signal a takeover attempt. See: Shark watcher.

Raider Individual or corporate investor who intends to take control of a company (often ostensibly for greenmail) by buying a controlling interest in its stock and installing new management. Raiders who accumulate 5% or more of the outstanding shares in the target company must report their purchases to the SEC, the exchange of listing, and the target itself. See: takeover.

Rainmaker A valuable employee, manager or subcontracted person who brings new business to a company.

Rally (recovery) An upward movement of prices. Opposite of reaction.

Reverse-annuity mortgages (RAM) Bank loan for an amount equal to a percentage of the appraisal value of the home. The loan is then paid to the homeowner in the form of an annuity.

Random variable A function that assigns a real number to each and every possible outcome of a random experiment.

Random walk Theory that stock price changes from day to day are accidental or haphazard; changes are independent of each other and have the same probability distribution. For a simple random walk, the best forecast of tomorrow's price is today's price. Related: Mean reversion.

Random walk with drift For a random walk with drift, the best forecast of tomorrow's price is today's price plus a drift term. One could think of the drift as measuring a trend in the price (perhaps reflecting long-term inflation). Given the drift is usually assumed to be constant. Related: Mean reversion.

Randomized strategy A strategy of introducing into the decision-making process a chance element that is designed to confound the information content of the decision-maker's observed choices.

Range The high and low prices, or high and low bids and offers, recorded during a specified time.

Range forward A forward exchange rate contract that places upper and lower bounds on the future cost of foreign exchange.

Rate anticipation swaps An exchange of bonds in a portfolio for new bonds that will achieve the target portfolio duration, given the investor's assumptions about future changes in interest rates.

Rate base The value of a regulated public utility

and its operations as defined by its regulators and on which the company is allowed to earn a particular rate of return.

Rate covenant A provision governing a municipal revenue project financed by a revenue bond issue, which establishes the rates to be charged users of the new facility.

Rate lock An agreement between the mortgage banker and the loan applicant guaranteeing a specified interest rate for a designated period, usually 60 days.

Rate-lock selling Underwriters of corporate bonds sell Treasuries to hedge what they will take in a planned underwriting.

Rate of interest The rate, as a proportion of the principal, at which interest is computed.

Rate of return Calculated as the (value nowminus value at time of purchase) divided by value at time of purchase. For equities, we often include dividends with the value now. See also: Return, annual rate of return.

Rate of return ratios Ratios that measure the profitability of a firm in relation to various measures of investment in the firm.

Rate risk In banking, the risk that profits may drop or losses occur because a rise in interest rates forces up the cost of funding fixed-rate loans or other fixed-rate assets.

Ratings An evaluation of credit quality of a company's debt issue by Thomson Financial BankWatch, Moody's, S&P, and Fitch Investors Service. Investors and analysts use ratings to assess the riskness of an investment. Ratings can also be an evaluation a country's creditworthiness or ability to repay, taking into consideration its estimated percentage default rate and political risk.

Ratio analysis A way of expressing relationships between a firm's accounting numbers and their trends over time that analysts use to establish values and evaluate risks.

Ratio Calendar Combination A strategy consisting of a simultaneous position of a ratio calendar spread using "calls" and a similar position using puts, where the striking price of the "calls" is greater that the striking price of the "puts".

Ratio Calendar Spread Selling more near-termoptions than longer-term ones purchased, all with the same strike; either puts or calls.

Ratio Spread Constructed with either puts or calls, the strategy consists of buying a certain amount of options and then selling a larger quantity of more out-of-the-money options.

Ratio Strategy A strategy in which one has an unequal number of long secruities and short sercurities. Normally, it implies a preponderance of short options over either long options or long stock.

Ratio writer An option writer who does not own the number of shares required to cover the call options he or she writes.

Rational expectations The idea that people rationally anticipate the future and respond today to what they see ahead. This concept was pioneered by Nobel Laureate, Robert E. Lucas, Jr.

Raw material Materials a manufacturer converts into a finished product.

Raw material supply agreement As used in connection with project financing, an agreement to furnish a specified amount per period of a specified raw material.

Reachback The ability of a tax shelter or limited partnership to deduct certain costs and expenses at the end of the year that were incurred throughout the entire year.

Reaction A decline in prices following an advance. Opposite of rally.

Reading the tape Judging the performance of stocks by monitoring changes in price as they are displayed on the ticker tape.

Real Used in the context of general equities.

(1) natural,

(2) not dividend roll-or program trading-related;

(3) not tax-related. "Real" indications have three major repercussions:

a) pricing will be more favorable to the other side of the trade since an investment bank is not committing any capital;

b) price pressure will be stronger if real since a natural buyer/seller may have information leading to his decision or more behind it, and,

c) an uptick may be required for the trader to transact if the indication is not real and the trader has no long position.

Real assets Identifiable assets, such as land and buildings, equipment, patents, and trademarks, as distinguished from a financial investment.

Real appreciation/depreciation A change in the purchasing power of a currency.

Real body On a candlestick line, it is the broad part consisting of the difference between opening and closing prices.

Real capital Wealth that can be represented in financial terms, such as savings account balances, financial securities, and real estate.

Real cash flow Income expressed in current purchasing power terms.

Real Currency The purchasing power in today's currency of future nominal currency to be disbursed or received.

Real estate A piece of land and whatever physical property is on it.

Real estate appraisal An estimate of the value of property using various methods.

Real estate broker An intermediary who receives a commission for arranging and facilitating the sale of a property for a buyer or a seller.

Real Estate Investment Trust (REIT) REITs invest in real estate or loans secured by real estate and issue shares in such investments. A REIT is similar to a closed-end mutual fund.

Real Estate Mortgage Investment Conduit (REMIC) A pass-through tax entity that can hold mortgages secured by any type of real property and can issue multiple classes of ownership interests to investors in the form of pass-through certificates, bonds, or other legal forms. A financing vehicle created under the Tax Reform Act of 1986.

Real exchange rates Exchange rates that have been adjusted for the inflation differential between two countries.

Real gain or loss A gain or loss adjusted for increasing prices by an inflation index such as the CPI.

Real GDP Inflation-adjusted measure of Gross Domestic Product.

Real income The income of an individual, group, or country adjusted for inflation.

Real interest rate The rate of interest excluding the effect of expected inflation; that is, the rate that is earned in terms of constant-purchasing-power dollars. Interest rate expressed in terms of real goods, i.e. nominal interest rate adjusted for expected inflation.

Real market The bid and offer prices at which a dealer could execute the desired quantity of shares. Quotes in the brokers market.

Real option An option or option-like feature embedded in a real investment opportunity.

Real property Land plus all other property that is in some way attached to the land.

Real rate of return The percentage return on some investments that has been adjusted for inflation.

Real return The actual payback on an investment after removing the effect of inflation.

Real time A real-time stock or bond quote is one that states a security's most recent offer to sell or bid (buy). Different from a delayed quote, which shows the same bid and ask prices 15 minutes and sometimes 20 minutes after a trade takes place.

Realistic on price In trading, and indication that the size under consideration requires price give, especially with illiquid stocks. See: Takes price.

Realized compound yield Yield assuming that coupon payments are invested at the going market interest rate at the time of their receipt and held thus until the bond matures.

Realized profit (or loss) A capital gain or loss on securities held in a portfolio that has become actual by the sale or other type of surrender of one or many securities.

Realized return The return that is actually earned over a given time period.

Realized volatility Sometimes referred to as the historical volatility, this term usually used in the context of derivatives. While the implied volatility refers to the market's assessment of future volatility, the realized volatility measures what actually happened in the past. The measurement of the volatility depends on the particular situation. For example, one could calculate the realized volatility for the equity market in March of 2003 by taking the standard deviation of the daily returns within that month. One could look at the realized volatility between 10:00AM and 11:00AM on June 23, 2003 by calculating the standard deviation of one minute returns.

Realized yield The holding-period return actually generated from an investment in a bond.

Realtor A specific designation given to members of real estate firms affiliated with the National Association of Realtors (NAR) who are trained and licensed to assist clients in buying and selling real estate.

Rebalancing Realigning the proportions of assets in a portfolio as needed.

Rebate Negotiated return of a portion of the interest earned by the lender of stock to a short seller. When a stock is sold short, the seller borrows stock from an owner or custodian and delivers it to the buyer. The proceeds are delivered to the lender. The borrower, who is short, often wants a rebate of the interest earned on the proceeds under the lender's control, especially when the stock can be borrowed from many sources. Note: The seller must pay the lender any dividends paid out or, in the case of bonds, interest that accrues daily during the term of the loan.

Recalculation method A method of calculating required minimum distributions from a retirement plan using life expectancy tables. Unisex data tables allow a plan holder to determine the applicable life expectancy each year a distribution is required.

Recapitalization proposal Often used in risk arbitrage. Plan by a target company to restructure its capitalization (debt and equity) in a way to ward off a hostile or potential suitor.

Recapture A provision in a contract that allows one party to recover (recapture) some degree of possession of an asset, such as a share of the profits derived from some property.

Receipts Funds collected from selling land, capital, or services, as well as collections from the public (budget receipts), such as taxes, fines, duties, and fees.

Receive fixed counterparty The transactor in an interest rate swap who receives payments based on the fixed rate and makes payments based on the floating rate.

Receive floating counterparty The transaction in an interest rate swap who receives payments based on the floating rate and makes payments based on the fixed rate.

Receive versus payment An instruction that only cash will be accepted in exchange for delivery of securities.

Receivables balance fractions The percentage of a month's sales that remains uncollected (and part of accounts receivable) at the end of succeeding months.

Receivables turnover ratio Total operating revenues divided by average receivables. Used to measure how effectively a firm is managing its accounts receivable.

Received for Shipment Bill of Lading A document issued by a carrier that looks like a bill of lading as evidence of receipt of goods for shipment. This type of document is issued prior to the vessel loading and is therefore not an on board bill of lading.

Receiver A bankruptcy practitioner appointed by secured creditors to oversee the repayment of debts.

Receiver's certificate A debt instrument issued by a receiver and serving as a lien on the property, which provides funding to continue operations or to protect assets in receivership.

Recession A temporary downturn in economic activity, usually indicated by two consecutive quarters of a falling GDP.

Recharacterization The reversal of a traditional IRA contribution or conversion into a Roth IRA, or vice versa.

Reciprocal marketing agreement A strategic alliance in which two companies agree to comarket each other's products. Production rights may or may not be transferred.

Reclamation A claim for the right to return or the right to demand the return of a security that has been previously accepted as a result of bad delivery or other irregularities in the delivery and settlement process.

Record date Date by which a shareholder must officially own shares in order to be entitled to a dividend. For example, a firm might declare a dividend on Nov. 1, payable Dec. 1 to holders of record Nov. 15. Once a trade is executed, an investor becomes the "owner of record" on settlement, which currently takes five business days for securities and one business day for mutual funds. Stocks trade ex-dividend the fourth day before the record date, since the seller will still be the owner of record and is thus entitled to the dividend. (2) The date that determines who is entitled to payment of principal and interest due to be paid on a security. The record date for most MBS is the last day of the month, although the last day on which an MBS may be presented for the transfer is the last business day of the month. The record dates for CMOs and asset-backed securities vary with each issue.

Recordholder The individual or institution listed on the corporation's books as a securityholder as of a specified record date.

Record Owner The stockholder of record as distinguished from the beneficial owner.

Recourse Term describing a type of loan. If a loan is with recourse, the lender has a the ability has the ability to fall back to the guarantor of the loan if the borrower fails to pay. For example, Bank A has a loan with Company X. Bank A sells the loan to Bank B with recourse. If Company X defaults, Bank B can demand Bank A fulfill the loan obligation.

Recovery The use of depreciation of assets to offset costs; or a new period of rising securities prices after a period of declining security values.

Redemption date The date on which a bond matures or is redeemed.

Redemption fee A fee some mutual funds charge when an investor sells shares within a specified short period of time.

Red herring A preliminary prospectus providing information required by the SEC. It excludes the offering price and the coupon of the new issue.

Redeemable Eligible for redemption under the terms of an indenture.

Redemption Repayment of a debt security or

preferred stock issue, at or before maturity, at par or at a premium price.

Redemption charge The commission a mutual fund charges an investor who is redeeming shares. For example, a 2% redemption charge (also called a back end load) on the sale of shares valued at $1000 will result in payment of $980 (or 98% of the value) to the investor. This charge may decline or be eliminated as shares are held for longer time periods.

Redemption cushion The percentage by which the conversion value of a convertible security exceeds the redemption price (strike price).

Redemption or call Right of the issuer to force holders on a certain date to redeem their convertibles for cash. The objective usually is to force holders to convert into common prior to the redemption deadline. Typically, an issue is not called away unless the conversion price is 15%-25% below the current level of the common. An exception might occur when an issuer's tax rate is high, and the issuer could replace it with debt securities at a lower after-tax cost.

Rediscount To discount short-term negotiable debt instruments for a second time, after they have been discounted with a bank.

Red-lining Illegal discrimination in making loans, insurance coverage, or other financial services available to people or property in certain areas because of poor economic conditions, high levels of fraudulent transaction, or frequent defaults.

Reduction-Option Loan (ROL) A hybrid of a fixed-rate and adjustable-rate mortgage. An ROL matches the borrower to the current mortgage rate, which then becomes fixed for the rest of the term. This reduction is usually allowed if rates drop more than 2% in a year.

Reference rate A benchmark interest rate (such as LIBOR) used to specify conditions of an interest rate swap or an interest rate agreement.

Refinancing An extension and/or increase in amount of existing debt.

Reflation Government monetary action that causes a reversal of deflation.

Refund To retire existing bond issues through the sale of a new bond issue, usually to reduce the interest rate being paid.

Refundable Eligible for refunding under the terms of a bond indenture.

Refunded bond Also called a prerefunded bond, a bond that originally may have been issued as a general obligation or revenue bond but that is now secured by an escrow fund consisting entirely of direct U.S. government obligations that are sufficient for paying the bondholders.

Refunding Redeeming a bond with proceeds received from issuing lower-cost debt obligations with ranking equal to or superior to the debt to be redeemed.

Refunding Escrow Deposits (REDs) A financial instrument involving a forward purchase contract that obligates investors to bonds at a certain rate when issued. The future date coincides with the first optional call date on an existing high-rate bond. In the interim, investors' money is invested in secondary market Treasury bonds. The Treasuries mature around the call date on the existing bonds, providing the money to buy the new issue and redeem the old one.

Regional bank A bank operating in a specific region of the country, taking deposits and offering loans.

Regional Check Processing Center (RCPC) A Federal Reserve check processing operation that clears checks drawn on depository institutions located within a specified area. RCPCs expedite collection and settlement of checks within the area on an overnight basis.

Regional fund A mutual fund that invests in a specific geographic area overseas, such as Asia or Europe.

Registered bond A bond whose issuer records ownership and interest payments. Differs from a bearer bond, which is traded without record of ownership and whose possession is the only evidence of ownership.

Registered check A check issued and guaranteed by a bank for a customer who provides funds for payment of the check.

Registered company A company that is listed with the SEC after submission of a required statement and compliance with disclosure requirements.

Registered competitive market maker An NASD-registered dealer who acts as a market maker for a designated over-the-counter stock by buying and selling that stock to maintain stability.

Registered equity market maker Member firm of the American Stock Exchange registered as a trader to make stabilizing trades for its own account in particular securities.

Registered investment adviser SEC-registered individual or firm that substantiates completion of education and work experience in the field, and pays an annual membership fee.

Registered investment company An investment firm which is registered with the SEC and complies with certain stated legal requirements.

Registered options trader An American Stock Exchange specialist who monitors a certain group of options to help maintain a fair and orderly market.

Registered Owner An individual or organization to whom certificates are directly issued and who, as a result, is recorded on the corporation's securityholder records (as maintained by the transfer agent).

Registered Retirement Savings Plan (RRSP) Tax-sheltered retirement plan for Canadian citizens, much like an American IRA.

Registered representative A person registered with the CFTC who is employed by and solicits business for a commission house or futures commission merchant.

Registered secondary offering A reoffering of a large block of securities, previously publicly issued, by the holder of a large portion of some corporation through an investment firm.

Registered security Used in the context of general equities. Securities whose owner's name is recorded on the books of the issuer or the issuer's agent, called a registrar.

Registered Shares Shares that are issued in a shareholder's name as the holder of record.

Registered trader A member of the exchange who executes frequent trades for his or her own account.

Registrar Financial institution appointed to record issue and ownership of company securities.

Registration In the securities market describes process set up pursuant to the Securities Exchange Acts of 1933 and 1934 whereby securities that are to be sold to the public are reviewed by the SEC.

Registration statement A legal document filed with the SEC to register securities for public offering that details the purpose of the proposed public offering. The statement outlines financial details, a history of the company's operations and management, and other facts of importance to potential buyers. See: Registration.

Regression A mathematical technique used to explain and/or predict. The general form is $Y = a + bX + u$, where Y is the variable that we are trying to predict; X is the variable that we are using to predict Y, a is the intercept; b is the slope, and u is the regression residual. The a and b are chosen in a way to minimize the squared sum of the residuals. The ability to fit or explain is measured by the R-square.

Regression analysis A statistical technique that can be used to estimate relationships between variables.

Regression coefficient Term yielded by

regression analysis that indicates the sensitivity of the dependent variable to a particular independent variable. See: Parameter.

Regression equation An equation that describes the average relationship between a dependent variable and a set of explanatory variables.

Regression toward the mean The tendency that a random variable will ultimately have a value closer to its mean value.

Regressive tax A tax system that provides that average tax rates decrease with increases in individuals' income brackets.

Regular settlement Transaction in which a stock contract is settled and delivered on the fifth full business day following the date of the transaction (trade date). In Japan, regular settlement occurs three business days following the trade date; in London, two weeks following the trade date (at times, three weeks); in France, once per month.

Regular way settlement In the money and bond markets, the standard basis on which some security trades are settled is that the delivery of the securities purchased is made against payment in Fed funds on the day following the transaction.

Regulated commodities The group of registered commodity futures and options contracts traded on organized U.S. futures exchanges.

Regulated investment company An investment company allowed to pass capital gains, dividends, and interest earned on fund investments directly to its shareholders so that it is taxed only at the personal level, and double taxation is avoided.

Regulations Rules specifying the appropriate behavior of agencies, organizations or individuals in the securities industry.

Regulatory accounting procedures (RAP) Accounting principles required by the FHLB that allow S&Ls to elect annually to defer gains and losses on the sale of assets and amortize these deferrals over the average life of the asset sold.

Regulatory pricing risk Risk that arises when insurance companies are subject to regulation of the premium rates that can they charge.

Regulatory surplus The surplus as measured using regulatory accounting principles (RAP), which may allow the nonmarket valuation of assets or liabilities and which may be materially different from economic surplus.

Rehypothecation Pledging to banks by securities brokers of the amount in customers' margin account as collateral for broker loans, which are used to cover margin loans to customers for margin purchases and selling short.

Reimbursement Payment made to someone for out-of-pocket expenses has incurred.

Reinstatement The restoration of an insurance policy after it has lapsed for nonpayment of premiums.

Reinsurance The spreading of risk and division of client premiums among insurance companies allowing the sharing of the burden of a large risk.

Reinvestment Use of investment income to buy additional securities. Many mutual fund companies and investment services offer the automatic reinvestment of dividends and capital gains distributions as an option investors.

Reinvestment date The date on which an investment's dividend or capital gains income is reinvested, if requested by the shareholder, to purchase additional shares. Also known as the ex-dividend date.

Reinvestment effect The impact of a change in interest rates on the reinvestment rate.

Reinvestment privilege A shareholder's right

to reinvest dividends and buy more shares in the corporation or mutual fund.

Reinvestment rate The rate at which an investor assumes interest payments made on a debt security can be reinvested over the life of that security.

Reinvestment risk The risk that proceeds received in the future may have to be reinvested at a lower potential interest rate.

Reinvoicing center A central financial subsidiary an MNC uses to reduce transaction exposure by billing all home country exports in the home currency and reinvoicing to each operating affiliate in that affiliate's localcurrency. It can also be used as a netting center.

Rejection Refusal by a bank to grant credit, usually because of the applicants financial history, or refusal to accept a security presented to complete a trade, usually because of a lack of proper endorsements or violation of rules of a firm.

Relative form of purchasing power parity Theory that the rate of change in the prices of products should be somewhat similar, but not absolutely the same when measured in a common currency, as long as transportation costs and trade barriers are unchanged.

Relative PE A firm or industry's price to earnings ratio divided by the market price to earnings ratio.

Relative purchasing power parity (RPPP) Idea that the rate of change in the price level of commodities in one country relative to the price level in another determines the rate of change of the exchange rate between the two countries' currencies.

Relative strength Movement of a stock price over the past year as compared to a market index (like the S&P 500). A value below 1.0 means the stock shows relative weakness in price movement (underperformed the market); a value above 1.0 means the stock shows relative strength over the one-year period. Equation for Relative Strength: [current stock price/year-ago stock price] divided by [current S&P 500/year-ago S&P 500]. Note: this can be a misleading indicator of performance because it does not take risk into account.

Relative strength index Used in technical analysis, it is a measure of the number of days a stock increases in value relative to the number of days it decreases in value. The rule of thumb is that values over 70 suggest overvaluation and hence selling where as values around 30 suggest undervaluation or buying. Of course, this indicator completely ignores all fundamental information about the firm's prospects and, hence, is problematic to use as a stand-alone indicator for an investment strategy.

Relative value The attractiveness measured in terms of risk, liquidity, and return of one instrument relative to another, or, for a given instrument, of one maturity relative to another.

Relative yield spread The ratio of the yield spread to the yield level. Used for bonds.

Release Relieve party to a trade of any previously made obligation concerning that trade, hence allowing the would-be transactor to show the inquiry/order to a new broker.

Release clause A mortgage provision that releases a pledged asset after a certain portion of the total payments has been made.

Reload Stock Option A replacement stock option granted by some companies to optionees upon a stock swap. The number of reload shares granted is equal to the number of shares delivered to exercise the option plus, in some cases, any shares withheld for tax withholding obligations. The exercise price of the new option is the current market price. The option generally expires on the same date that the original option would have.

Remainderman One who receives the principal of a trust when it is dissolved.

Remaining maturity The length of time remaining until a bond comes due

Remaining principal balance The amount of principal dollars remaining to be paid under a mortgage as of a given time.

Remargining Putting up additional cash or securities after a margin call on a brokerage customer's margin account so that it meets minimum maintenance requirements.

Rembrandt market The foreign market in the Netherlands.

Remit To pay for purchases by cash, check, or electronic transfer.

Remote disbursement Technique that involves writing checks drawn on banks in remote locations so as to maximize disbursement float.

Renegotiable rate A type of variable rate involving a renewable short-term "balloon" note. The interest rate on the loan is generally fixed during the term of the note, but when the balloon comes due, the lender may refinance it at a higher rate. In order for the loan to be fully amortized, periodic refinancing may be necessary.

Renewal Placement of a day order identical to one not completed on the previous day.

Renewable term life insurance A policy for a stated period that may be renewed if desired at the end of the term.

Rent Regular payments to an owner for the use of some leased property.

Rent control Municipal regulation restricting the amount of rent that a building owner can charge.

Reoffering yield In a purchase and sale, the yield to maturity at which an underwriter offers to sell bonds to investors.

Reopen an issue The Treasury, when it wants to sell additional securities, will occasionally sell more of an existing issue (reopen it) rather than offer a new issue.

Reopening Treasury offerings of additional amounts of outstanding issues, rather than an entirely new issue. A reopened issue will always have the same maturity date, CUSIP number, and interest rate as the original issue.

Reorg (or Corporate Action or Reorganization) Any transaction involving the issuance of stock or cash, or the cancellation of stock tendered by a shareholder, such as in the case of a merger, acquisition or tender offer.

Reorganization Creation of a plan to restructure a debtor's business and restore its financial health.

Reorganization bond A bond issued by a company undergoing a reorganization process.

Repatriation The return from abroad of the financial assets of an organization or individual.

Replacement Chain A concept that views a capital investment as an indefinite commitment to a specific type of technology. The replacement chain concept can be used to allow the comparison of mutually exclusive investments with unequal lives.

Replacement cost Cost to replace a firm's assets.

Replacement cost accounting An accounting method that includes as part of depreciation the difference between the original purchase price of an asset and the current replacement cost.

Replacement cost insurance Insurance that pays out the full amount required to replace damaged property with new property, without taking into account the depreciated value of the property.

Replacement cycle The frequency with which an asset is replaced by an equivalent asset.

Replacement value Current cost of replacing the firm's assets.

Replacement-chain problem Idea that future replacement decisions must be taken into account in selecting among projects.

Replicating portfolio A portfolio constructed to match an index or benchmark.

Repo An agreement in which one party sells a security to another party and agrees to repurchase it on a specified date for a specified price. See: Repurchase agreement.

Report Written or oral confirmation that all or part of one's order has been executed, including the price and size parameters of the trade being reported; often followed by a fresh picture.

Report of Condition and Income Financial report that all banks, bank holding companies, savings, and loan associations, Edge Act and agreement corporations, and certain other types of organizations must file with a federal regulatory agency. Informally termed a call report.

Reported factor The pool factor as reported by the bond buyer for a given amortization period.

Reporting currency The currency in which the parent firm prepares its own financial statements; that is, US dollars for a US company.

Representations In the context of project financing, a series of statements about a project, a sponsor, or the obligations under the project contracts or the financing agreements.

Repricing To change the price of an asset. In derivatives, it sometimes refers to the exchange of options of with different strike prices.

Reproducible assets A tangible asset with physical properties that can be matched or duplicated, such as a building or machinery.

Repurchase agreement An agreement with a commitment by the seller (dealer) to buy a security back from the purchaser (customer) at a specified price at a designated future date. Also called a repo, it represents a collateralized short-term loan for which, where the collateral may be a Treasury security, money market instrument, federal agency security, or mortgage-backed security. From the purchaser's (customer's) perspective, the deal is reported as a reverse repo.

Repurchase of stock Technique to pay cash to firm's shareholders that provides more preferential tax treatment for shareholders than dividends. Treasury stock is the name given to previously issued stock that has been repurchased by the firm. A repurchase is achieved through either a Dutch auction, open market, purchase, or tender offer.

Required Rate of Return (RRR) The minimum expected yield by investors require in order to select a particular investment.

Required reserves The dollar amounts, based on reserve ratios, that banks are required to keep on deposit at a Federal Reserve Bank.

Required return The minimum expected return you would need in order to purchase an asset, that is, to make the investment.

Required yield Generally referring to bonds; the yield required by the marketplace to match available expected returns for financial instruments with comparable risk.

Rescaled Range (R/S) Analysis The analysis developed by H.E. Hurst to determine long-memory effects and fractional Brownian motion. Rescaled range analysis measures how the distance covered by a particle increases as we look at longer and longer time scales. For Brownian motion, the distance covered increases with the square root of time. A series which increases at a different rate is not random. See: Anti-persistence, Fractional Brownian Motion, Hurst Exponent, Persistence, Joseph Effect, Noah Effect.

Rescheduled loans Bank loans that are usually altered to have longer maturities in order to assist the borrower in making the necessary repayments.

Rescind To cancel a contract because of misrepresentation, fraud, or illegal procedure.

Research and development limited partnership A partnership whose investors put up money to finance new product R&D in return for profits generated from the products.

Research department The office in an institutional investing organizations that analyzes markets and securities.

Research portable Service offered to clients that transmits investment bank research electronically by computers.

Reserve An accounting entry that properly reflects contingent liabilities.

Reserve account A separate amount of cash or letter of credit to service a payment requirement such as debt service or maintenance.

Reserve currency A foreign currency held by a central bank or monetary authority for the purposes of exchange intervention and the settlement of intergovernmental claims.

Reserve ratios Specified percentages of deposits, established by the Federal Reserve Board, that banks must keep in a noninterest-bearing account at one of the twelve Federal Reserve Banks.

Reserve requirements The percentage of different types of deposits that member banks are required to hold on deposit at the Fed.

Reservation price The price below or above which a seller or purchaser is unwilling to go.

Reset bonds Bonds that allow the initial interest rates to be adjusted on specific dates in order that the bonds trade at the value they had when they were issued.

Reset frequency The frequency with which the floating rate changes.

Residential mortgage Mortgage on a residential property, tax-deductible for individuals up to $1 million.

Residential property Property that consists of homes, apartments, townhouses, and condominiums.

Residual assets Assets that remain after sufficient assets are dedicated to meet all senior debtholders' claims in full.

Residual claim Related: Equity claim

Residual cover The cash flow remaining after a project financing has been repaid, expressed as a percentage of the original loan. Residual cushion.

Residual dividend approach An approach that suggests that a firm pay dividends if and only if acceptable investment opportunities for those funds are currently unavailable.

Residual method A method of allocating the purchase price for the acquisition of another firm among the acquired assets.

Residual Return Return independent of the benchmark. The residual return is the return relative to beta times the benchmark return. To be exact, an asset's residual return equals its excess return minus beta times the benchmark excess return.

Residual risk Related: Unsystematic risk

Residuals Part of stock returns not explained by the explanatory variable (the market index return). Residuals measure the impact of firm-specific events during a particular period. (2) Remainder cash flows generated by pool collateral and those needed to fund bonds supported by the collateral.

Residual value Usually refers to the value of a lessor's property at the time the lease expires.

Resiliency Speed with which new orders respond to a change in prices.

Resistance An effective upper bound on prices achieved because of many willing sellers at that price level.

Resistance level A price level above which it is supposedly difficult for a security or market to rise. Price ceiling at which technical analysts note persistent selling of a commodity or security. Antithesis of support level.

Resolution A document that records a decision or action by a Board of Directors, or a bond resolution by a government entity authorizing a bond issue.

Restricted Placed on a list that dictates that the trader may not maintain positions, solicit business, or provide indications in a stock, but may serve as broker in agency trades after being properly cleared. Traders are so restricted due to investment bank involvement with the company on nonpublic activity (i.e., mergers and acquisitions defense), affiliate ownership, or underwriting activities; signified on the Quotron by a flashing "R." A restricted list and the stocks on it should never be conveyed to anyone outside of the trading areas, much less outside the firm.

Restricted account A margin account without enough equity to meet the initial margin requirement that is restricted from any purchases until the requirement is fulfilled.

Restricted surplus A portion of retained earnings not allowed by law to be used for the payment of dividends.

Restricted stock Stock that must be traded in compliance with special SEC regulations concerning its purchase and resale. These restrictions generally result from affiliate ownership, M&A activity, and underwriting activity. Many firms are now using restricted stock as a reward for employees. The advantages to restricted stock are: employees get dividends, employees usually get voting rights, and employee gets something even if the stock price drops over the vesting period (whereas an option would be worthless).

Restricted Stock Award Grants of shares of stock subject to restrictions on sale and risk of forfeiture until vested by continued employment. Restricted stock typically vests in increments over a period of several years. Dividends or dividend equivalent rights may be paid, and award holders may have voting rights, during the restricted period.

Restricted stock units Similar to restricted stock. However, the unit represents a promise that employees will receive stock in the future. The units do not pay dividends until the stock is vested.

Restrictive covenants Provisions that place constraints on the operations of borrowers, such as restrictions on working capital, fixed assets, future borrowing, and payment of dividends.

Restrictive endorsement An endorsement signature on the back of a check that specifies the conditions under which the check can be transferred or paid out.

Restructuring The reorganization of a company in order to attain greater efficiency and to adapt to new markets. Major corporate restructuring include mergers, acquisitions, tender offers, leveraged buyouts, divestitures, spin-offs, equity carve-outs, liquidations and reorganizations.

Resyndication limited partnership The sale of existing properties to new limited partners, so that they can receive the tax advantages that are no longer available to the old partners.

Retail Individual and institutional customers as opposed to dealers and brokers.

Retail credit Credit granted by a firm to consumers for the purchase of goods or services. See: consumer credit.

Retail house A brokerage firm that caters to individual customers rather than large institutions.

Retail investors Small individual investors who commit capital for their personal account rather than on behalf of another company.

Retail price The total price charged for a product sold to a customer, which includes the manufacturer's cost plus a retail markup.

Retained earnings Accounting earnings that are retained by the firm for reinvestment in its operations; earnings that are not paid out as dividends.

Retained earnings statement A statement of all transactions affecting the balance of a company's retained earnings account.

Retention The number of units allocated to an underwriting syndicate member less the units held back by the syndicate manager for facilitating institutional sales and for allocation to nonmember firms. In the context of construction contracts, an amount retained from construction contract payments (5-15% of the contract price) to ensure the contractor completes the construction before the retention is returned.

Retention rate The percentage of present earnings held back or retained by a corporation, or one minus the dividend payout rate. Also called the retention ratio.

Retire To extinguish a security, as in paying off a debt.

Retirement Removal from circulation of stock or bonds that have been reacquired or redeemed.

Retracement A price movement in the opposite direction of the previous trend.

Return The change in the value of a portfolio over an evaluation period, including any distributions made from the portfolio during that period.

Return if Exercised The return that a covered call writer would make if the underlying stock were called away.

Return of capital A cash distribution resulting from the sale of a capital asset, or securities, or tax breaks from depreciation.

Return on assets (ROA) Indicator of profitability. Determined by dividing net income for the past 12 months by total average assets. Result is shown as a percentage. ROA can be decomposed into return on sales (net income/sales) multiplied by asset utilization (sales/assets).

Return on capital employed (ROCE) Indicator of profitability of the firm's capital investments. Determined by dividing Earnings Before Interest and Taxes by (capital employed plus short-term loans minus intangible assets). The idea is that this ratio should at least be greater than the cost of borrowing.

Return on equity (ROE) Indicator of profitability. Determined by dividing net income for the past 12 months by common stockholder equity (adjusted for stock splits). Result is shown as a percentage. Investors use ROE as a measure of how a company is using its money. ROE may be decomposed into return on assets (ROA) multiplied by financial leverage (total assets/total equity).

Return on investment (ROI) Generally, book income as a proportion of net book value.

Return on sales A measurement of operational efficiency equalingnet pre-tax profits divided by net sales expressed as a percentage.

Return on total assets The ratio of earnings available to common stockholders to total assets.

Return-to-maturity expectations A variant of pure expectations theory that suggests that the return an investor will realize by rolling over short-term bonds to some investment horizon will be the same as holding a zero-coupon bond with a maturity that is the same as that investment horizon.

Reuters International news and quotation service based in London.

Revaluation An increase in the foreign exchange value of a currency that is pegged to other currencies or gold.

Revenues Sales or royalty proceeds. Quantity times price sold.

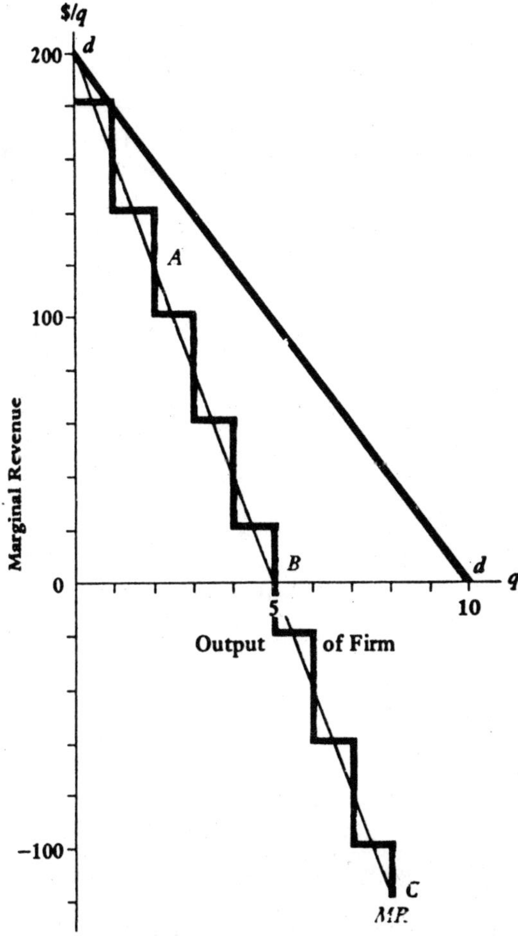

Fig. Revenue

Reversal Arbitrage A riskless arbitrage that involves selling the stock short, writing a put, and buying a call. The options have the same .

Revenue Anticipation Note (RAN) A short-term municipal debt issue that will be repaid with anticipated revenues, such as sales taxes, from the project.

Revenue bond A bond issued by a municipality to finance either a project or an enterprise in which the issuer pledges to the bondholders the revenues generated by the operation of the projects financed. Examples are hospital revenue bonds and sewer revenue bonds.

Revenue fund A fund accounting for all revenues from an enterprise financed by a municipal revenue bond.

Revenue Reconciliation Act of 1993 Legislation created to reduce the federal budget deficit by cutting spending and increasing taxes.

Revenue sharing The percentage split between the general partner and limited partners of profits and losses resulting from the operation of the involved business.

Reversal Turn, unwind. For convertible reversal, selling a convertible and buying the underlying common, usually effected by an arbitrageur. For market reversal, change in direction in the stock or commodity futures markets, as charted by technical analysts in trading ranges. For options reversal, closing the positions of each aspect of an options spread or combination strategy.

Reverse a swap Reswap of bonds to gain the advantage of a yield spread or tax loss and restore a bond portfolio to its position before the original swap.

Reverse conversion A technique in which brokerage firms earn interest on the stocks they hold for their customers by selling the short and investing the proceeds in money market accounts. The short positions are hedged to protect against adverse market conditions.

Reverse leverage Occurs when the interest on borrowings exceeds the return on investment of the funds that were borrowed.

Reverse leveraged buyout Bringing back into publicly traded status a company that had been privatized by way of a leveraged buyout.

Reverse mortgage A mortgage agreement allowing a homeowner to borrow against home equity and receive tax-free payments until the total principal and interest reach the credit limit of equity, and the lender is either repaid in full or takes the house.

Reverse price risk A type of mortgage pipeline risk that occurs when a lender commits to sell loans to an investor at rates prevailing at the time of mortgage application but sets the note rates when the borrowers closes. The lender is thus exposed to the risk of falling rates.

Reverse repo In essence, refers to a repurchase agreement. From the customer's perspective, the customer provides a collateralized loan to the seller.

Reverse stock split A proportionate decrease in the number of shares, but not the total value of shares of stock held by shareholders. Shareholders maintain the same percentage of equity as before the split. For example, a 1-for-3 split would result in stockholders owning one share for every three shares owned before the split. After the reverse split, the firm's stock price is, in this example, three times the pre-reverse split price. A firm generally institutes a reverse split to boost its stock's market price. Some think this supposedly attracts investors.

Reversion See: Mean reversion.

Reversing trade Entering the opposite side of a currently held futures position to close out the position.

Revised estimate The third estimate of GDP released about three months after the measurement period.

Revisionary trust An irrevocable trust that becomes a revocable trust after a certain amount of time.

Revocable letter of credit Assurance of funds issued by a bank that can be canceled at any time without prior notification to the beneficiary.

Revocable trust A trust that may altered as many times as desired in which income-producing property passes directly to the beneficiaries at the time of the grantor's death. Since the arrangement can be altered at any time, the assets are considered part of the grantor's estate and they are taxed as such.

Revolving credit agreement A legal commitment in which a bank promises to lend a customer up to a specified maximum amount during a specified period.

Revolving line of credit A bank line of credit on which the customer pays a commitment fee and can take and repay funds at will. Normally a revolving LOC involves a firm commitment from the bank for a period of several years.

Reward-to-volatility ratio Ratio of excess return to portfolio standard deviation.

Rich Term for a security whose price seems too high in light of its price history.

RICO Stands for Racketeer Influenced and Corrupt Organization Act. Legislation under/which inside traders may be convicted.

Rider A form accompanying an insurance policy that alters the policy's terms or coverage.

Riding the yield curve Buying long-term bonds in anticipation of capital gains as yields fall with the declining maturity of the bonds.

Riegle-Neal Interstate Banking and Branching Efficiency Act of 1994 Law permitting interstate banking in the U.S.

Rigged market Manipulation of prices in a market to attract buyers and sellers.

Right Privilege granted shareholders of a corporation to subscribe to shares of a new issue of common stock before it is offered to the public. Such a right, which normally has a life of two to four weeks, is freely transferable and entitles the holder to buy the new common stock below the public offering price. See: Warrant.

Right here Used in the context of general

equities. In-line, emphasizing that this is a customer inquiry that is ready to be executed and not distant on price. See: Tight.

Right of accumulation Allow investors to combine prior mutual fund purchases with current purchases in the same mutual fund or related mutual fund family to qualify for a breakpoint and obtain a lower sales charge.

Rights offering Issuance to shareholders that allows them to purchase additional shares, usually at a discount to market price. Holdings of shareholders who do not exercise rights are usually diluted by the offering. Rights are often transferable, allowing the holder to sell them on the open market to others who may wish to exercise them. Rights offerings are particularly common to closed-end funds, which cannot otherwise issue additional common stock.

Right of first refusal The right of a person or company to purchase some thing before the offering is made to others.

Right of redemption The right to recover property that has been attached by paying off the debt .

Right of rescission The right to void a contract without any penalty within three days as provided in the Consumer Credit Protection Act of 1968.

Rights Agreement (aka "Poison Pill") An anti-takeover arrangement often established by a company in anticipation of a hostile takeover attempt. The company appoints a Rights Agent who will issue Rights Certificates to each shareholder at the time of the takeover attempt. The shareholder may then exercise these rights to receive additional shares of stock and/or debentures, making the target company more expensive to acquire as a result of the additional shares outstanding, or the additional debt.

Rights Offering A popular means of raising capital by offering shareholders the opportunity to buy additional shares of the same stock at a price below the current market value.

Rights-on Shares trading with rights attached to them.

Rights of set-off An agreement defining each party's rights should one party default on its obligation. A setoff is common in parallel loan arrangements.

Rings Trading arenas located on the floor of an exchange in which traders execute orders. Sometimes called a pit.

"Ring the cash register" Used in the context of general equities. "Take a profit." See: Profit taking.

Rio de Janeiro Stock Exchange (Bolsa do Rio) Brazil's major securities market.

Rising bottoms Chart pattern showing an increasing trend in the daily low prices of a security or commodity.

Risk Often defined as the standard deviation of the return on total investment. Degree of uncertainty of return on an asset. In context of asset pricing theory. See: Systematic risk.

Risk-adjusted discount rate The rate established by adding a expected risk premium to the risk-free rate in order to determine the present value of a risky investment.

Risk-adjusted profitability A probability used to determine a "sure" expected value (sometimes called a certainty equivalent) that would be equivalent to the actual risky expected value.

Risk-adjusted return Often we subtract from the rate of return on an asset a rate of return from another asset that has similar risk. This gives an abnormal rate of return that shows how the asset performed over and above a benchmark asset with the same risk. We can also use the beta against the benchmark to calculate an alpha, which is also risk-adjusted performance.

Risk arbitrage Traditionally, the simultaneous

purchase of stock in a company being acquired and the sale of stock of the acquirer. Modern risk arbitrage focuses on capturing the spreads between the market value of an announced takeover target and the eventual price at which the acquirer will buy the target's shares.

Risk-averse Describes an investor who, when faced with two investments with the same expected return but different risks, prefers the one with the lower risk.

Risk-based capital ratio Bank requirement that there be a minimum ratio of estimated total capital to estimated risk-weighted asset.

Risk classes Groups of projects that have approximately the same amount of risk.

Risk controlled arbitrage A self-funding, self-hedged series of transactions that generally use mortgage securities (MBS) as the primary assets.

Risk factor In arbitrage pricing theory or the multibeta capital asset pricing model, the set of common factors that impact returns, e.g., market return, interest rates, inflation, or industrial production.

Risk-Free Interest Rate Describes return available to an investor in a security somehow guaranteed to produce that return. The risk-free interest rate compensataes the investor for the temporary sacrifice of consumption.

Risk indexes Categories of risk used to calculate fundamental beta, including.

(1) market variability,

(2) earnings variability,

(3) low valuation,

(4) immaturity and smallness, (5) growth orientation, and (6) financial risk.

Risk profile The slope of a line graphed according to the value of an underlying asset on the x-axis and the value of a position exposed to risk in the underlying asset on the y-axis. Also used with changes in value. See: Payoff profile.

Risk-return trade-off The tendency for potential risk to vary directly with potential return, so that the more risk involved, the greater the potential return, and vice versa.

Risk tolerance An investor's ability or willingness to accept declines in the prices of investments while waiting for them to increase in value.

Riskless arbitrage The simultaneous purchase and sale of the same asset to yield a profit.

Riskless or risk-free asset An asset whose future return is known today with certainty. The risk-free asset is commonly defined as short-term obligations of the US government.

Riskless rate The rate earned on a riskless investment, typically the rate earned on the 90-day US Treasury Bill.

Riskless rate of return The rate earned on a riskless asset.

Riskless transaction A transaction that is guaranteed a profit, such as the arbitrage of a temporary differential between commodity prices in two different markets. The evaluation of whether dealer markups and markdowns in OTC transactions are reasonable. According to NASD, markups or markdowns should not exceed 5%.

Risk lover A person willing to accept lower expected returns on prospects with higher amounts of risk.

Risk management The process of identifying and evaluating risks and selecting and managing techniques to adapt to risk exposures.

Risk-neutral Insensitive to risk.

Risk-prone Willing to pay money to assume risk from others.

Risk premium The reward for holding the risky market portfolio rather than the risk-free asset. The spread between Treasury and non-Treasury bonds of comparable maturity.

Risk premium approach A common approach for tactical asset allocation to determine the relative valuation of asset classes based on expected returns.

Risk profile A mapping of the change in value or profits and to which an organization has exposure.

Risk-return tradeoff The basic concept that higher expected returns accompany greater risk, and vice versa.

Risk-reward ratio Relationship of substantial reward corresponding to the amount of risk taken; mathematically represented by dividing the expected return by the standard deviation.

Risk seeker Investor who likes to take risk and is even willing to pay for it. Also called risk lover.

Risk transfer The shifting of risk through insurance or securitization of debt because of risk aversion.

Risky asset An asset whose future return is uncertain.

Risk-adjusted return Return earned on an asset normalized for the amount of risk associated with that asset.

Risk-free asset An asset whose future normal return is known today with certainty.

Risk-free rate The rate earned on a riskless asset.

Road show A promotional presentation by an issuer of securities to potential buyers about the desirable qualities of the investments.

Rotation An active asset management strategy that tactically overweighted and underweighted certain sectors, depending on expected performance. Sometimes called sector rotation.

Rocket scientist An employee of an investment firm (often having a Ph.D. in physics or mathematics) that works on highly mathmatic models of derivative pricing.

Roll down To move to an option position with a lower exercise price.

Roll forward To move to an option position with a later expiration date.

Roll, Richard Author of path-breaking work on asset pricing including the famous Roll critique. Finance professor at UCLA.

Roll order Dividend roll; (2) Replacement of a maturing position with an identical one in the new maturity; (3) Recognizition of capital gain or loss while reestablishing the position at the risk of the market.

Roll over To reinvest funds received from a maturing security in a new issue of the same or a similar security.

Roll up To move to an option position with a higher exercise price. In venture capital, refers to the venture capitalist forcing small firms to merge operations in order to reduce costs

Rolling of Futures As financial futures have short-term maturities, often 3-9 months, before or at maturity, the future must be sold and a new future (for the same asset but with a new maturity) must be repurchased.

Rollover Means that a loan is periodically repriced at an agreed spread over the appropriate, currently prevailing rate. Most term loans in the Euromarket are made on a rollover basis as to current LIBOR rate.

Rollover IRA A traditional individual retirement account holding money from a qualified plan or 403(b) plan. These assets, as long as they are not mixed with other contributions, can later be rolled over to another qualified plan or 403(b) plan. Also known as a conduit IRA.

Roll's Critique That the CAPM holds by construction when performance is measured against a mean-variance efficient index; otherwise, it holds not at all. Attributable to Richard Roll in 1977.

Ross, Stephen Developer of the Arbitrage Pricing Theory. Finance professor at MIT.

Roth IRA Individual Retirement Account that allows contributors to make annual contributions and to withdraw the principal and earnings tax-free under certain conditions. Maximum annual contributions are $3,000 per year (phasing up to $4,000 per year in 2005 and $5,000 per year in 2008.

Round lot A trading order typically of 100 shares of a stock or some multiple of 100. Related: odd lot.

Round-trip trade The purchase and sale of a security within a short period of time.

Round-trip transactions costs Costs of completing a transaction, including commissions, market impact costs, and taxes.

Round-turn Procedure by which the long or short position of an individual is offset by an opposite transaction or by accepting or making delivery of the actual financial instrument or physical commodity.

Royalty Payment for the right to use intellectual property or natural resources.

Rubber check A check that bounces for lack of funds.

R square (R^2) Square of the correlation coefficient. The proportion of the variability in one series that can be explained by the variability of one or more other series a regression model. A measure of the quality of fit. 100% R-square means perfect predictability.

Rule lOb-5 An SEC rule that prohibits trading by insiders on material nonpublic information. This is also the rule under which a company may be sued for false or misleading disclosure.

Rule 13-d Often used in risk arbitrage. Requirement under Section 13-d of the Securities Act of 1934 that a form must be filed with the SEC within ten business days of acquiring direct or beneficial ownership of 5% or more of any class of equity securities in a publicly held corporation. The purchaser of such stock must also file a 13-d with the stock exchange on which the shares are listed (if any) and the company itself. Required information includes the way the shares were acquired, the purchaser's background, and future plans regarding the target company. The law is designed to protect against insidious takeover attempts and to keep the investing public aware of information that could affect the price of their stock. See: Williams Act.

Rule 14-d Often used in risk arbitrage. Regulations and restrictions covering public tender offers and related disclosure requirements.

Rule 144 Restricts solicitation of buyers to complete the sell order of an insider (unless the firm is already a buyer); signified by a flashing "E" on Quotron.

Rule 144a SEC rule allowing qualified institutional buyers to buy and trade unregistered securities.

Rule 405 NYSE codification of "know your customer" rules, which require that a customer's situation is suitable for any investment being made.

Rule 415 Permits corporations to file a registration for securities they intend to issue in the future when market conditions are favorable. See: Shelf registration.

Rule of Absolute Priority A condition of bankruptcy proceedings under which junior (subordinated) claim holders can receive no payment until senior (priority) claim holders are paid in full.

Rule of 72 A formula used to determine the amount of time it will take for invested money to double at a given compound interest rate, which is 72 divided by the interest rate. The logic is as follows. The time for an amount A to double is given by 2A=A(1+i)^t where ^ represents exponent and i is the interest rate, e.g. .05 is 5%. The A term cancels from both sides of the question. Solve for t by taking the natural log of both sides of the equation. Hence, t= [ln(2) over {ln(1+i)}], which is approximately equal to 0.72 over i. Hence the rule of 72.

Rules of fair practice Rules established by the NASD that lay down guidelines for just and equitable principles of trade and business in securities markets.

Rumortrage A term combining the words "rumor" and arbitrage, used to describe trading that occurs on the basis of rumors of a takeover.

Rump Usually used in the context of a merger or acquisition. A group of shareholders who refuse to tender their shares for a merger or acquisition. In a merger of Company A and Company B for example, if a sufficient number of Company B shareholders do not tender their shares, the new company will not be able to access the cash flows of Company B.

Run A run consists of a series of bid and offer quotes for different securities or maturities. Dealers give and ask for runs from each other.

Rundown A summary of the amount and prices of a serial bond issue that is still available for purchase.

Running ahead The illegal practice of trading in a security for a broker's personal account before placing an order for the same security for a customer.

Runoff Used for listed equity securities. Series of trades printed on the ticker tape that occur on the NYSE before 4:00 p.m., but are not reported until afterwards due to heavy trading that makes the tape late.

Russell Indexes US equity index widely used by pension and mutual fund investors that are weighted by market capitalization and published by the Frank Russell Company of Tacoma, Washington. For example, the Russell 3000 index includes the 3,000 largest US companies according to market capitalization.

Russell 1000 A market capitalization-weighted benchmark index made up of the 1000 highest-ranking US stocks in the Russell 3000.

Russell 2000 A market capitalization-weighted benchmark index made up of the 2000 smallest US companies in the Russell 3000.

Russell 3000 A market capitalization-weighted benchmark index made up of the 3000 largest US stocks, which represent about 98% of the US equity market.

Russian Exchange Russia's major securities market.

Russian Trading System (RTS) An electronic system in Russia, like the Nasdaq system on which the majority of Russian equities trading is conducted.

S

Safe harbor Often used in risk arbitrage as a form of shark repellent. A target company acquires a business so onerously regulated that it makes the target less attractive, giving it, in effect, a safe harbor.

Safe harbor lease A lease to transfer tax benefits of ownership (depreciation and debt tax shield) from the lessee, if the lessee could not use them, to a lessor that could use them.

Safekeep Holding by a bank of bonds and money market instruments. For a fee, the bank clips coupons and presents for payment at maturity.

Safety cushion In a contingent immunization strategy, the difference between the initially available immunization level and the safety-net return.

Safety-net return The minimum available return that will trigger an immunization strategy in a contingent immunization strategy.

Salary Regular wages and benefits an employee receives from an employer.

Salary freeze A temporary halt to increases in salary due to financial difficulties experienced by a company.

Salary reduction plan A plan allowing employees to contribute pre-tax income to a tax-deferred retirement plan.

Salary Reduction Simplified Employee Pension Plan (SARSEP) A low-cost, no-frills version of a 401(k) employee savings plan available to companies with 25 or fewer employees. It allows employees to make pretax contributions to their IRAs through salary reduction each year. The Small Business Job Protection Act of 1996 replaced SARSEPs with SIMPLE (Savings Incentive Match Plan for Employees) plans. Existing SARSEPs were allowed to add new participants, but new plans could not be formed after December 31, 1996.

Sale An agreement between a buyer and a seller on the price to be paid for a security, followed by delivery.

Sale and lease-back Sale of an existing asset to a financial institution that then leases it back to the user. Related: Lease.

Sales charge The fee charged by a mutual fund at purchase of shares, usually payable as a commission to a marketing agent, such as a financial adviser, who is thus compensated for assistance to a purchaser. It represents the difference, if any, between the share purchase price and the share net asset value.

Sales completion In the context of project financing, the state in which the project has reached physical completion and has delivered product or generated revenues in satisfaction of a sales completion test.

Sales Contract Contract between a seller and buyer for the sale of goods, services, or both.

Sales forecast A key input to a firm's financial planning process. External sales forecasts are based on historical experience, statistical analysis, and consideration of various macroeconomic factors.

Sales literature Material written by an institution selling a product, which informs potential buyers of the product and its benefits.

Sales load See: Sales charge

Sales tax A percentage tax on the selling price of goods and services.

Sales-type lease The leasing out of a firm's own equipment, such as a printing company leasing its own presses, thereby competing with an independent leasing company.

Sallie Mae See: Student Loan Marketing Association

Salomon Brothers World Equity Index (SBWEI) A top-down, float capitalization-weighted index used to measure the performance of fixed-income and equity markets. It includes approximately 6000 companies in 22 countries.

Salomon Brothers Non-U.S. Dollar World Government Bond Index A benchmark index that includes institutionally traded bonds other than U.S. issues that have a fixed rate and a remaining maturity of one year or longer.

Salvage value Scrap value of plant and equipment.

Same-Day Funds Settlement (SDFS) A method of settlement used in trading between well-collateralized parties in good-the-same-day federal funds used by the Depository Trust Company for transactions in US government securities, short-term municipal notes, medium-term commercial paper notes, CMOs, and other instruments.

Same-day substitution Offsetting changes in a margin account during the day that result in no overall change in the balance of the account.

Samurai bond A yen-denominated bond issued in Tokyo by a non-Japanese borrower. Related: Bulldog bond and Yankee bond.

Samurai market The foreign market in Japan.

Santa Claus Rally Seasonal rise in stock prices in the last week of the calendar year, between Christmas and New Year's Day.

Sao Paulo Stock Exchange See: Bolsa de Valores de Sao Paulo

S&P Standard & Poor's Corporation.

S&P 500 Composite Index Index of 500 widely held common stocks that measures the general performance of the market.

S&P phenomenon Tendency of stocks newly added to the S&P composite index to rise in price due to a large number of buy orders as S&P-related index funds add the stock to their portfolios.

S&P Rating Rating service provided by S&P that indicates the amount of risk involved with different securities.

Sarbanes Oxley Act of 2002 Legislation passed largely as a result of a number of accounting scandals. Among the many features is the creation of the Public Company Accounting Oversight Board. This board is charged to: The Board shall: 1) register public accounting firms; 2) establish, or adopt, by rule, auditing, quality control, ethics, independence, and other standards relating to the preparation of audit reports for issuers; (3) conduct inspections of accounting firms; (4) conduct investigations and disciplinary proceedings, and impose appropriate sanctions; (5) perform such other duties or functions as necessary or appropriate; (6) enforce compliance with the Act, the rules of the Board, professional standards, and the securities laws relating to the preparation and issuance of audit reports and the obligations and liabilities of accountants with respect thereto; (7) set the budget and manage the operations of the Board and the staff of the Board.

Saturday night special Often used in risk arbitrage. Sudden attempt by one company to take over another by making a public tender offer.

Saucer Technical chart pattern depicting a security whose price has reached bottom and is moving up.

Savings Association Insurance Fund (SAIF) A government organization that replaced the Federal Savings and Loan Insurance Corporation as the provider of deposit insurance for thrift institutions.

Savings bank An institution that primarily accepts consumer savings deposits and to make home mortgage loans.

Savings bond A government bond issued in face value denominations from $50 to $10,000, with local and state tax-free interest and semiannually adjusted interest rates.

Savings deposits Accounts that pay interest, typically at below-market interest rates, that do not have a specific maturity, and that usually can be withdrawn upon demand.

Savings element Used in the context of life insurance, the cash value built up in a policy, which equals the amount of premium paid minus the cost of protection. This excess is invested by the insurance company, and the returns are tax-deferred inside the policy.

Savings Incentive Match Plan for Employees (SIMPLE) 401(k) plan A tax-deferred retirement savings plan similar to a conventional 401(k) plan, redesigned with specific rules to meet the needs of small employers. The Small Business Job Protection Act of 1996 created these plans for companies with fewer than 100 employees. An employee's contributions are indexed for inflation, and employers must make annual annual matching contributions.

Savings and loan association National- or state-chartered institution that accepts savings deposits and invests the bulk of the funds thus received in mortgages.

Savings rate Personal savings as a percentage of disposable personal income.

Say's Law of Markets A law which was formulated by the French economist J.B. Say. This law states that every increase in the supply of goods is a corresponding increase in the demand for goods so long as producers have directed thier production in accordance with each other's wants. Hence, it was argued, a general over-production of goods cannot take place.

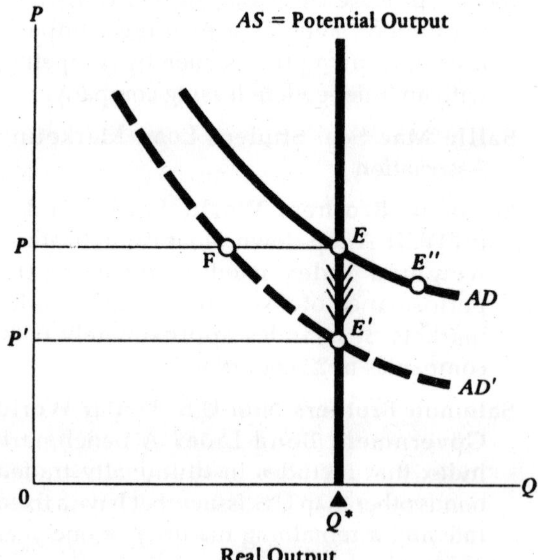

Fig. Say's Law

Scalability The ability to increase the size of a business while either maintaining or increasing its profit margin.

Scale Payment of different rates of interest on CDs of varying maturities. A bank is said to "post a scale." Commercial paper dealers also post scales.

Scale-enhancing Describes a project that is in the same risk class as the whole firm. That is, the project allows the firm to grow larger in the context of their current business rather than diversify into new businesses.

Scale in Gradually taking a position in a security or market over time.

Scale order Order to buy (sell) a security that specifies the total amount to be bought (sold) and the amount to be bought (sold) at successively decreasing (increasing) price intervals; often placed in order to average the price.

Scaling How the characteristics of an object change as you change the size of your measuring device. For a three dimensional object, it could be the volume of an object covered as you increase the radius of a covering sphere. In a times series, it could be the change in the amplitude of the time series as you increase the increment of time.

Scalp To trade for small gains. Scalping normally involves establishing and liquidating a position quickly, usually within the same day.

Scalping Buying up the good IPOs.

Scattered Used for listed equity securities. Unconcentrated buy or sell interest.

Scenario analysis The use of horizon analysis to project total returns under different reinvestment rates and future market yields.

Schedule C Describes membership requirements and procedures of NASD, in its bylaws.

Schedule 13d Disclosure form required when more than 5% of any class of equity securities in a publicly held corporation is purchased.

Scheduled cash flows The mortgage principal and interest payments due to be paid under the terms of the mortgage, not including possible prepayments.

Scorched-earth policy Often used in risk arbitrage. Any technique a company that has become the target of a takeover attempt uses to make itself unattractive to the acquirer. For example, it may agree to sell off its crown jewels, or schedule all debt to become due immediately after a merger.

S Corporation A corporation that elects not to be taxed as a corporation. That is, the corporation does not directly pay federal income tax on its earnings. Similar to a partnership, it passes its income or losses and other tax items on to its shareholders.

Screen stocks To analyze various stocks in search of stocks that meet predetermined criteria. For example, a simple value screen would sort all stocks by their price-to-book ratio and pick the stocks with the lowest ratios as candidates for the value portfolio.

Scrip A temporary document that represents a portion of a share of stock, often issued after a stock split or spin-off.

Scripophily Collecting stock and bond certificates for their scarcity, rather than for their value as securities.

Search costs Costs associated with locating a counterparty to a trade, including explicit costs (such as advertising) and implicit costs (such as the value of time). Related: Information costs.

Seasonally adjusted Mathematically adjusted by moderating a macroeconomic indicator (e.g., oil prices/imports) so that relative comparisons can be drawn from month to month all year.

Seasoned In the case of equity, having gained a reputation for quality with the investing public and enjoying liquidity in the secondary market; in the case of convertibles, having traded for at least 90 days after issue in Europe, and thus available for sale legally to U.S. investors.

Seasoned datings Extended credit for customers who order goods in periods other than peak seasons.

Seasoned issue Issue of a security for which there is an existing market. Related: Unseasoned issue.

Seasoned new issue A new issue of stock after

the company's securities have previously been issued. A seasoned new issue of common stock can be made using a cash offer or a rights offer.

Seat Position of membership on a securities or commodity exchange, bought and sold at market prices.

Sec fee Small fee the SEC charges to sellers of equity securities on an exchange.

Second market The OTC market.

Second pass regression A cross-sectional regression of portfolio returns on betas. The estimated slope is the measurement of the reward for bearing systematic risk during the period analyzed.

Second-preferred stock Preferred stock issue that has less priority in claiming dividends and assets in liquidation than another issue of preferred stock.

Second round Stage of venture capital financing following the start-up and first round stages and before the mezzanine level stage.

Second-to-die insurance Insurance policy that, on the death of the spouse dying last, pays a death benefit to the heirs that is designed to cover estate taxes.

Secondary distribution/offering Public sale of previously issued securities held by large investors, usually corporations or institutions, as distinguished from a primary distribution, where the seller is the issuing corporation. The sale is handled off the NYSE, by a securities firm or a group of firms, and the shares are usually offered at a fixed price related to the current market price of the stock.

Secondary issue Procedure for selling blocks of seasoned issues of stocks.

(2) More generally, sale of already issued stock.

Secondary Offering An IPO in which privately held shares in a corporation are sold to the public.

Secondary market The market in which securities are traded after they are initially offered in the primary market. Most trading occurs in the secondary market. The New York Stock Exchange, as well as all other stock exchanges and the bond markets, are secondary markets. Seasoned securities are traded in the secondary market.

Secondary mortgage market Buying and selling existing mortgage loans, which are often pooled and traded as mortgage-backed securities.

Secondary stocks Stocks with smaller market capitalization, less quality and more risk than blue chip issues that behave differently than larger corporations' stocks.

Second mortgage lending Loans secured by real estate previously pledged in a first mortgage.

Secert Ballot In the context of corporate governance, this is also known as confidential voting. An independent third party or employees sworn to secrecy are used to count proxy votes, and the management usually agrees not to look at individual proxy cards. This can help eliminate potential conflicts of interest for fiduciaries voting shares on behalf of others, or can reduce pressure by management on shareholder-employees or shareholder-partners.

Section 16(a) Provision of the Securities Exchange Act of 1934 that requires company insiders to file periodic reports disclosing their holdings and changes in beneficial ownership of the company's equity securities.

Section 16(b) Provision of the Securities Exchange Act of 1934 that requires that any profit realized by a company insider from the purchase and sale, or sale and purchase, of the company's equity securities within a period of less than six months must be returned to the company. It is also known as the "short-swing profit" rule.

Section 83(b) Election A tax filing within 30 days of grant that allows employees granted stock to pay taxes on the grant date instead of on the date restrictions lapse. If an employee files the election, taxes are based on the fair market value on the grant date, with any future appreciation taxed as a capital gain. If the employee does not file an election, taxes are based on the fair market value on the date the restrictions lapse, which will be higher assuming the stock has appreciated in value.

Section 423 The government agency responsible for the supervision and regulation of the securities industry and markets, as well as public securities offerings and the ongoing disclosure obligations of public companies.

Section 482 US Department of Treasury regulations governing transfer prices.

Sector Used to characterize a group of securities that are similar with respect to maturity, type, rating, industry, and/or coupon.

Sector allocation Investment of certain proportions of a portfolio in certain sectors. See: Industry allocation.

Sector diversification Constituting of a portfolio of stocks of companies in each major industry group.

Sector fund A mutual fund that concentrates on a relatively narrow market sector. These funds can experience higher share price volatility than some diversified funds because sector funds are subject to common market forces specific to a given sector.

Sector rotation An active asset management strategy certain sectors, that tactically overweights and underweights depending on expected performance. Sometimes called rotation.

Secular Long-term time frame (10-50 years or more).

Secured bond A bond backed by the pledge of collateral, a mortgage, or other lien, as opposed to an unsecured bond, called a debenture .

Secured debt Debt that has first claim on specified assets in the event of default.

securities Paper certificates (definitive securities) or electronic records (book-entry securities) evidencing ownership of equity (stocks) or debt obligations (bonds).

Securities Act of 1933 First law designed to regulate securities markets, requiring registration of securities and disclosure.

Securities Acts Amendments of 1975

Legislation to encourage the establishment of a national market system together with a system for nationwide clearing and settlement of securities transactions.

Securities analysts Related: Financial analysts

Securities and commodities exchanges Exchanges on which securities, options, and futures contracts are traded by members for their own accounts and for the accounts of customers.

Securities & Exchange Commission (SEC) A federal agency that regulates the US financial markets. The SEC also oversees the securities industry and promotes full disclosure in order to protect the investing public against malpractice in the securities markets.

Securities and Exchange Commission Rules Rules enacted by the SEC to assist in the regulation of US financial markets.

Securities Exchange Act of 1934 Legislation that created the SEC, outlawing dishonest practices in the trading of securities.

Securities Exchange of Thailand (SET) The only stock market in Thailand, based in Bangkok.

Securities Industry Association (SIA) An association of broker-dealers who sell taxable securities, which lobbies the government, records industry trends, and keeps records of broker profits.

Securities Industry Committee on Arbitration (SICA) A private group that provides mediation services in case of customer complaints against securities firms.

Securities Investor Protection Corporation (SIPC) A nonprofit corporation that insures customers' securities and cash held by member brokerage firms against the failure of those firms.

Securities loan The loan of securities between brokers, often to cover a client's short sale; or a loan secured by marketable securities.

Securities markets Organized exchanges plus over-the-counter markets in which securities are traded.

Securitization Creating a more or less standard investment instrument such as the mortgage pass-through security, by pooling assets to back the instrument. Also refers to the replacement of nonmarketable loans and/or cash flows provided by financial intermediaries with negotiable securities issued in the public capital markets.

Security Piece of paper that proves ownership of stocks, bonds, and other investments.

Security characteristic line A plot on a graph of the excess return on a security over the risk-free rate as a function of the excess return on the market. The slope of this line is the security's beta.

Security deposit (initial) Synonymous with the term margin. A cash amount that must be deposited with the broker for each contract as a guarantee of fulfillment of the futures contract. It is not considered as part payment or purchase. Related: Margin.

Security deposit (maintenance) Related: Maintenance margin

Security Industry Automated Corporation (SIAC) Entity that executes automated DOT orders.

Security interest The creditor's right to take property or a portion of property offered as security.

Security market line Line representing the relationship between expected return and market risk or beta. The slope of this line is the risk premium for beta.

Security Market Line The linear relationship between expected asset returns and betas posited by the Capital Asset Pricing Model.

Security market plane A plane that shows the relationship between expected return and the beta coefficient of more than one factor.

Security ratings Commercial rating agencies' assessment of the credit and investment risk of securities.

Security selection See: Security selection decision

Security selection decision Choosing the particular stocks or bonds or other investment instruments to include in a portfolio.

Seed money The first contribution by a venture capitalist toward the financing of a new business, often using a loan or purchase of convertible bonds or preferred stock. See: Mezzanine level and second round.

Seek a market Search for a securities buyer or seller.

Segmented Market A market in which there are impediments to the free flow of labor, capital, and information.

Segregation of securities SEC rules to dictate how customers' securities may be used by broker-dealers in broker loans.

Seigniorage The amount of goods and services that the government obtains by printing new money in a given period. Often we consider this in real terms, by dividing the new money by the price level.

Select ten portfolio A unit investment trust that buys and holds for one year the ten stocks in the Dow Jones Industrial Average with the highest dividend yields.

Selective hedging Protecting investments during some time periods and not during others.

Selected dealer agreement The set of rules governing the selling group in an underwriting.

Self-amortizing mortgage Mortgage whose entire principal is paid off in a specified period of time with regular interest and principal payments.

Self-directed IRA An IRA that the account holder can after appointing a custodian manager to carry out investment instructions.

Self-employed income Taxable income of a person involved in a sole proprietorship or other sort of free-lance work.

Self-employment tax A tax self-employed people must pay to qualify them to receive Social Security benefits at retirement.

Self-liquidating loan Loan to finance current assets. The sale of the current assets provides the cash to repay the loan.

Self-regulatory organization (SRO) Organizations that enforce fair, ethical, and efficient practices in the securities and commodity futures industries, including all national securities and commodities exchanges and the NASD.

Self-selection Consequence of a contract that induces only one group to participate.

Self-Similar When small parts of an object are qualitatively the same, or similar to the whole object. In certain deterministic fractals, like the Sierpinski Triangle, small pieces look the same as the entire object. In random fractals, small increments of time will be statistically similar to larger increments of time. See: Fractal.

Self-supporting debt Bonds sold to finance a project that will produce enough revenue through tolls or other charges to retire the debt . See: revenue bond.

Self Tender A company buys back a certain percentage of its own shares through a tender offer.

Self-tender offer A company that tenders for its own shares.

Sell the book Used for listed equity securities. Order to a broker by the holder of a large quantity of shares of a security to sell all that can be absorbed at the current bid price. The term derives from the specialist's book - the record of all the buy and sell orders members have placed in the stock one handles. In this scenario, the buyers potentially include those in the specialist's book, the specialist for its own account, and broker-dealers.

Sell hedge Related: short hedge.

Sell limit order Conditional trading order that indicates that a security may be sold at the designated price or higher. Related: Buy limit order.

Sell off Sale of securities under pressure. See: Dumping.

Sell order An order that may take many different forms by an investor to a broker to sell a particular stock, bond, option, future, mutual fund, or other holding.

Sell out Liquidation of a margin account after a customer has failed to bring an account to a required level by producing additional equity after a margin call. The selling of securities by a broker when a customer fails to pay for them. The complete sale of all securities in a new issue.

Sell plus order Market or limit order to sell a stated amount of stock provided that the price to be obtained is not lower than the last sale if the last sale was a plus, or zero plus tick, and is not lower than the last sale plus the minimum fractional change in the stock if the last sale was a minimum or zero minimum tick. (In a limit order, sale cannot be lower than the limit, regardless of tick.)

Sell price See: Redemption price

Sell-side analyst A financial analyst who works for a brokerage firm and whose recommendations are passed on to the brokerage firm's customers. Also called Wall Street analyst.

Seller financing Funding a purchase by a seller's loan to the buyer, the buyer takes full title to the property when the loan is fully repaid.

Seller's market Market in which demand exceeds supply. As a result, the seller can dictate the price and the terms of sale.

Seller's option Delayed settlement/delivery in a transaction.

Seller's points In reference to a loan, seller's points consist of a lump sum paid by the seller to the buyer's creditor to reduce the cost of the loan to the buyer. This payment is either required by the creditor or volunteered by the seller, usually in a loan to buy real estate. Generally, one point equals one percent of the loan amount.

Selling climax A sudden drop in security prices as sellers dump their holdings.

Selling concession The discount underwriters offer the selling group on securities in a new issue.

Selling dividends Inducing a prospective customer tobuy shares in order to profit from a dividend scheduled in the near future.

Selling, general, and administrative (SG&A) expenses Expenses such as salespersons' salaries and commissions, advertising and promotion, travel and entertainment, office payroll and expenses, and executives' salaries.

Selling on the good news A strategy of selling stock shortly after a company announces good news and the stock price rises. Investors believe that the price is as high as it can go and is on the brink of going down.

Selling group All banks involved in selling or marketing a new issue of stock or bonds.

Selling short Selling a stock not actually owned. If an investor thinks the price of a stock is going down, the investor could borrow the stock from a broker and sell it. Eventually, the investor must buy the stock back on the open market. For instance, you borrow 1000 shares of XYZ on July 1 and sell it for $8 per share. Then, on Aug. 1, you purchase 1000 shares of XYZ at $7 per share. You've made $1000 (less commissions and other fees) by selling short.

Selling short against the box Selling short stock that is actually owned by the seller but held in the box, meaning it is held in safekeeping. The seller borrows securities needed to cover as the stock in the box may be inaccessible, or the seller may not wish to disclose ownership. The traditional motive for this transaction was to defer capital gains taxes. However, this method became infeasible under the Taxpayer Relief Act of 1997.

Selling the spread A spread whose option to be sold is trading at a higher premium than the option to be bought.

Selling Syndicate A group of underwriters that issues a firm's securities by buying them from the issuing firm and reselling them to a group of smaller brokerage firms for eventual sale to individual investors.

Semistrong-form efficiency A form of pricing efficiency that profits the price of a security fully reflects all public information (including, but not limited to, historical price and trading patterns). Compare weak-form efficiency and strong-form efficiency.

"Send it in" Market language: "I bought your stock - 'send it in' (and possibly more)."

Senior debt Debt whose terms in the event of bankruptcy, require it to be repaid before subordinated debt receives any payment.

Senior mortgage bond A bond that, in the event of bankruptcy, will be redeemed before any other bonds are repaid.

Senior refunding Replacement by the issuer of securities with 5-to 12-year maturities with securities of 15-year or longer maturities, in order to delay, reduce, or consolidate payment.

Senior security A security that, in the event of bankruptcy, will be redeemed before any other securities.

Seniority The order of repayment. In the event of bankruptcy, senior debt must be repaid before subordinated debt is repaid.

Sensitive market A market that reacts to a great extent to good or bad news.

Sensitivity analysis Analysis of the effect on a project'sprofitability of changes in sales, cost, and so on.

Sentiment indicators The general feeling of investors about the state of the market, such as whether they are bullish or bearish.

Separate customer Method of allocating insurance by the Securities Investor Protection Corporation. Each account that is under the name of a different person or group of people is entitled to maximum protection.

Separate tax returns Tax returns of married persons who choose to file their returns individually, usually because this approach produces lower overall tax payments.

Separate Trading of Registered Interest and Principal Securities (STRIPS) Long-term notes and bonds divided into principal and interest-paying components, which may be transferred and sold in amounts as small as $1000. STRIPS are sold at auction at a minimum par amount, varying for each issue. The amount is an arithmetic function of the issue's interest rate.

Separation property The property that portfolio choice can be divided into two independent tasks: (1) Determination of the optimal risky portfolio, which is a purely mathematical problem, and (2) the personal choice of the best mix of the optimal risky portfolio and the risk-free asset, which depends on a person's degree of risk aversion.

Separation theorem Theory that the value of an investment to an individual is not dependent on consumption preferences. That is, investors will want to accept or reject the same investment projects by using the NPV rule, regardless of personal preference.

Serial bonds Corporate bonds arranged so that specified principal amounts become due on specified dates. Related: Term bonds.

Serial covariance The covariance between a variable and the lagged value of the variable; the same as autocorrelation.

Serial entrepreneur Business person that successfully starts (does not kill) a number of different businesses.

Serial redemption The redemption of a serial bond.

Series Options: All option contracts of the same class that also have the same unit of trade, expiration date, and exercise price. Stocks: shares that have common characteristics, such as rights to ownership and voting, dividends, or par value. In the case of many foreign shares, one series may be owned only by citizens of the country in which the stock is registered.

Series bond Bond that may be issued in several series under the same indenture document.

Series E bond A local and state tax-free bond issued by the U.S. government from 1941 to 1979, which was then replaced by Series HH bonds.

Series EE bond See: Savings bond.

Series HH bond See: Savings bond.

Service charge A component of some finance charges, such as the fee for triggering an overdraft checking account into use.

Set-aside A percentage of a municipal or corporate bond underwriting that is allocated for handling by a minority-owned broker/dealer firm.

Set of contracts perspective View of corporation as a set of contracting relationships among individuals who have conflicting objectives, such as shareholders or managers. The corporation is a legal construct that serves as the nexus for the contracting relationships.

Set up Applies mainly to convertible securities. Arbitrage involving going long the convertible and short a certain percentage of the underlying common. Antithesis of Chinese hedge.

Setoff Money held on behalf of a borrower that may be applied to repay the loan, but usually without the permission of the borrower.

Settle price An average of the trading prices in the futures market during the last few minutes of trading.

Settlement When payment is made for a trade.

Settlement date The date on which payment is made to settle a trade. For stocks traded on US exchanges, settlement is currently three business days after the trade. For mutual funds, settlement usually occurs in the US the day following the trade. In some regional markets, foreign shares may require months to settle.

Settlement options The various possibilities open to a beneficiary under a life insurance policy as to how the benefit will be paid out.

Settlement price A figure determined by the closing range that is used to calculate gains and losses in futures market accounts. Settlement prices are used to determine gains, losses, margin calls, and invoice prices for deliveries. Related: Closing range.

Settlement rate The rate suggested in Financial Accounting Standards Board (FASB) 87 for discounting the obligations of a pension plan. The rate at which the pension benefits could be effectively settled if the company sponsoring the pension plan wishes to terminate its pension obligation.

Settlement risk The risk that one party will deliver and the counterparty will not be able to pay and vice versa.

Severally but not jointly An agreement between members of an underwriting group buy a new issue (severally), but not to assume joint liability for shares left unsold by other members.

Severance A settlement received after being released from a corporation. In the context of corporate governance, an agreement that assures high-level executives of their postions or some compensation and are not contingent upon a change in control.

Segmented market A market that is partially or wholly isolated from other markets by one or more market imperfections.

Shadow calendar A backlog of securities issues registered with the SEC, awaiting the determination of an offer date.

Shadow stock First, a public company may create a stock that strips out the market wide movements for the purpose of rewarding managers. That is, the management might have done a great job - but the traded stock plummets because the market as a whole plummets. A second interpretation of shadow stock is a phantom stock that is created by a private company (i.e. that does not have stock traded either on exchange or over the counter) again for the purpose of performance evaluation and rewards.

Shadows The thin lines above and below the real body on a candlestick line.

Shakeout A dramatic change in market conditions that forces speculators to sell their positions, often at a loss.

Sham A business transaction, such as a limited partnership, that is entered into for the sake of avoiding tax.

Shanghai Stock Exchange One of two major securities markets in China.

Share broker A discount broker who charges per share traded, and reduces the per unit charge as the number of shares traded increases, as opposed to a dealer who charges a percentage of the dollar amount of the trade.

Share repurchase Program by which a corporation buys back its own shares in the open market. It is usually done when shares are undervalued. Since repurchase reduces the number of shares outstanding and thus increases earnings per share, it tends to elevate the market value of the remaining shares held by stockholders.

Shared Appreciation Mortgage (SAM) A mortgage with a low rate of interest, offset by giving the lender some portion of the appreciation in the value of the underlying property.

Shareholder Person or entity that owns shares or equity in a corporation.

Shareholders' equity This is a company's total assets minus total liabilities. A company's net worth is the same thing.

Shareholders' letter A section of an annual report where one can find general overall discussion by management of successful and failed strategies. Provides guidance for looking at specific parts of the report.

Shares Certificates or book entries representing ownership in a corporation or similar entity.

Shares authorized The maximum number of shares of stock of a company allowed in the articles of incorporation, which may be changed only by a shareholder vote. See: Issued and outstanding.

Shark repellant Often used in risk arbitrage. Examples are golden parachutes, poison pills, safe harbor, and scorched-earth policy. Porcupine provision. Amendment to company charter intended to protect it against takeover.

Shark watcher Often used in risk arbitrage. Firm specializing in the early detection of takeover activity. Such a firm, whose primary business is usually the solicitation of proxies for client corporations, monitors trading patterns in a client's stock and attempts to determine the identity of parties accumulating shares.

Sharpe benchmark A statistically created benchmark that adjusts for a manager's index-like tendencies. Named after William Sharpe, Nobel Laureate, and developer of the capital asset pricing model.

Sharpe ratio A measure of a portfolio's excess return relative to the total variability of the portfolio. Related: Treynor index. Named after William Sharpe, Nobel Laureate, and developer of the capital asset pricing model.

Shelf offering Offering of registered securities covered by a prospectus whose distribution is not underwritten on a firm commitment basis. The shares may be sold in one block or in small amounts from time to time in agency or principal transactions. See: Rule 415.

Shelf registration A procedure that allows firms to file one registration statement covering several issues of the same security. SEC Rule 415, adopted in the 1980s, allows a corporation to comply with registration requirements up to two years prior to a public offering of securities. With the registration "on the shelf," the corporation, by simply updating regularly filed annual, quarterly, and related reports to the SEC, can go to the market as conditions become favorable with a minimum of administrative preparation and expense.

Shell corporation An incorporated company with no significant assets or operations, often formed to obtain financing before beginning actual business, or as a front tax evasion.

Shenzhen Stock Exchange One of two major securities markets in China.

Shipper's Export Declaration (SED) Document required by the U.S. Department of Commerce for exports of certain controlled items, and/or shipments to certain countries, and/or shipments anywhere that exceed certain dollar amounts. This document is used to monitor shipments of controlled goods.

Shipping Documents A generic term for the various typesof forms required for overseas shipments, such as commercial invoices, transport documents, packing lists, origin certificates, etc.

Shirking The tendency to do less work when the return is smaller. Owners may have more incentive to shirk if they issue equity as opposed to debt, because they retain less ownership interest in the company and therefore may receive a smaller return. Thus, shirking is considered an agency cost of equity.

Shock absorbers See: Circuit breakers

Shogun bond Dollar bond issued in Japan by a nonresident.

Shootout Venture capital jargon. Refers to two or more venture capital firms fighting for the startup.

Shop Wall Street slang for a firm.

Shopped stock Sell inquiry that has been seen by or shown to other dealers before coming to an investment bank.

Shopping Seeking to obtain the best bid or offer available by calling a number of dealers and/or brokers.

Short One who has sold a contract to establish a market position and who has not yet closed out this position through an offsetting purchase; the opposite of a long position. Related: Long.

Short against the box A short sale of a stock is where the seller actually owns the stock, but does not want to close out the position.

Short Bias In the context of hedge funds, a style of management where part or all of the fund consists of short sales.

Short bonds Bonds with short (not much time to maturity) current maturities.

Short book See: Unmatched book.

Short coupon A bond payment covering less than six-months' interest, because the original issue date is less than six months from the first scheduled interest payment. A bond with a short time to maturity, usually two years or less.

Short covering Used in the context of general equities. Actual purchase of securities by a short seller to replace those borrowed at the time of a short sale.

Short exempt Used for listed equity securities. A special trading situation where a short sale is allowed on a minustick. The owners of a convertible trading at parity can sell the equivalent amount of common short on a minus tick, assuming they have the firm intention to convert.

Short hedge The sale of futures contracts to eliminate or lessen the possible decline in value of an approximately equal amount of the actual financial instrument or physical commodity. Related: Long hedge.

Short interest Total number of shares of a security that investors have sold short and that have not been repurchased to close out the short position. Usually, investors sell short to profit from price declines. As a result, the short interest is often an indicator of the amount of pessimism in the market about a particular security, although there are other reasons to short that are not related to pessimism. For example, hedging strategies for mergers and acquisition as well as derivative positions may involve short sales.

Short interest theory The theory that a large interest in short positions in stocks will precede a rise in the market prices, because the short positions must eventually be covered by purchases of the stock.

Short-Form Registration A procedure that allows a firm to condense its registration statement and prospectus by referencing financial data already on file with the SEC.

Short position Occurs when a person sells stocks he or she does not yet own. Shares must be borrowed, before the sale, to make "good delivery" to the buyer. Eventually, the shares must be bought back to close out the transaction. This technique is used when an investor believes the stock price will drop.

Short ratio(or short interest ratio) Number of shares of a security that investors have sold short divided by average daily volume of the security (measured over 30 days or 90 days). There are various interpretations of this ratio. When people short, it is usually (but not always) because they are pessimistic about the security's future performance. Shorting involves buying at at some point however. Hence, some would interpret a high short ratio as an indicator that there will be some buying pressure on the security that would increase its price.

Short-run operating activities Events and decisions concerning the short-term finance of a firm, such as how much inventory to order and whether to offer cash terms or credit terms to customers.

Short sale Selling a security that the seller does not own but is committed to repurchasing eventually. It is used to capitalize on an expected decline in the security's price.

Short-sale rule An SEC rule requiring that short sales be made only in a market that is moving upward; this means either on an uptick from the last sale, or showing no downward movement.

Short selling Establishing a market position by selling a security one does not own in anticipation of the price of that security falling.

Short settlement Trade settlement made prior to the standard five-day period due to customer request.

Short-short test A repealed IRS restriction, that used to limit profits from short-term trading, which three months, to 30% of gross income. The penalty for exceeding this limit would be the loss of certain tax-free benefits.

Short squeeze When a lack of supply tends to force prices upward. In particular, when prices of a stock or commodity futures contracts start to move up sharply and many traders with short positions are forced to buy stocks or commodities in order to cover their positions and prevent (limit) losses. This sudden surge of buying leads to even higher prices, further aggravating the losses of short sellers who have not covered their positions.

Short straddle A straddle involves both purchase and sale. In short straddle one put and one call are sold.

Short-term capital gain A profit on the sale of a security or mutual fund share that has been held for one year or less. A short-term capital gain is taxed as ordinary income.

Short-term interest rates Interest rates on loan contracts-or debt instruments such as Treasury bills, bank certificates of deposit or commerical paper-having maturities of less than one year. Often called money market rates.

Short-term reserves Investments in interest-bearing bank deposits, money market instruments, U.S. Treasury bills, and short-term bonds.

Short tender Practice prohibited by SEC that involves the use of borrowed stock to respond to a tender offer.

Short-term Any investments with a maturity of one year or less.

Short-term bond fund A bond mutual fund holding short to intermediate-term bonds that have maturities of three to five years.

Short-term debt Debt obligations, recorded as current liabilities, requiring payment within the year.

Short-term financial plan A financial plan that covers the coming fiscal year.

Short-term gain (or loss) A profit or loss realized from the sale of securities held for less than a year that is taxed at normal income tax rates if the net total is positive.

Short-term investment services Services that assist firms in making short-term investments.

Short-term solvency ratios Ratios used to judge the adequacy of liquid assets for meeting short-term obligations as they come due, including (1) the current ratio, (2) the acid test ratio, (3) the inventory turnover ratio, and (4) the accounts receivable turnover ratio.

Short-term tax exempts Short-term securities issued by states, municipalities, and quesi-government entities such as local housing and urban renewal agencies.

Short-term trend Erratic price movements that last less than three weeks.

Shortage cost Costs that fall with increases in the level of investment in current assets.

Shortfall risk The risk of falling short of any investment target.

Show me buyer/seller Used in the context of general equities. Customer who has not placed a firm order to buy stock but has requested that the salesperson propose available stock for sale or purchase, along with the asking/bid price. See: Bidding buyer.

Show stopper A legal barrier, such as a scorched-earth policy or shark repellant system, that firms use to prevent a takeover.

Show and tell list Used in the context of general equities. Block list which is full of real customer indications (rather than profile).

Shrinkage Discrepancy between a firm's actual inventory and its recorded inventory due to theft, deterioration, loss, or clerical problems.

Shut out the book Used for listed equity securities. Exclude a public bid or offer from participation in a print.

Side effects Effects of a proposed project on other parts of the firm.

Side-by-side trading Trading a security and an option on the same security on the same exchange.

Sidelines Hypothetical position referring to noninvolvement in a stock; merely watching.

Sideways market See: Horizontal price movement

Sight draft Demand for immediate payment.

Sight Letter of Credit A letter of credit made payable to a beneficiary upon presentation to the opener of conforming documents.

Signal To convey information through a firm's actions. The more costly it is to provide a signal, the more credibility it has. For example, to call a press conference and tell everyone that the firm's prospects have improved is less effective than saying the same thing and raising the dividend.

Signaling approach Notion that insiders in a firm have information that the market does not have, and that the choice of capital structure by insiders can signal information to outsiders and change the value of the firm. This theory is also called the asymmetric information approach.

Signaling approach (on dividend policy) The argument that dividend changes are important signals to investors about changes in management's expectation about future earnings.

Signature guarantee The authentication of a signature in the form of a stamp, seal, or written confirmation by a bank or member of a domestic stock exchange (or other acceptable guarantor). A notary public cannot provide a signature guarantee. A signature guarantee is a common requirement when transferring or redeeming shares or changing the ownership of an account.

Signature loan A good faith loan that is unsecured and requires only the borrower's signature on the loan application.

Signatures on Proxies The basic rule of

acceptability is that if the signature reads as the proxy is printed, it is acceptable. If an individual signs on behalf of another individual and states a legal representation, it is acceptable. Examples: executor, guardian, power of attorney; but not husband, wife, next of kin, etc. On corporate registrations, a manual signature in the name of the corporation is acceptable. A facsimile signature is also acceptable, but a rubber-stamp signature with a signature line is acceptable only if signed on that line. With joint tenancy, one signature is sufficient, as in the case of one trustee signing for two or more.

Significant influence The holding of a large portion of the equity of a corporation, usually at least 20%, which gives the holder a significant amount of control over the corporation. This degree of holding must be recorded in a firm's financial statements.

Significant order An order to buy or sell a large enough quantity of securities that the price of the security may be affected. Institutional investors usually spread out such an order over a few days or weeks to avoid adverse pressures on the buy or sell price.

Significant order imbalance A large number of buy or sell orders for a stock that cause an abnormally wide spread between bid and offer prices, and often causes the exchange to halt the sale of the stock until significant balance has been reestablished.

Silent partner A partner in a business who has no role in management but shares in the liability, tax responsibility, and cash flow.

Silver Parachutes These provisions are similar to Golden Parachutes in that they provide severance payments upon a change in corporate control, but unlike Golden Parachutes, a large number of a firm's employees are eligible for these benefits.

Single-buyer policy Ex-Im Bank practice allows the exporter to insure certain transactions selectively.

Single European Act Act intended to eliminate barriers on trade and capital flows between and among European countries.

Simple compound growth method Calculating a growth rate by relating terminal value to initial value and assuming a constant percentage annual rate of growth between the two values.

Simple interest Interest calculated as a simple percentage of the original principal amount. Compare to compound interest.

Simple IRA A salary deduction plan for retirement benefits provided by some small companies with no more than 100 employees.

Simple linear regression A regression analysis between only two variables, one dependent and the other explanatory.

Simple linear trend model An extrapolative statistical model that asserts that earnings have a base level and grow at a constant amount each period.

Simple moving average The mean, calculated at any time over a past period of fixed length.

Simple prospect An investment opportunity in which only two outcomes are possible.

Simple rate of return The return from investments figured by dividing income plus capital gains by the amount of capital invested. The effect of compounding is not taken into account.

Simplified Employee Pension (SEP) plan A pension plan in which both the employee and the employer contribute to an individual retirement account. Also available to the self-employed.

Simulation The use of a mathematical model to imitate a situation many times in order to estimate the likelihood of various possible outcomes. See: Monte Carlo simulation.

Singapore International Monetary Exchange (SIMEX) A leading futures and options exchange in Singapore.

Single-country fund A mutual fund that invests in individual countries outside the United States.

Single-factor model A model of security returns that acknowledges only one common factor. The single factor is usually the market return. See: Factor model.

Single-index model A model of stock returns that decomposes influences on returns into a systematic factor, as measured by the return on the broad market index, and firm specific factors. Related: Market Model

Single life annuity An annuity covering one person. A straight life annuity provides payments until death, while a life annuity with a guaranteed period provides payments until death or continues payments to a beneficiary for a guaranteed term, such as ten years.

Single option A single put option or call option, as opposed to a spread or straddle, which involves multiple puts and calls.

Single-payment bond A bond that makes only one payment of principal and interest.

Single-Premium Deferred Annuity (SPDA) An IRA-like annuity into which an investor makes a lump-sum payment that is invested in either a fixed-return instrument or a variable-return portfolio, which is taxed only when distributions are taken.

Single-premium life insurance A whole life insurance policy requiring one premium payment, which accrues cash value much more quickly than a policy paid in installments.

Single-state municipal bond fund A mutual fund investing only in government obligations within a single state, with state tax-free dividends, but taxed capital gains.

Sinker A bond with interest and principal payments coming from the proceeds of a sinking fund.

Sinking fund A fund to which money is added on a regular basis that is used to ensure investor confidence that promised payments will be made and that is used to redeem debt securities or preferred stock issues.

Sinking fund requirement A condition included in some corporate bond indentures that requires the issuer to retire a specified portion of debt each year. Any principal due at maturity is called the balloon maturity.

Sit tight Directive from the trader to the customer to be patient, emphasizing that one's piece of business will be executed.

Size Refers to the magnitude of an offering, an order, or a trade. Large as in the size of an offering, the size of an order, or the size of a trade. Size is relative from market to market and security to security. "I can buy size at 102-22," means that a trader can buy a significant amount at 102-22. Small is <10,000 shares. Medium is 15,000-25,000 shares. Good is 50,000 shares. Size is 100,000 shares. Good six-figure size is 200,000-300,000 shares. Multiple six-figure size is >300,000 shares. Size of the market is actual number of shares represented in one's market, or bid and offering; unless specified, assumed to be at least 500 to 1000 shares, depending on the stock.

Size out the book Overt action to exclude a public bid or offer from participation in a print through trading a larger size in the book. Can never size out a market order. See: Priority, shut out the book.

Skewed distribution Probability distribution in which an unequal number of observations lie below (negative skew) or above (positive skew) the mean.

Skewness Negative skewness means there is a substantial probability of a big negative return. Positive skewness means that there is a greater-than-normal probability of a big positive return.

Skill The ability to accurately forecast returns. We measure skill using the information coefficient.

Skip-day settlement Settling a trade one business day beyond what is normal.

Skip-payment privilege A mortgage contract clause giving borrowers the right to skip payments if they are ahead of schedule.

Skort-Swing Transaction Any purchase and sale, or sale and purchase, of the issuer's equity securities by an insider within a period of less than six months, See: Section 16(b) above.

SLD last sale Shortened version of "sold last sale," which shows up on the consolidated tape when a large change (one point for lower priced securities and two points for higher-priced securities) occurs between transactions.

Sleeper Stock in which there is little investor interest but that has significant potential to gain in price once its attractions are recognized. Antithesis of high flyer.

Sleeping beauty Often used in risk arbitrage. Potential takeover target that has not yet been approached by an acquirer. Such a company usually has particularly attractive features, such as a large amount of cash, or undervalued real estate or other assets.

Slippage The difference between estimated transactions costs and actual transactions costs. The difference usually represents revisions to price difference or spread and commission costs.

Slump A temporary fall in performance, often describing consistently falling security prices for several weeks or months.

Small business policy Insurance coverage available to new exporters and small businesses.

Small-cap A stock with a small capitalization, meaning a total equity value of less than $500 million.

Small-capitalization (small-cap) fund A mutual fund that invests primarily in stocks of companies whose market value is less than $1 billion. Small-cap stocks historically have been more volatile than large-cap stocks, and often perform differently from the overall market.

Small-capitalization (small cap) stocks The stocks of companies whose market value is less than $1 billion. Small-cap companies tend to grow faster than large-cap companies and typically use any profits for expansion rather to pay dividends. They also are more volatile than large-cap companies, and have a higher failure rate.

Small-firm effect The tendency of small firms (in terms of total market capitalization) to outperform the stock market (consisting of both large and small firms).

Small investor An individual person investing in small quantities of stock or bonds. This group of investors makes up a minimal fraction of total stock ownership.

Small issues exemption Securities issues that involve less than $1.5 million are not required to file a registration statement with the SEC. Instead, they are governed by Regulation A, for which only a brief offering statement is needed.

Small Order Execution System (SOES) Three-tiered system of automatic execution of an order at the best price. Size is either 200, 500, or, most often, 1000 shares.

Smart money Investors who make consistent profits in the market, regardless of the investing environment, by making wise, educated moves.

Smidge Small amount of price, usually +/- 1/8 or 1/4.

Smithsonian Agreement A revision to the Bretton Woods international monetary system that was signed at the Smithsonian Institution in Washington, D.C., in December 1971. Included were a new set of par values, widened bands to +/- 2.25% of par, and an increase in the official value of gold to US$38.00 per ounce.

Snake Arrangement established in 1972, that ties European currencies to each other within specified limits.

Snowballing Used in the context of general equities. Process by which the exercise of stop orders in a declining or advancing market causes further downward or upward pressure on prices, thus triggering more stop orders and more price pressure, and so on.

Social Security benefits Monthly government payments to retired workers or their families who have paid Social Security taxes for a total of 40 quarters or 10 years.

Social Security Disability Income Insurance Program financed by the Social Security tax to provide assistance to disabled individuals with disabilities expected to last at least one year, to compensate for lost income.

Socially conscious mutual fund A mutual fund that does not invest in companies that have interests in socially unacceptable markets or produce harmful products or by-products, such as high levels of environmental pollution.

Society for Worldwide Interbank Financial Telecommunications (SWIFT) A dedicated computer network to support funds transfer messages internationally between over 900 member banks world-wide.

"Soft" capital rationing Constraints on spending that under certain circumstances can be violated or even viewed as constituting targets rather than absolute limits.

Soft currency The money of a country that is expected to drop in value relative to other currencies.

Soft dollars The value of research services that brokerage houses supply to investment managers "free of charge" in exchange for the investment manager's business commissions.

Soft landing A term describing a growth rate high enough to keep the economy out of recession, but also slow enough to prevent high inflation and interest rates.

Soft market A buyer's market in which supply exceeds demand, causing little trading activity and wide bid-ask spreads.

Soft spot Stocks or groups of stocks that remain weak in a strong market.

Softs Tropical commodities such as coffee, sugar, and cocoa.

Sold away Refers to over-the-counter trading. Having sold stock to another dealer before making the present offering.

Sold-out market Unavailability of a futures contract in a particular commodity or maturity date because of contract executions and limited offerings.

Sole proprietorship A business owned by a single individual. A sole proprietor pays no corporate income tax but has unlimited liability for business debts and obligations.

Solvency Ability to meet obligations.

Sour bond A bond issue that has defaulted on interest or principal payments, and will thus trade at a large discount and a poor credit rating.

Source of funds seller Customer seller of stock for the purpose of raising cash for other purchases. Such a seller will sell only at advantageous prices, and not aggressively.

Sources and applications of funds statement See: Statement of cash flows

South African Futures Exchange (SAFEX) Electronic futures and options exchange based in South Africa.

Sovereign risk The risk that a central bank will impose foreign exchange regulations that will reduce or negate the value of foreign exchange contracts. Also refers to the risk of government default on a loan made to a country or guaranteed by it. The government's part of political risk.

Span To cover all contingencies within a specified range.

SPDRs SPDRs (Spiders) are designed to track the value of the Standard & Poor's 500 Composite Price Index. Stands for Standard & Poor's Depositary Receipt. They trade on the American Stock Exchange under the symbol SPY. SPDRs are similar to closed-end funds but are formally known as, a unit investment trust. One SPDR unit is valued at approximately one-tenth (1/10) of the value of the S&P 500. Dividends are disbursed quarterly, and are based on the accumulated stock dividends held in trust, less any expenses of the trust. See: Mid-cap SPDR.

Special arbitrage account A margin account with lower cash requirements, reserved for transactions that are hedged by an offsetting position in futures or options.

Special assessment bond A municipal bond with interest paid by the taxes of the community benefiting from the bond-funded project.

Special bid A method of purchasing a large block of stock on the NYSE by advertising a client's large buy order, and matching it up with a number of other traders' smaller sell orders.

Special bond account A special broker margin account used only for transactions in US government bonds, municipals, and eligible listed and unlisted non-convertible corporate bonds.

Special Claim on Residual Equity (SCORE) A certificate that entitles the owner to the capital appreciation of an underlying security, but not to the dividend income from the security.

Special dividend Also referred to as an extra dividend. Dividend that is unlikely to be repeated.

Special Drawing Rights (SDR) A form of international reserve assets, created by the IMF in 1967, whose value is based on a portfolio of widely used currencies.

Special Meeting Refers to a meeting of shareholders outside the usual annual general meeting. In the context of corporate governance, some limitations either increase the level of shareholder support required to call a special meeting beyond that specified by state law or eliminate the ability to call one entirely. Such provisions add an extra time delay to many proxy fights, since bidders must wait until the regularly scheduled annual meeting to replace board members or dismantle takeover defenses.

Special-Purpose Entity A financing technique in which a company decreases its risk by creating separate partnerships, rather than subsidiaries, for certain holdings and solicits outside investors to take on the risk. In order to qualify as a special-purpose entity, whose financial results are not carried on the company's books, the unit must meet strict accounting guidelines. Compare to subsidary.

Specialist On an exchange, the member firm that is designated as the market maker (or dealer for a listed common stock). Member of a stock exchange who maintains a "fair and orderly market" in one or more securities. Only one specialist can be designated for a given stock, but dealers may be specialists for several stocks. In contrast, there can be multiple market makers in the OTC market. Major functions include executing limit orders on behalf of other exchange members for a portion of the floor broker's commission, and buying or selling for the specialist's own account to counteract temporary imbalances in supply and demand and thus prevent wide swings in stock prices.

Specialist block purchase and sale Purchase of a large number of securities by a specialist for himself or to pass on to another floor trader or block buyer.

Specialist market Market in a stock made solely by the specialist, as no public orders, and henceforth no depth, exist in the market.

Specialist unit A specialist who maintains a stable market by acting as a principal and agent for other brokers in one or many stocks.

Specialist's book Chronological record maintained by a specialist that includes the specialist's own inventory of securities, market orders to sell short, and limit orders and stop orders that other stock exchange members have placed with the specialist.

Specialist's short-sale ratio The percentage of the total short sales of stock sold short by specialists.

Specific issues market The market in which dealers reverse in securities they wish to short.

Specific Return The part of the excess return not explained by common factors. The specific return is independent of (uncorrelated with) the common factors and the specific returns to other assets. It is also called the idiosyncratic return.

Specific risk See: Unique risk

Spectail A dealer doing business with retail but concentrating more on acquiring and financing its own speculative positions.

Speculation Purchasing risky investments that present the possibility of large profits, but also pose a higher-than-average possibility of loss. A profitable strategy over the long term if undertaken by professionals who hedge their portfolios to control the amount of risk.

Speculative Securities that involve a high level of risk.

Speculative demand (for money) The need for cash to take advantage of investment opportunities that may arise.

Speculative-grade bond Bond rated Ba or lower by Moody's, or BB or lower by S&P, or an unrated bond.

Speculative motive A desire to hold cash in order to be poised to exploit any attractive investment opportunity requiring a cash expenditure that might arise.

Speculative stock Very risky stock.

Speculator One who attempts to anticipate price changes and, through buying and selling contracts, aims to make profits. A speculator does not use the market in connection with the production, processing, marketing, or handling of a product. See: Trader.

Speed Related: Prepayment speed

Spider See: SPDRs

Spike Order ticket that shows the stock, price, number of shares, type, and account of the order. Origin: Practice of placing the ticket on a metal spike upon execution or cancellation. Spike is also a sudden, drastic increase in a company's share price.

Spin-off A company can create an independent company from an existing part of the company by selling or distributing new shares in the so-called spin-off.

Spinning In investment banking, the practice of an investment bank setting aside portions of a corporation's Initial Public Offering for senior management of that corporation. Ethically questionable practice which appears to be a form of bribery.

SPINs Stands for Standard & Poor's 500 Index Subordinated Notes.

Split Sometimes companies split their outstanding shares into more shares. If a company with 1 million shares executes a two-for-one split, the company would have 2 million shares. An investor with 100 shares before the split would hold 200 shares after the split. The investor's percentage of equity in the company remains the same, and the share price of the stock owned is one-half the price of the stock on the day prior to the split.

Split commission A commission shared between a broker and a financial adviser or other professional who brought the customer to the broker.

Split-coupon bond A bond that begins as a zero-coupon bond paying no interest and converts to an interest paying bond on a future date.

Split-fee option An option on an option. The buyer generally executes the split fee with first an initial fee, with a window period at the end of which (upon payment of a second fee) the original terms of the option may be extended to a later predetermined final notification date.

Split offering A municipal bond issue that is made up of serial bonds and term maturity bonds.

Split order A large securities transaction that is divided into smaller orders that are spread out over some period of time to avoid large fluctuations in the market price.

Split print Block trade printed at two different prices. Often used in dividend rolls to get an average price equal to the dividend.

Split-rate tax system A tax system that taxes retained earnings at a higher rate than earnings that are distributed as dividends.

Split rating Two different ratings given to the same security by two important rating agencies.

Split stock Purchases or sales shared with others. (2) Division of the outstanding shares of a corporation into a large number of shares. Ordinarily, splits must be proposed by directors and approved by shareholders.

Spoken for Amount of opposite demand (placement) or supply (availability) the trader has in efforts to cross the stock. Not open.

Sponsor An underwriting investment company that offers shares in its mutual funds, or an influential institution that highly values a particular security and thus creates additional demand for the security. In the context of project financing, a developer of the project or a party poviding financial support.

Spontaneous Current Liabilities Short-term obligations that automatically increase and decrease in response to financing needs, such as accounts payable.

Spontaneous Liabilities Obligations that arise automatically in the course of operating a business when a firm buys goods and services on credit.

Spot commodity A commodity that is traded with the expectation of actual delivery, as opposed to a commodity future that is usually not delivered.

Spot exchange rates Exchange rate on currency for immediate delivery. Related: Forward exchange rate.

Spot futures parity theorem Describes the theoretically correct relationship between spot and futures prices. Violation of the parity relationship gives rise to arbitrage opportunities.

Spot interest rate Interest rate fixed today on a loan that is made today. Related: Forward interest rates.

Spot lending Originating mortgages by processing applications taken directly from prospective borrowers.

Spot markets Related: Cash markets

Spot month The nearest delivery month on a futures contract.

Spot price The current market price of the actual physical commodity. Also called cash price. Current delivery price of a commodity traded in the spot market, in which goods are sold for cash and delivered immediately. Antithesis of futures price.

Spot rate The theoretical yield on a zero-coupon Treasury security.

Spot rate curve The graphical depiction of the relationship between the spot rates and maturity.

Spot secondary Secondary distribution that may not require an SEC registration statement and may be attempted without delay. An underwriting discount is normally included in these offerings.

Spot trade The purchase and sale of a foreign currency, commodity, or other item for immediate delivery.

Spot transaction A foregin exchange transaction in which each party promises to pay a certain amount of currency to the other on the same day or within one or two days.

Spousal IRA An individual retirement account in the name of an unemployed spouse.

Spousal remainder trust A fixed-term trust from which income is distributed to the beneficiary (such as a child of the grantor) to take advantage of a lower tax bracket, and that at the end of the term passes to the grantor's spouse.

Spread The gap between bid and ask prices of a stock or other security. (2) The simultaneous purchase and sale of separate futures or options contracts for the same commodity for delivery in different months. Also known as a straddle. (3) Difference between the price at which an underwriter buys an issue from a firm and the price at which the underwriter sells it to the public. (4) The price an issuer pays above a benchmark fixed-income yield to borrow money.

Spread income Also called margin income, the difference between income and cost. For a depository institution, the difference between the assets it invests in (loans and securities) and the cost of its funds (deposits and other sources).

Spread option A position consisting of the purchase of one option and the sale of another option on the same underlying security with a different exercise price and/or expiration date.

Spread order An order listing the series of options that the customer wants to buy and sell and the desired spread between the premiums paid and received for the options.

Spread position The status of an account after a spread order has been carried out.

Spread strategy A strategy that involves a position in one or more options so that the cost of buying an option is funded entirely or in part by selling another option in the same underlying. Also called spreading.

Spreadsheet A computer program that organizes numerical data into rows and columns in order to calculate and make adjustments based on new data.

Sprinkling trust A trust in which the trustee decides how to distribute trust income among a group of designated people.

SPX Applies to derivative products. Symbol for the S&P 500 index.

Squeeze Period when stocks or commodities futures increase in price and investors who have sold short must cover their short positions to prevent loss of large amounts of money.

SS1 Securities sales speaker box that transmits to all investment banks' regional trading and sales desks.

Stabilization The action undertakes a country when it buys and sells its own currency to protect its exchange value. Actions registered competitive traders undertake by on the NYSE to meet the exchange requirement that 75% of their traded be stabilizing, meaning that sell orders follow a plus tick and buy orders a tick. Actions a managing underwriter undertake so that the market price does not fall below the public offering price during the offering period.

Stable Paretian, or Fractal Hypothesis In the characteristic function of the fractal family of distributions, the characteristic exponent alpha can range between one and two. See: Alpha, Fractal Distributions, Gaussian.

Stability The relative steadiness or safety of a security or fund compared to the market as a whole. For example, money market funds and other short-term investments offer more stability than funds that invest in growth stocks.

Stag Speculator who buys and sells stocks to hold for short intervals to make quick profits.

Stagflation A period of slow economic growth and high unemployment with rising prices (inflation).

Staggered board of directors Occurs when a portion of directors are elected periodically, instead of all at once. Board terms are often staggered in order to thwart unfriendly takeover attempts, since potential acquirers would have to wait longer before they could take control of a company's board through the normal voting procedure.

Staggering maturities Hedging against interest rate movements by investment in short-, medium-, and long-term bonds.

Stagnation A period of slow economic growth, or, in securities trading, a period of inactive trading.

Stakeholders All parties that have an interest, financial or otherwise, in a firm-stockholders, creditors, bondholders, employees, customers, management, the community, and the government.

Stale price An old price of the asset that does not reflect the most recent information.

Stale price arbitrage For a number of assets, the most recent transaction price at 4PM ET does not fully reflect all available market information. One example is international equities that trade on exchanges that are located in different time zones and close 2-15 hours before U.S. markets. In addition, domestic small-capitization equities and high-yield and convertible bonds often trade infrequently and have wide bid-ask spreads. This can cause the most recent transaction price to be much different from the price that one would see in a liquid market at 4 PM, even for assets that trade on exchanges that are open at that time. Investors can take advantage of mutual funds that calculate their NAVs using stale closing prices by trading based on recent market movements. For example, if the U.S. market has risen since the close of overseas equity markets, investors can expect that overseas markets will open higher the following morning. Investors can buy a fund with a stale-price NAV for less than its current value, and they can likewise sell a fund for more than its current value on a day that the U.S. market has fallen. Similar opportunities exist when the values of infrequently or illiquidly-traded domestic assets have recently changed. Also referred to as Net Asset Value Arbitrage or NAV Arbitrage.

Stalking horse In bankruptcy proceedings, this refers to the company that first bids for the companies assets.

Stalking horse bid In bankruptcy proceedings, this refers to first bid for the companies assets. This is the bid to beat. If there are multiple bids, often there is a bankruptcy auction.

Stamp duty Applies mainly to international equities. Taxes on foreign transactions, usually a percentage of total transaction amount, that can be unilateral or bilateral in nature.

Stamp tax Tax on a financial transaction.

Stand-alone principle Investment approach that advocates a firm should accept or reject a project by comparing it with securities in the same risk class.

Standby Letter of Credit Documents evidencing failure of the bank's customer (the applicant) to pay an obligation when due.

Stand up to Make a good-sized market in the trader's own bid and offering prices. Hence, "standing up" to the bid signifies the trader's willingness to buy size (i.e., 50m) volume at the advertised bid, even if the customer buyer/seller falls down.

Standard deduction The IRS-specified amount by which a taxpayer is entitled to reduce income an alternative to itemizing deductions.

Standard deviation The square root of the variance. A measure of dispersion of a set of data from its mean.

Standard error In statistics, a measure of the possible error in an estimate. Plus or minus 2 standard errors usually provides a 95% confidence interval.

Standard Industrial Classification (SIC) A code system that designates a unique business activity classified by industry.

Standard & Poor's MidCap 400 Index A market capitalization-weighted benchmark index made up of 400 securities with market values between $200 million and $5 billion.

Standard & Poor's SmallCap 600 Index A small-capitalization benchmark index made up of 600 domestic stocks chosen for market size, liquidity, and industry group representation.

Standardized normal distribution A normal distribution with a mean of 0 and a standard deviation of 1.

Standardized value Also called the normal deviate, the distance of one data point from the mean, divided by the standard deviation of the distribution.

Standby agreement In a rights issue, agreement that the underwriter will purchase any stock not purchased by investors.

Standby commitment An agreement between a corporation and investment firm that the firm will purchase whatever part of a stock issue that is offered in a rights offering that is not subscribed to in the two- to four- week standby period.

Standby fee Amount paid to an underwriter who agrees to purchase any stock that is not purchased by public investors in a rights offering.

Standby letter of credit Agreement to guarantee invoice payments to a supplier; a standby LOC promises to pay the seller if the buyer fails to pay.

Standing Level of priority in the trading crowd.

Standstill agreement Contract by which the bidding firm in a takeover attempt agrees to limit its holdings of another firm.

Start-up The earliest stage of a new business venture.

State bank A bank authorized in a specific state by a state-based charter, with generally the same functions as a national bank.

State and local government series (SLUGs) Special nonmarketable certificates, notes, and bonds offered to state and local governments as a means to invest proceeds from their own tax-exempt financing. Interest rates and maturities comply with IRS arbitrage provisions. Slugs are offered in both time deposit and demand deposit forms. Time deposit certificates have maturities of up to one year. Notes mature in one to ten years and bonds mature in more than ten years. Demand deposit securities are one-day certificates rolled over with a rate adjustment daily.

State tax-exempt income fund A mutual fund that seeks current income exempt from federal and a specific state's income taxes.

Stated annual interest rate The interest rate expressed as a per year percentage, by which interest payments are determined. See: Annual percentage rate.

Stated conversion price At the time of issuance of a convertible security, the price the issuer effectively grants the securityholder to purchase the common stock, equal to the par value of the convertible security divided by the conversion ratio.

Stated maturity For the CMO tranche, the date the last payment would occur at zero CPR.

Stated value A monetary worth figure that bears no relation to market value that is assigned, for accounting purposes, to stock for use instead of par value.

Statement of Additional Information (SAI) A document provided as a supplement to a mutual fund prospectus. It provides more detailed information about fund policies, operations, and risks. Also known as a Part B prospectus.

Statement billing Billing method in which the sales for a period such as a month (for which a customer also receives invoices) are collected into a single statement, and the customer must pay all the invoices represented on the statement.

Statement of Cash Flows A financial statement showing a firm's cash receipts and cash payments during a specified period.

Statement-of-Cash-Flows Method A method of cash budgeting that is organized along the lines of the statement of cash flows.

Statement of condition A document describing the status of assets, liabilities, and equity of a person or business at a particular time.

Statement of Financial Accounting Standards No. 8 The is a currency translation standard once used by U.S. accounting firms. See: Statement of Accounting Standards No. 52.

Statement of Financial Accounting Standards No. 52 The currency translation standard currently used by US firms. It mandates the use of the current rate method. See: Statement of Financial Accounting Standards No. 8.

"Static" Return The return that an investor would make on a particular position if the underlying stock were unchanged in price at the expiration of the options in the position.

Static theory of capital structure Theory that the firm's capital structure is determined by a trade-off of the value of tax shields against the costs of bankruptcy.

Stationary time series A longitudinal measure in which the process generating returns is identical over time.

Statistical Arbitrage In the context of hedge funds, a style of management that employs complex statistical models that try to capture small abnormalities in a security's intraday return.

Statistical tracking error Used in the context of general equities. Standard deviation of the difference between the portfolio return and the desired investment benchmark return.

Statutory debt limit The cap that Congress imposes on the amount of public debt that may be outstanding whether temporary or permanent. When this limit is reached, the Treasury may not sell new debt issues until Congress raises the limit. For a detailed listing of changes in the limit since 1941, see Budget of the United States Government. See: Debt outstanding subject to limitation.

Statutory investment An investment that a trustee is authorized to make under state law.

Statutory merger A merger in which one corporation remains as a legal entity, instead of a new legal entity being formed.

Statutory surplus The surplus of an insurance company determined by the accounting treatment of both assets and liabilities as established by state statutes.

Statutory voting The standard rule in most corporations that there is one vote per share in elections of the Board of Directors.

Staying power The ability of an investor to stay in the market and not to sell out of a position when an investment has fallen in value.

Steady state As an MBS pool ages, or four to six months after component mortgages have passed at least once the threshold for refinancing, the prepayment speed tends to stabilize within a fairly steady range.

Steenth 1/16 (0.0625) of one full point in price.

Often used in negotiations to compromise an eighth difference, and in options trading.

Steepening of the yield curve A change in the yield curve where the spread between the yield on a long-term and short-term Treasury has increased. Compare flattening of the yield curve and butterfly shift.

Step aside Allow a block to trade at a price at which you do not care to participate in the trade.

Step-down note A floating-rate note whose interest rate declines after a specified period of time.

Step up To increase, as in step up the tax basis of an asset.

Step-up bond A bond that pays a lower coupon rate for an initial period, and then increases to a higher coupon rate. Related: Deferred-interest bond, payment-in-kind bond.

Step-up swap An interest rate swap on which the notional principal increases according to a predetermined schedule.

Sterilized intervention Foreign exchange market activity by which monetary authorities insulate their domestic money supplies from the foreign exchange transactions with offsetting sales or purchases of domestic assets.

Sticky deal A new securities issue that may be difficult to sell because of problems in the market or underlying problems with the corporation.

Stochastic models Liability-matching models that assume that the liability payments and the asset cash flows are uncertain. Related: Deterministic models.

Stochastics index A computerized tool measuring overbought and oversold conditions in a stock over a certain period.

Stock Ownership of a corporation indicated by shares, which represent a piece of the corporation's assets and earnings.

Stock ahead When two or more orders for a stock at a certain price arrive about the same time, and the exchange's priority rules take effect. NYSE rules stipulate that the bid made first should be executed first, or, if two bids come in at once, the bid for the larger number of shares receives priority. The bid that is not executed is then turned to the broker, who informs the customer that the trade was not completed because there was "stock ahead.".

Stock bonus plan A plan used as an incentive that rewards employee performance with stock in the company.

Stockbroker See: Registered representative

Stock Appreciation Right (SAR) A contractual right, often granted in tandem with an option that allows an individual to receive cash or stock of a value equal to the appreciation of the stock from the grant date to the date the SAR is exercised.

Stock buyback A corporation's purchase of its own outstanding stock, usually in order to raise the company's earnings per share.

Stock certificate A document representing the number of shares of a corporation owned by a shareholder.

Stock dividend Payment of a corporate dividend in the form of stock rather than cash. The stock dividend may be additional shares in the company, or it may be shares in a subsidiary being spun off to shareholders. Stock dividends are often used to conserve cash needed to operate the business. Unlike a cash dividend, stock dividends are not taxed until sold.

Stock Exchange Automated Quotation System (SEAQ) London's Nasdaq system.

Stock Exchange of Hong Kong (SEHK) Only stock exchange located in Hong Kong.

Stock Exchange, Mumbai (BSE) Formerly the Bombay stock exchange, the BSE accounts for more than one-third of Indian trading volume.

Stock Exchange of Singapore (SES) The only stock exchange in Singapore.

Stock Appreciation Rights An incentive scheme for employees similar to stock options. The employee get the increase in the stock price from the date of the grant to the date of the exercise. However, in contrast to options, there is no dillutive effect. That is, no shares are issued. Similar to options, if the company's stock falls in value, the appreciation right is worthless.

Stock Exchange of Thailand The major securities market of Thailand.

Stock exchanges Formal organizations, approved and regulated by the Securities and Exchange Commission (SEC), that are made up of members who use the facilities to exchange certain common stocks. The two major national stock exchanges are the New York Stock Exchange (NYSE) and the American Stock Exchange (ASE or AMEX). Five regional stock exchanges include the Midwest, Pacific, Philadelphia, Boston, and Cincinnati. The Arizona Stock Exchange is an after-hours electronic marketplace where anonymous participants trade stocks via personal computers.

Stock index Index like the Dow Jones Industrial Average that tracks a portfolio of stocks.

Stock Index Future A security that uses composite stock indexes to allow investors to speculate on the performance of the entire market, or to hedge against losses in long or short positions. The settlement of the contracts is in cash.

Stock index option An option in which the underlying is a common stock index.

Stock index swap A swap involving a stock index. The other asset involved in a stock index swap can be another stock index (a stock-for-stock swap), a debt index (a debt-for-stock swap), or any other financial asset or financial price index.

Stock insurance company An insurance company owned by a group of stockholders, who are not necessarily policyholders.

Stock jockey A stock broker who frequently buys and sells shares in a client's portfolios.

Stock list The department within a stock exchange that oversees compliance with listing requirements and exchange regulations.

Stock market Also called the equity market, the market for trading equities.

Stock option An option whose underlying asset is the common stock of a corporation.

Stock power A power of attorney form giving ownership of a security to another person, brokerage firm, bank, or lender after it has been sold or pledged to that party.

Stock purchase plan A plan allowing employees of a company to purchase shares of the company, often at a discount or with matching employer funds.

Stock rating An evaluation by a rating agency of the expected financial performance or inherent risk of common stocks.

Stock record The accounting a brokerage firm keeps of all securities held in inventory.

Stock replacement strategy A strategy for enhancing a portfolio's return, used when the futures contract is expensive according to its theoretical price. The strategy involves a swap between the futures and a Treasury bill and stock portfolio.

Stock repurchase A firm's repurchase of outstanding shares of its common stock.

Stock right Another terminology for a stock option.

Stock selection An active portfolio management technique that focuses on advantageous selection of particular stock rather than on broad asset allocation choices.

Stock split Occurs when a firm issues new shares of stock and in turn lowers the current market price of its stock to a level that is proportionate to pre-split prices. For example, if IBM trades at $100 before a two-for-one split, after the split it will trade at $50, and holders of the stock will have twice as many shares as they had before the split. See: Split.

Stock symbol See: Ticker symbol

Stock ticker A letter designation assigned to securities and mutual funds that trade on US financial exchanges.

StockWatch A stock surveillance program offered by proxy solicitation firms, and selected transfer agents, to track and monitor sales and purchases of a corporation's shares and provide valuable information at the beneficial owner level.

Stock watcher (NYSE) A computerized service that monitors and investigates trading activity on the NYSE in order to identify any unusual activity or security movement that might be caused by rumors or illegal activities.

Stockholder See: Shareholder.

Stockholder books Set of books kept by firm management for its annual report that follows Financial Accounting Standards Board rules. The tax books follow IRS tax rules.

Stockholder equity Balance sheet item that includes the book value of ownership in the corporation. It includes capital stock, paid-in surplus, and retained earnings.

Stockholder of record Stockholder whose name is registered on the books of a corporation and thus will receive dividends from the corporation.

Stockholder's equity The residual claims that stockholders have against a firm's assets, calculated by subtracting all current liabilities and debt liabilities from total assets.

Stockholder's report The annual report and other reports given to stockholders to inform them of the company's financial standing and developments.

Stockholm Stock Market (Stockholm B&#ouml;rsen) The major securities market of Sweden.

Stockout Running out of inventory.

Stop basis Refers to over-the-counter trading. Method of entering an OTC trade into the trader's position without reporting the trade on the OTC tape.

Stop-limit order A stop order that designates a price limit. Unlike the stop order, which becomes a market order once the stop is reached, the stop-limit order becomes a limit order.

Stop-loss order An order to unwind a position when the price moves against you. For example, you had purchased a stock, the stop-loss order would be to sell the stock when the price falls to a specified level. If you were short the asset, the stop-loss would trigger a purchase.

Stop order (or stop) An order to buy or sell at the market when a definite price is reached, either above (on a buy) or below (on a sell) the price that prevailed when the order was given.

Stop-out price The lowest auction price at which Treasury bills are sold.

Stop payment An order given a depository institution not to pay out cash for a check; often used when the check has been stolen or lost.

Stop Transfer A block placed against a security reported lost or stolen (an adverse claim), so it cannot be transferred.

Stopped Guaranteed a specific price on the customer's working order while the dealer tries to obtain a better one. Stopped against one's self involves a customer order and a firm's own account, not two customers. One can cancel an order even after being stopped by another party.

Stopped out A purchase or sale that is executed under a stop order at the stop price specified by the customer.

Stopping curve A curve showing the refunding rates for different times at which the expected value of refunding immediately equals the expected value of waiting to refund.

Stopping curve refunding rate A refunding rate that falls on the stopping curve.

Story stock/bond A highly complex security that requires a long "story" so that investors may understand the corporation and be persuaded of its merits.

Straddle Purchase or sale of an equal number of puts and calls with the same terms at the same time. Related: Spread.

Straight Direct telephone line, compared to an outside line that requires a telephone number to be dialed.

Straight Bill of Lading A bill of lading that is cosigned to a specific party and is therefore non negotiable.

Straight Discount The rate applied to the face value of the promissory note to calculate present value without compounding. For example, a note with a face value in three years of 100, with a straight discount of 10% per annum has a present value of 70.

Straight-line depreciation Amortizing or apportioning an equal dollar amount of depreciation in each accounting period.

Straight term insurance policy Term life insurance policy providing a fixed-amount death benefit over a certain number of years.

Straight value Also called investment value, the value of a convertible security without the conversion option.

Straight voting Allows shareholder to cast all of the shareholder's votes for each candidate for the Board of Directors.

Strange Attractor An attractor in phase space, where the points never repeat themselves, and orbits never intersect, but they stay within the same region of phase space. Unlike limit cycles or point attractors, strange attractors are non-periodic, and generally have a fractal dimension. They are a picture of a non-linear, chaotic system. See: Attractor, Chaos, Limit Cycle, Point Attractor.

Strangle Buying or selling an out-of-the-money put option and call option on the same underlying instrument, with the same expiration. Profits are made only if there is a drastic change in the underlying instrument's price.

Strategic alliance Collaboration between two or more companies designed to achieve some corporate objective. May include international licensing agreements, management contracts, or joint ventures.

Strategic buyout Acquisition of another firm in order to realize some operational benefits which will result in increased earnings.

Strategy The general or specific approach to investing that an individual, institution, or fund manager employs.

Stratified equity indexing A method of constructing a replicating portfolio that classifies the stocks in the index into strata, and represents each stratum in the portfolio.

Stratified sampling approach to indexing Dividing an index into cells, each representing a different characteristic of the index, such as duration or maturity.

Stratified sampling bond indexing A method of bond indexing that divides the index into cells, each cell representing a different characteristic, and that buys bonds to match those characteristics.

Stray Not a member of the participating party in the trade at hand; (2) not a meaningful indication of a customer's desire to take a sizable position or be involved in a stock.

Street Means Wall Street financial community; brokers, dealers, underwriters, and other knowledgeable participants.

Street name Registration under which

securities maybe held by a broker on behalf of a client but be registered in the name of the Wall Street firm.

Strike index For a stock index option, the index value at which the buyer of the option can buy or sell the underlying stock index. The strike index is converted to a dollar value by multiplying by the option's contract multiple. Related: Strike price.

Strike price The stated price per share for which underlying stock may be purchased (in the case of a call) or sold (in the case of a put) by the option holder upon exercise of the option contract.

Striking price The price at which an option can be exercised. See: Exercise price.

Striking Price Intercal The distance between striking prices on a particular underlying security. Normally, the interval is 2-1/2 points for stocks under $25, 5 points for stocks selling over $25 per share, and 10 points (or greater) is acceptable for stocks over $200 per share. There are, however, exceptions to this general guideline.

Strip Variant of a straddle. A strip is two puts and one call on a stock. A strap is two calls and one put on a stock. The puts and calls have the same strike price and expiration date. See: Strap.

Strip mortgage participation certificate (strip PC) Ownership interests in specified mortgages purchased by Freddie Mac from a single seller in exchange for separate instruments representing interests in the same mortgages.

Stripped bond Bond that can be subdivided into a series of zero-coupon bonds.

Stripped mortgage-backed securities (SMBS) Securities that redistribute the cash flows from the underlying generic MBS collateral into the principal and interest components of the MBS to enhance their attractiveness to different groups of investors.

Stripped yield Applies mainly to convertible securities. Return on the debt portion of a bond/warrant unit after subtracting the value of the issued warrant segment.

Strong Currency A currency whose value compared to other currencies is improving, as indicated by a decrease in the direct exchange rates for the currency.

Strong dollar When the dollar can be exchanged for a large amount of foreign currency, benefiting travelers but hurting exporters.

Strong-form efficiency A form of pricing efficiency, that posits that the price of a security reflects all information, whether or not it is publicly available. Related: Weak-form efficiency, semi-strong form efficiency.

Strong form of the EMT Theory that market prices reflect all relevant publicly and privately available information. Defined by Eugene F. Fama in 1970.

Structural Adjustment Loan Facility (SAL) World Bank program established in 1980 to enhance a country's long-term economic growth through financing projects.

Structure The description of how a project financing is drawdown, repaid, and collateralized secured.

Structured arbitrage transaction A self-funding, self-hedged series of transactions that usually use mortgage-backed securities (MBS), commercial mortgage backed securities, and collateralized debt obligations as the primary assets.

Structured Asset Trust Unit Repackagings A synthetic security linked or weak-linked to underlying collateral. Ratings usually reflect the credit quality of the underlying securities.

Structured debt Debt that has been customized for the buyer, often by incorporating unusual options.

Structured finance Often refers to a group within an investment bank that deals with mortgage-backed securities (MBS), commercial mortgage backed securities, and collateralized debt obligations,and real estate.

Structured note A derivative investment that will change in value with movements of an underlying index; or a note whose issuer makes swap arrangements to alter its required cash flows.

Structured portfolio strategy Designing a portfolio to achieve a level of performance that matches some predetermined liabilities that must be paid out in the future.

Structured settlement An agreement in settlement of a lawsuit involving specific payments made over a period of time. Property and casualty insurance companies often buy life insurance products to pay the costs of such settlements.

Stub Often used in risk arbitrage. Piece of equity security left over from a major cash or security distribution from a recapitalization.

Student Loan Marketing Association (SLMA) A publicly traded corporation established by federal action that increases availability of educational loans by guaranteeing student loans traded in the secondary market. Also known as Sallie Mae.

Subaccount A term used in bookkeeping. For example, the insurance expense account may have various different subcategories such as building and property insurance, auto/fleet insurance, general liability, environmental, professional liability, law enforcement, and other insurance.

Subchapter M An IRS regulation dealing with investment companies and real estate investment trusts that avoid double taxation by distributing interest, dividends, and capital gains directly to shareholders, who are taxed individually.

Subchapter S IRS regulation that gives a corporation with 35 or fewer shareholders the option of being taxed as a partnership to escape corporate income taxes.

Subject Refers to a bid or offer that cannot be executed without confirmation from the customer. In other words, not firm, but a bid/offer that needs additional information/confirmation before becoming firm and is therefore still negotiable.

Subject market Quote in which prices are subject to confirmation. See: Fast market.

Subject to a (NY) can Contingent upon trader's ability to cancel an order (on the indicated exchange).

Subject to opinion An auditor's opinion reflecting acceptance of a company's financial statements subject to pervasive uncertainty that cannot be adequately measured, such as information relating to the value of inventories, reserves for losses, or other matters open to judgment.

Subject to a print/execution/trading Contingent on execution of a trade because the picture in the stock has not been materially altered.

Subjective probabilities Probabilities that are determined subjectively (for example, on the basis of judgment rather than statistical sampling).

Subordinated A claim ranked lower in priority than other claims. Common stock claims are always subordinated to debt.

Subordinated bonds Securities that fall after others in priority of claims on the entity in the case of financial distress.

Subordinated debenture bond An unsecured bond that ranks after secured debt, after debenture bonds, and often after some general creditors in its claim on assets and earnings. Related: Debenture bond, mortgage bond, collateral trust bonds.

Subordinated debt Debt over which senior debt takes priority. In the event of bankruptcy, subordinated debtholders receive payment only after senior debt claims are paid in full.

Subordination clause A provision in a bond indenture that restricts the issuer's future borrowing by subordinating future lenders' claims on the firm to those of the existing bondholders.

Subpart F Special category of foreign-source "unearned" income that is currently taxed by the IRS whether or not it is remitted to the US

Subperiod return The return of a portfolio over a shorter period of time than the evaluation period.

Subrogation An insurance process whereby a company that has paid out to a policyholder for a loss incurred recovers the amount of the loss from the party that is legally liable.

Subscription Agreement to buy new issue of securities.

Subscription agreement An application reviewed by the general partner to join a limited partnership.

Subscription price Price that current shareholders pay for a share of stock in a rights offering.

Subscription privilege The right of current shareholders of a corporation to buy newly issued shares before they are available to the public.

Subscription right See: Subscription privilege

Subscription warrant Applies to derivative products. Type of security, usually issued with another security, such as a bond or stock, that entitles the holder to buy a proportionate amount of common stock at a specified price, usually higher than the market price at the time of issuance. Warrant.

Subsidiary A wholly or partially owned company that is part of a large corporation. A foreign subsidiary is a separately incorporated entity under the host country's law. A subsidiary's financial results are carried on the parent company's books.

Subsidized financing Funding provided by a government or other entity that is available at a below-market interest rate.

Substitute sale A method for hedging price risk that uses debt market instruments, such as interest rate futures, or that involves selling borrowed securities as the primary assets.

Substitution swap A swap in which a money manager exchanges one bond for another bond that is similar in terms of coupon, maturity, and credit quality, but that offers a higher yield.

Substantially equal periodic payments (SEPP) A method of distribution from IRA account assets that under certain conditions is not subject to the IRS's 10% premature withdrawal penalty for those under age 59-1/2.

Success tax A 15% excise tax on "excess" distributions from tax-deferred retirement plans that was repealed by the Taxpayer Relief Act of 1997. In essence, the tax had penalized "successful" investors who accumulated large retirement accounts and took distributions that exceeded an annual limit deemed excessive by the tax code.

Suicide pill A hostile takeover prevention tactic that could destroy the target company. Taking on a large amount of debt to prevent the takeover might cause bankruptcy, for example.

Suitability A requirement that any investing strategy fall within the financial means and investment objectives of an investor.

Suitability rules Policies and guidelines that brokers must use to ensure that investors have the financial means to assume risks that they wish to undertake. These are enforced by the NASD and other self-regulatory organizations.

Suitable Describing a strategy or trading philosophy in which the investor is operating in accordance with his(her) financial means and investment objectives.

Summary plan description (SPD) A document that explains the fundamental features of an employer's defined benefit or defined contribution plan, including eligibility requirements, contribution formulas, vesting schedules, benefit calculations, and distribution options. ERISA requires that the SPD be easy to understand and that each participant receive a copy within 90 days of joining the plan.

Sum-of-the-years'-digits depreciation Method of accelerated depreciation.

Sunk costs Costs that have been incurred and cannot be reversed.

Sunrise industries Growth industries in an economy that may become leaders in the market in the future.

Super Bowl indicator A theory that if a team from the old American Football League pre-1970 wins the Super Bowl, the stock market will decline during the coming year. If a team from the old pre-1990 National Football League wins the Super Bowl, stock prices will increase in the coming year.

Super DOT Super DOT provides faster execution than regular DOT and focuses on large-size trades and baskets. See: Program trading.

Super Majority A proposal requiring more than a simple majority of the votes eligible to be cast at an annual or special meeting. A super majority is often a 2/3 (66.66%) vote, but it can be as high as 3/4 (75%) or 4/5 (80%).

Super message See: Autex

Super sinker bond Usually a home financing bond, but also any other bond that has long-term coupons but short maturity; the mortgages may be prepaid, and the holders may receive the long-term yield after a short period of time.

Supermajority Provision in a company's charter requiring a majority of, say, 80% of shareholders to approve certain changes, such as a merger.

Supermajority amendment Often used in risk arbitrage. Corporate amendment requiring that a substantial majority (usually 67% to 90%) of stockholders approve important transactions, such as mergers.

Supervisory analyst An analyst who is qualified to approve publicly distributed research reports on the NYSE.

Supervisory board The board of directors that represents stakeholders in the governance of the corporation.

Supplemental Security Income A Social Security program established to help the blind, disabled, and poor.

Supplier credit Self-financing of a supplier's operations. Also the agreement of a supplier of goods or services to deferred repayment terms.

Supply risk The risk associated with a change in raw materials or input to a project from those assumed or projected. In the context of a resources production project, this is called reserves risk.

Supply shock An event that influences production capacity and costs in an economy.

Supply-side economics A theory of economics that reductions in tax rates will stimulate investment and in turn will benefit the entire society.

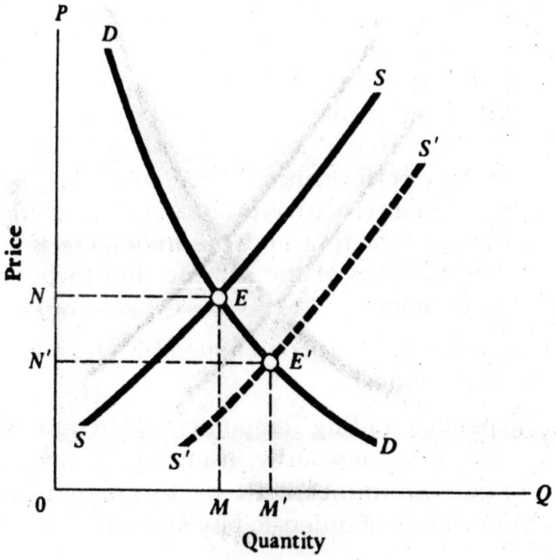

Fig. Supply

Support An effective lower bound on prices supported because of many willing buyers at that price level.

Support level A price level below which it is supposedly difficult for a security or market to fall. That is, the price level at which a security tends to stop falling because there is more demand than supply; can be identified on a technical basis by seeing where the stock has bottomed out in the past.

Surcharge An additional levy added to some charge.

Surety An individual or corporation that guarantees the performance or actions of another.

Surplus funds Cash flow available after payment of taxes in a project.

Surplus management Related: Asset management

Surtax A tax added to the normal tax paid by corporations or individuals who have earned income above a certain level.

Surveillance department of exchanges A department that monitors trading activity on an exchange in order to identify any unusual activity that may indicate illegal practices.

Survivorship bias Usually pertaining to fund manager or individual investor performance. Suppose we examined the performance over the last ten years of a group of managers that exist today. This performance is biased upwards because we are only considering those that survived for 10 years. That is, some dropped out because of poor performance. Hence, in evaluating performance, one has to be careful to include both the current and the managers that dropped out of the sample due to poor performance.

Sushi bond A Eurobond issued by a Japanese corporation.

Suspended trading Temporary halt in trading in a particular security, in advance of a major news announcement or to correct an imbalance of orders to buy and sell.

Suspense account An account used temporarily to record receipts and disbursements that have yet to be classified.

Sustainable growth rate Maximum rate of growth a firm can sustain without increasing financial leverage.

Swap An arrangement in which two entities lend to each other on different terms, e.g., in different currencies, and/or at different interest rates, fixed or floating.

Swap arrangements Short-term reciprocal lines of credit between the Federal Reserve and 14 foreign centeral banks as well as the Bank for International Settlements. Through a swap transactions, the Federal Reserve can, in effect, borrow foreign currency in order to purchase dollars in the foreign exchange market. In doing so, the demand for dollars and the dollar's foreign exchange value are increased. Similarly, the Federal Reserve can temporarily provide dollars to foreign central banks through swap arrangments.

Swap assignment Related: Swap sale

Swap book A swap bank's portfolio of swaps, usually arranged by currency and maturity.

Swap buy back The sale of an interest rate swap by one counterparty to the other, effectively ending the swap.

Swap fund See: Exchange fund

Swap option See: Swaption. Related: Quality option.

Swap rate The difference between spot and forward rates expressed in points, e.g., $0.0001 per pound sterling.

Swap reversal An interest rate swap designed to end a counterparty's role in another interest rate swap, accomplished by counterbalancing the original swap in maturity, reference rate, and notional amount.

Swap sale Also called a swap assignment, a transaction that ends one counterparty's role in an interest rate swap by substituting a new counterparty whose credit is acceptable to the other original counterparty.

Swaption Options on interest rate swaps. The buyer of a swaption has the right to enter into an interest rate swap agreement by some specified date in the future. The swaption agreement will specify whether the buyer of the swaption will be a fixed-rate receiver or a fixed-rate payer. The writer of the swaption becomes the counterparty to the swap if the buyer exercises.

Sweat equity An increase in equity created by the labor of the owner.

Sweep The act of using all available cash flow for the repayment of debt service.

Sweep account Account providing that a bank invest all the excess available funds at the close of each business day for the firm.

Sweetener A feature of a security that makes it more attractive to potential purchasers.

Swing Trading Refers to a type of short term (one day to a couple of weeks) trading, triggered by technical analysis, for example, momentum. Swing trading is distinguished by the notion thatthe trades are executed while the assets is moving in upward or downward momentummomentum.

Swingline facility Bank borrowing facility to provide finance while the firm replaces US commercial paper with eurocommercial paper.

Swiss Electronic Bourse (EBS) Computer linking system between the former stock exchange trading floors in Zurich, Geneva, and Basel, Switzerland so that trades can be carried out among traders on all three of the trading floors.

Swiss Options and Financial Futures Exchange (SOFFEX) The Swiss derivatives market with the first fully electronic trading system in the world, now called Eurex Zurich AG.

Swiss Exchange The major securities market of Switzerland.

Swissy Slang for the Swiss franc.

Switch order Order for the purchase (sale) of one stock and the sale (purchase) of another stock at a stipulated price difference. Contingent order, swap.

Switching Liquidating a position and simultaneously reinstating a position in another futures contract of the same type.

Switching options A sequence of transactions in which exercise of one option creates one or more additional options. Investment-disinvestment, entry-exit, expansion-contraction, and suspension-reactivation decisions are switching options.

Sydney Futures Exchange (SFE) The derivatives market of Australia.

Symbol Letters used to identify companies on the consolidated tape and other locations.

Symbol book special Illiquid, inactively traded stock not familiar market

Symmetric cash matching An extension of cash flow matching that allows for the short-term borrowing of funds to satisfy a liability prior to the liability due date, reducing the cost of funding liabilities.

Synchronous data Information available at the same time. To test option-pricing models, the price of the option and of the underlying should be synchronous and reflect the same moment in the market.

Syndicate A group of banks that acts jointly, on a temporary basis, to loan money in a bank credit (syndicated credit) or to underwrite a new issue of bonds.

Syndicate manager See: Managing underwriter

Syndicated Eurocredit loans Funding provided by a group (or syndicate) of banks in the Eurocredit market.

Syndicated Loan A large Eurocurrency loan from a group of international banks.

Syndication The selling of a project finance to

a group of prospective participants, the syndicate.

Synergistic effect A violation of value-additivity in that the value of a combination is greater than the sum of the individual values.

Synergy Describes a combination whose value is greater than the sum of the separate individual parts.

Synthetic convertible Combination of usable bonds and warrants (that expire on or after the bonds' maturity) that resembles convertible bond.

Synthetic forward position A forward position constructed through borrowing in one currency, lending in another currency, and offsetting these transactions in the spot exchange market.

Synthetic Lease When a company creates a special-purpose entity to arrange for a loan to purchase property, and then leases the property from the entity.The synthetic lease therefore keeps the loan off the company's balance sheet, while the company provides enough income to the special-purpose entity to cover the interest rate on the loan.

Synthetic put A strategy equivalent in risk to purchasing a put option where an investor sells stock short and buys a call.

Synthetic stock An option strategy that is equivalent to the underlying stock. A long call and a short put is synthetic long stock. A long put and a short call is sythetic short stock.

Synthetics Customized hybrid instruments created by blending an underlying price on a cash instrument with the price of a derivative instrument. It is a combination of security holdings that mimics the price movement of another single security (i.e., synthetic call: long position in a stock combined with a put on that position; a protected long sale; synthetic put: short position in a stock combined with a call on that position; a protected short sale).

System Noise See: Dynamical Noise

Systematic Common to all businesses.

Systematic investment plan An approach involving regular investments in order to take advantage of dollar-cost averaging.

Systematic Return The part of the return dependent on the benchmark return. We can break excess returns into two components: systematic and residual. The systematic return is the beta times the benchmark excess return.

Systematic risk Also called undiversifiable risk or market risk.

Systematic risk principle Only the systematic portion of risk matters in large, well-diversified portfolios. Thus, expected returns must be related only to systematic risks.

Systematic withdrawal plan A provision of certain mutual funds to pay out to the shareholder specified amounts after specified periods of time.

Systemic Risk Risk common to a particular sector or country. Often refers to a risk resulting from a particular "system" that is in place, such as the regulator framework for monitoring of financial_institutions.

T

Tabulation Report A proxy tally report detailing the current quorum and vote figures on each proposal.

TAC bonds See: Targeted amortization class bond.

Tactical Asset Allocation (TAA) Portfolio strategy that allows active departures from the normal asset mix according to specified objective measures of value. Often called active management. It involves forecasting asset returns, volatilities, and correlations. The forecasted variables may be functions of fundamental variables, economic variables, or even technical variables.

Tail The remaining reserves after a project financing has been repaid. Sometimes refers to the residual value.

Tailgating Purchase of a security by a broker after the broker places an order for the same security for a customer. The broker hopes to profit either because of information which the customer has or because the customer's purchase is of sufficient size to affect security prices. This is an unethical practice.

Taiwan Stock Exchange (TSEC) Established in 1961, the only centralized securities market in Taiwan.

Take To agree to buy. A dealer or customer who agrees to buy at another dealer's offered price is said to take the offer.

(2) Euro bankers speak of taking deposits rather than buying money.

Take a bath To sustain a loss on either a speculation or an .

Take a flier To speculate on highly risky securities.

"Take it down" Reduce the offering price or hit others' bids to such an extent as to lower the inside market.

"Take me along" "Allow me to participate in the side of a particular trade.

Take off A sharp increase in the price of a stock, or a positive movement of the market as a whole.

Take the offer Buy stock by accepting a floor broker's (listed) or dealer's (OTC) offer at an agreed-upon volume. Antithesis of hit the bid.

Take-out A cash surplus generated by the sale of one block of securities and the purchase of another, e.g., selling a block of bonds at 99 and buying another block at 95. Also, a bid made to a seller of a security that is designed (and generally agreed) to take the seller out of the market.

Take-and-pay contract An agreement that obligates the purchaser to take any product that is offered (and pay the cash purchase price) and pay a specified amount if the product is not taken.

Take a position To buy or sell short; that is to own or to owe some amount on an asset or derivative security.

Take a powder Temporarily cancel an order or indication in a stock, while unrepresented interest still exists. See: Back on the shelf, sidelines.

Take a swing Execute a trade at a price that the trader feels is higher or more risky than would normally be acceptable, in order to gain market share in the institutional arena.

Takedown The share of securities of each participating investment banker in a new or a secondary offering, or the price at which the securities are distributed to the different members of an underwriting group.

Takeout A financing to refinance or take out another loan.

Takeover General term referring to transfer of control of a firm from one group of shareholders to another group of shareholders. Change in the controlling interest of a corporation, either through a friendly acquisition or an unfriendly, hostile, bid. A hostile takeover (with the aim of replacing current existing management) is usually attempted through a public tender offer.

Takeover target A company that is the object of a takeover attempt, friendly or hostile.

Take-up fee A fee paid to an underwriter in connection with an underwritten rights offering or an underwritten forced conversion. Represents compensation for each share of common stock the underwriter obtains and must resell upon the exercise of rights or conversion of bonds.

Takes a call Requires a phone call to an account in order for a trade to be completed. See: Show me.

Takes price Requiring some price movement or concession on behalf of the initiating party before a trade can be consummated. See: Price give.

Taking delivery When the buyer actually assumes possession from a seller of assets agreed upon in a forward contract or a futures contract.

Taking a view A London expression; means forming an opinion as to where market prices are headed and acting on it.

Tandem programs Ginnie Mae mortgage funds provided at below-market rates to residential MBS buyers with FHA Section 203 and 235 loans and to developers of multifamily projects with Section 236 loans initially and later with Section 221(d)(4) loans.

Tangible asset An asset whose value depends on particular physical properties. These include reproducible assets such as buildings or machinery and non-reproducible assets such as land, a mine, or a work of art. Also called real assets. Converse of: Intangible asset

Tangible net worth Total assets minus intangible assets, which include patents and copyrights, and total liabilities.

Tangibility Characteristic that an assets can be used as collateral to secure debt.

Tape Service that reports prices and sizes of transactions on major exchanges-ticker tape. (2) Dow Jones and other news wires. See: Consolidated tape.

Tape is late When the trading volume is so heavy that trades appear on the tape more than a minute behind the timer they actually take place.

Tare Weight The weight of an empty container and any packaging materials used in the container.

Tariff A tax on imports or exports.

Target cash balance Optimal amount of cash

for a firm to hold, considering the trade-off between the opportunity costs of holding too much cash and the trading costs of holding too little cash.

Target company Often used in risk arbitrage. Firm chosen as an attractive takeover candidate by a potential acquirer. The acquirer may buy up to 5% of the target's stock without public disclosure, but it must report all transactions and supply other information to the SEC, the exchange the target company is listed on, and the target company itself once the 5% threshold is hit. See: Raider.

Target firm A firm that is the object of a takeover by another firm.

Target investment mix The percentage mix of stocks, bonds, and short-term reserves that an investor considers appropriate based on his/her personal objectives, time horizon, risk tolerance, and financial resources.

Target Leverage Ratio The ratio of the market value of debt to the total market value of the firm that management seeks to maintain.

Target payout ratio A firm's long-run dividend-to-earnings ratio. The firm's policy is to attempt to pay out a certain percentage of earnings, but it pays a stated dollar dividend and adjusts it to the target as base line increases in earnings occur.

Target price In the context of takeovers, the price at which an acquirer aims to buy a target firm. In the context of options, the price of the underlying security at which an option will become in the money. In the context of stocks, the price that an investor hopes a stock will reach in a certain time period.

Target zone arrangement A monetary system under which countries pledge to maintain their exchange rates within a specific margin around agreed-upon, fixed central exchange rates.

Target zones Implicit boundaries on exchange rates established by central banks.

Targeted registered offerings Securities issues sold to "targeted" foreign financial_institutions according to U.S. Securities and Exchange Commission guidelines. These foreign institutions then maintain a secondary market in the foreign market.

Targeted repurchase Buying back of a firm's stock from a potential acquirer, usually at a substantial premium, to forestall a takeover attempt. Related: Greenmail.

Targeted Amortization Class (TAC) bonds Bonds offered as a tranche class of some CMOs, according to a sinking fund schedule. They differ from PAC bonds whose amortization is guaranteed as long as prepayments on the underlying mortgages do not exceed certain limits. A TAC's schedule is met at only one prepayment rate.

Tax anticipation bills (Tabs) Special bills that the Treasury occasionally issues that mature on corporate quarterly income tax dates and can be used at face value by corporations to pay their tax liabilities.

Tax Anticipation Notes (Tans) Notes issued by states or municipalities to finance current operations in anticipation of future tax receipts.

Tax arbitrage Trading that takes advantage of a difference in tax rates or tax systems as the basis for profit.

Tax audit Audit by the IRS or other tax-collecting agency to determine whether a taxpayer has paid the correct amount of tax.

Tax avoidance Minimizing tax burden through legal means such as tax-free municipal bonds, tax shelters, IRA accounts, and trusts. Compare with tax evasion.

Tax base The assessed value of the taxable property, assets, and income within a specific geographic area.

Tax basis In the context of finance, the original cost of an asset less depreciation that is used to determine gains or losses for tax purposes. In the context of investments, the price of a stock or bond plus the broker's commission.

Tax books Records kept by a firm's management that follow IRS rules. The books follow Financial Accounting Standards Board rules.

Tax bracket The percentage of tax obligation for a particular taxable income.

Tax clawback agreement An agreement to contribute as equity to a project the value of all previously realized project-related tax benefits not already clawed back. Exercised to the extent required to cover any cash deficiency of the project.

Tax clientele Categories of investors who have specific preferences for debt or equity because of differences in their personal tax rates.

Tax credit A direct dollar-for-dollar reduction in tax allowed for expenses such as child care and R&D for building low-income housing. Compare tax deduction.

Tax-deductible The effect of creating a tax deduction, such as charitable contributions and mortgage interest.

Tax deduction An expense that a taxpayer is allowed to deduct from taxable income.

Tax-deferred income Dividends, interest, and unrealized capital gains on investments in an account such as a qualified retirement plan, where income is not subject to taxation until a withdrawal is made.

Tax deferral option Allowing the capital gains tax on an asset to be payable only when the gain is realized by selling the asset.

Tax-deferred retirement plans Employer-sponsored and other plans that allow contributions and earnings to be made and accumulate tax-free until they are paid out as benefits.

Tax differential view (of dividend policy) The view that shareholders prefer capital gains over dividends, and hence low payout ratios, because capital gains are effectively taxed at lower rates than dividends.

Tax Equity and Fiscal Responsibility Act of 1982 (TEFRA) Legislation to increase tax revenue by eliminating various taxation loopholes and instituting tougher enforcement procedures in collecting taxes.

Tax-equivalent yield The pre-tax yield required from a taxable bond in order to equal the tax-free yield of a municipal bond.

Tax evasion Illegal by reducing tax burden by underreporting income, overstating deductions, or using illegal tax shelters.

Tax-exempt bond A bond usually issued by municipal, county, or state governments whose interest payments are not subject to federal and, in some cases, state and local income tax.

Tax-exempt income Dividends and interest not subject to federal and, in some cases, state and local income taxes.

Tax-exempt income fund A mutual fund that seeks income that is exempt from federal and, in some cases, state and local income taxes.

Tax-exempt money market fund A money market fund that invests in short-term tax-exempt municipal securities.

Tax-exempt sector The municipal bond market where state and local governments raise funds. Bonds issued in this sector are exempt from federal income taxes.

Tax-exempt security An obligation whose interest is tax-exempt, often called a municipal bond, offered by a country, state, town, or any political district.

Tax free acquisition A merger or consolidation in which (1) the acquirer's tax basis on each asset whose ownership is transferred in the transaction is generally the same as the acquiree's, and (2) each seller who receives only stock does not have to pay any tax on the gain realized until the shares are sold.

Tax haven A nation with a moderate level of taxation and/or liberal tax incentives for undertaking specific activities such as exporting or investing.

Tax haven affiliate A wholly owned entity in a low-tax jurisdiction that is used to channel funds to and from a multinational's foreign operations. The tax benefits of tax haven affiliates were largely removed in the US by the Tax Reform Act of 1986.

Tax holiday A reduced tax rate that a government provides as an inducement to foreign direct investment.

Tax liability The amount in taxes a taxpayer to the government.

Tax lien The right of the government to enforce a claim against the property of a person owing taxes.

Tax and loan account An account at a private bank, held in the name of the district Federal Reserve Bank, which holds operating cash for the business of the US Treasury.

Tax loss carryback, carryforward A tax benefit that allows business losses to be used to reduce tax liability in previous and or following years.

Tax-neutrality Characteristic that taxes do not interfere with the natural flow of capital toward its most productive use.

Tax planning Devising strategies throughout the year in order to minimize tax liability, for example, by choosing a tax filing status that is most beneficial to the taxpayer.

Tax preference item Items that must be included when calculating the alternative minimum tax.

Tax preparation services Firm that prepare tax returns for a fee.

Tax rate The percentage of tax paid for different levels of income.

Tax Reduction Strategy A source of competitive advantage that depends on differences in the tax rates imposed in different locations.

Tax Reform Act of 1976 Legislation aimed at tightening provisions relating to taxation, including changes in the capital gains tax laws.

Tax Reform Act of 1984 Legislation enacted as part of the Deficit Reduction Act of 1984 to reduce the federal budget deficit. Among its provisions are a decrease in the minimum holding period for assets to qualify for long-term capital gains treatment from one year to six months.

Tax Reform Act of 1986 A 1986 law involving a major overhaul of the US tax code.

Tax Reform Act of 1993 See: Revenue Reconciliation Act of 1993

Tax refund Money back from the government when too much tax has been paid or withheld from a salary.

Tax schedules Tax forms used to report itemized deductions, dividend and interest income, profit or loss from a business, capital gains and losses, supplemental income and loss, and self-employment tax.

Tax selling Selling of securities to realize losses that will offset capital gains and reduce tax liability. See: Wash sale.

Tax shelter Legal methods taxpayers can use to reduce tax liabilities. An example is the use of depreciation of assets.

Tax-sheltered annuity A type of retirement plan under Section 403(b) of the Internal Revenue Code that permits employees of public educational organizations or tax-exempt organizations to make before-tax contributions via a salary reduction agreement to a tax-sheltered retirement plan. Employers are also allowed to make direct contributions on behalf of employees.

Tax shield The reduction in income taxes that results from taking an allowable deduction from taxable income.

Tax software Computer software designed to assist taxpayers in filling out tax returns and minimizing tax liability.

Tax status election The decision of the status under which to file a tax return. For example, a corporation may file as a C corporation or an S corporation.

Tax straddle Technique used in futures and options trading to create tax benefits. For example, an investor with a capital gain takes a position creating an artificial offsetting loss in the current tax year and postponing a gain from the position until the next tax year.

Tax swap Swapping two similar bonds to receive a tax benefit.

Tax-timing option The option to sell an asset and claim a loss for tax purposes or not sell the asset and defer the capital gains tax.

Tax umbrella Tax loss carryforwards from previous business losses that form a tax shelter for profits earned in current and future years.

Taxpayer Relief Act of 1997 Legislation forming part of a larger act designed to balance the federal budget. Some of the legislation's provisions included tax credits for taxpayers supporting children, an increase in the amount that could be excluded from estate taxes, and a lower capital gains tax rate.

Taxable acquisition A merger or consolidation that is not a acquisition. The selling shareholders are treated as having sold their shares.

Taxable equivalent yield The return from a higher-paying but taxable investment that would equal the return from a tax-free investment. This depends on the investor's tax bracket.

Taxable estate That portion of a deceased person's estate that is subject to transfer tax.

Taxable event An event or transaction that has a tax consequence, such as the sale of stock holding that is subject to capital gains taxes.

Taxable income Gross income less a variety of deductions.

Taxable municipal bond Taxed private-purpose bonds issued by the state or local government to finance prohibited projects such as sports stadiums.

Taxable transaction Any transaction that is not tax-free to the parties involved, such as a taxable acquisition.

Taxable year The 12-month period an individual uses to report income for income tax purposes. For most individuals, their tax year is the calendar year.

Tear sheet A page from an S&P stock that provides information on thousands of stocks, often sent to prospective purchasers.

Teaser rate A low initial interest rate on an adjustable-rate mortgage to entice borrowers, that is later eliminated and replaced by a market-level rate.

Technical analysis Security analysis that seeks to detect and interpret patterns in past security prices.

Technical analysts Also called chartists or technicians, analysts who use mechanical rules to detect changes in the supply of and demand for a stock, and to capitalize on the expected change.

Technical condition of a market Demand and supply factors affecting price, in particular, the net position, either long or short, of the dealer community.

Technical descriptors Variables that are used to describe the market in terms of patterns in historical data.

Technical forecasting A forecasting method that uses historical prices and trends.

Technical Information Information related to the momentum of a particular variable. In market analysis, technical information is information related to market dynamics and crowd behavior only.

Technical insolvency Default on a legal obligation of the firm. Technical insolvency occurs when a firm doesn't pay a bill on time.

Technical rally Short rise in securities or commodities futures prices in the face of a general declining trend. Such a rally may result because investors are bargain hunting or because analysts have noticed a particular support level at which securities usually bounce up. Antithesis of correction.

Technical sign A short-term trend in the price movement of a security that analysts recognize as significant.

Technician Related: Technical analysts

TED spread Difference between US Treasury bill rate and Eurodollar rate; used by some traders as a measure of investor/trader anxiety or credit quality.

Teeny 1/16 or 0.0625 of one full point in price. Steenth.

Tel Aviv Stock Exchange Israel's only stock exchange.

Telephone switching Moving one's assets from one mutual fund or variable annuity to another by telephone.

Temporal method A currency translation method under which the choice of exchange rate depends on the underlying method of valuation. Assets and liabilities valued at historical cost (market cost) are translated at the historical (current market) rate.

Temporary Assets That portion of a firm's current assets that fluctuates in response to seasonal or anticipated short-term.

Temporary Financing The sum of negotiated current liabilities and temporary spontaneous current liabilities.

Temporary investment A short-term investment, such as a money market fund, Treasury bills, or short-term CD, which is usually held a year or less.

Ten largest holdings The percentage of a portfolio's total net assets or equity holdings in its ten largest securities positions. As this percentage rises, a portfolio's returns are likely to be more volatile because they are more dependent on the fortunes of fewer companies.

10% guideline The standard analysts' principle that funded debt over 10% of the assessed valuation of taxable property for a municipality is excessive.

10-K Annual report required by the each year. Provides a comprehensive overview of a company's state of business. Must be filed within 90 days after fiscal year-end. A 10-Q report is filed quarterly.

10-Q Quarterly report required by the SEC each quarter. Provides a comprehensive overview of a company's state of business.

1040 form The standard individual tax return form of the IRS.

1099 A statement sent to the IRS and taxpayers by the payers of dividends and interest and by issuers of taxable original issue discount securities.

1099 B The tax statement used for reporting proceeds resulting from the sale, redemption or liquidation of shares.

1099 DIV The tax statement used for reporting dividends paid to registered shareholders.

Ten-Day Rule The New York Stock Exchange rule permitting member firms (brokers) to vote in favor of management ten days or less before the meeting, provided that the member firm mailed proxy material to beneficial owners at least 15 business days before the meeting. The rule allows many shares to be voted, which would otherwise not be, to reach a quorum, approve the choice of directors and auditors and handle other routine matters. This rule does not apply to banks, their nominees or their depository positions, nor to non-routine proposals such as approval for the corporation to issue more shares.

Tenant A partial owner of a security, or the holder of some property. See: Lessee.

Tenants by Entireties (TEN ENT) Joint ownership of property or securities by a husband and wife where, upon the death of one, the property goes to the survivor.

Tenants in common Account registration in which two or more individuals own a certain proportion of an account. Each tenant's proportion is distributable as part of the owners estate, so that if one of the account holders dies, that owner's heirs are entitled to that proportional share of the account.

Tenbagger A stock that grows in value ten-fold.

Tender To offer for delivery against futures.

Tender offer General offer made publicly and directly to a firm's shareholders to buy their stock at a price well above the current value market price.

Tender offer premium The premium offered above the current market price in a tender offer.

Tenor The length of time until a loan is due. For example, a loan is taken out with a two year tenor. After one year passes, the tenor of the loan is one year.

Term The period of time during which a contract is in force.

Term bonds Bonds whose principal is payable at maturity. Often referred to as bullet-maturity bonds or simply bullet bonds. Related: Serial bonds.

Term certificate A certificate of deposit with a longer time to maturity.

Term Fed funds Fed funds sold for a period of time longer than overnight.

Term insurance Provides a death benefit only, no build up of cash value.

Term life insurance A contract that provides a death benefit but no cash build up or investment component. The premium remains constant only for a specified term of years, and the policy is usually renewable at the end of each term.

Term loan A bank loan, typically with a floating interest rate, for a specified amount that matures in between one and ten years, and requires a specified repayment schedule.

Term to maturity The time remaining on a bond's life, or the date on which the debt will cease to exist and the borrower will have completely paid off the amount borrowed. See: Maturity.

Term premiums Excess of the yields to maturity on long-term bonds over those of short-term bonds.

Term repo A repurchase agreement with a term of more than one day.

Term structure of interest rates Relationship between interest rates on bonds of different maturities, usually depicted in the form of a graph often called a yield curve. Harvey shows that inverted term structures (long rates below short rates) have preceded every recession over the past 30 years.

Term trust A closed-end fund that has a fixed termination or maturity date.

Terminal value The value of a bond at maturity, typically its par value, or the value of an asset (or an entire firm) on some specified future valuation date. Usually, a perpetuity formula is used. For example, suppose we forecast cash flows through year 10. We make an assumption that year 11 and beyond will be no growth (except for inflation). If the cash flow forecast for year 11 is 100, the firm's discount rate is 12%, and inflation is expected to be 2%, we use the formula $V_{10} = CF_{11}/(\text{disc rate-inflation})$. Hence, the value is 100/(0.12 - 0.02) that is 1,000. This cash flow needs to be brought back to present value using the formula $1000/(1.12)^{10}$, which is 321.97. Note the importance of the inflation assumption.

Terms of Delivery The part of a sales contract that indicates the point at which title and risk of loss of merchandise pass from the seller to the buyer. See: Incoterms.

Terms of sale Conditions under which a firm proposes to sell its goods or services for cash or credit.

Terms of trade The weighted average of a nation's export prices relative to its import prices.

Territorial tax system A tax system that taxes domestic income but not foreign income. Territorial tax regimes are found in Hong Kong, France, Belgium, and the Netherlands.

Test The event of a price movement that approaches a support level or a resistance level established earlier by the market. A test is passed if prices do not go below the support or resistance level, and the test is failed if prices go on to new lows or highs.

Testamentary trust A trust created by a will, that is scheduled to occur after the maker's death.

The Curb Another name for the American Stock Exchange (AMEX).

The Desk The trading desk at the Federal REserve Bank of New York through which open market purchases and sales of government and federal agancy securities are made. The desk maintains direct telephone communication with major government securities dealers. A "foreign desk" at the Federal Reserve Bank of New York conducts transactions in the foregin exchange market.

Theoretical futures price The equilibrium futures price. Also called the fair price.

Theoretical spot rate curve A curve derived from theoretical considerations as applied to the yields of actually traded Treasury debt securities, because there are no zero-coupon Treasury debt issues with a maturity greater than one year. Like the yield curve, this is a graphic depiction of the term structure of interest rates.

Theoretical value Applies to derivative products. Mathematically determined value of a derivative instrument as dictated by a pricing model such as the Black-Scholes model.

Theta The ratio of the change in an option price to the decrease in time to expiration. Also called time decay.

Thin market A market in which trading volume is low, and consequently bid and asked quotes are wide and the instrument traded is not very liquid. Very little stock to buy or sell.

Thinly traded Infrequently traded.

Third market Exchange-listed securities trading in the OTC market.

Thirty-day visible supply The total volume in dollars of municipal bonds with maturities of 13 months or more that should reach the market within 30 days.

Thirty-day wash rule IRS rule stating that losses on a sale of stock may not be used as tax shelter if equivalent stock is purchased 30 days or less before or after the sale of the stock.

Three-phase DDM A version of the dividend discount model that applies a different expected dividend rate depending on a company's life-cycle phase: growth phase, transition phase, or maturity phase.

Three steps and a stumble rule A rule predicting that stock and bond prices will fall following three increases in the discount rate by the Federal Reserve. This is a result of increased costs of borrowing for companies and the increased attractiveness of money market funds and CDs over stocks and bonds as a result of the higher interest rates.

Threshold for refinancing The point when the weighted-average coupon of an MBS is at a level to induce homeowners to prepay the mortgage in order to refinance to a lower-rate mortgage, generally reached when the weighted-average coupon of the MBS is 2 percentage points or more above currently available mortgage rates.

Thrift institution An organization formed as a depository for primarily consumer savings. Savings and loan associations and savings banks are thrift institutions.

Thrift Institution Advisory Council (TIAC) A council, established following the passage of the Monetary Control Act of 1980, whose purpose is to provide information and views on the special needs and problems of thrifts. The group is comprised of representatives of savings banks, savings and loan associations, and creditor unions.

Thrift plan A defined contribution plan in which an employee contributes, usually on a before-tax basis, toward the ultimate benefits that will be provided. The employer usually agrees to match all or a portion of the employee's contributions.

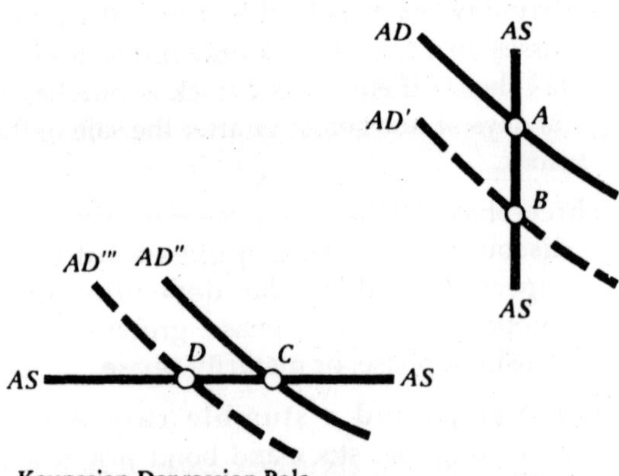

Fig. Thrift

Throughput agreement An agreement to put a specified amount of product per period through a particular facility. An example is an agreement to ship a specified amount of crude oil per period through a particular pipeline.

Tick Refers to the minimum change in price a security can have, either up or down. Related: Point.

Tick indicator A market indicator based on the number of stocks whose last trade was an uptick or a downtick. Used as an indicator of market sentiment or psychology to try to predict the market's trend.

Tick-test rules SEC-imposed restrictions on when a short sale may be executed, intended to prevent investors from destabilizing the price of a stock when the market price is falling. A short sale can be made only when either (1) the sale price of the particular stock is higher than the last trade price (referred to as an uptick trade) or (2) if there is no change in the last trade price of the particular stock, the previous trade price must be higher than the trade price that preceded it (referred to as a zero uptick).

Ticker symbol An abbreviation assigned to a security for trading purposes.

Ticker tape Computerized device that relays to investors around the world the stock symbol and the latest price and volume on securities as they are traded.

Ticket An abbreviation of order ticket.

Tier 1 and Tier 2 Descriptions of the capital adequacy of banks. Tier 1 refers to core capital while Tier 2 refers to items such as undisclosed resources.

Tight In line with or extremely close to the inside market or last sale in a stock (+/- 1/8). On the money.

Tight market A market in which volume is high, trading is active and highly competitive, and consequently spreads between bid and ask prices are narrow.

Tight money When a restricted money supply makes credit difficult to secure. The antithesis of tight money is easy money.

Tiki Tick of Dow Jones Industrial Average component issues.

Tilted portfolio An indexing strategyy that is linked to active management through the emphasis of a particular industry sector, selected performance factors such as earnings momentum, dividend yield, price-earnings ratio, or selected economic factors such as interest rates and inflation.

Time decay Related: Theta

Time deposit Interest-bearing deposit at a savings institution that has a specific maturity. Related: Certificate of deposit.

Time draft Demand for payment at a stated future date.

Time horizon The period, usually expressed in years, for which an investor expects to hold an investment.

Time Letter of Credit See: Usance Letter of Credit.

Time to maturity The time remaining until a financial contract expires. Also called time until expiration.

Time order Order that becomes a market or limited price order or is canceled at a specific time.

Time premium Also called time value, the amount by which an option price exceeds its intrinsic value. The value of an option beyond its current exercise value representing the optionholder's control until expiration, the risk of the underlying asset, and the riskless return.

Time-series analysis Assessment of relationships between two or among more variables over periods of time.

Time series models Systems that examine series of historical data; sometimes used as a means of technical forecasting, by examining moving averages.

Time spread strategy Buying and selling puts and calls with the same exercise price but different expiration dates, and trying to profit from the different premiums of the options.

Time until expiration The time remaining until a financial contract expires. Also called time to maturity.

Time value Applies to derivative products. Portion of an option price that is in excess of the intrinsic value, due to the amount of volatility in the stock; sometime referred to as premium. Time value is positively related to the length of time remaining until expiration.

Time value of money The idea that a dollar today is worth more than a dollar in the future, because the dollar received today can earn interest up until the time the future dollar is received.

Time value of an option The portion of an option's premium that is based on the amount of time remaining until the expiration date of the option contract, and the idea that the underlying components that determine the value of the option may change during that time. Time value is generally equal to the difference between the premium and the intrinsic value. Related: In the money.

Time value permium The amount by which an option's total premium exceeds its intrinsic value.

Times-interest-earned ratio Earnings before interest and tax, divided by interest payments.

Time-weighted rate of return Related: Geometric mean return

Time-Zone Arbitrage A form of stale price arbitrage where the pricing discrepencies are due to the primary markets for the underlying securities being closed at the times that the fund is traded. Note that time zone arbitrage is sometimes mistakenly used if it were a pure synonym for stale price arbitrage. These are not synonyms since stale prices can also be due to illiquid stocks or bonds that are not traded frequently.

Timeliness A source of competitive advantage that depends on being the first to enter a given market with a product or service.

Timing See: Market timing

Timing option The seller's choice of when in the delivery month to deliver. A Treasury Bond or note futures contract.

Tip Information given by one trader to another, which is used in making buy or sell decisions but is not available to the general public.

Tired Has been strong for a while and will probably fall due to increased supply at current price level (due to e.g. profit taking, technical analysis). Heavy.

Title insurance Insurance policy that protects a policyholder from future challenges to the title claim a property that may result in loss of the property.

To be announced (TBA) A contract for the purchase or sale of an MBS to be delivered at an agreed-upon future date but does not include a specified pool number and number of pools or precise amount to be delivered.

Tobin's Q Market value of assets divided by replacement value of assets. A Tobin's Q ratio greater than 1 indicates the firm has done well with its investment decisions. Named after James Tobin, Yale University economist.

Toehold purchase Often used in risk arbitrage. Accumulation by an acquirer of less than 5% of the shares of a target company. Once 5% is acquired, the acquirer must file with the SEC and other agencies to explain its intentions and notify the acquiree. See: Rule 13d.

Tokyo Commodity Exchange (TOCOM) Tokyo exchange for trading futures on gold, silver, platinum, palladium, rubber, cotton yarn, and woolen yarn.

Tokyo International Financial Futures Exchange Exchange that trades Euroyen futures and options, and futures on the one-year Euroyen, three-month eurodollar, and US dollar/Japanese yen currency.

Tokyo Stock Exchange (TSE) The largest stock exchange in Japan with the some of the most active trading in the world.

Toll revenue bond A municipal bond that is repaid with revenues from tolls that are paid by users of the public project built with the bond revenue.

Tolling agreement An agreement to put a specified amount of raw material per period through a particular processing facility. For example, an agreement to process a specified amount of alumina into aluminum at a particular aluminum plant.

Tom next Means to "tomorrow next". In the interbank market in Eurodollar deposits and the foreign exchange market, the value (delivery) date on a tom next transaction is the next business day.

Tombstone Advertisement listing the underwriters of a security issue.

Ton $100 million in bond trader's terms.

too-big-too-fail Government practices that protect large banking organizations from the normal discipline of the marketplace because of concerns that such institutions are so important to markets and their positions so intertwined with those of other banks that their failure would be unaccrptably disruptive, financially and economically.

Top Indicates the higher price one is willing to pay for a stock in an order; implies a not held order.

Top-down approach A method of security selection that starts with asset allocation and works systematically through sector and industry allocation to individual security selection.

Top-down equity management style Investment style that begins with an assessment of the overall economic environment and makes a general asset allocation decision regarding various sectors of the financial markets and various industries. The bottom-up manager, in contrast, selects specific securities within the particular sectors.

Top-heavy At a price level where supply is exceeding demand. See: Resistance level.

Topline growth Growth in revenues. Also see: Bottomline growth.

Topping out Denoting a market or a security that is at the end of a period of rising prices and can now be expected to stay on a plateau or even to decline.

Toronto Stock Exchange (TSE) Canada's largest stock exchange, trading approximately 1,200 company stocks and 33 options.

Total Complete amount of buy or sell interest, as opposed to having more behind it. See: Partial.

Total asset turnover The ratio of net sales to total assets.

Total capitalization The total long-term debt and all types of equity of a company that constitutes its capital structure.

Total cost The price paid for a security plus the broker's commission and any accrued interest that is owed to the seller (in the case of a bond).

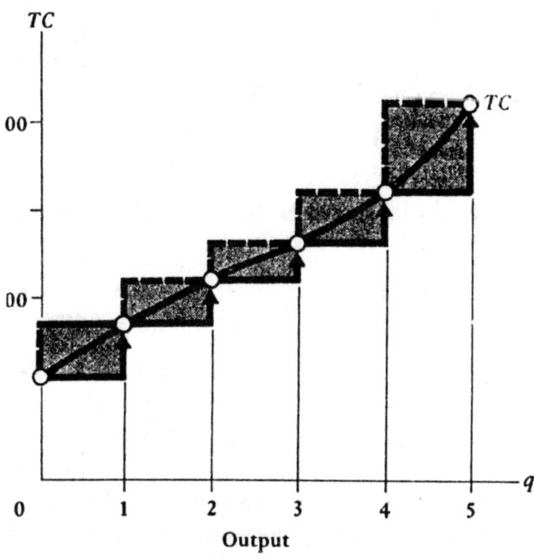

Fig. Total Coast

Total debt-to-equity ratio A capitalization ratio comparing current liabilities plus long-term debt to shareholders' equity.

Total dollar return The dollar return on a nondollar investment, which includes the sum of any dividend/interest income, capital gains or losses, and currency gains or losses on the investment. See also: Total return.

Total Market Capitalization The total market value of all of a firm's outstanding securities.

Total return In performance measurement, the actual rate of return realized over some evaluation period. In fixed income analysis, the potential return that considers all three sources of return (coupon interest, interest on coupon interest, and any capital gain/loss) over some investment horizon.

Total return for calendar year The profit or loss realized by an investment at the end of a specified calendar year, stated as the percentage gained or lost per dollar invested on January 1.

Total revenue Total sales and other revenue for the period shown. Known as "turnover" in the U.K.

Total risk The sum of systematic and unsystematic risk.

Total volume The total number of shares or contracts traded on national and regional exchanges in a stock, bond, commodity, future, or option on a certain day.

Touch, the Mainly applies to international equities. Inside market in London terminology.

Tough on price Firm price mentality at which one wishes to transact stock, often at a discount/premium that is not available at the time.

Tout To promote a security in order to attract buyers.

Toxic Convertible Used by companies that are in such bad shape, that there is no other way to get financing. This instrument is similar to a convertible bond, but convertible at a discount to the share price at issuance and for a fixed dollar amount rather than a specific number of shares. The further the stock falls, the more shares you get. Popular in the mid to late 1990s. Also known as death spiral convertibles or floorless convertibles.

Tracers Refers to investment trusts which are populated by corporate bonds. In October 2001, Morgan Stanley's Tradable Custodial Receipts (Tracers) was launched. Tracers contain a number of coporate bonds and credit default swaps which are selected for liquidity and diversity. Lehman Brothers launched a similar product, Targeted Return Index Securities (Trains) in January 2002. Both contain investment grade bonds. If a bond falls out of the investment grade category, it is either liquidated from the trust or delivered to the investor. Both Tracers and Trains are 144a trust structures and are only available to qualified buyers because they are considered private securities due to the trust structure.

Tracking error In an indexing strategy, the standard deviation of the difference between the performance of the benchmark and the replicating portfolio.

Tracking stock Best defined with an example. Suppose Company A purchases a business from Company B and pays B with 1 million shares of A's stock. The agreement provides that B cannot sell the 1 million shares for 60 days, and also prohibits B from hedging by purchasing put options on A's shares or short-selling A's shares. B is worried that the market may fall in the next 60 days. B could hedge by purchasing put options or selling the futures on the S&P 500. However, it is possible that A's business is much more cyclical than the S&P 500. One solution to this problem is to find a tracking stock. This is a stock that has high correlation with A. Let us call it Company C. The solution is to sell short or buy protective put options on this tracking stock C. This protects B from fluctuations in the price of A's stock over the next 60 days. Because the degree of the protection is related to the correlation of A and C's stock, it is extremely unlikely that the protection is perfect. Multidivisional firms have used a form of restructuring called tracking stock since 1984 to segment the performance of a particular division — similar to a spin-off or carve-out, except that the parent firm does not relinquish control of the tracked division. Previously, this was known as alphabet stock, but the technically correct name is tracking stock (e.g., EDS traded for years as a tracking stock of GM). This is a way to reward managers for good divisional performance with an equity that is tied to their division-rather than potentially penalizing them compensation for bad performance in a division they have no control over.

Trade An oral (or electronic) transaction involving one party buying a security from another party. Once a trade is consummated, it is considered "done" or final. Settlement occurs 1-5 business days later.

Trade acceptance Written demand that has been accepted by an industrial company to pay a given sum at a future date. Related: Banker's acceptance.

Trade away Trade execution by another broker/dealer.

Trade balance Overall result of a country's exports.

Trade credit Credit one firm grants to another firm for the purchase of goods or services.

Trade date The date that the counterparties in an interest rate swap commit to the swap. Also, the day on which a security or a commodity future trade actually takes place. Trades generally settle (are paid for) 1-5 business days after a trade date. With stocks, settlement is generally 3 business days after the trade. The settlement date usually follows the trade date by five business days, but varies depending on the transaction and method of delivery used.

Trade debt Accounts payable.

Trade deficit or surplus The difference in the

value of a nation's imports over exports (deficit) or exports over imports (surplus).

Trade draft A draft addressed to a commercial enterprise. See: Draft.

Trade flat For convertibles, trade without accrued interest. Preferred stock always "trades flat," as do bonds on which interest is in default or is in doubt. In general, trade in and out of a position at the same price, neither making a profit nor taking a loss.

Trade house A firm that deals in actual commodities.

Trade Lanes The direction of trade, e.g. US to Europe.

"Trade me out" Work out of one's long position (usually created by committing firm principal to complete a trade block trade) by selling stock. Antithesis of "buy them back."

Trade on the wire Immediately give a bid or offer to a salesperson without checking the floor conditions (listed), dealer depth (OTC) or customer interest. An aggressive trading posture.

Trade on top of Trade at a narrow speed or no spread in basis points relative to some other bond yield, usually Treasury bonds.

Trade reporting Dealer: In a trade between two registered Market Participants (MP), only the sell side reports the trade. Auction: In a trade between two member firms, only the sell side reports the trade. Dealer: In a trade between a registered MP and a non-registered MP (Market Maker not registered in a particular stock, an ECN, etc.), the registered MP reports the trade as a buy or sell. Auction: Trading can occur ONLY between two member firms. (Thus, a buy is never reported.)

Trade Surplus A nation's excess of exports over imports during a given time frame.

Trade-weight value of the dollar The value of the dollar pegged to, a market basket of selected foreign currencies. The Federal Reserve calculates a trade-weighted value of the dollar based on the weighted-average exchange value of the dollar against the currencies of 10 industrial countries.

Trademark A distinctive name or symbol used to identify a product or company and build recognition. Trademarks may be registered with the US Patent and Trademark Office.

Traders Individuals who take positions in securities and their derivatives with the objective of making profits. Traders can make markets by trading the flow. When they do this, their objective is to earn the bid/ask spread. Traders can also take proprietary positions in which they seek to profit from the directional movement of prices or spread positions.

Trades by appointment A stock that is very difficult to trade to because of illiquidity.

Trading Buying and selling securities.

Trading Ahead A New York Stock Exchange rule violation. Basically, in this situation the specialist puts their firm's interest ahead of the investor's interest. Consider an example. Suppose that the specialist simultaneously receives orders from two investors, one to sell 5,000 shares of XYZ and one to buy 5,000 shares of XYZ. Normally, these orders are matched. However, suppose that the specialist substitutes (matches) her own firm's 5,000 shares of XYZ. That is, the firm's own shares are sold instead of the order that came in previously. This disadvantages the buyer because the very next transaction will be the order to sell 5,000 shares of XYZ (which will likely put downward pressure on the price). Notice that the firm has bailed out of XYZ at a higher price than if the order was reversed (the specialist's firm selling afterwards). Trading ahead is part of what is known as negative obligation. Trading ahead should not be confused with front_running.

Trading authorization A document (power of attorney) a customer gives to a broker in order that the broker may buy and sell securities on behalf of the customer.

Trading costs Costs of buying and selling marketable securities and borrowing. Trading costs include commissions, slippage, and the bid/ask spread. See: Transactions costs.

Trading desk (dealing desk) Personnel at an international bank who trade spot and forward foreign exchange.

Trading dividends Maximizing a firm's revenues by purchasing stock in other firms in order to collect the maximum amount of dividends of which 70% is tax-free.

Trading halt When trading of a stock, bond, option or futures contract is stopped by an exchange while news is being broadcast about the security. See: Suspended trading.

Trading limit The exchange-imposed maximum daliy price change that a futures contract or futures option contract can undergo.

Trading paper CDs purchased by accounts that are likely to resell them. The term is commonly used in the Euromarket.

Trading pattern Long-range direction of a security or commodity futures price, charted by drawing one line connecting the highest prices the security has reached and another line connecting the lowest prices at which the security has traded over the same period. See: Technical analysis.

Trading posts The positions on the floor of a stock exchange where the specialists stand and securities are traded.

Trading price The price at which a security is currently selling.

Trading profit The profit earned on short-term trades of securities held for less than one year, subject to tax at normal income tax rates.

Trading range The difference between the high and low prices traded during a period of time; for commodities, the high/low price limit an exchange establishes for a specific commodity for any one day's trading.

Trading symbol See: Ticker symbol

Trading unit The number of shares of a particular security that is used as the acceptable quantity for trading on the exchanges.

Trading variation The increments to which securities prices are rounded up or rounded down.

Trading volume The number of shares transacted every day. As there is a seller for every buyer, one can think of the trading volume as half of the number of shares transacted. That is, if A sells 100 shares to B, the volume is 100 shares.

Traditional IRA A tax-deferred individual retirement account that allows annual contributions of up to $2000 for each income earner. Contributions are fully deductible for all individuals who are not active participants in employer-sponsored plans or for plan participants within certain income ranges.

Traditional view (of dividend policy) An argument that, "within reason," investors prefer higher dividends to lower dividends because the dividend is sure but future capital gains are uncertain.

Trailing earnings Past earnings. Often used in the context of the price earnings ratio. This ratio is usually distinguished as price to trailing earnings (today's price divided by the most recent 12 months of earnings) versus price to prospective earnings (today's price divided by consensus forecast earnings for the next 12 months).

Trailing sales Past sales. Often used in the valuation of companies that have negative cash flows or earnings. The company is said to be valued at some multiple of past sales - usually, the last 12 months sales.

Trains Refers to investment trusts which are populated by corporate bonds. In October 2001, Morgan Stanley's Tradable Custodial Receipts (Tracers) was launched. Tracers contain a number of coporate bonds and credit default swaps which are selected for liquidity and diversity. Lehman Brothers launched a similar product, Targeted Return Index Securities (Trains) in January 2002. Both contain investment grade bonds. If a bond falls out of the investment grade category, it is either liquidated from the trust or delivered to the investor. Both Tracers and Trains are 144a trust structures and are only available to qualified buyers because they are considered private securities due to the trust structure.

Tranche One of several related securities offered at the same time. Tranches from the same offering usually have different risk, reward, and/or maturity characteristics.

Transaction The delivery of a security by a seller and its acceptance by the buyer.

Transaction account A checking or similar account from which transfers can be made to third parties. Demand-deposit accounts, negotiable order of withdrawal NOW accounts, automatic transfer service (ATS) accounts, and credit union share draft accounts are examples of transaction accounts at banks and other depository institutions.

Transactions costs The time, effort, and money necessary, including such things as commission fees and the cost of physically moving the asset from seller to buyer. Transcations costs should also include the bid/ask spread as well as price impact costs (for example a large sell order could lower the price). Related: Round-trip transactions costs, information costs, search costs.

Transaction demand (for money) The money needed to accommodate a firm's expected cash transactions.

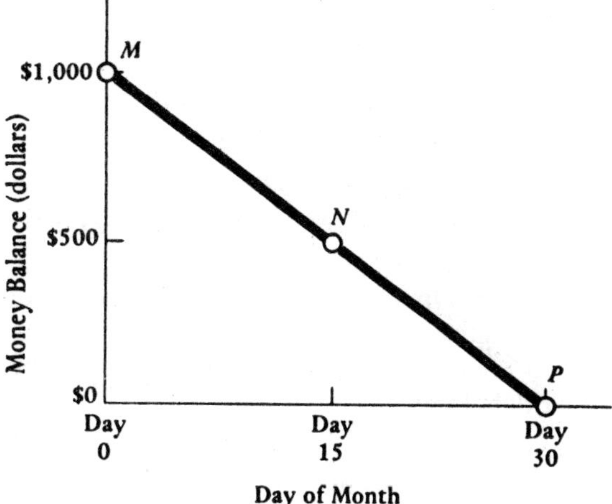

Fig. Transactions demand for money

Transaction exposure Risk to a firm with known future cash flows in a foreign currency, that arises from possible changes in the exchange rate. Related: Translation exposure.

Transaction fee A charge an intermediary, such as a broker-dealer or a bank, assesses for assisting in the sale or purchase of a security.

Transaction Insured Trade Acceptance Locator (TITAL) A trade acceptance through an insurance entity (rather than a bank) which is conditional upon exporter performance.

Transaction loan A loan extended by a bank for a specific purpose. Lines of credit and revolving credit agreements involve by contrast loans that can be used for various purposes.

Transaction tax Applies mainly to international equities. Levies on a deal that foreign governments sometimes charge.

Transaction risk The risk of changes in the home currency value of a specific future foreign currency cash flow.

Transactions motive A desire to hold cash in order to conduct cash-based transactions.

Transcript of Account A listing of all prior and present registered securityholder account information.

Transfer A change of ownership from one person or party to another.

Transfer agent Individual or institution a company appoints to look after the transfer of securities.

Transfer On Death (TOD) The process of changing title of a security from one name to another upon the death of one of the titleholders.

Transfer payments Payments from a government to its citizens, such as welfare and other government benefits.

Transfer price The price at which one unit of a firm sells goods or services to another unit of the same firm.

Transfer risk The risk associated with the possibility of a currency not being able to be sent out of the country, usually due to central bank restrictions or a national debt rescheduling.

Transfer tax A small federal tax on the movement of ownership of all bonds (except obligation of the US, foreign governments, states, and municipalities) and all stocks.

Transferable letter of credit Document that allows the first beneficiary on a standby bank assurance of funds to transfer all or part of the original letter of credit to a third party.

Transferable Stock Options Options that provide by their terms that they may be transferred by the optionee, generally only to a family member or to a trust, limited partnership or other entity for the benefit of family members, or to a charity.

Transferable put right An option issued by a firm to its shareholders to sell the firm one share of its common stock at a fixed price (the strike price) within a stated period (the time to maturity). The put right is "transferable" because it can be traded in the capital markets.

Transferee The party who has received the benefits of a letter of credit by action of a transfer.

Transferor The beneficiary of a transferable letter of credit who causes a bank to transfer the credit to another party.

Transshipment The passing goods from one ocean vessel to another.

Transition phase A stage of development when a company begins to mature and its earnings decelerate to the rate of growth of the economy as a whole. Related: Three-phase DDM.

Translation exposure Risk of adverse effects on a firm's financial statements that may arise from changes in exchange rates. Related: Transaction exposure.

Translation Risk The risk of changes in the reported home currency accounting results of foreign operations due to changes in currency exchange rates.

Transmittal letter A letter describing the contents and purpose of a transaction delivered with a security that is changing ownership.

Travel and entertainment expense Funds spent on business travel and entertainment that qualify for a tax deduction of 50% of the amount claimed.

Treasurer The corporate officer responsible for designing and implementing a firm's financing and investing activities.

Treasurer's check A check issued by a bank to make a payment. Treasurer's checks outstanding are counted as part of a bank's reservable deposits and as part of the money supply.

Treasuries Related: Treasury securities

Treasury US Department of the Treasury, which issues all Treasury bonds, notes, and bills as well as overseeing agencies. Also, the department within a corporation that oversees its financial operations including the issuance of new shares.

Treasury bills Debt obligations of the US Treasury that have maturities of one year or less. Maturities for T-bills are usually 91 days, 182 days, or 52 weeks. Treasury bills are sold at a discount from face value and do not pay interest before maturity. The interest is the difference between the purchase price of the bill and the amount that is paid to you either at maturity (this amount is the face value) or when you sell the bill prior to maturity.

Treasury bonds Debt obligations of the US Treasury that have maturities of more than 10 years.

Treasury certificates From 1963 to 1975, the Treasury issued something called a "Treasury Certificates". It was a nonmarketable, public issue with a short maturity, usually three months and never more than a one year. They were issued once or twice every month with odd interest rates (such as 5.471% and 6.053%) and sold at par.

Treasury direct A system allowing an individual investor to make a noncompetitive bid on US Treasury securities and thus avoid broker-dealer fees.

Treasury Inflation-Indexed Securities (TIIS) Refers to a broad range of U.S. Treasury securities that are inflation indexed. The most popular are the TIPS. The index for measuring the inflation rate is the non-seasonally adjusted U.S. City Average All Items Consumer Price Index for All Urban Consumers (CPI-U), published monthly by the Bureau of Labor Statistics (BLS).

Treasury Inflation-Protected Security (TIPS) First issued by the U.S. Treasury in 1997, these Treasury bonds attempt to protect investors against fluctuations in inflation by linking the principal amount to the consumer price index. Each year, the principal is adjusted by the inflation rate during the previous year. The index for measuring the inflation rate is the non-seasonally adjusted U.S. City Average All Items Consumer Price Index for All Urban Consumers (CPI-U), published monthly by the Bureau of Labor Statistics (BLS). These bonds are taxable. Indeed, one must pay tax on both the interest and the increase in principal. TIPS are one of two types of inflation-indexed securities sold by the U.S. Treasury; the other type is Series I Savings Bonds.

Treasury Investors Growth Receipt (TIGER) US government-backed bonds without coupons, meaning that the bondholders do not receive the periodic interest payments. The principal of the bond and the individual coupons are sold separately.

Treasury notes Debt obligations of the US Treasury that have maturities of more than one year, but not more than 10 years.

Treasury securities Securities issued by the US Department of the Treasury.

Treasury Shares Shares issued in the name of the corporation. The shares are considered issued, but not outstanding.Usually refers to stock that was once traded in the market but has since been repurchased by the corporation. Treasury stock not considered when calculating dividends or earnings per share.

Treasury stock Common stock that has been repurchased by the company and held in the company's treasury.

" Treat me subject " In the equities market, a conditional bid or offer. "My bid or offer is not firm, but is subject to confirmation between other parties and to market changes."

Trend The general direction of the market.

Trend Ratio Analysis The comparison of the successive values of each ratio for a single firm over a number of years.

Trendline A technical chart line that depicts the past movement of a security and that is used in an attempt to help predict future price movements.

Treynor Index A measure of the excess return per unit of risk, where excess return is defined as the difference between the portfolio's return and the risk-free rate of return over the same evaluation period and where the unit of risk is the portfolio's beta. Named after Jack Treynor.

T-Rex Fund A large venture capital fund (over one billion dollars). Such funds are known for imposing strong discipline on the firms they fund.

Triangular arbitrage Striking offsetting deals among three markets simultaneously to obtain an arbitrage profit.

Trickle down An economic theory that the support of businesses that allows them to flourish will eventually benefit middle- and lower-income people, in the form of increased economic activity and reduced unemployment.

TRIN Name derived from TRading INdex. Also known as an ARMS index. The index is usually calculated as the number of advancing issues divided by the number of declining issues. This, in turn, is divided by the advancing volume divided by the declining volume. If there is considerably more advancing volume relative to declining volume this will tend to reduce the index (i.e. increase the denominator). Hence, a value less than 1.0 is bullish while values greater than 1.0 indicate bearish demand. The index often is smoothed with a simple moving average.

Triple net lease A lease providing that the tenant pay for all maintenance expenses, plus utilities, taxes, and insurance. This results in lower risk for investors, who usually form a limited partnership.

Triple tax-exempt Municipal bonds featuring federal, state, and local tax-free interest payments.

Triple witching hour The four times a year that the S&P futures contract expires at the same time as the S&P 100 index option contract and option contracts on individual stocks. It is the last trading hour on the third Friday of March, June, September, and December, when stock options, futures on , and options on these futures expire concurrently. Massive trades in index futures, options, and underlying stock by hedge strategists and arbitrageurs cause abnormal activity (noise) and volatility.

Trough The transition point between economic recession and recovery.

True interest cost For a security such as commercial paper that is sold on a discount basis, true interest cost is the coupon rate required to provide an identical return assuming a coupon-bearing instrument of like maturity that pays interest in arrears.

True lease A contract that qualifies as a valid lease agreement under the Internal Revenue Code.

Trust A fiduciary relationship calling for a trustee to hold the title to assets for the benefit of the beneficiary. The person creating the trust, who may or may not also be the beneficiary, is called the grantor.

Trust company An organization that acts as a fiduciary and administers trusts.

Trust deed Agreement between trustee and borrower setting out terms of a bond.

Trust fund transaction An intra budgetary financial arrangement in which both payments and receipts occur within the same trust fund group.

Trust Indenture Act of 1939 A law that requires all corporate bonds and other debt securities to be issued subject to indenture agreements and comply with certain indenture provisions approved by the SEC.

Trust receipt Receipt for goods that are to be held in trust for the lender.

Trustee Agent of a bond issuer who handles the administrative aspects of a loan and ensures that the borrower complies with the terms of the bond indenture.

Trustee in bankruptcy An appointed trustee who supervises and administers the affairs of a bankrupt company or individual.

TSE 300 (Toronto Stock Exchange 100 index) Canadian form of a S&P 500.

Truth in lending law Legislation governing the granting of credit, that requires lenders to disclose the true cost of loans and the actual interest rates and terms of the loans in a manner that is easily understood.

TT&L account Treasury tax and loan account at a bank.

Turkey A losing investment.

Turn In the equities market, a reversal; unwind.

Turnaround Securities bought and sold for settlement on the same day. Also describes a firm that has been performing poorly, but changes its financial course and improves its performance.

Turnaround time Time available or needed to effect a turnaround.

Turnkey construction contract A type of construction contract under which the construction firm is obligated to complete a project according to prespecified criteria for a price that is fixed at the time the contract is signed.

Turnover For mutual funds, a measure of trading activity during the previous year, expressed as a percentage of the average total assets of the fund. A turnover rate of 25% means that the value of trades represented one-fourth of the assets of the fund. For finance, the number of times a given asset, such as inventory, is replaced during the accounting period, usually a year. For corporate finance, the ratio of annual sales to net worth, representing the extent to which a company can grow without outside capital. For markets, the volume of shares traded as a percent of total shares listed during a specified period, usually a day or a year. For Great Britain, total revenue. Percentage of the total number of shares outstanding of an issue that trades during any given period.

Turnover rate Measures trading activity during a particular period. Portfolios with high turnover rates incur higher transaction costs and are more likely to distribute capital gains, which are taxable to nonretirement accounts.

12B-1 fees The percent of a mutual fund's assets used to defray marketing and distribution expenses. The amount of the fee is stated in the fund's prospectus. The SEC has recently proposed that 12B-1 fees in excess of 0.25% be classed as a load. A true no load fund has neither a sales charge nor a 12b-1 fee.

12B-1 funds Mutual funds that do not charge an up-front or back-end commission, but instead take out up to 1.25% of average daily fund assets each year to cover the costs of selling and marketing shares, an arrangement allowed by the SEC's Rule 12B-1 (passed in 1980).

Twenty bond index A benchmark indicator of the level of municipal bond yields. It consists of the yields on 20 general obligation municipal bonds with 20-year maturities with an average rating equivalent to a1l.

Twenty-day period The period during which the SEC inspects registration statement and preliminary prospectus prior to a new issue or secondary distribution.

20% cushion rule Guideline that revenues from facilities financed by municipal bonds should exceed the operating budget plus maintenance costs and debt service by at least 20% to allow for unforeseen expenses.

25% rule The guidelines that bonded debt over

25% of a municipality's annual budget is excessive.

Twisting Convincing a customer that trades are necessary in order to generate a commission. This is an unethical practice.

Two dollar broker Floor broker of the NYSE, who executes orders for other brokers having more business at that time than they can handle with their own private floor brokers or who do not have their exchange member on the floor.

Two-factor model Usually, Fischer Black's zero-beta version of the capital asset pricing model. It may also refer to another type of model whereby expected returns are generated by any two factors.

Two-fund separation theorem The theoretical result that all investors will hold a combination of the risk-free asset and the market portfolio.

Two-sided market A market in which both bid and asked prices, good for the standard unit of trading, are quoted. When customers or market makers are lined up on both sides (buy and sell) of a stock.

Two-state option pricing model A pricing equation allowing an underlying asset to assume only two possible (discrete) values in the next time period for each value it can take on in the preceding time period. Also called the binomial option pricing model.

Two-tier bid Takeover bid in which the acquirer offers to pay more for the shares needed to gain control than for the remaining shares, or to pay the same price but at different times in the merger period; contrasts with any-or-all bid.

Two-tier tax system Taxation system that results in taxing the income going to shareholders twice.

Type The classification of an option contract as either a put or a call.

U

Ultra-short-term bond fund A mutual fund that invests in bonds with very short maturity periods, usually one year or less.

Umbrella personal liability policy A liability insurance policy that provides protection against damages not covered by standard liability policies, such as large jury awards in lawsuits.

Umbrella policy Insurance for exports of an exporter whose issuer handles all administrative requirements.

Unamortized bond discount Par value of a bond less the proceeds received from the sale of the bond, less whatever portion has been amortized.

Unamortized premiums on investments The unexpensed portion of the difference between the price paid for a security and its par value.

Unbiased expectations hypothesis Theory that forward exchange rates are unbiased predictors of future spot rates. See Forward parity.

Unbiased predictor A theory that spot prices at some future date will be equal to today's forward rates.

Unbundling Separation of a multinational firm's transfers of funds into discrete flows for specific purposes. See: Bundling.

Uncollected funds The amount of bank deposits in the form of checks that have not yet been paid by the banks on which the checks are drawn.

Uncollectible account An account which cannot be collected by a company because the customer is not able to pay or is unwilling to pay.

Unconfirmed Letter of Credit A letter of credit which has not been guaranteed or confirmed by any bank other than the bank that opened it. The advising bank merely informs the beneficiary of the letter of credit terms and conditions.

Uncovered call A short call option position in which the writer does not own shares of underlying stock represented by the option contracts. Uncovered calls are much riskier for the writer than a covered call, where the writer of the uncovered call owns the underlying stock. If the buyer of a call exercises the option to call, the writer would be forced to buy the asset at the current market price. Also called a "naked" asset.

Uncovered call writing A short call option position in which the writer does not own an equivalent position in the underlying security represented by his option contracts.

Uncovered options See: Naked options

Uncovered put A short put option position in which the writer does not have a corresponding short stock position or has not deposited, in a cash account, cash or cash equivalents equal to the exercise value of the put. The writer has pledged to buy the asset at a certain price if the buyer of the option chooses to exercise it. Uncovered put options limit the writer's risk to the value of the stock (adjusted for premium received.) Also called "naked" puts.

Uncovered Put writing A short put option position in which the writer does not have a corresponding short position in the underlying security or has not deposited, in a cash account,

Under the belt Long position in a stock.

Underbanked When an originating investment banker cannot find enough firms to underwrite a new issue.

Underbooked Describes limited interest by prospective buyers in a new issue of a security during the preoffering registration period.

Undercapitalized A business has insufficient capital to carry out its normal functions.

Underfunded pension plan A pension plan that has a negative surplus (i.e., liabilities exceed assets).

Underinvestment problem The mirror image of the asset substitution problem, in that stockholders refuse to invest in low-risk assets to avoid shifting wealth from themselves to debtholders.

Underlying What supports the security or instrument that parties agree to exchange in a derivative contract.

Underlying asset The security or property or loan agreement that an option gives the option holder the right to buy or to sell.

Underlying debt Municipal bonds issued by government entities but under the control of larger government entities and for which the larger entity shares the credit responsibility.

Underlying futures contract A futures contract that supports an option on that future, which is executed if the is exercised .

Underlying security For options, the security that is subject to purchase or sold upon exercise of an option contract. For example, IBM stock is the underlying security for IBM options. For Depository receipts, the class, series, and number of the foreign shares represented by the depository receipt.

Undermargined account A margin account that no longer meets minimum maintenance requirements, requiring a margin call on the investor.

Underperform In general, this means to do worse than some particular benchmark. Mutual Fund XYZ is said to underperform the S&P500 if its return falls short of the S&P500 return. However, this language does not take risk into account. That is, one might have a lower return than the benchmark in a particular year because of lower risk exposure. Underperform is also a term used by analysts to describe the prospects of a particular company. Usually, this means that the company will do worse than its industry average. Related: outperform.

Underpricing Issuing securities at less than their market value.

Undervalued A stock price perceived to be too low or cheap, as indicated by a particular valuation model. For instance, some might consider a particular company's stock price cheap if the company's price-earnings ratio is much lower than the industry average. To refer to undervaluation or overvaluation implicitly assumes some model of valuation. It is always possible that the security is valued correctly and that model applied is wrong.

Undervalued security A security selling below its market value or liquidation value.

Underweight Usually refers to recommendation that leads an investor to reduce their investment in a particular security or asset class. The reduction is usually with respect to a benchmark. Suppose that U.S. equities compose 40% of the benchmark portfolio. If one thinks the U.S. will underperform, the investor may reduce the exposure to U.S. equity to less than 40%.

Underwithholding When a taxpayer has withheld too little tax from salary and will therefore owe tax when filing a return.

Underwrite To guarantee, as to guarantee the issuer of securities a specified price by entering into a purchase and sale agreement. To bring securities to market.

Underwriter A firm, usually an investment bank, that buys an issue of securities from a company and resells it to investors. In general, a party that guarantees the proceeds to the firm from a security sale, thereby in effect taking ownership of the securities

Underwriter's discount See: Gross spread

Underwriting Acting as the underwriter in the issue of new securities for a firm.

Underwriting agreement The contract between a corporation issuing new publicly offered securities and the managing underwriter as agent for the underwriting group. Compare to agreement among underwriters.

Underwriting Commission The fee investment bankers charge for underwriting a security issue.

Underwriting fee The portion of the gross underwriting spread that compensates the securities firms that underwrite a public offering for their services.

Underwriting income For an insurance company, the difference between the premiums earned and the costs of settling claims.

Underwriting spread The income that is generated by the underwriting syndicate and the selling group, which is essentially the difference between the amount paid to the issuer of securities in a primary distribution and the public offering price.

Underwriting syndicate A group of investment banks that work together to sell new security offerings to investors. The underwriting syndicate is led by the lead underwriter. See also: Lead underwriter.

Underwritten offering A purchase and sale.

Undigested securities Newly issued securities that are not purchased because of lack of demand during the initial public offering.

Undiversifiable risk Related: Systematic risk

Unearned income (revenue) Income received in advance of the time at which it is earned, such as prepaid rent.

Unearned interest Interest that has been received on a loan, but that cannot be treated as a part of earnings yet, because the principal of the loan has not been outstanding long enough.

Unemployment rate The percentage of the people classified as unemployed as compared to the total labor force.

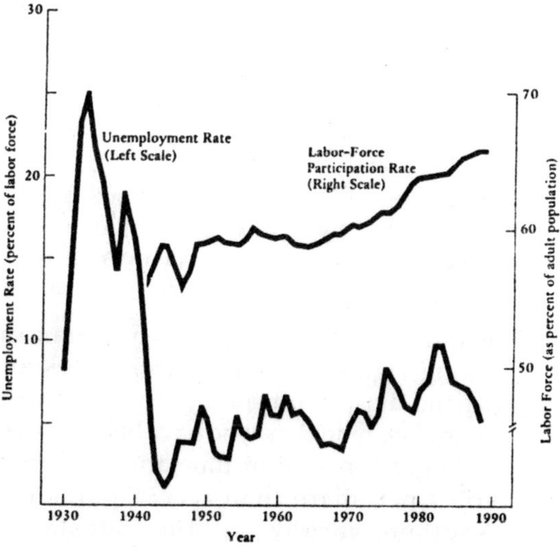

Fig. Labor force and unemployment

Unemployment Trap Refers to a situation in which middle and older aged persons retrenched during a recession have ben passed over during the subsequent economic recovery, employers tending to hire the younger short-term unemployers tending to hire the younger short-term unemployed and new job seekers; the predicament of those caught in the unemployment trap worsens not withstanding considerable experience and competence previously greatly valued.

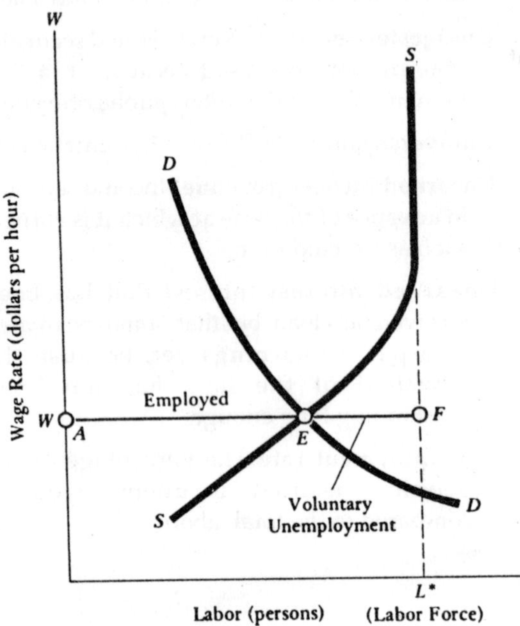

Fig. Unemployment Trap

Unencumbered Property that is not subject to any claims by creditors. For example, securities bought with cash instead of on margin and homes with mortgages paid off.

Unequal Voting These provisions limit the voting rights of some shareholders and expand those of others. Under time-phased voting, shareholders who have held the stock for a given period of time are given more votes per share than recent purchases. Another variety is the substantial shareholder provision, which limits the voting power of shareholders who have exceeded a certain threshold of ownership.

Unfavorable Balance of Trade The value of a nation's imports in excess of the value of its exports.

Unfunded debt Debt maturing within one year (short-term debt). See: Funded debt.

Unfunded pension plan Provides for the employer to pay out amounts to retirees or beneficiaries as and when they are needed. There is no money put aside on a regular basis. Instead, it is taken out of current income.

Unified tax credit A federal tax credit that reduces tax liability, dollar for dollar, on lifetime gifts and asset transfers at death.

Uniform Commercial Code (UCC) Collection of laws dealing with commercial business.

Uniform Customs and Practices (Brochure 500) International Chamber of Commerce rules (commonly referred to as UCP 500 or ICC 500), that are used for Letters of credit. These letters then become legally binding when written into the text of the letter.

Uniform Gifts to Minors Act (UGMA) Legislation that provides a tax-effective manner of transferring property to minors without the complications of trusts or guardianship restrictions.

Uniform practice code Standards of the NASD prescribing procedures for handling over-the-counter securities transactions, such as delivery, settlement date, and ex-dividend date.

Uniform Rules for Collections International Chamber of Commerce rules on the handling of documentary and clean collections.

Uniform securities agent state law examination A test required in some states for registered representatives who are employees of member firms of the NASD or over-the-counter brokers.

Uniform Transfers to Minors Act (UTMA) A law similar to the Uniform Gifts to Minors Act that extends the definition of gifts to include real estate, paintings, royalties, and patents.

Unilateral transfers Items in the current account of the balance of payments of a country's accounting books that correspond to gifts from foreigners or pension payments to foreign residents who once worked in the particular country.

Unincorporated joint venture A joint venture in which the legal means of dividing the project's equity is by shareholdings in a company.

Uninsured motorist insurance Insurance that covers the policyholder and family if they are injured by a hit-and-run or uninsured motorist, assuming the other driver is at fault.

Unique Diversification Benefit Reduction in the likelihood of financial distress for a conglomerate firm that comes with its diversified investments.

Unique risk Also called unsystematic risk or idiosyncratic risk. Specific company risk that can be eliminated through diversification. See: Diversifiable risk and unsystematic risk.

Uninvested Usually refers to cash that could be invested but is being held in reserve.

Unissued stock Shares authorized in a corporation's charter, but not issued.

Unit More than one class of securities traded together (e.g., one common share and three subscription warrants).

Unit benefit formula Method used to determine a participant's benefits in a defined benefit plan. Involves multiplying years of service by the percentage of salary.

Unit investment trust Money invested in a portfolio whose composition is fixed for the life of the fund. Shares in a unit trust are called redeemable trust certificates, and they are sold at a premium to net asset value.

Unit Share Investment Trust (USIT) A unit investment trust comprising one unit of prime and one unit of score.

Unit of trading See: Trading unit.

Unit trust In the United Kingdom and other foreign markets, an open-end mutual fund.

United States Customs Service An agency of the Treasury Department charged with enforcing laws relative to imports.

United States government securities Debt issues of the U.S. government, as distinguished from government-sponsored agency issues.

Universal life A whole life insurance product whose investment component pays a competitive interest rate rather than the below-market crediting rate.

Universe of securities A group of stocks having a common feature, such as similar outstanding market capitalization or same product line.

Unleveraged beta The beta of an unleveraged required return (i.e., no debt) on an investment when the investment is financed entirely by equity.

Unleveraged program The use of borrowed funds to finance less than 50% of a purchase of assets. In a leveraged program borrowed funds are used to finance more than 50%.

Unleveraged required return The required return on an investment when the investment is financed entirely by equity (i.e., no debt).

Unlevered cost of equity The discount rate appropriate for an investment that it is financed with 100% equity.

Unlimited liability Full liability for the debt and other obligations of a legal entity. The general partners of a partnership have unlimited liability.

Unlimited marital deduction An Internal Revenue Service provision that allows an individual to transfer an unlimited amount of assets to a spouse, during life or at death, without incurring federal estate or gift tax.

Unlimited tax bond A municipal bond secured by the pledge to levy taxes until full repayment at an unlimited rate.

Unlisted security A security traded in the over-the-counter market that is not listed on an orgaized exchange.

Unlisted trading Trading in unlisted securities that occurs on an organized exchange to accommodate members. This practice is not permitted at the NYSE.

Unloading Selling securities or commodities whose prices are dropping to minimize loss.

Unmargined account A cash account held at a brokerage firm.

Unmatched book If the average maturity of a bank's liabilities is shorter than that of its assets, it is said to be running an unmatched book. The term is commonly used with the Euromarket. Also refers to entering into OTC derivatives contracts and not hedging by making trades in the opposite direction to another financial intermediary. In this case, the firm with an unmatched book usually hedges its net market risk with futures and options. Related expressions: Open book and short book.

Unpaid dividend A dividend declared by the directors of a corporation that has not yet been paid.

Unqualified opinion An independent auditor's opinion that a company's financial statements comply with accepted accounting procedures. Antithesis of qualified opinion.

Unrealized capital gain/loss An increase/decrease in the value of a security that is not "real" because the security has not been sold. Once a security is sold by the portfolio manager, the capital gains/losses are "realized" by the fund, and any payment to the shareholder is taxable during the tax year in which the security is sold.

Unrelated Business Tax Income Income earned by a tax-exempt entity that does not result from tax-exempt activities. The entity may owe taxes on this income.

Usance The time allowed for settlement of a draft.

Usance Draft See: Time Draft

Usance Letter of Credit A letter of credit payable at a determined future date after presentation of conforming documents.

Unseasoned issue Issue of a security for which there is no existing market. See: Seasoned issue.

Unsecured debt Debt that does not identify specific assets that the debtholder is entitled to in case of default.

Unsterilized intervention Foreign exchange market intervention in which the monetary authorities have not insulated their domestic money supplies from the foreign exchange transactions.

Unsystematic risk Also called the diversifiable risk or residual risk. The risk that is unique to a company such as a strike, the outcome of unfavorable litigation, or a natural catastrophe that can be eliminated through diversification. Related: Systematic risk.

Unwind a trade Reverse a securities transaction through an offsetting transaction in the market.

Up Market indication; willingness to go both ways (buy or sell) at the mentioned volume and market. Print; up on the ticker tape, confirming that the trade has been executed.

Up tick Plus tick.

Up volume When a stock closes increases in value on a particular day, the volume in that stock is considered up volume. Related: Down volume.

Upgrading Raising the quality rating of a security because of new optimism about the prospects of a firm due to tangible or intangible factors. This can increase investor confidence and push up the price of the security.

Upset price The minimum price at which a seller of property will accept a bid at an auction.

Upside potential The amount by which analysts or investors expect the price of a security may increase.

Upstairs market A network of trading desks for the major brokerage firms and institutional investors, which communicate with each other by means of electronic display systems and telephones to facilitate block trades and program trades.

Upstairs order Used for listed equity securities. Off-floor order.

Upswing An upward turn in a security's price after a period of falling prices.

Uptick rule SEC rule that selling short is allowed only on an up tick.

Uptick trade A transaction that takes place at a higher price than the preceding transaction involving the same security. Related: Tick test rules.

Useful life The expected period of time during which a depreciating asset will be productive.

US Treasury bill US government debt with a maturity of less than a year.

US Treasury bond US government debt with a maturity of more than 10 years.

US Treasury note US government debt with a maturity of one to 10 years.

U.S. Treasury securities Interest-bearing obligations if the U.S. government issued by the U.S. Department of the Treasury as a means of borrowing money to meet government expenditures not covered by tax revenues. There are three types of marketable Treasury securities-bills, notes and bonds.

Usury laws Laws limiting the amount of interest that can be charged on loans.

Utility A power company that owns or operates facilities used for the generation, transmission, or distribution of electric energy, which is regulated at state and federal levels.

Utility function A mathematical expression that assigns a value to all possible choices. In portfolio theory, the utility function expresses the preferences of economic entities with respect to perceived risk and expected return.

Utility revenue bond A municipal bond issued to finance the construction of public utility services. These bonds are repaid from the operating revenues the project produces after the utility is finished.

Utility value The welfare a given investor assigns to an investment with a particular expected return and risk.

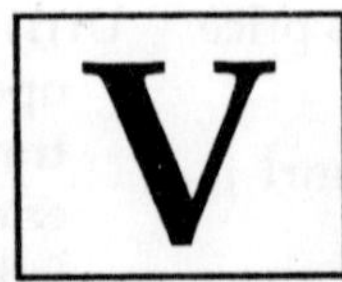

Valuation Determination of the value of a company's stock based on earnings and the market value of assets.

Valuation Clause Stipulates a fixed sum for insured property in the event of loss when included in a marine cargo insurance policy.

Valuation Opportunity Cost The potential increase in firm value associated with investments that are for gone due to capital rationing.

Valuation reserve An allowance to provide for changes in the value of a company's assets, such as depreciation.

Value Added Value added is the risk adjusted return generated by an investment strategy: the return of the investment strategy minus the return of the benchmark.

Value-added tax Tax added onto a product during each step of production, from raw material to finished good.

Value additivity principal When the value of a whole group of assets exactly equals the sum of the values of the individual assets that make up the group of assets. Or, the principle that the net present value of a set of independent projects is just the sum of the net present values of the individual projects.

Value broker A discount broker whose rates are a percentage of the dollar value of each transaction.

Value date In the market for Eurodollar deposits and foreign exchange, the delivery date of funds traded. For spot transactions, it is normally on spot transactions two days after a transaction is agreed upon. In the case of a forward foreign exchange trade, it is the future date.

Value dating When value or credit is given for funds transferred between banks.

Value investing In the context of asset management, mutual funds, and hedge funds, the a style of investment that focuses on securities with low price to earnings ratios or low price to book ratios. Some of these securities are deemed cheap and are viewed by manager as having a lot of profit potential.

Value Line investment survey A proprietary service that ranks stocks for timeliness and safety.

Value manager A manager who seeks to buy stocks that are at a discount to their "fair value" and to sell them at or in excess of that value. Often a value stock is one with a low price-to-book value ratio. Opposite of to growth stock.

Value Maximization Increases in owners'

wealth achieved by maximizing of the value of a firm's common stock.

Value-at-risk model (VaR) Procedure for estimating the probability of portfolio losses exceeding some specified proportion based on a statistical analysis of historical market price trends, correlations, and volatilities.

Value stocks Stocks with low price/book ratios or price/earnings ratios. Historically, value stocks have enjoyed higher average returns than growth stocks (stocks with high price/book or P/E ratios) in a variety of countries.

Value stock fund A mutual fund that emphasizes stocks of companies whose growth opportunities are generally regarded as subpar by the market. A value stock company often pays regular dividend income to shareholders and sells at relatively low prices in relation to its earnings or book value.

Vancouver Stock Exchange (VSE) A securities and options exchange in Vancouver, British Columbia, (Canada), specializing in venture capital companies.

Vanilla issue A security issue that has no unusual features.

Variable An element in a model. For example, in the model $RS\&P_{t+1} = a + b\ Tbill_t + e_t$, where $RS\&P_{t+1}$ is the return on the S&P in month $_{t+1}$ and Tbill is the Tbill return at month $_t$, both RS&P and Tbill are "variables" because they change through time; i.e., they are not constant.

Variable annuities Investment contracts whose issuer pays a periodic amount linked to the investment performance of an underlying portfolio.

Variable cost A cost that is directly proportional to the volume of output produced. When production is zero, the variable cost is equal to zero.

Variable interest rate See: Adjustable rate

Variable life insurance policy A whole life insurance policy that provides a death benefit dependent on the insured's portfolio market value at the time of death. Typically the company invests premiums in common stocks, so variable life policies are referred to as equity-linked policies.

Variable Plan A plan in which either the number of shares and/or the price at which they will be issued is not known on the grant date.

Variable-price security A security that sells at a fluctuating market-determined price stocks and bonds are example.

Variable-rate A varible-rate agreement, as distinguished from a fixed-rate agreement, calls for an interest rate that may fluctuate over the life of the loan. The rate is often tied to an index that reflects changes in market rates of interest. A fluctuation in the rate causes changes in either the payments or the length of the loan term. Limits are often placed on the degree to which the interest rate or the payments can vary.

Variable-rate CDs Short-term certificate of deposits that pay interest periodically on roll dates. On each roll date, the coupon on the CD is adjusted to reflect current market rates.

Variable-rate demand note A note that is payable on demand and bears interest tied to a money market rate.

Variable-rate loan Loan made at an interest rate that fluctuates depending on a base interest rate, such as the prime rate or LIBOR.

Variable rated demand bond (VRDB) Floating-rate bond that periodically can be sold back to the issuer.

Variable Ratio Write An option strategy in which the investor owns 100 shares of the underlying security and writes two call options against it, each option having a different striking price.

Variance A measure of dispersion of a set of data points around their mean value. The mathematical expectation of the average squared deviations from the mean. The square root of the variance is the standard deviation.

Variance-minimization approach to tracking An approach to bond indexing that uses historical data to estimate the variance of the tracking error.

Variance rule Specifies the permitted minimum or maximum quantity of securities that can be delivered to satisfy a TBA trade. For Ginnie Mae, Fannie Mae, and Freddie Mac pass-through securities, the accepted variance is plus or minus 2.499999 % per million of the par value of the TBA quantity.

Variation margin An additional required deposit to bring an investor's equity account up to the margin level when the balance falls below the maintenance margin requirement.

Vault cash Cash kept on hand in a depository institution's vault to meet day-to-day business needs, such as cashing checks for customers; can be counted as a portion of the institution's required reserves.

Vega A term that describes the sensitivity of the option price to a one-percent change in volatility.

Velda Sue Stands for Venture Enhancement and Loan Development Administration for Smaller Undercapitalized Enterprises. A federal agency that buys and pools small business loans made by banks, and then issues securities that are bought by large institutional investors.

Velocity The number of times a dollar is spent, or turns over, in a specific period of time. Velocity affects the amount of economic activity generated by a given money supply.

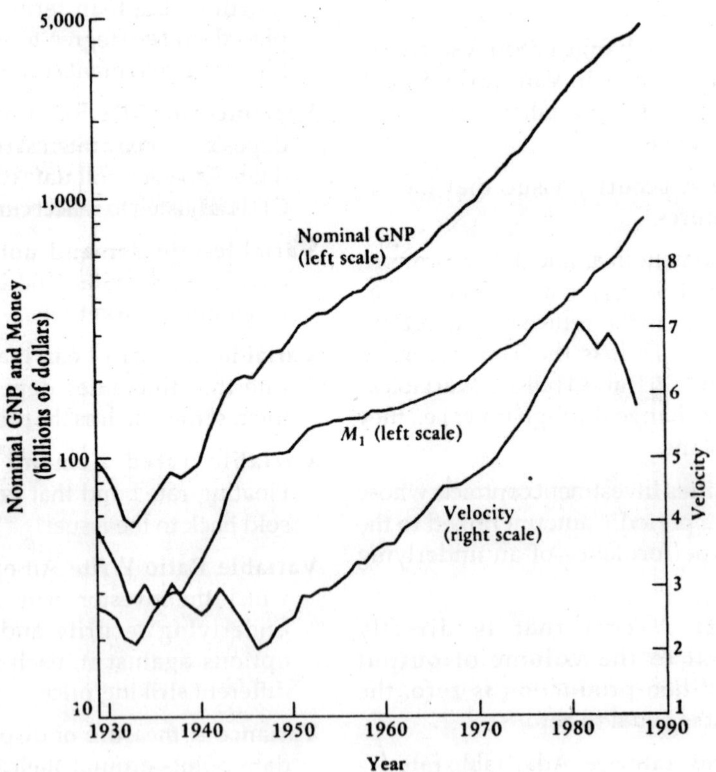

Fig. Income Velocity

Vendor Seller or supplier.

Venture capital An investment in a start-up business that is perceived to have excellent growth prospects but does not have access to capital markets. Type of financing sought by early-stage companies seeking to grow rapidly.

Venture capital limited partnership A partnership between a startup company and a brokerage firm or entrepreneurial company that provides capital for the new business in return for stock in the company and a share of the profits.

Vertical acquisition Buying or taking over a firm in the same industry in which the acquired firm and the acquiring firm represent different steps in the production process.

Vertical analysis Dividing each expense item in the income statement of a given year by net sales to identify expense items that rise more quickly or more slowly than a change in sales.

Vertical line charting A form of technical charting that shows the high, low, and closing prices of a stock or a market on each day on one vertical line with the closing price indicated by a short horizontal mark.

Vertical merger When one firm acquires another firm that is in the same industry but at another stage in the production cycle. For example, the firm being acquired serves as a supplier to the firm doing the acquiring.

Vertical spread Simultaneous purchase and sale of two options that differ only in their exercise price. See: Horizontal spread.

Vessel A conveyance for the transport of goods by water.

Vest Become applicable or exercisable. A term mainly used on the context of employee stock ownership or option programs. Employees might be given equity in a firm but they must stay with the firm for a number of years before they are entitled to the full equity. This is a vesting provision. It provides incentive for the employee to perform.

Vesting Nonforfeitable ownership (or partial ownership) by an employee of the retirement account balances or benefits contributed on the employees behalf by an employer. The Tax Reform Act of 1986 established minimum vesting rights for employees based on their years of service—full vesting in five years or 20% vesting per year starting by the end of the third year.

Vesting Schedule Schedule setting forth when, and to what extent, options become exercisable or restricted stock or stock units are no longer subject to forfeiture (for example, 20% per year over five years).

Veterans Administration (VA) mortgage A home mortgage loan granted by a lending institution to U.S. veterans and guaranteed by the Veterans Administration.

V formation A technical chart pattern that follows a letter V form, indicating that the security price has bottomed out, and is now in a bullish trend.

Vienna Convention Common name for the United Nations Convention on Contracts for the International Sale of Goods. They are a body of law governing the international sale of goods between parties domiciled in member countries.

Vienna Stock Exchange (VSX) One of the world's oldest exchanges, which accounts for approximately 50% of Austrian stock transactions; the balance are traded OTC.

Vignette A symbol or pictorial representation of the corporation on a stock certificate. Usually a complicated and artistic design, it is meant to make the counterfeiting of stock certificates as difficult as possible.

Virtual currency option A new option contract introduced by the PHLX in 1994 that is settled in US dollars rather than in the underlying currency. These options are also called 3-Ds (dollar-denominated delivery).

Visible supply New muni bond issues scheduled to come to market within the next 30 days.

VIX The implied volatility on the S&P 100 (OEX) option. This volatility is meant to be a forward looking volatility. It is calculated from both calls and puts that are near the money. The VIX is a popular measure of market risk.

Volatility A measure of risk based on the standard deviation of the asset return. Volatility is a variable that appears in option pricing formulas, where it denotes the volatility of the underlying asset return from now to the expiration of the option. There are volatility indexes. Such as a scale of 1-9; a higher rating means higher risk.

Volume counting The SEC dictates how volume is counted. Thus, volume is counted in the same manner on all markets based on the above reporting structure. Any time money changes hands (or any time capital is risked), it must be counted as a trade. Examples: 1) One registered market participant on Nasdaq buys 100 shares into inventory from another registered market participant or from one of its clients. In either case, it is counted as 100 shares. 2) One member firm on the NYSE or Amex buys 100 shares from another member firm. The Specialist matches the order between the two firms and it is counted as 100 shares. 3)The Specialist sells 100 shares from his inventory to a member firm on the NYSE. It is counted as 100 shares. 4) A Market Maker receives an order to buy 100 shares from it's client. It does not have 100 shares in its inventory. It must go buy 100 shares from someone else. It then sells these 100 shares to the client. Thus, there are two trades in this example for a total of 200 shares.

Volume deleted A note appearing on the consolidated tape when the tape is running behind under heavy trading, meaning that only the stock symbol and price will be shown for trades under 5000 shares.

Volume discount A reduction in price based on the purchase of a large quantity.

Voluntary accumulation plan Arrangement allowing shareholders of a mutual fund to purchase shares over a period of time on a regular basis, and in so doing take advantage of dollar cost averaging.

Voluntary bankruptcy The legal proceeding that follows a petition of bankruptcy.

Voluntary liquidation Liquidation proceedings that are supported by a company's shareholders.

Voluntary plan A pension plan supported partially by the employee by pension contributions deducted from each paycheck.

Volatility risk The risk in the value of options portfolios due to the unpredictable changes in the volatility of the underlying asset.

Voting Instruction Card The voting card sent to participants in an employee plan giving the trustee of the plan the authority to vote the shares as indicated on a proxy card.

Volume This is the daily number of shares of a security that change hands between a buyer and a seller. Also known as volume traded. Also see Up volume and Down volume.

Up volume When a stock closes increases in value on a particular day, the volume in that stock is considered up volume. Related: Down volume.

Voting certificate Certificates issued by a voting trust to stockholders in exchange for their common stock, which represent all the rights of common stock except voting rights.

Voting rights The right to vote on matters that are put to a vote of security holders. For example the right to vote for directors.

Voting stock The shares in a corporation that entitle the shareholder to vote.

Voting trust certificate A trust in which control of a corporation is given to a few individuals, usually to support reorganization of a corporation without interference.

VXN The implied volatility on the Nasdaq 100 (NPX) option. This volatility is meant to be a forward looking volatility. It is calculated from both calls and puts that are near the money.

Wage Any regular payment to an employee of a business for his or her labour by the hour, week, month, or some other period, or by units of output.

Waiting period Time during which the Securities and Exchange Commission (SEC) studies a firm's registration statement. During this time the firm may distribute a preliminary prospectus.

Waiver of premium A provision in an insurance policy that allows payment of insurance premiums to be permanently or

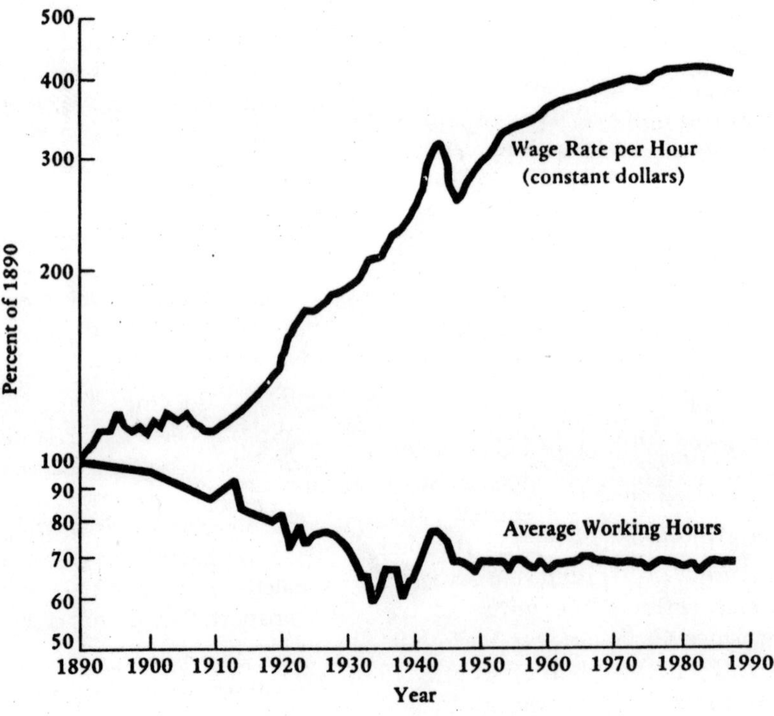

Fig. Wage versus working hours

temporarily stopped in the event the policyholder becomes incapacitated.

Walk away To take and maintain a position in a stock after going to the floor to consummate a trade. Antithesis of trade me out, buy them back.

Wall Street Generic term for the securities industry firms that buy, sell, and underwrite securities.

Wall Street analyst Related: Sell-side analyst

Wallflower Stock that has fallen out of favor with investors; stock that tends to have a low P/E (price-to-earnings ratio).

Wallpaper A security with no monetary value.

Wanted for cash A statement displayed on market tickers indicating that a bidder will pay cash for same-day settlement of a block of a specified security.

War babies Slang term for the stocks ana bonds of corporations in the defense industry.

War chest Cash kept aside for a takeover or for defense against a takeover bid.

War Risk Insurance Separate insurance coverage against loss or damage due to acts of war (including objects left over from previous wars).

Warehouse receipt Evidence that a firm owns goods stored in a warehouse.

Warehousing The interim holding period from the time of the closing of a loan to its subsequent marketing to capital market investors.

Warrant A security entitling the holder to buy a proportionate amount of stock at some specified future date at a specified price, usually one higher than current market price. Warrants are traded as securities whose price reflects the value of the underlying stock. Corporations often bundle warrants with another class of security to enhance the marketability of the other class. Warrants are like call options, but with much longer time spans-sometimes years. And, warrants are offered by corporations, while exchange-traded call options are not issued by firms.

Warranty A guarantee by a seller to a buyer that if a product requires repair or remedy of a problem within a certain period after its purchase, the seller will repair the problem at no cost to the buyer.

Warsaw Stock Exchange The major securities market of Poland.

Wash Gains equal losses.

Wash sale Purchase and sale of a security either simultaneously or within a short period of time, often in order to recognize a tax loss without altering one's position. See: Tax selling.

Wasting asset An asset that has a limited life and thus decreases in value (depreciates) over time. Also applies to consumed assets, such as oil or gas, and termed "depletion."

Watch list A list of securities selected for special surveillance by a brokerage, exchange, or regulatory organization; firms on the list are often takeover targets, companies planning to issue new securities, or stocks showing unusual activity.

Watered stock A stock representing ownership in a corporation that is worth less than the actual invested capital, resulting in problems of low liquidity, inadequate return on investment, and low market value.

Waybill A document (that looks like a bill of lading) issued by a carrier that describes the goods to be transported and that details the shipping particulars. Waybills are issued by both air carriers (air waybills) and ship lines (sea waybills). They merely indicate that the stated goods were received by the carrier for transport, they do not convey title.

Weak dollar A depreciated dollar with respect to other currencies, meaning that more dollars are needed to buy a unit of foreign currency. Antithesis of strong dollar.

Weak-form efficiency A pricing theory that the price of a security reflects the past price and trading history of the security. Theory implies that security prices follow a random walk. Related: Semistrong-form efficiency, strong-form efficiency.

Weak market A market with few buyers and many sellers and a declining trend in prices.

Wedge A chart pattern composed of two converging lines connecting peaks and troughs. In the case of falling wedges, the pattern indicates temporary interruptions of upward price rallies. In the case of rising wedges, indicates interruptions of a falling price trend.

Weekend effect The common recurrent low or negative average return from Friday to Monday in the stock market.

Weight Either Gross Weight, Net Weight, or Tare Weight.

Weighted average cost of capital (WACC) Expected return on a portfolio of all a firm's securities. Used as a hurdle rate for capital investment. Often the weighted average of the cost of equity and the cost of debt The weights are determined by the relative proportions of equity and debt in a firm's capital structure.

Weighted average Coupon The weighted average of the gross interest rates of mortgages underlying a pool as of the pool issue date; the balance of each mortgage is used as the weighting factor.

Weighted average life See: Average life

Weighted average maturity The weighted average maturity of an MBS is the weighted average of the remaining terms to maturity of the mortgages underlying the collateral pool at the date issue, using as the weighting factor the balance of each of the mortgages as of the issue date.

Weighted average portfolio yield The weighted average of the yield of all the bonds in a portfolio.

Weighted average remaining maturity The average remaining term of the mortgages underlying a MBS.

Well-diversified portfolio A portfolio that includes a variety of securities so that the weight of any security is small. The risk of a well-diversified portfolio closely approximates the systematic risk of the overall market, and the unsystematic risk of each security has been diversified out of the portfolio.

When distributed When issued.

When issued (W.I.) Refers to a transaction made conditionally, because a security, although authorized, has not yet been issued. Treasury securities, new issues of stocks and bonds, stocks that have split, and in-merger situations after the time the proxy has become effective but before completion are all traded on a when-issued basis. With ice.

Whipsawed Buying stocks just before prices fall and selling stocks just before prices rise in a volatile market, often as the result of misleading signals.

Whisper number or forecast An unofficial earnings estimate of a company given to clients by a security analyst if there is more optimism or pessimism about earnings than shown in the published number. These are often found on the Internet.

Whisper stock A stock rumored to be the target of a takeover bid, drawing speculators who hope to make a profit after the takeover is completed.

Whistle blower A person who has knowledge of fraudulent activities inside a firm or government agency, who is protected from the employer's retribution by federal law.

White knight A friendly potential acquirer sought out by a target firm that is threatened by a less welcome suitor.

White Noise The audio equivalent of Brownian motion. Sounds that are unrelated and sound like a hiss. The video equivalent of white noise is "snow" in television reception.

White sheets Lists of prices published by the National Quotation Bureau for Market Makers.

White-shoe firm Broker-dealer firms that disdain practices such as hostile takeovers.

White squire White knight who buys less than a majority interest.

White's rating A rating of municipal securities, that uses market factors rather than credit considerations to find appropriate yields.

Whitemail Sale of a large amount of stock by a company that is the target of a takeover bid to a friendly party at below-market prices, so that the raider is forced to buy more of highly priced shares to accomplish the takeover.

Whole life insurance A contract with both insurance and investment components: (1) It pays off a stated amount upon the death of the insured, and (2) it accumulates a cash value that the policyholder can redeem or borrow against.

Whole loan A term that distinguishes an investment representing an original mortgage loan from a loan representing a participation with one or more lenders.

Wholesale mortgage banking The purchasing of loans originated by others, for the acquisition of the servicing rights.

Wholesaler An underwriter or a broker-dealer who trades with other broker-dealers, rather than with the retail investor.

Wholly owned subsidiary A subsidiary whose parent company owns virtually 100% of its common stock.

Whoops A nickname for the Washington Public Power Supply System, which in the 1970s raised billions of dollars through municipal bond offerings, the projects that never materialized. WPPSS defaulted on the payments to bondholders.

WI WI Come from when issued. Treasury bills trade on a WI basis between the day they are auctioned and the day settlement is made. Bills traded before they are auctioned are said to be traded WI WI

Wide opening Abnormally wide spread between the and asked prices of a security at the opening of a trading session.

Widow-and-orphan stock A stock paying high dividends with a low beta and noncyclical business, that is an extremely safe investment.

Wiener B&#ouml;rse (Austrian Stock Exchange) Established in 1771, the major securities market of Austria.

Wild card option The right of the seller of a Treasury bond futures contract to give notice of intent to deliver at or before 8:00 p.m. Chicago time after the closing of the exchange (3:15 p.m. Chicago time) when the futures settlement price has been fixed. Related: Timing option.

Williams Act Federal legislation enacted in 1968 (and now constituting Rules 13d and 14d of the Security Exchange Act of 1934) that imposes requirements with respect to public tender offers.

Wilshire indexes Widely followed performance measurement indexes measuring performance of all U.S.-headquartered equity securities with readily available price data, created by Wilshire Associates, Inc.

Windfall profit A sudden unexpected profit uncontrolled by the profiting party.

Window A brokerage firm's cashier department, where delivery of securities and settlement of transactions take place.

Window contract A guaranteed investment contract purchased with deposits over some future designated time period (the "window"), usually between 3 and 12 months. All deposits made are guaranteed the same credit rating. Related: Bullet contract.

Window dressing Trading activity near the end of a quarter or fiscal year that is designed to improve the appearance of a portfolio to be presented to clients or shareholders. For example, a portfolio manager may sell losing positions so as to display only positions that have gained in value.

Winnipeg Commodity Exchange Canada's only agricultural futures and options exchange, located in Manitoba.

Winner's curse Problem faced by uninformed bidders. For example, in an initial public offering uninformed participants are likely to receive larger allotments of issues that informed participants know are overpriced.

Wire house A firm operating a private wire to its own branch offices or to other firms, commission houses, or brokerage houses.

Wire room A department within a brokerage firm that receives customers' orders and transmits the orders to the exchange floor or the firm's trading department.

Wire transfer Electronic transfer of funds; usually involves large dollar payments.

With Average (W.A.) Marine cargo insurance coverage providing for partial loss or damage to goods, either with or without a deductible. Also called With particular average.

With dividend Purchase of shares that entitle the buyer to the forthcoming dividend. Related: Ex-dividend.

With ice When issued.

With rights Shares sold accompanied by entitlement the buyer to buy additional shares in the company's rights issue.

Withdrawal plan Agreement that a mutual fund will disburse automatic periodic redemptions to the investor.

Withholding Used in the context of securities, the illegal practice of a public offering participant keeping some shares in a private account or with a family member, employee, or dealer to profit from the higher market price of a hot issue. Used in the context of taxes, the withholding by an employer of a certain amount of an employee's income in order to cover the employee's tax liability. Also used to refer to the withholding by corporations and financial institutions of a flat 10% of interest and dividend payments due to security holders.

Withholding tax A tax levied by a country of source on income paid, usually on dividends remitted to the home country of the firm operating in a foreign country.

Without Indicates a one-way market if 70 were bid in the market and there was no offer, the quote would be "70 bid without.".

With Particular Average (WPA) See: With Average

Without recourse Giving the lender no right to seek payment or seize assets in the event of nonpayment from anyone other than the party specified in the debt contract (such as a special-purpose entity).

Without Recourse Financing Financing in which the right of recourse to the party receiving funds is forfeited to the party advancing funds. This may be evidenced by conditions added to the endorsement of a draft being sold by an exporter in order to protect the exporter, if the instrument is not paid at maturity by the original obligor.

Woody Slang to describe a market moving strongly upward, as in, "This market has a woody."

Working Attempting to complete the remaining part of a trade, by finding either buyers or sellers for the rest.

Working away Transacting with another broker/dealer.

Working capital Defined as the difference between current assets and current liabilities. There are some variations in how working capital is calculated. Variations include the treatment of short-term debt. In addition, current assets may or may not include cash and cash equivalents, depending on the company.

Working capital management The deployment of current assets and current liabilities so as to maximize short-term liquidity.

Working capital ratio Working capital expressed as a percentage of sales.

Working control Control of a corporation by a shareholder or shareholders having less than 51% voting interest because of the wide dispersion of share ownership.

Working order Standing order in the marketplace, through which a broker bids or offers to fill the order in a series of lots at opportune times in hopes of obtaining the best price.

Workout Informal repayment or loan forgiveness arrangement between a borrower and creditors.

Workout market Market indicating prices at which it is believed a security can be bought or sold within a reasonable length of time.

Workout period Realignment of a temporarily misaligned yield relationship that sometimes occurs in fixed income markets.

World Bank A multilateral development finance agency created by the 1944 Bretton Woods, (New Hampshire) negotiations. It makes loans to developing countries for social overhead capital projects that are guaranteed by the recipient country. See: International Bank for Reconstruction and Development.

World Equity Benchmark Series (WEBS) The World Equity Benchmark Series are similar to SPDRs. WEBS trade on the AMEX, and track the Morgan Stanley Capital International (MSCI) country indexes. WEBS are available for: Australia, Austria, Belgium, Canada, France, Germany, Hong Kong, Italy, Japan, Malaysia Free, Mexico, the Netherlands, Singapore, Spain, Sweden, Switzerland, and the United Kingdom.

World investible wealth The part of world wealth that is traded and is therefore accessible to investors.

World Trade Organization (WTO) A multilateral agency that administers world trade agreements, fosters trade relations among nations, and solves trade disputes among member countries.

Wrap account An investment consulting relationship for management of a client's funds by one or more money managers, that bills all fees and commissions in one comprehensive fee charged quarterly.

Wraparound A financing device that permits an existing loan to be refinanced and new money to be advanced at an interest rate between the rate charged on the old loan and the current market interest rate. The creditor combines or "wraps" the remainder of the old loan with the new loan at the intermediate rate.

Wraparound annuity An investment that allows the annuitant the choice of underlying investments tax-deferred.

Wraparound mortgage A second mortgage that leaves the original mortgage in force. The wraparound mortgage is held by the lending institution as security for the total mortgage debt. The borrower makes payments on both loans to the wraparound lender, which in turn makes payments on the original senior mortgage.

Wrinkle A feature of a new product or security intended to entice a buyer.

Write Sell an option. Applies to derivative products.

Write-down Reducing the book value of an asset if its is overstated compared to current market values.

Write-off Charging an asset amount to expense or loss, such as through the use of depreciation and amortization of assets.

Write out The procedure used when a specialist makes a trade involving his own inventory, on one hand, and a floor broker's order, on the other. The broker must first complete the trade with the specialist, who then transacts a separate trade with the customer.

Writer The seller of an option, usually an individual, bank, or company that issues the option and consequently has the obligation to sell the asset (if a call) or to buy the asset (if a put) on which the option is written if the option buyer exercises the option.

Writing cash-secured puts An option strategy to avoid using a margin account. Instead of depositing margin with a broker, a put writer can deposit a cash balance equal to the option exercise price, and can avoid additional margin calls.

Writing puts to acquire stock Selling a put option at an exercise price that would represent a good investment by an option writer who believes a stock's value will fall, so that the writer cannot lose. If the stock price unexpectedly goes up, the option will not be exercised and the writer is at least ahead the amount of the premium received. If the stock loses value, as expected, the option will be exercised, and the writer has the stock at what he had earlier decided was originally a good buy, and he has the premium income in addition.

Written-down value The book value of an asset after allowing for depreciation and amortization.

Wrong-way risk This type of risk occurs when exposure to a counterparty is adversely correlated with the credit quality of that counterparty. There are two types of wrong-way risk. Specific wrong way risk arises through poorly structured transactions, for example, those collateralized by own or related party shares. General or conjectural wrong way risk arises where the credit quality of the counterparty may for non-specific reasons be held to be correlated with a macroeconomic factor which also affects the value of derivatives transactions. An example of conjectural wrong way risk is that fluctuations in the interest rate causes changes in the value of the derivative transactions but could also impact the credit worthiness of the counterparty. Another example might occur with an emerging-market counterparty, where there is country and possibly currency risk associated with the counterparty (however creditworthy it might otherwise be).

XD or X/D The abbreviation for ex-divident.

Xerography A dry electrostatic process for making photocopies and printing computer output with a laser printer. In a plain-paper photocopier an electrostatic image is formed on a seleniumcoated plate or cylinder. This is dusted with a resinous toner, which sticks, selectively to the charged areas, the image so formed being transferred to a sheet of paper, on which it is fixed by heating.

XR or Xr, or Ex-Right A method of quoting shares in a Stock Exchange, whereby the current purchaser is not entitled to a right of issue by the comany which is contemplated or annonced to be issued.

Xu A monetary unit of Vietnam, worth one hundredth of a dông.

Y

Yankee bonds Foreign bonds denominated in USRs. issued in the United States by foreign banks and corporations. These bonds are usually registered with the SEC. For example, bonds issued by originators with roots in Japan are called Samurai bonds.

Yankee market The foreign market in the United States.

Yard Slang for one billion currency units. Used particularly in currency trading, e.g., for Japanese yen since one billion yen equals approximately USRs.10 million. It is clearer to say, "I'm a buyer of a yard of yen," than to say, "I'm a buyer of a billion yen," which could be misheard as "I'm a buyer of a million yen."

Year-end dividend A special dividend declared at the end of a fiscal year that usually represents distribution of higher-than-expected company profits.

Yellow sheets Sheets published by the National Quotation Bureau that detail bid and ask prices, plus those firms that are making a market in over-the-counter corporate bonds.

Yield advantage The advantage gained by purchasing convertible securities instead of common stock, which equals the difference between the rates of return of the convertible security and the common shares.

Yield burning A municipal bond financing method. Underwriters in advance refundings add large markups on US Treasury bonds bought and held in escrow to compensate investors while waiting for repayment of old bonds after issuance of the new bonds. Since bond prices and yields move in opposite directions, when the bonds are marked up, they "burn down" the yield, which may violate federal tax rules and diminishes tax revenues.

Yield curb Applies mainly to convertible securities. Difference in current yield between the convertible and the underlying common.

Yield curve option-pricing models Models that can incorporate different volatility assumptions along the yield curve, such as the Black-Derman-Toy model. Also called arbitrage-free option-pricing models.

Yield curve strategies Investments that position a portfolio to capitalize on expected changes in the shape of the Treasury yield curve.

Yield curve The graphic depiction of the relationship between the yield on bonds of the same credit quality but different maturities. Related: Term structure of interest rates. Harvey (1991) finds that the inversions of the yield curve (short-term rates greater than long term rates) have preceded the last five US recessions. The yield curve can accurately forecast the turning points of the business cycle.

Yield differential/pickup Mainly applies to convertible securities. Graph showing the term structure of interest rates by plotting the yield of all bonds of the same quality with maturities ranging from the shortest to the longest available.

Yield equivalence The interest rate at which a tax-exempt bond and a taxable security of similar quality give the investor the same rate of return.

Yield ratio The quotient of two bond yields.

Yield spread strategies Investments that position a portfolio to capitalize on expected changes in yield spreads between sectors of the bond market.

Yield spread The difference in yield between different security issues usually securities of different credit quality.

Yield The percentage return paid on a stock in the form of dividends, or the effective rate of interest paid on a bond or note.

Yield to average life A yield calculation in which bonds are retired routinely during the life of the issue. Since the issuer buys its own bonds on the open market because of sinking fund requirements, if the bonds are trading below par, this action provides automatic price support for these bonds and they will usually trade on a yield to average life basis.

Yield to call The percentage rate of a bond or note if the investor buys and holds the security until the call date. This yield is valid only if the security is called prior to maturity. Generally bonds are callable over several years and normally are called at a slight premium. The calculation of yield to call is based on coupon rate, length of time to call, and market price.

Yield to maturity The percentage rate of return paid on a bond, note, or other fixed income security if the investor buys and holds it to its maturity date. The calculation for YTM is based on the coupon rate, length of time to maturity, and market price. It assumes that coupon interest paid over the life of the bond will be reinvested at the same rate.

Yield to warrant expiration Applies mainly to convertible securities. Effective yield of usable convertible bonds determined by the expiration date of the applicable warrants.

Yield to worst The bond yield computed by using the lower of either the yield to maturity or the yield to call on every possible call date.

Yo-yo stock A highly volatile stock that moves up and down like a yo-yo.

Yuan The standard monetary unit of the People's Republic of China, divided into 10 jiao or 100 fen.

Yuppie A successful and ambitious young person, especially one from the world of business. The word is formed from young upwardly-mobile professional.

Z

Z bond A bond on which interest accrues but is not currently paid to the investor but rather is added to the principal balance of the Z bond and becoming payable upon satisfaction of all prior bond classes.

Z score Statistical measure that quantifies the distance (measured in standard deviations) a data point is from the mean of a data set. Separately, z score is the output from a credit-strength test that gauges the likelihood of bankruptcy.

Z zaïre The standard monetary unit of Zaïre, divided into 100 makuta (singular, likuta).

ZBB Abbreviation for zero-base budget.

Zebra A discounted zero-coupon bond, in which the accrued income is taxed annually rather than on redemption.

Zero-base budget (ZBB) A cash-flow budget in which the manager responsible for its preparation is required to prepare and justify the budgeted expenditure from a zero base, i.e. Assuming that initially there is no commitment to spend on any activity.

Zero-coupon bond A type of bond that offers no interest payments. In effect, the interest is paid at maturity in the redemption value of the bond. Investors sometimes prefer zero-coupon bonds as they may confer more favourable tax treatment. *See also*deepdiscount bond; TIGR; Zebra.

Zero defects A component of total quality management aimed at changing workers' attitudes to quality by stressing the goal of error-free performance. In practice, error-free performance is neither economically nor practically realizable. Experience suggests that, while short-term improvements may be achieved by these programs, performance usually returns to previous levels quite quickly.

Zabara Applies mainly to international equities. Japanese securities transactions conducted on the principal of auction, i.e., (1) price priority in which the selling (buying) order with the lowest (highest) price takes precedence over other orders, and (2) time priority in that an earlier order takes precedence over other orders at the same price.

Zaibatsu Large family-owned conglomerates that controlled much of the economy of Japan prior to World War II.

Zero coupon bond Such a debt security pays an investor no interest. It is sold at a discount to its face price and matures in one year or longer.

Zero prepayment assumption The assumption of payment of scheduled principal and interest with no payments.

Zero uptick Related: tick-test rules.

Zero-balance account (ZBA) A checking account in which zero balance is maintained by transfers of funds from a master account in an amount only large enough to cover checks presented.

Zero-base budgeting (ZBB) Budgeting method that disregards the previous year's budget in setting a new budget, since circumstances may have changed. Each and every expense must be justified in this system.

Zero-beta portfolio A portfolio constructed to have zero systematic risk, similar to the risk-free asset, that is, having a beta of zero.

Zero-coupon bond A bond in which no periodic coupon is paid over the life of the contract. Instead, both the principal and the interest are paid at the maturity date.

Zero-coupon convertible security A zero-coupon bond convertible into the common stock of the issuing company after the stock reaches a certain price, using a put option inherent in the security. Also refers to zero-coupon bonds, which are convertible into an interest bearing bond at a certain time before maturity.

Zero-investment portfolio A portfolio of zero net value established by buying and shorting component securities, usually in the context of an arbitrage strategy.

Zero-minus tick Sale that takes place at the same price as the previous sale, but at a lower price than the last different price. Antithesis of zero-plus tick.

Zero-one integer programming An analytical method that can be used to determine the solution to a capital rationing problem.

Zero-plus tick Used for listed equity securities. Transaction at the same price as the preceding trade, but higher than the preceding trade at a different price.

Zero-sum game A type of game wherein one player can gain only at the expense of another player.

Zioty The standard monetary unit of Poland, divided into 100 groszy.

Zone pricing A pricing strategy in which a company delineates two or more zones. All the customers within a zone pay the same price for a product; the more distant the zone from the company's headquarters or warehouse, the higher the price.

Z-score A single statistic that attempts to measure the susceptibility of a business to failure. It is computed by applying beta coefficients to a number of selected ratios taken from an organization's final accounts using the technique of multiple discriminant analysis.